A. N. ZAVARITSKII and V. S. SOBOLEV

THE PHYSICOCHEMICAL PRINCIPLES OF IGNEOUS PETROLOGY

Israel Program for Scientific Translations
Jerusalem 1964

Published in the U. S. A. by:
DANIEL DAVEY & CO., INC.
257 Park Avenue South, New York, N. Y.

This book is a translation of
FIZIKO-KHIMICHESKIE OSNOVY PETROGRAFII
IZVERZHENNYKH GORNYKH POROD

Gosgeoltekhizdat
Gosudarstvennoe Nauchno-tekhnicheskoe
Izdatel'stvo Literatury po Geologii i Okhrane Nedr
Moskva 1961

Translated by J. Kolodny and R. Amoils, B. Sc.

IPST Cat. No. 2056

Printed in Israel by S. Monson, Jerusalem

All cross-references in this book allude to the original Russian page numbers indicated marginally

TABLE OF CONTENTS

Page

Chapter IV

Page

Page

Chapter X

Chapter XIV

Explanatory List of Abbreviations

Abbre-viation	Full name (transliterated)	Translation
AN SSSR	Akademiya Nauk SSSR	Academy of Sciences of the USSR
IL	Inostrannaya Literatura	Foreign Literature
TsNIGRI	Tsentral'nyi Nauchno-Issledovatel'skii geologo-razvedochnyi Institut	Central Scientific Research Institute of Geological Survey
IGN	Institut Geologicheskikh Nauk	Institute of Geological Sciences
MGK	Mezhdunarodnyi Geofizicheskii Komitet	International Geophysical Committee
ONTI	Ob"edinenie Nauchno-Tekhnicheskikh Izdatel'stv	United Scientific and Technical Publishing Houses

3 PREFACE

A. N. Zavaritskii's book "Physicochemical Principles of the Petrography of Igneous Rocks"†, which was published in 1926, was the first systematic textbook on this subject to appear in world literature. It was written in conjunction with a course of lectures delivered by the author over a period of many years (both before and after its publication) at the Leningrad Mining Institute, developing the trend initiated by E. S. Fedorov.

As this book is not only a systematic review, but also reflects several personal ideas of the author himself, it has played a major role in the development of petrology. A. N. Zavaritskii himself continued investigations in this field and presented his results and consideration of various petrological problems in several separate publications. He was also engaged in the revision of the book itself.

A. N. Zavaritskii's work on this book may be divided into two periods. During the first period, approximately up to 1940, the author made numerous remarks, on pages pasted into a special copy of the book, as to the amendments which should be made (with references to literature), and in some cases he even inserted some corrections and amendments. After 1940, A. N. Zavaritskii started a complete revision of the book, consisting of three parts (a total of 15 printer's sheets): Part I — "The chemical composition of igneous rocks and methods of comparison of chemical compositions" was used by the author as a basis for his monograph "Introduction to the Petrochemistry of Igneous Rocks" (31 printer's sheets), which has already been published in the USSR in two editions (1944, 1950), and has been translated into several foreign languages.

After a certain interruption, A. N. Zavaritskii started preparations of physicochemical principles proper of petrography, which were included in Part II of the 1926 edition — "Physicochemical processes occurring during the crystallization of magma". This work was, however, not completed by the author before his death; he prepared the introduction, Chapters I and II, the larger part of Chapter III, a part of Chapter IV and Chapter VII.

It is evident from the manuscript that A. N. Zavaritskii had intended to carry out a thorough revision of the book, making maximum utilization of the extensive material which had accumulated in science during the interim period. This revision was aimed at extending the scope of the book far beyond the limits of the earlier remarks. On the other hand, it is evident from the comparison of the rewritten introduction to the one of 1926 that
4 the author had no intension of including both Part I, which was already included in the above-mentioned monograph, and Part III, which was dedicated to problems of petrogenesis, probably intending to return to it later. In the course of the text, A. N. Zavaritskii somewhat broadened

† [T. N. In our translation: "Physicochemical Principles of Igneous Petrology"].

his considerations of individual questions of the direct application of physical chemistry to petrology.

All material left after the decease of A. N. Zavaritskii was transferred to me by E. P. Zavaritskaya, who had copied and brought to order the remarks in the private copy of the book. I have attempted to complete the work according to the author's intentions. In addition to the manuscripts, use was also made during the revision of materials especially related to the considered problems. These were contained in one of the chapters of "Introduction to the Petrochemistry of Igneous Rocks" and a paper dedicated to systems with volatile components (see p. 283). The greater amendments were made in Chapters VIII, IX, X, XI and XIV, in which is presented extensive accumulated material dealing with experimental investigations of the most important silicate systems.

In order to distinguish, where possible, the original text of A. N. Zavaritskii, a great part of which is published for the first time, the titles of the sections have the following markings: those written by A. N. Zavaritskii are not marked; sections which were substantially revised and enlarged, are marked by a single asterisk; where possible, the co-author's text has been indicated by square brackets; sections written by V. S. Sobolev are indicated by two asterisks.

While preparing this book for print, great assistance was rendered by N. S. Vartanova, E. A. Kostyuk and E. N. Ushakova. The Academicians A. G. Betekhtin and D. S. Korzhinskii kindly agreed to read the manuscript, offering valuable advice. I take this opportunity of extending to them my deepfelt gratitude.

V. Sobolev

5 # INTRODUCTION

Petrology is the study of rocks, i. e., of mineral aggregates forming the Earth's crust as definite, geologically independent constituents. Petrology does not only cover the study of those extraordinary accumulations of matter including rare elements of the crust, which form metalliferous and non-metalliferous ore deposits, although it should be mentioned that in some cases the limits between a rock and an ore are merely conventional.

Attempts were made to determine the difference between the terms "rock" and "ore", on the basis of geological criteria. However, geological criteria in this case do not present the possibility of drawing a distinct boundary between these terms. From the geological point of view, the main characteristic feature of both rocks and ores is the mode of their formation. In the majority of cases, the mode of formation of rocks and ores are truly different, but for certain ores, e. g., iron or aluminum ores, the boundaries between ore and rock disappear; both in the mode of their formation and their size, the mentioned ore bodies differ little from masses of rocks. It is preferable to distinguish rocks from ores on the basis of geochemical properties. Among all chemical elements composing the Earth's crust, one may distinguish several groups, which differ from one another by the inner structure of atoms. One of these groups, the so-called group of petrogenic elements (most widely distributed in the Earth's crust), include the elements which constitute the rocks. Elements of the so-called iron group are also essential participants in the composition of rocks. One may assume that iron and probably also other elements of this group play a most significant role in the structure of the globe. The important role which they play in the formation of rocks is only a reflection of their significance in the structure of the globe as a whole.

As in any other branch of science, the first task of petrology is to determine similarities and differences between the objects of study — in our case between rocks. In other words, one should start with the description and determination of rocks. When doing this, concepts of distinct species and varieties of rocks are formed, being designated by special terms — the names of these species and varieties. The determination of each of these concepts is the main task of that branch of petrology which is called petrography, or descriptive petrography. Petrography, being a branch of petrology, is also a geological science. It not only regards rocks as mineral aggregates, but considers them as parts of the Earth's crust — as those materials from which the Earth's crust is
6 comprised. In this respect it may be defined as the histology of the Earth's crust. Therefore, the great role played by the study of textures

of rocks in petrography becomes evident. If the study of a rock does not extend beyond the limits of study of properties of the constituent minerals, this study is related to mineralogy. A study becomes petrographic only when it not only answers the question of the nature of the minerals and in what amounts they contribute to rock formation, but when it also clarifies the genesis of these minerals.

Rocks, however, should not only be considered as mineral aggregates, which make up the constituents of the Earth's crust; as mentioned above, they are also the products of certain geological processes. As they act as physicochemical systems during these processes, they have reached their present state through a series of changes and inversions of other systems. Physicochemical, mechanical, or, eventually, inversions of organic origin, take place under definite geological conditions, and are interconnected by the general history of geological development of the area upon which the rocks were formed. Thus, the concept of the genetic connection between rocks arises, as well as of the relations following from this connection, and of the very origin of rocks, as various manifestations of the general evolution of the earth. These questions are dealt with in the theoretical part of petrology (mainly in the study of petrogenesis).

Petrology may consider a rock from two points of view. On the one hand, each rock is, as stated above, a physicochemical system, i. e., the outcome of certain physicochemical processes, either involving the cooling and crystallization of molten masses, the precipitation of dissolved substances, the chemical inversions connected with the activity of organisms, the mechanical processes of transport, sorting, and deposition of material forming the rock, or, lastly, the processes of alteration and inversion of one type of rock to another. The formation of rocks is governed by general laws of physics and chemistry. The presentation of these general laws, as applied to petrogenesis, should obviously constitute an essential part of petrology. On the other hand, petrology, being a branch of geology, considers rocks as products of geological processes, and studies their mutual relations and connections, regularities of formation in time and space, relation of rocks to the structure of the crust (tectonics) and conditions of their emplacement; this is also a substantial part of petrology as a science.

In the first part, we closely approach these problems, the solution of which is the task of physical chemistry and theoretical mineralogy. In the other we come in contact with the theory of vulcanism in its broad sense — when speaking of igneous rocks, and the theory of facies — when dealing with sediments.

Petrology considers rocks as products of various processes in the lithosphere and on the Earth's surface. All these processes (all changes occurring in the Earth's crust and on the surface) are united in the general concept of the evolution of our planet. This concept may also be applied to petrogenetic phenomena. From this point of view various rocks are not isolated objects of study. They are products of a general evolutionary process and are in one way or another interconnected genetically. The idea of evolution (as in biological sciences) enables one to determine mutual relations between rocks and to place single species and groups of species in
7 a definite order. In other words, it enables a natural classification of

rocks. Though we are yet far from possessing such a classification, this goal is nevertheless attainable; even at present it is possible to trace its general scheme.

The principal law of evolution, which governs all processes of petrogenesis, is reflected in all physicochemical processes occurring in the Earth's crust and on its surface — the law of dissipation of energy or the law of entropy. This is self-evident, as in petrogenesis we deal with various physicochemical and physical processes. In particular, the idea of evolution leading to the formation of various species of rocks from a primary magma has existed for some time in petrology; but up to relatively recent times this has only been a general idea; the acceptance of this idea could not lead to any concrete conclusions as to the trend of evolution, and one could not discover the forms in which its motive force — the phenomenon of dissemination of energy — is manifest.

In petrology one may observe a process similar to that in the biological sciences, where the theory of evolution now serves as their basis. This idea appeared in petrology at a much earlier date than the formulation of Darwin's theory, but it is only since Darwin's time that we regard natural selection as the principal mechanism of the evolutionary process. The causes and regularities of the evolution in petrogenesis of igneous rocks began to become clearer as from the time that laws of physical chemistry, which were earlier studied on metallic melts and melts of salts, were applied to petrogenesis. The significance of the study of eutectics, the cooling of melts of isomorphous mixtures, etc, as well as experimental investigations of simple systems of the main, rock-forming silicates are of no less importance in formulating the evolutionary concept explaining the origin of igneous rocks than is the theory of Darwin in biology. In both cases, the unsubstantiated acceptance of the existence, development, subsequent appearance and genetic connection of separate examined forms is replaced by an explanation of how this process of development takes place, and why such or other inversion, the results of which we study, take place.

In all these inversions we see one common property — the common principle of evolution; in each single process we are confronted with the reestablishment of disturbed equilibrium between the considered form, the considered system and the environment. The series of changes which they undergo constitute a series of subsequent adaptations to the changing external conditions. During the development, contradictions naturally arise between the states of the system and the changing conditions of the environment with which it is in interaction. These contradictions are either overcome by large scale phenomena such as volcanic eruptions, or in barely evident processes of the appearance of crystals in the molten magma, which crystals nevertheless indicate the appearance of matter in a new form, being constructed of atomic lattices instead of the loose linking of atoms as in the mobile state of the liquid.

Each new state of the system corresponds to new conditions of its physicochemical equilibrium. On changing, it becomes better adapted to the conditions of its existence. This continuous change of states of physicochemical systems from which rocks are formed, the subsequent reestablishment of the disturbed equilibruim and the subsequent "adaptation" of the systems to new conditions, may, to a certain degree, be compared

to the survival of the fittest in the organic world. In petrogenesis, as in the inorganic world, the overall trend of evolution is determined by the
8 interaction between the appearing forms and the external environment. Naturally, the mechanism of "survival of the fittest" is different in the inorganic world from that in the organic one, but "selection", as a principle, remains valid in both cases: in each of the rock species we may find the form which is "most stable" — most "adapted to existence" under the conditions causing its formation.

The cooling and solidification of magma causes its differentiation and the formation of a series of igneous rocks; mechanical reworking of the solid material constituting the Earth's crust, by means of the energy of a mobile environment (water, air), leads to the formation of clastic sedimentary rocks; changes in rocks under the influence of the internal heat energy of the Earth and of pressure prevailing in the Earth's crust convert them into metamorphic rocks. Systems more stable under the new set of conditions are obtained in all these processes. Studying these processes of petrogenesis, we should regard the initial material from which a rock is formed (e. g. , magma, aqueous solution, mud), as well as the external influences on it, which cause the changes (cooling of magma, vaporization of solution, coagulation of mud) and which result in the formation of a rock. However, in most cases, we may only very approximately assess the nature of the initial material, using the composition of the resulting rock as our basis.

The study of modes of formation of rocks offers a means of arriving at an explanation of mutual relations between various rocks, the establishment of genetic connection between single types and the subsequent clarification of the relations between the formation of these rock types and the entire geological history of the investigated area. This calls for the study of both the morphology of the rock masses and of the morphological relations between them, as the conditions of formation and inversion of rocks are connected with the conditions of emplacement. From this aspect, the study of occurrence of rocks is also an essential part of petrology, although it is mainly the subject of structural geology or the study of tectonics.

As in other branches of Natural Sciences, e. g. , biology, three types of problems are encountered in petrology. The first of these are the problems of physiography or morphology of rocks. This is the task of descriptive petrography. Secondly, our science is confronted with questions on regularities in processess occurring during the formation of rocks(in analogy to physiology in biological sciences). Lastly, there arise problems as to the natural history of rocks.

Great advances have been made in the field of descriptive petrography as the result of the use of the polarizing microscope in studying rocks. Comparatively recently (as from the end of the last century), a marked impetus was given to the study of processes of petrogenesis by the application of physical chemistry for solving petrological problems, and especially by recent experimental works, which placed the study of petrogenesis of igneous rocks on a firm footing. Recently, a wide development of petrological ideas, connecting the processes of petrogenesis with the particular geological conditions under which they occur, has been noted. This was possible owing to the former development of physiographic and physicochemical petrology on one hand, and of the accumulation of geological observations in the field on the other. Thus, from a purely descriptive science (as was the case until fairly recently), petrology has developed into the natural history of rocks.

9 Only one group of problems pertaining to igneous petrology is dealt with in this book. These problems concern the physicochemical laws, which govern the inversions leading to the formation of a rock or a series of rocks from an igneous magma. The problems are basic for the study of igneous rocks and should naturally have much attention devoted to them. These problems are an essential part of petrology, but far from cover its entire range. The material presented in this book is equally related to physical chemistry and to the study of rocks. Only the basic concepts of physical chemistry which are necessary for the elucidation of processes of petrogenesis of igneous rocks are presented, but no all-comprehensive consideration of those processes constituting the main field of petrology is provided.

Part I

PHYSICOCHEMICAL DIAGRAMS OF CONDENSED SYSTEMS

10 CHAPTER I

THERMODYNAMICS OF IGNEOUS PROCESSES AND THE PHASE RULE

The processes occurring in the igneous-liquid magma during the formation of igneous rocks are only one of the manifestations of that evolution of the Earth which finds its expression in all endogenetic geological phenomena resulting from the cooling of our planet. The foremost is the physicochemical process of igneous rock formation, which obeys the general rules governing such processes. All the changes occurring in the magma, and resulting finally in the formation of rocks, are due to that energy reserve which the magma derived from the volcanic chamber. The magmatic melt should be considered as a system with a certain energy reserve. Certain processes, which might be summarized as phenomena of differentiation and crystallization, occur at the expense of this energy, and result in the formation of igneous rocks.

The composition of the igneous rocks differs from that of the parent magma from which they are derived. During the formation of a rock from the magma, constituents of two kinds might be distinguished in the magma: the fixed constituents which are transferred to the rock and the fugitive constituents, or as they are usually called — the volatiles. The latter constitute a comparatively small part by weight, but their presence in the magma exerts enormous influence upon the rock-forming processes. These are the components of the magma which give rise to all volcanic phenomena; the occurrence of ores which accompany masses of igneous rocks is associated with them; their presence in the magma is essential to the formation of the rocks themselves.

The magma is a melt of a rather complex composition, which crystallizes mainly as a mineral aggregate of silicates. In order to understand the processes which occur during the formation of rocks and of their associations from such a melt, we should first regard the general features of these physicochemical processes in general, and later divide this complex phenomenon into its component parts and regard the peculiarities of these simpler phenomena in greater detail.

As already stated in the introduction, during the evolution of rocks, several changes are observed, which tend to bring the systems from a state of inequilibrium caused by changing conditions (a state which is in contradiction to the conditions of their stable existence) to a state of equilibrium. The resolution of this contradiction, which occurs naturally during evolution, makes up the history of the rock.

11 Every rock is defined by the paragenesis of its constituent minerals; this paragenesis may correspond to various cases of physicochemical equilibrium relative to the conditions in which the given rock exists.

1. A rock may be in a state of true equilibrium of a completed process. One may speak of equilibrium either in relation to the conditions of formation of the rock, or in relation to the conditions in which it is observed. In the first case, the paragenesis of the minerals correspond to conditions which existed at the time of formation of the rock and which differ from the conditions in which it is now found. These parageneses were, however, preserved because the rate of inversion, subsequent to the initial formation of the rock, is very slow. This is usually the case in plutonic rocks. Thanks to the slowness of the reaction of decomposition of rocks upon the Earth surface, we may study the conditions of their formation.

The products of decomposition of igneous and other rocks upon the Earth surface are also a case of true equilibrium; this is a case of equilibrium in relation to the conditions in which these formations are observed, but such an equilibrium only gives rise to difficulties in the study of igneous rocks.

2. During the crystallization of a magma, extrusion may interrupt the process at a point when true equilibrium conditions (corresponding to the termination of the process) have not yet been attained. Thus that state of equilibrium, reached at the moment of eruption, will be represented by a volcanic rock consisting of crystals fixed in a glass matrix. The study of such states which succeed one another consecutively, and which were preserved despite subsequent influences, is of special interest, as it enables one to reconstruct the process of rock formation.

3. In an igneous rock (plutonic or volcanic) post-formational processes occur, under different conditions, which cause the appearance of new suites of minerals. The initial minerals may yet be preserved, but they are no longer in true equilibrium with the new conditions and the new associations. Various paragenetic associations may exist in the rock, which correspond to various conditions of equilibrium. Thus one may speak of partial equilibrium in such rocks. Among these are also included the metamorphosed igneous rocks, which are often encountered in nature, but these will not be considered by us here.

4. A complete absence of equilibrium is characteristic of the paragenesis of minerals in clastic sedimentary rocks, including to a certain degree pyroclastic rocks. Here we are confronted with completely different forms of petrogenesis, which cannot be included among the igneous magmatic processes.

Processes which occur during the formation of igneous rocks are mainly connected with the change of two physical factors — temperature and pressure — and consist in chemical changes, and changes in state of the matter. The study of these phenomena is most generally related to thermodynamics, and involves the study of phases. We shall start with the presentation of these.

The process of formation of igneous rocks, as well as the changes which occur in them, are changes in heterogeneous systems consisting of several phases. Even the initially homogeneous magma, on being converted into a rock consisting of several minerals, become a heterogeneous system consisting of several physically homogeneous, but individually differing bodies (crystals, melt), which are termed as the phases of the system. Even when the homogeneous magma cools to form a glass, it liberates gases, thus becoming a heterogeneous system.

A phase which is included in a heterogeneous system is a physically uniform part of the system, differing by its physical properties from the other parts, and is capable of exchanging its constituents with the other parts when equilibrium is disturbed. Phases of a heterogeneous system may be divided mechanically, i. e., they possess boundaries which separate them.

Every mineral species which forms homogeneous crystals in a solidifying melt or in a cooling rock presents a separate phase, as it differs from other minerals and from the melt. The melt, or the glass which it forms on cooling, is also such a phase. The bubbles of gas in the foaming lava also constitute a separate phase. If, under certain conditions the melt decomposes into two or several immiscible fluids, each of them presents a separate phase, but a system which is in equilibrium may contain only one gaseous phase, as all gases are miscible in all proportions, and no boundaries exist between them.

The number of phases in a heterogeneous system may differ, but under conditions of equilibrium it complies with a rule — the Phase Rule, which will be mentioned below.

A phase system is in equilibrium when matter does not pass from one phase to another. This may be a state of true equilibrium or a state of false equilibrium. We have already seen that one may only speak of equilibrium between mineral constituents of a rock with a certain reservation, relating this equilibrium to the conditions of formation of the rock. As an example of false equilibrium, even in this sense, one may cite the vitreous or semi-vitreous condition of the groundmass in volcanic rocks, which undergoes recrystallization, being unstable at low temperatures. We know that in ancient, Paleozoic volcanic rocks the glass very rarely remains fresh, becoming dull and replaced by cryptocrystalline aggregates, e. g., silica-rich glasses are converted into microfelsite.

Further on, when considering the general regularities in the changes of heterogeneous systems, we shall term as equilibrium only the case of true equilibrium. The mass of every single phase in a system which is in such an equilibrium does not influence the equilibrium. If, to a saturated solution in which crystals of some substance were formed, a quantity of the same crystals should be added, equilibrium will not be disturbed. Similarly, the equilibrium will not be disturbed by the increase of the quantity of the same solution. Equilibrium is, however, influenced by surface properties of the substance, if its granulation is exceedingly fine. We shall not consider this particular case.

When equilibrium is disturbed, some parts of the system pass from one phase to another. When for instance water freezes, a part of it passes
13 from the liquid phase into the solid, and on evaporation a part of liquid water passes into the gaseous phase. On the crystallization of a salt from a solution a part of the substance passes into the solid phase, etc.

These constituents of a system, which, as it changes, may individually pass from one phase to another, are called components. The content of a component in a phase does not depend on the content of other components in the same phase.

Assuming we consider a melt of anorthite $CaAl_2Si_2O_8$ and of diopside $CaMgSi_2O_6$, then on crystallization either anorthite, or diopside, or both will crystallize out. These (anorthite and diopside) are the components of the system. They pass from the liquid to the solid phase or vice versa (on melting). But should we consider nepheline $NaAlSiO_4$ and diopside $CaMgSi_2O_6$ in various ratios, then after melting and crystallization of the melt, not only these substances will crystallize, but also melilite, monticellite and olivine. The components of such a system are not nepheline and diopside, but the oxides or constituents into which these compounds decompose, and which pass independently from the liquid to the solid phase on crystallization, and in the reverse direction on melting. Should we consider diopside and labradorite, then, on melting and crystallization of their mixture, the components will be diopside, albite and anorthite. In this case, it is not the constituents of the minerals, i. e., their oxides, Na_2O, CaO, MgO, Al_2O_3, SiO_2, — which are the components of the system, but their compounds: albite $NaAlSi_3O_8$, anorthite $CaAl_2Si_2O_8$ and diopside $CaMgSi_2O_6$. These compounds, and not the oxides, pass individually from one phase to another.

The determination of the components on the inversion in a given system is an important problem inherent in its study.

BASIC PRINCIPLES OF THERMODYNAMICS AND CONDITIONS FOR EQUILIBRIUM IN A HETEROGENEOUS SYSTEM †

At every inversion the system tends toward equilibrium. The deduction of the conditions for equilibrium is therefore an important theoretical problem. The conditions for equilibrium of a heterogeneous physicochemical system are deduced from the general laws of thermodynamics. It is therefore necessary to mention these general laws, and the deduction of conditions for equilibrium.

The first law of thermodynamics — the law of conservation of energy:

$$dQ = dE + PdV,$$

where dQ — is an infinitesimal amount of heat which has been introduced into the system;

dE — a minute increase in the internal energy of the system;

† For more details, see textbooks on thermodynamics: Karapet'yants, 1953; Leontovich, 1950; Epstein, 1948, as well as Tammann, 1935.

PdV — an infinitesimal amount of work done by the system during the process (here P — is pressure, dV — increase of volume).

The second law of thermodynamics — the law of entropy.

$$dS \geqslant \frac{dQ}{T},$$

i. e., the increase of entropy (dS) is equal to or greater than the amount of heat obtained by the system (dQ), divided by the absolute temperature of the system (T); the sign of equality applies to reversible processes, that of inequality ($>$) to irreversible ones; the first case corresponds to the equilibrium of a system. The last formula might be rewritten differently

$$dQ \leqslant TdS,$$

14 i. e., an infinitesimal amount of heat which has been added to the system is equal to or less than the product of absolute temperature and change of entropy of the system; the sign of equality corresponds to conditions of equilibrium.

Entropy, as is known, is a function which is dependent only on parameters of the state of the system, and therefore characterizes a particular state.

In principle these two laws of thermodynamics are a mathematical expression, as applied to heat processes, of two theses which summarize the total of human experience in the understanding of nature. The first law of thermodynamics — the law of conservation of energy — states that energy cannot be created or destroyed†. The most comprehensible expression of the second law is the statement that heat may flow only from a warmer body to a colder one, but not in the reverse direction. Some "trend" exists in the course of natural processes, and the state of systems passes from one level to another in a manner similar to the downward flow of a stream.

From the two formulas which express the first and second laws of thermodynamics it follows that:

$$dE - TdS + PdV \leqslant 0,$$

where, as in the formulas cited above, the sign of equality applies to reversible processes, i. e., to the case of equilibrium, whereas the sign of inequality applies to irreversible processes.

For constant values of T and P this expression becomes

$$d(E - TS + PV) \leqslant 0,$$

and as all functions in the brackets are functions of state —

$$Z = E - TS + PV$$

is also a function of state. This function is termed the thermodynamic potential††. Therefore it might be stated that during any irreversible changes in a system, at constant T and P, the thermodynamic potential decreases. This can be expressed by

$$dZ < 0.$$

† [T. N. Russian text states that "energy cannot be obtained from nothing"].

†† [T. N. In Western literature, commonly designated by F or G, and known as the Gibb's free energy].

If, however, stable equilibrium is achieved and changes cease, the decrease of the potential will also cease, and

$$dZ=0.$$

The function Z will then attain a minimum value.

In a heterogeneous system in equilibrium, T and P are equal for the entire system, where E, S, Q and V of the entire system are equal to the sum of these same functions for each individual phase; therefore Z of the entire system equals the total of Z_i of the individual phases, i. e.,

$$Z=\sum Z_i.$$

Moreover, for every Z_i of a phase with the mass M

$$Z_i=M_i\zeta_i,$$

where ζ_i is obviously the thermodynamic potential for unit mass of this phase (the so-called specific thermodynamic potential). For the entire system

$$Z=\sum M_i\zeta_i.$$

If the heterogeneous systems are in equilibrium, then

$$dE=TdS-PdV.$$

By differentiating the equation

$$Z=E-TS+PV,$$

we obtain

$$dZ=dE-TdS-SdT+PdV+VdP.$$

On substituting the expression for dE for the case of equilibrium in the above equation we find that

$$dZ=-SdT+VdP.$$

Taking T and P as independent variables, we obtain the partial derivatives

$$\frac{\partial Z}{\partial T}=-S \text{ and } \frac{\partial Z}{\partial P}=V.$$

As entropy is always positive, the partial derivative $\frac{\partial Z}{\partial T}$ is negative, i. e., maintaining all other parameters of the system constant, its potential decreases when temperature increases.

Let us now determine the conditions for equilibrium of a heterogeneous system consisting of p phases and c components. As mentioned above, the potential of the entire system must then attain the minimum value; hence it is necessary that dZ be zero or

$$dZ_1 + dZ_2 + \ldots\ldots + dZ_p = 0,$$

where $Z_1, Z_2, \ldots, Z_p$ — are the potentials of the phases 1, 2, . . . , p.

The variable parameters which characterize the state of every phase will be the temperature T, pressure P, and the mass of each component included in each phase (in other words — the concentration of the components in the various phases); designating the mass of the component "k" in the phase i by the sign m_i^k, we may write the conditions for equilibrium as follows:

$$\frac{\partial Z_1}{\partial m_1'} dm_1' + \frac{\partial Z_1}{\partial m_1''} dm_1'' + \ldots + \frac{\partial Z_1}{\partial m_1^{(c)}} dm_1^{(c)} +$$
$$+ \frac{\partial Z_2}{\partial m_2'} dm_2' + \frac{\partial Z_2}{\partial m_2''} dm_2'' + \ldots + \frac{\partial Z_2}{\partial m_2^{(c)}} dm_2^{(c)} +$$
$$+ \ldots\ldots\ldots\ldots\ldots\ldots\ldots\ldots\ldots\ldots +$$
$$+ \frac{\partial Z_p}{\partial m_p'} dm_p' + \frac{\partial Z_p}{\partial m_p''} dm_p'' + \ldots + \frac{\partial Z_p}{\partial m_p^{(c)}} dm_p^{(c)} = 0.$$

Every line corresponds to one of the phases and contains as many terms as there are components (c); the number of lines is equal to the number of phases (p); if any phase i does not contain the component k, then $\frac{\partial Z_i}{\partial m_i^k} = 0$.

The mass of every component in the entire system remains constant in spite of the transfer of a certain amount of components from one phase to another (closed system). This is expressed by a set of equations, the number of which is c — the number of components:

$$dm_1' + dm_2' + dm_3' + \ldots + dm_p' = 0;$$
$$dm_1'' + dm_2'' + dm_3'' + \ldots + dm_p'' = 0;$$
$$\ldots\ldots\ldots\ldots\ldots\ldots\ldots\ldots\ldots\ldots;$$
$$dm_1^{(c)} + dm_2^{(c)} + dm_3^{(c)} + \ldots + dm_p^{(c)} = 0.$$

Conditions for equilibrium in a closed system demand that, in addition to the former equations, the last-mentioned equations should also be kept equal. This condition might be obtained in another form on the introduction
16 of the values μ', μ'',...., μ^c according to the number of components, and each of which characterizes one of these components. Considering these values first as certain undefined multipliers, we shall multiply every equation of the above-mentioned system by each of these multipliers in the sequence μ', μ'', ..., μ^c and subtract it from the first equation which expresses the conditions for equilibrium.

We shall then obtain the following new equation:

$$\left(\frac{\partial Z_1}{\partial m_1'}-\mu'\right)dm_1'+\left(\frac{\partial Z_1}{\partial m_1''}-\mu''\right)dm_1''+\ldots+\left(\frac{\partial Z}{\partial m_1^{(c)}}-\mu^{(c)}\right)dm_1^{(c)}+$$
$$+\left(\frac{\partial Z_2}{\partial m_2'}-\mu'\right)dm_2'+\left(\frac{\partial Z_2}{\partial m_2''}-\mu''\right)dm_2''+\ldots+\left(\frac{\partial Z}{\partial m_2^{(c)}}-\mu^{(c)}\right)dm_2^{(c)}+$$
$$+\ldots\ldots\ldots\ldots\ldots\ldots\ldots\ldots\ldots\ldots\ldots\ldots+$$
$$+\left(\frac{\partial Z_p}{\partial m_p'}-\mu'\right)dm_p'+\left(\frac{\partial Z_p}{\partial m_p''}-\mu''\right)dm_p''+\ldots+\left(\frac{\partial Z_p}{\partial m_p^{(c)}}-\mu^{(c)}\right)dm_p^{(c)}=0.$$

The differentials dm_1', dm_1'', ..., dm_1^c, dm_2', dm_2'', ..., dm_2^c, dm_p', dm_p'', ..., dm_p^c may be of any infinitesimal value, the equation being kept equal for all these values. Obviously this is possible only if every one of the expressions in brackets of the form $(\frac{\partial Z_i}{\partial m_i^k}-\mu_i^{(k)})$ equals zero; whence conditions for equilibrium might be written as follows:

$$\frac{\partial Z_1}{\partial m_1'}=\frac{\partial Z_2}{\partial m_2'}=\frac{\partial Z_3}{\partial m_3'}=\ldots=\frac{\partial Z_p}{\partial m_p'}=\mu';$$
$$\frac{\partial Z_1}{\partial m_1''}=\frac{\partial Z_2}{\partial m_2''}=\frac{\partial Z_3}{\partial m_3''}=\ldots=\frac{\partial Z_p}{\partial m_p''}=\mu'';$$
$$\ldots\ldots\ldots\ldots\ldots\ldots\ldots\ldots$$
$$\frac{\partial Z_1}{\partial m_1^{(c)}}=\frac{\partial Z_2}{\partial m_2^{(c)}}=\frac{\partial Z_3}{\partial m_3^{(c)}}=\ldots=\frac{\partial Z_p}{\partial m_p^{(c)}}=\mu^{(c)}.$$

The equations obtained show that under equilibrium conditions the change of potential during the transfer of an infinitesimal amount of a component from one phase to another is of the same value in both phases. As one phase gains a certain infinitesimal amount of the component and the other phase loses it, the signs of the changes of mass are opposite, and the total value of Z does not change.

The partial derivatives $\frac{\partial Z_i}{\partial m_i}$ for a given component in all phases at equilibrium should be equal.

The value $\mu_i^{(k)}=\frac{\partial Z_i}{\partial m_i^k}$, which characterizes the change of potential of any phase, as dependent on the transfer of an infinitesimal amount of a component from this phase to another, is called the chemical potential of the given component **k** in the given phase **i**. It is necessary for equilibrium that the chemical potentials of every component in all phases should be equal.

We see that the chemical potential is a variable factor, similar to temperature or pressure. For equilibrium, not only temperatures and pressures in the various phases should remain constant, but also the chemical potentials. Just as heat cannot be transferred without a difference in temperatures, and mechanical work done — without differences in pressures, chemical reactions are possible only if a difference of chemical potentials exists. Chemical reactions, which essentially consist of the transfer of
17 component matter from one phase to another, call for a difference of chemical potentials of the considered components in those phases.

From the equality (at equilibrium) of chemical potentials of any component in all phases, it follows directly that if two phases, which belong to the same system, are individually in equilibrium with a third phase, they are also in equilibrium with one another.

PHASE RULE

The derivation of the Phase Rule

From thermodynamic conditions of equilibrium of a heterogeneous system the Phase Rule can be derived. This defines how many independently variable factors determine the state of a heterogeneous system in equilibrium. These factors are temperature, pressure, and the composition of all phases which form the system.

If, from the number of variable factors which determine the state of a system, we subtract the number of equations which connect these variables and answer the conditions of equilibrium, the result will be the number of independently variable factors which may change their value without disturbing the state of the system. This number is called the d e g r e e o f f r e e d o m o f t h e s y s t e m which is in equilibrium, i. e., in such a state that no matter is transported from one phase to another within the heterogeneous system.

The composition of any phase i is determined by the concentration of the components in this phase x_i', x_i'', x_i''', ..., x_i^c, where c is the number of components; obviously

$$x_i' + x_i'' + x_i''' + \dots + x_i^{(c)} = 1.$$

Thus, the composition of every phase is determined by the number of variables $c-1$, and the composition of all phases p, by $(c-1)\,p$ variables. Adding the temperature and pressure, we obtain the total number of variables —

$$p\,(c-1)+2.$$

The number of equations might be arrived at by rewriting the set of equations of chemical potentials of every component in all phases in the form of equations of partial derivatives, and thus providing the following set of equations:

$$\frac{\partial Z_1}{\partial m_1'} = \frac{\partial Z_2}{\partial m_2'} = \frac{\partial Z_3}{\partial m_3'} = \dots = \frac{\partial Z_p}{\partial m_p'};$$

$$\frac{\partial Z_1}{\partial m_1''} = \frac{\partial Z_2}{\partial m_2''} = \frac{\partial Z_3}{\partial m_3''} = \dots = \frac{\partial Z_p}{\partial m_p''};$$

$$\dots\dots\dots\dots\dots\dots\dots\dots$$

$$\frac{\partial Z_1}{\partial m_1^{(c)}} = \frac{\partial Z_2}{\partial m_2^{(c)}} = \frac{\partial Z_3}{\partial m_3^{(c)}} = \dots = \frac{\partial Z_p}{\partial m_p^{(c)}}.$$

In such a table the number of lines is c, every line containing $(p-1)$ equations (or equality signs). Thus the table contains $(p-1)c$ equations. The number of degrees of freedom f is therefore:

$$f=[p(c-1)+2]-[(p-1)c];$$
$$f=c-p+2,$$

18 i. e., "the number of degrees of freedom is equal to the number of components minus the number of phases plus two". This expression is the famous Phase Rule of Willard Gibbs (Gibbs, 1876—1878).

The number 2, which is included in the expression of the Phase Rule, is determined on the assumption that the state of the considered system is completely determined by the variables P and T. If, in a system of condensed phases at moderate pressures (such as in dry silicate melts at normal pressure), the pressure is of no great importance, we may consider the state of the system as sufficiently determined by temperature only, and the Phase Rule attains the form of

$$f=c-p+1.$$

On the other hand, if the change of state is dependent on other forms of energy, e. g., the gravitational energy, the number 2 increases by the number of such independent forms of energy.

General remarks on the Phase Rule

The above derivation shows that the Phase Rule is a necessary conclusion drawn from the basic laws of thermodynamics, and not an empirical rule which generalizes data of experiments, as it sometimes appears even in textbooks. The Phase Rule is the result of the experience of several generations of scientists who investigated nature. This experience resulted in the establishment of such laws of thermodynamics as the law of conservation of energy, and the law of entropy. This is of great significance to the petrologist, who becomes convinced of the basic nature of the principles which he is using on investigating complex phenomena.

The derivation of the Phase Rule also shows its limitations. From this derivation we accept such ideal concepts as "true equilibrium", "closed system", and others. The phenomenon of chemical inversion is described as an ideal process of transfer of a component from one phase to another. The treatment of phenomena in the thermodynamic theory of phases is analogous to the method of calculus of infinitesimal values in mathematics; this also determines the mathematical form of the conclusions which result from our considerations.

The application of the Phase Rule to natural phenomena, involving the formation of heterogeneous systems (rocks), is important, first of all, because it explains the nature of the components involved in the process, thus clarifying the nature of the resultant transformations. Phases might be determined here by observation, which may also give some idea of the degrees of freedom. This may be done in a manner similar to the observation of laboratory experimental results. The chemistry of the entire system, and sometimes also of individual phases can be revealed by analysis. Hence, one would consider the solution of the problem of the system's components as always being attainable. However, a critical point of view is necessary here, taking into account all those factors which make the phenomena more complex, and sometimes appear to be simpler than they really are. Incomplete equilibrium, partially closed systems, vagueness in the determination of the components, etc, may give rise to difficulties in the application of the Phase Rule to natural phenomena, such as are investigated by the petrologist. These difficulties, however, do not detract from the importance of the Phase Rule in our considerations of petrogenesis.

19 Phase Rule and PT-Diagrams

As shown, the factors of equilibrium of a system are pressure, temperature, and concentration, or in other words, chemical potential. The chemical potential is dependent on the composition of the components. Temperature and pressure are different as they are external factors of the physical state of the system. Their specific status in relation to other factors allows one to describe the Phase Rule conveniently in the form of a diagram, which is called the phase diagram or the PT-diagram.

Let us plot temperature (T) — as abscissa, and pressure (P) — as ordinate on a set of coordinate axes. Every point on such a diagram represents certain physical conditions of the considered system (Figure 1).

According to the Phase Rule, in a system of n components, the external conditions (temperature and pressure) may change without causing transformations in the system, with the formation of new phases or the disappearance of existent ones, only in cases when the number of phases does not exceed the number of components. In these cases the number of degrees of freedom will be not less than two. The stability fields of such (with the formation or disappearance of phases) heterogeneous systems are presented upon the diagram as certain areas. If the number of phases equals $n+1$, then temperature and pressure cannot be varied independently, as is possible within the limits of the mentioned fields. In a system of $n+1$ phases they remain invariable only if a change in temperature is accompanied by a change in pressure, one of them being the function of the other. Here we see only one independently variable factor of equilibrium, all others being functions of it. If either temperature or pressure are altered, then for equilibrium to be maintained, a corresponding alteration in the other factor is necessary. Upon the diagram, the possible equilibrium states of a system $n+1$ phase are represented by points upon lines.

Finally, if a system of n components consists of $n+2$ phases, then it may exist only at one definite value both of temperature and pressure. This particular value is described upon the diagram as a point for these values of P and T. Such a point corresponds to an invariable state of the system, both in the number and composition of the phases, and it is termed an invariant point. The line of coexistence of $n+1$ phases makes possible the variance of only one factor of equilibrium — it is a univariant line. The areas of the diagram correspond to equilibrium states, in which two factors of equilibrium may be varied — these are bivariant fields.

At the invariant point, i. e., at constant P and T which determine this point, the system consists of the maximum number of phases $n+2$. If there is one phase less, then a simultaneous alteration of both temperature and pressure is possible along the univariant curve which issues from the invariant point. The maximum number of such curves is obviously $n+2$, corresponding to the number of phases at the invariant point. However if some phases do not participate in the reactions of inversion on the univariant curves, then the number of the curves is correspondingly reduced.

We shall now turn to the invariant point, which represents the conditions of coexistence of $n+2$ phases. By decreasing the number of phases by one, the system, as just mentioned, moves from the intersection point, along one of several univariant lines, which meet at this point. A further decrease in the number of phases by one leads the system into the area of bivariant fields, limited by the univariant lines. In each of these fields several combinations of n phases are possible, depending on the total composition of the system. On the other hand, on increasing the number of phases, the bivariant system which consists of n phases is converted into a univariant system of $n+1$ phases, and further on into an invariant system of $n+2$ phases, passing from the field to the line, and moving along the line towards the invariant point.

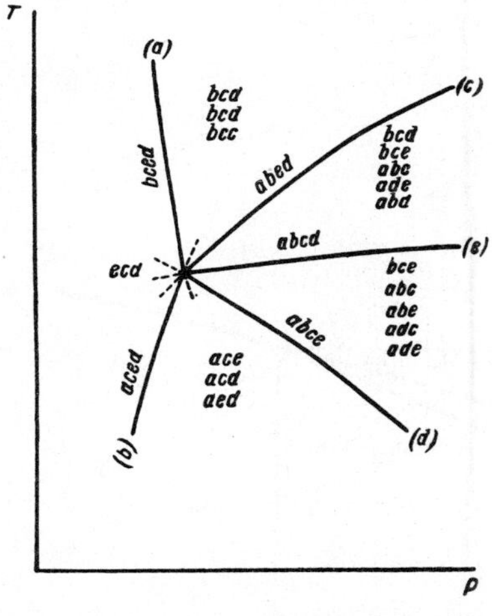

FIGURE 1. An example of a phase diagram (PT diagram) of a multicomponent system

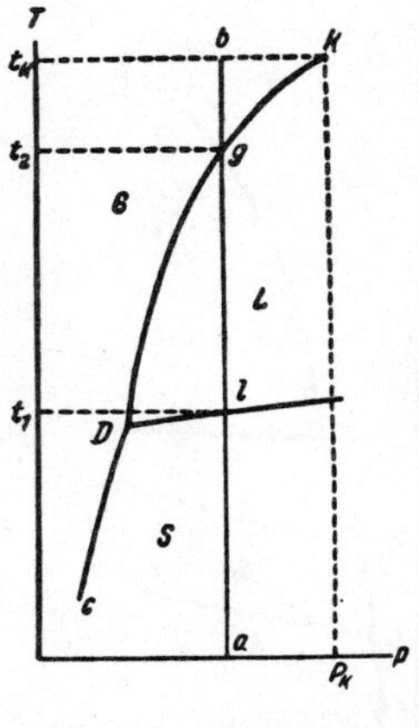

FIGURE 2. Phase diagram of a unary system

Let us take for example one of the possible cases of a ternary system. At the invariant point, five phases coexist which will be denoted by the letters *a, b, c, d, e* (see Figure 1). The univariant lines represent combinations of four phases†, and within the fields between these lines, various

† Each such line is usually marked by the symbol of the phase which does not participate in the considered equilibrium. Within every field the possible combinations are shown — the parageneses of three phases for the given variance of the system. For details on the construction of such diagrams see Schreinemakers (1957).

combinations of three phases may exist, the difference between them being dependent on the general composition of the system (see Figure 1).

If a system consists of one component, which forms various phases, then the *PT*-diagram will attain the well-known form of the phase diagram for the substance (Figure 2).

21 The area G is the stability field of a single gaseous phase; the area L — of the liquid, and the area S — of the solid. The line Dk is the line of coexistence of the gaseous and liquid phases; this is the line of vaporization. It represents the connection between P and T, showing the curve of increase of the temperature of boiling versus the pressure (or on the other side, the curve of the saturated vapor pressure versus temperature). The curve Dl represents the coexistence of the solid and liquid phase; this is the line of melting as dependent on pressure. The line Dc — is the line of sublimation of the solid substance, representing the coexistence of solid and gaseous phases.

At a certain constant pressure, which corresponds to the line ab, the point l represents the melting point, which corresponds with the temperature t_1, and the point g — the boiling point with the temperature t_2. The curve Dk terminates in the critical point k, where the boundary between liquid and gaseous phases disappears. This point is determined by constant temperature t_k and pressure P_k. Some authors also presume the existence of a critical point for the Dl branch, at very high pressures†.

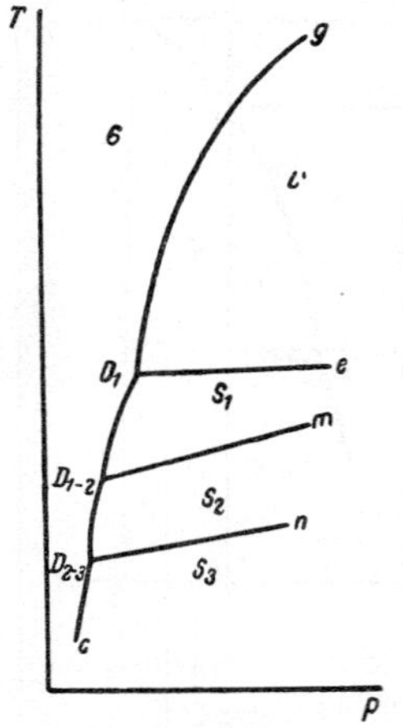

FIGURE 3. Phase diagram of a unary system with enantiotropic polymorphous transformations

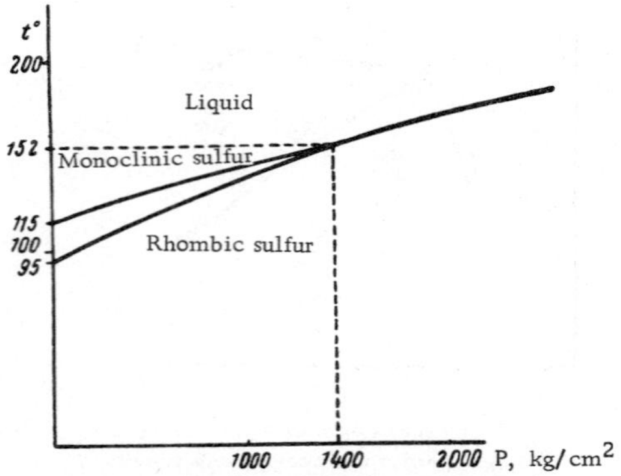

FIGURE 4. Phase diagram of a unary system — sulfur. In the given scale, the line of inversion of liquid into gas (Dk in Figure 2) is practically identical with the ordinate

† [There is little possibility for the existence of a critical point upon this curve, as the transition from the solid phase to the liquid demands the destruction of the crystal lattice. This makes the possibility of a gradual transition doubtful, as it should have taken place at the critical point or before reading].

Should the substance in the solid crystalline state exist in two or more modifications, then each of them presents a single phase (Figure 3); besides the triple point D_1, there are also other triple points D_{1-2}, D_{2-3}, upon the D_1c branch. At each of these points two solid phases and one gaseous phase coexist. The lines $D_{1-2}m$ and $D_{2-3}n$ are lines of corresponding polymorphous transformations of one modification into another — S_1 into S_2 and S_2 into S_3 (see Figure 3). Sometimes these lines intersect, as for instance on the phase diagram of sulfur (Figure 4).

From the melt, at a pressure of 1 kg/cm^2 and 115°C temperature monoclinic sulfur crystallizes, which inverts to the rhombic form at the same pressure and at 95.3°C. The curve of transformation of rhombic sulfur into monoclinic intersects with the curve of melting at the point which corresponds with a temperature of 152° and a pressure of 1,400 kg/cm^2. This point is a triple point, where two solid modifications of sulfur are coexistent: the rhombic and monoclinic, together with liquid sulfur melt.

ON PHYSICOCHEMICAL DIAGRAMS IN GENERAL

The concept of the thermodynamic potential, which combines both the first and second laws of thermodynamics, offers the possibility of forecasting the transformations which will occur in systems when factors are changed. In our case the factors are temperature and pressure, which factors determine the state of the system. Such conclusions may be most conveniently drawn, if the geometrical method of investigation is applied to the consideration of transformation phenomena, as was done by van der Waals (1936), van-Rijn-van-Alkemade (1893), Roozeboom (1891, 1899), Schreinemakers (1948)†, and others. Further on we shall only consider the most important cases, beginning with the simple ones and passing on to the more complicated.

The factors governing equilibrium are temperature (*T*) and pressure (*P*). First we shall deal with phenomena of crystallization and melting, in which pressure is of no importance. Here we deal with cases of melts which do not contain volatile components, capable of passing to a gaseous phase when the system undergoes transformation. Furthermore, the process occurs at constant pressure.

As is obvious from the above mentioned, if two phases are in equilibrium, their specific thermodynamic potentials ζ_1 and ζ_2 must be equal. If they are not equal, a transformation takes place, the system tending toward the equilibrium, which corresponds to the lower value of the thermodynamic potential, thus determining the direction of the process.

We have noted above that the partial derivative $\frac{\partial Z}{\partial T} = -S$; hence the potential increases when temperature is lowered. At the melting point of a substance, its solid phase is in equilibrium with its liquid phase; hence their specific thermodynamic potentials are equal. After plotting the change of potentials against temperature for both phases, we shall find that at the melting point these two curves intersect. The slope of the curves, as is evident from the formula, is determined by the value of the entropy. By integrating

$$dS = \frac{dQ}{T},$$

we obtain:

$$S_2 - S_1 = \int_1^2 \frac{dQ}{T}.$$

†[T.N. The date here seems to apply to the year of appearance of the Russian translation].

At constant temperature this equation takes the form of:

$$S_2 - S_1 = \frac{1}{T}\int_1^2 dQ.$$

23 When a substance melts (T remains constant), the amount of heat which is measured by the value of the integral in the last equation represents that heat which is absorbed by the melting crystals at constant temperature, i. e., it is the latent heat of fusion. Hence, the entropy of a substance in the liquid state is higher than that of the solid, causing the slope of the potential curve of the liquid to have steeper gradient than that of the solid (Figure 5). It is evident from the curves that above the melting point (T_1) the potential of the liquid is lower than that of the solid; hence the liquid is more stable above this temperature than the solid. At temperatures lower than the melting point, the potential of the liquid phase is higher, and the solid is therefore the more stable phase.

If the solid crystalline phase forms two modifications (bimorphism), then we shall obtain differing curves of potential versus temperature variation for each of them. These two curves intersect at a point which corresponds to the transformation of one modification into the other. This point may be located either to the left or to the right of the curve of the thermodynamic potentials of the liquid phase (Figure 6). The first case corresponds to an enantiotropic change, the second to a monotropic change. One can imagine a case where the curves of potential of the solid phases do not intersect; with respect to stability this will be a case of monotropic change. As is obvious, in the case of monotropy, one solid phase is always unstable, and may exist only in unstable equilibrium. On changing the pressure, however, monotropy may pass into enantiotropy.

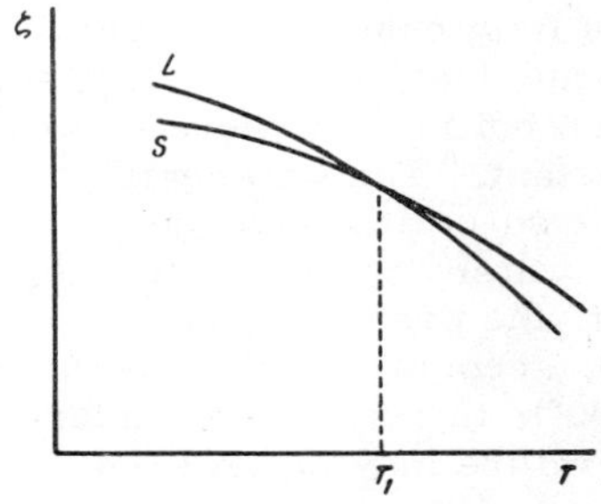

FIGURE 5. The variation with temperature of thermodynamic potential of the liquid (L) and the solid (S) phase of a unary system

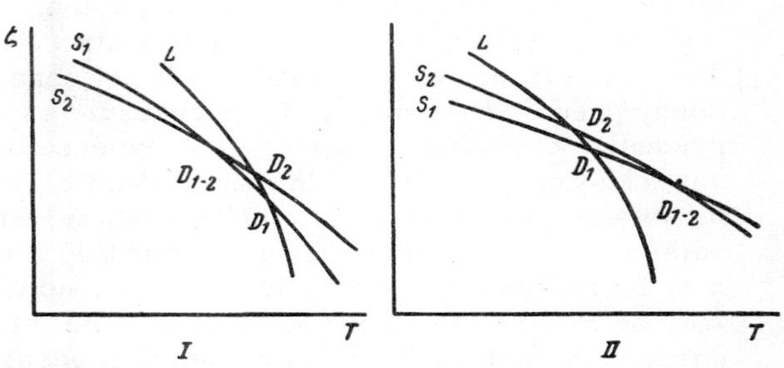

FIGURE 6. The variation of thermodynamic potentials of two solid phases (S_1 and S_2) and one liquid phase (L) of a unary system: I — temperature of transformation below the temperature of melting (enantiotropic change); II — temperature of transformation above the temperature of melting (monotropic change).

polymorphous modifications, which differ from one another in their microstructure, i. e., in the location of the atoms within the crystalline lattice. These polymorphous transformations are of special importance in mineralogy in general and in the mineralogy of igneous rocks in particular. They deserve a more detailed consideration.

In order to understand the polymorphous transformations, the possibility of delay in phase transformations should be mentioned, i. e., the possibility of a certain disturbance of equilibrium. The possibility of such delay is even evident in some systems at the transition from the liquid to the solid state. Under certain conditions, on arriving at the crystallization curve,
25 crystals do not form, and a so-called supercooled liquid is obtained. This phenomenon is of special importance in the crystallization of silicate melts (formation of glasses). It will be discussed in greater detail below.

Delay in transformations is more frequently observed at the transition boundary of two solid phases. In many cases a supercooled polymorphous modification may be obtained, which is stable at higher temperatures. In several cases a superheated modification may also be obtained, which is stable at lower temperatures. The rate of polymorphous transformations, and the possibility of a delay of transformation depend on two factors: the crystallographic character of the transformation, and the stability of the crystalline lattice of the considered phase.

Two types of transformations may be distinguished: 1) polymorphous transformations which take place by means of changing the position of atoms within the crystalline lattice, without causing disturbances in the latter; 2) polymorphous transformations in which the crystal lattice of one of the phases is disturbed, and the other phase crystallizes subsequently. Changes of the first type can readily occur, as a rule, in both directions. Changes of the second type are often quite restricted and supercooling or even superheating are obtained easily.

In order to destroy the crystal lattice, it is necessary to overcome the so-called energy barrier. This entails a certain expenditure of energy which is later compensated for on the formation of a new crystalline phase. The greater the stability of the lattice, the more the energy expended on its destruction, and the greater the difficulties encountered on polymorphous transformations of the second type. Thus, in halide salts, where the stability of the lattice is relatively low (low hardness, great solubility, relatively low melting point, etc), these changes take place relatively easily. Cesium chloride and ammonium fluoride may serve as examples of such salts. Crystal lattices of silicates are usually extremely stable, and delays in their transformation are quite characteristic.

The relative stability of phases is, as we have seen, well illustrated by the thermodynamic potential curve (see Figure 6). Upon the ***PT*** diagram the consideration of relative stability is most convenient by the extension of the two-phase curves (solid + gas, and liquid + gas) into the area of unstable equilibrium where they are usually represented by dotted lines. In one-component systems it has been proved that at a constant temperature (and low pressures) the stable phase is the one with the lowest vapor pressure. By constructing such curves (Figure 7), one may obtain new, invariant points for unstable, or metastable phases, e. g., the melting point of the low-temperature phase (D_2), (Figure 7, I and II), which is located in the area of the high-temperature modification. This point corresponds to the corresponding point D_2 on Figure 6 where the thermodynamic potential curves intersect. From this invariant point a hypothetical curve of the equilibrium solid S_2+ liquid, i. e., the melting curve of the unstable phase, tends towards the field of high pressures.

The diagrams I and II of Figure 7 represent the enantiotropic change, i. e., a change occurring in the region where each polymorphous modification is stable at normal pressures over a certain temperature range. In the area of high pressures two alternatives are possible. According to the first (Figure 7, I), the transformation curve intersects the melting curve at high pressures (E), as is shown in Figure 4 in the case of sulfur. Here the field of the high temperature modification wedges out. It has been
26 proved that the hypothetical curve D_2E must also pass through this point of intersection. To the right of the intersection point, this line D_2E has real values (equilibrium between the low temperature modification and the liquid), whereas the two former lines become hypothetical. According to the second alternative (Figure 7, II), the two curves diverge as pressure rises, i. e., both modifications remain stable.

As has already been mentioned, during monotropic changes one of the modifications is always unstable at low pressures, and therefore the transformation may occur only in one direction. The equilibrium curve S_2+G has a configuration such that the vapor pressure of the unstable phase is always higher than that of the other phase. This curve either intersects the corresponding curve of the stable phase S_1+G above the melting point
27 (Figure 7, III, point $D_{1\text{-}2}$), which corresponds with the location of the potential curves on Figure 6, II, or it does not intersect at all — the intersection point is below the absolute zero (Figure 7, V, VI). In the first case, there are two hypothetical invariant points on the diagram: S_2+L+G — the point D_2 and 2) S_2+S_1+G — the point $D_{1\text{-}2}$. From these points corresponding hypothetical dotted lines may be drawn into the field of high pressures: S_2+L and S_1+S_2 (see Figure 7, I). Obviously, here too there are two alternatives. If the mentioned curves intersect in the area of high pressures, they will take on real values to the right of the intersection point, and the phase S_2 is stable at high pressures. Its field of stability wedges out in the area of low pressures (see Figure 7, III). In the second case (see Figure 7, IV), the mentioned hypothetical curves diverge as pressure rises, i. e., there is no field of stability of S_2.

As regards diagrams with the intersection point S_1+S_2+G below the absolute zero (see Figure 7, V and VI), here, one may continue the curves and construct a diagram analogous to those shown on Figure 7, III and IV. As the curve S_1+S_2 rises usually from left to right (see below), the modification S_2 usually becomes stable at high pressures; in some cases the field of S_1 may wedge out (see Figure 7, V).

When the monotropic modification is stable at high pressures then, at low pressures, in analogy to the supercooled phase, it forms a solid modification which persists in an unstable state after the change of external equilibrium conditions. In order to achieve delayed equilibrium, a certain impetus is necessary to overcome the energy barrier. In most cases the rise in temperature serves as such an impetus, as it causes an increased mobility of the atoms. However, in this case, there is no need to attain any specific temperature, corresponding to the boundaries of the stability fields, as is the case with enantiotropic changes. Here heating plays the same role as in the ignition of wood or the explosion of gunpowder.

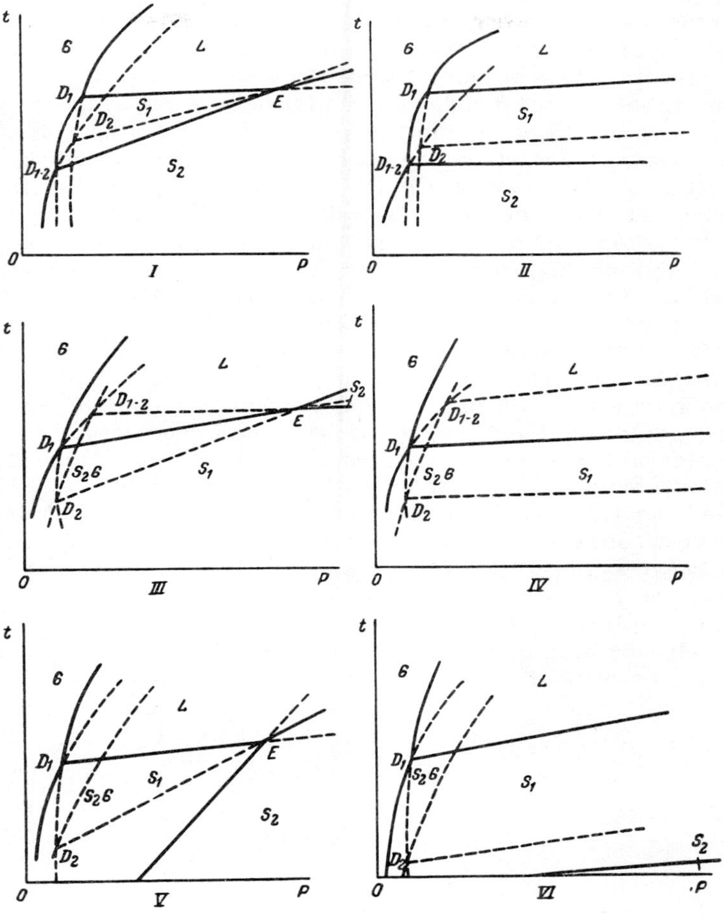

FIGURE 7. Possible types of diagrams of unary systems showing two polymorphous modifications

Naturally, the problem of the conditions of formation of monotropic phases, which do not possess fields of stability, arises. Such a possibility is explained by Oswald's rule which states that, in general, the transition of a system from an unstable to a stable state of equilibrium does not occur instantaneously, but in steps, through states which are intermediate with regard to their stability. In the unary system such intermediate steps are represented by phases, which possess a lower vapor pressure than the unstable phase, but which is nevertheless higher than that of the most stable phase. Thus, if a supercooled liquid appears in a system (see Figure 7, III and IV) below the imaginary point D_2, on the inception of crystallization the stable phase S_1 is not necessarily the first to form and may be preceded by phase S_2, which is more stable under these conditions than the supercooled liquid. If the transformation of S_2 to S_1 should occur with considerable delays, phase S_2 may remain in such an unstable state for a considerable time duration (entire geological periods). In more complicated systems, where there are more than two polymorphous modifications, several steps of transition are possible (see Chapter VIII).

The last problem upon which we shall touch here concerns the position of the equilibrium curves and their analytical determination. Most equilibrium curves rise toward the region of high pressures. Hence, in order to arrive at a curve one may lower the pressure instead of raising the temperature, and vice versa. This is due to the fact that in most physicochemical transformations, the rise in temperature and pressure, or their
28 corresponding lowering, have an opposing effect on the system (A. F. Kapustinskii). This, however, is not a law, but merely a rule which applies to all these normal cases, when the phase, stable at higher temperatures, is the one with lower density. In rare systems, where the high temperature phase is the denser, one is obliged to make exceptions to this rule. Such exceptions are known, although rare for melting curves (ice, bismuth, cast iron), and are utilized both in industry and mineralogy.

Phase diagrams are usually constructed on the basis of experimental investigations, which will be considered in brief further on.

However, equilibrium curves may equally be constructed analytically, being derived from the principal thermodynamic constants of the phases of the investigated system. These equations are illustrative of the system's properties, and in several cases they are of great importance for the solution of practical problems.

As has already been mentioned, the condition for equilibrium of two phases is the equality of their thermodynamic potentials, i.e.,

$$E_1 - TS_1 + PV_1 = E_2 - TS_2 + PV_2;$$
$$\Delta E - T\Delta S + P\Delta V = 0.$$

As shown on page 15

$$dZ = -SdT + VdP.$$

Applying this equation to our case:

$$-\Delta SdT + \Delta VdP = 0, \text{ i. e. } \frac{dT}{dP} = \frac{\Delta V}{\Delta S}.$$

According to the second law of thermodynamics

$$\Delta S = \frac{\Delta Q}{T},$$

The derived equation may also be written in the following form:

$$\frac{dT}{dP} = \frac{T\Delta V}{\Delta Q}.$$

The above-derived equation, which is known as the Clausius-Clapeyron equation, reflects several features of two-phase curves. Thus, for instance, it is evident that the slope of the curve is dependent on the sign of the latent heat of transformation, and on the difference of the specific volumes; a phase which is stable at higher temperature is formed with absorption of heat, i. e., ΔQ is positive. The entropy of the high temperature phase is always the higher. The specific volume of the high temperature phase is, in most cases, also the higher (i. e., ΔV is positive, and the gradient of the tangent to the curve is also positive).

The principal clue to the consideration of physicochemical equilibria is the so-called vant-Hoff law, which follows directly from the principle of LeChatelier: *the increase of temperature promotes reactions which involve the absorption of heat, and the increase of pressure promotes reactions which involve a decrease in volume.* As seen from the above-cited equation, this may be expressed not only qualitatively but also quantitatively. In cases where gaseous phases are absent, the difference in specific volumes may be neglected. If the scale of the diagram is chosen so that 1° corresponds
29 to 1 atm, the tangent of the angle of slope usually does not exceed 0.1 and varies correspondingly with the change of the scale. When the latent heat of transformation increases, the tangent of the angle of slope decreases.

If one of the phases is a gas, then ΔV is quite large, as is reflected by the form of the curve. On the same scale, such curves become vertical lines which coincide with the axis of the ordinate (see Figure 4).

Whereas for the equilibrium curves $L+G$ and $S+G$, ΔV varies rapidly with change of temperature or pressure, this value remains almost constant for the equilibrium $L+S$, and especially for S_1+S_2. Thus in the two latter cases $\frac{dT}{dP}$ = const., and the equilibrium curve becomes a straight line. As the practical performance of polymorphous inversions is sometimes quite difficult, e. g., the graphite-diamond inversion, the theoretical determination of the equilibrium line is of great importance. The slope of this line can be readily determined, if the specific volumes of the phases (or their specific weights) and the entropy are known. The entropy is usually determined from data on the heat capacity of the phases.

In order to locate the position of the equilibrium line, it is essential to know both its slope and the value of the constant member of the equation, which determines the intercept on the axis, i. e., ΔE in the following equation:

$$\Delta E - T\Delta S + P\Delta T = 0.$$

This calls for the experimental determination of the heat of transformation of the investigated phases at constant temperature and pressure. Obviously the value $\Delta Q = \Delta E + P\Delta V$, which can easily be calculated from the above-mentioned equations, does not provide us with any new information, as, if the position of the equilibrium line is unknown, the pressure to which the given ΔQ applies also remains unknown.

Further on (see p. 186 and 365) some examples of such calculations will be considered, both for the determination of the equilibrium line of polymorphous transformations, and for reactions involving the liberation of a gaseous phase.

The physicochemical diagrams with which we shall deal further on are geometrical models which enable us to assess the conditions of the system, i. e., its phases and their relative quantities, provided the factors which determine this state are given. These factors are, as we have seen, the composition of the system (if there is more than one component), and the conditions of its existence — temperature and pressure. In the most simple cases, with which we shall start our review, the factors to be dealt with are composition and temperature. Each state of the system, which is determined by its composition and temperature, should, in such a case, be represented by a definite point on the geometrical model (our diagram). The diagram should also show the boundaries between the fields or areas, the points of which represent states of the system, characterized by the same number of identical phases. These are areas showing equal qualitative phase conditions of the system. Various areas of such a diagram contain points which represent conditions of systems differing from one another by the phases which form them.

We should firstly know how to simply represent on the physicochemical diagram the composition both of the entire system as well as of the phases which comprise it.

The most natural method for depicting compositions is the so-called
30 barycentric system of coordinates. The principle of this method, its advantages and its limitations are evident from the following.

Let the composition of a certain substance be described on the diagram by the point *A*, and the composition of another substance by point *B*. If we mix the compositions represented by *A* and *B*, we may obtain an infinite number of mixtures, the composition of each of which will be completely dependent on the initial quantities of *A* and *B* taken. Each of the new compositions can also be represented by a point, and the location of this point is determined by the ratio of substances *A* and *B* taken. Should we take equal quantities of these, the resulting point, representing the composition of such a mixture, should naturally lie in the middle of the two initial points. A composition in which the *A* component is predominant should naturally be described by a point which lies nearer to the point *A*.

The composition of a mixture of two components is usually represented by a point upon a line. The distance from this point to the points which represent the pure components (*A* and *B*) is inversely proportional to the quantities of the components in the mixture. Such a system is called a barycentric system of coordinates. The compositional point, formed by the mixture of the components *A* and *B*, is determined here as the center of gravity of weights which are attached at these points, and which are proportional to the quantities of the components taken.

If we accept such a method of representing the composition of a two-component (binary) system in all ratios, we must obviously apply the same method for describing the compositions of a ternary system. Such a composition will be described by a point which is the center of gravity of weights attached at three points, which might be considered as apices of a triangle. Three different components cannot be colinear, as no single one of them can be obtained by a mixture of the other two, whereas, according to the accepted method of construction, a point upon a line, connecting two other points of compositions, describes a composition formed by the mixture of the said two compositional points.

Let us now consider three components, represented by points, A, B, C, their quantities being a, b and c respectively. The point which represents the composition of the mixture can be located by the following construction. Taking first the components A and B in the quantity of $a+b$, we obtain a composition represented by point D, which will be at the center of gravity of the weights a and b, on the AB side of the triangle ABC; on mixing the mixture of composition D with component C, we obtain a composition represented by the point M. This point will fall on the straight line CD, and according to the rule of barycentric coordinates $MD:MC=c:(a+b)$. Thus, in order to locate the point M, it is necessary to carry out all the operations necessary for the determination of the center of gravity of the weights a, b, and c, as applied at the points A, B, C, respectively.

Following our considerations, we arrive at the conclusion that the composition of a four component mixture should be represented by a point which is the center of gravity of four weights a, b, c and d, which are equal (or proportional) to the quantities of the components taken, the compositions of which are represented by points A, B, C and D respectively. These points cannot be coplanar, as none of the components can be obtained by a mixture of any of the others. Hence, in order to describe a substance composed of four components according to the rule of barycentric coordinates, one must resort to three-dimensional representation. The points which describe the composition of the pure components are located at the apices of a
31 tetrahedron, and all points within this tetrahedron represent all compositions formed by the mixing of these four components. The points upon the faces of the tetrahedron will represent all the compositions formed by three of the components, the points on the edges — compositions formed by two components.

In order to describe systems composed of more than four components, the method of barycentric coordinates — a visual method of representation in three dimensions — is no longer applicable. In order to achieve such a representation, more dimensions are needed, namely: in order to describe, by means of points, a system composed of n components, $(n-1)$ dimensions are necessary.

The colossal importance of the geometric method of representation consists, among other points, in its great visual clarity, which facilitates the revealing of regularities, when they are depicted on the diagrams. The present book is restricted to diagrams which do not extend beyond the boundaries of visual, three-dimensional space, although this strictly limits the number of components which may thus be represented.

SOME PROPERTIES OF BARYCENTRIC COORDINATES

Let the quantities of two components A and B of a system be respectively a and b (Figure 8, I). The composition represented by point M should fulfill the condition $MB:MA=a:b$; hence $MB:AB=a:(a+b)$ and $MA:AB=b:(a+b)$, i. e., the relative quantities of a and b in the mixture are proportional to the ratios of the lengths of the opposite segments (MB for the component A and

MA for the component B) to the length of the entire distance AB. In order to determine the percentage of A and B, the line AB should be taken as equalling 100 %. Hence, MB and MA will serve as a direct measure of the percentages A and B.

From the rule for determination of the point of composition of a binary system, as a center of gravity, the following may also be derived. Let M and N be compositions formed by the components A and B. We shall form from them the composition Q, taking them in the ratio $m:n=QN:QM$. The composition of Q could be formed by A and B, if these were taken in the ratio of $QB:QA$ (see Figure 8, II).

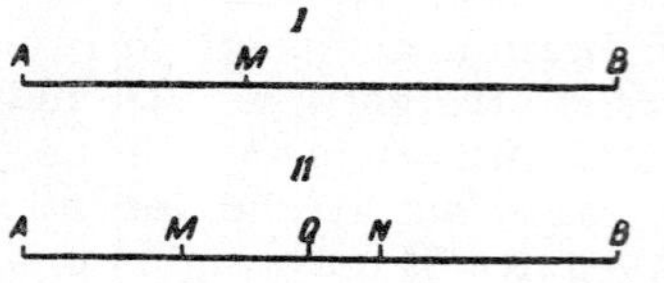

FIGURE 8. Representation of a binary system with barycentric coordinates

If a reaction occurs between M and N, and they are converted into a substance of composition Q, then this reaction will be complete, without any residuum, if $QM:NQ=m:n$; the location of points M, Q and N may thus represent the reaction $M+N\leftrightarrows Q$.

In a ternary system (Figure 9, I) composed of the components A, B and C, taken in the quantities a, b and c, the composition which they form together will be represented by point M which corresponds to the center of gravity of the weights a, b and c applied respectively from the points A, B and C. The weight which is applied at point M is $a+b+c$. Let us resolve it into two components; the weight $c+b$ applied at the point K and the weight a applied at A. Then $MK:MA=a:(b+c)$, and the relative amount of the component will be found from $a:(a+b+c)=MK:AK$. Hence the following rule is derived: in order to find the relative content of any component A in the composition of M, point A should be connected with M, and this line continued up to its intersection with the side BC of the triangle ABC, which
32 is opposite the corner A. The ratio of the segment MK to the entire length AK is the relative content of A in M. This rule holds for any shape of the triangle.

The weight applied at point K and equal to $b+c$, might be resolved into two weights, b and c, so that $b:c=KC:KB$. Through point M let us draw lines parallel to the sides of the triangle ABC (Figure 9, II). Then from the similarity of triangles we shall obtain:

$$\frac{PQ}{BC}=\frac{MK}{AK}=\frac{a}{a+b+c}. \tag{1}$$

On the other hand, $LM=BP$ and $MN=QC$; $LM:MN=c:b$; hence $BP:QC=c:b$. It follows from the last relation that:

and

$$\frac{b}{b+c}=\frac{QC}{BP+QC} \tag{2}$$

$$\frac{c}{b+c}=\frac{BP}{BP+QC}. \tag{3}$$

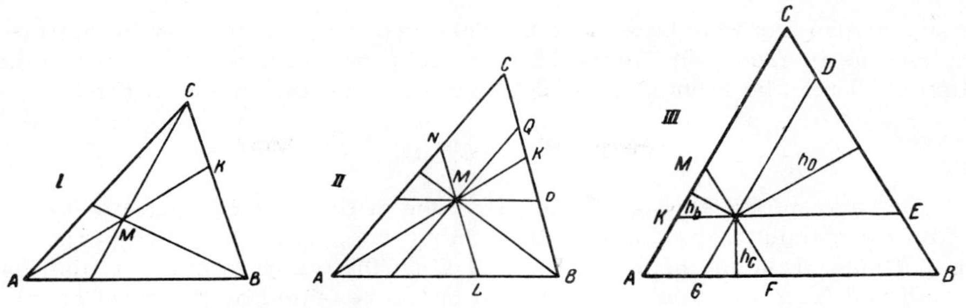

FIGURE 9. Representation of a ternary system with barycentric coordinates

From (1):

$$\frac{BC-PQ}{BC}=\frac{a+b+c-a}{a+b+c}=\frac{b+c}{a+b+c}\,. \tag{4}$$

As $BC-PQ=BP+QC$, we may multiply (2) by (3) and (4) and divide the second by the fourth, and the third by the fourth, obtaining:

$$\frac{BP}{BC}=\frac{c}{a+b+c}\ \text{and}\ \frac{QC}{BC}=\frac{b}{a+b+c}\,.$$

On drawing lines parallel to sides AB and AC through point M we shall divide the side BC into three parts, each of which is proportional to the relative quantities of components C, A and B.

$$c:a:b=BP:PQ:QC.$$

The property which is expressed by this equation is also independent of the
33 form of the triangle which we choose, the apices of which represent the components A, B and C. If this triangle is equilateral, on applying the construction mentioned above, we shall note the following property (Figure 9, III).

The content of every component (a, b and c) may, according to the above mentioned, be determined by the relations: $a=DE:CB$; $b=KM:AC$; $c=GF:AB$. Hence, in equilateral triangles ABC $CB=AB=AC$, and substituting the ratios of the sides (bases) of equilateral triangles by ratios of their heights, we obtain:

$$a:b:c=h_a:h_b:h_c,$$

i.e., the amounts of components are proportional to the normals drawn from the points of composition to the sides of the equilateral triangle opposite these points. If a triangle is so chosen that its height is equal to 100, these normals will show directly the percentage of the components.

The location of three points of composition in a binary system determines the direction of the reaction, during inversion processes, with the formation or disappearance of phases. Similarly, in a ternary system the location of four points determines the direction of the reaction between the corresponding four substances. Here two main cases may be distinguished, as is shown on Figure 10. In the first (Figure 10, I), the composition of D may be formed by A, B, C, and the reaction is one of the form:

$$A + B + C \rightleftarrows D.$$

In the second case (Figure 10, II), none of the four compositions can be formed without a residuum of the other three, but taking two of them, e. g. , D and B on one hand, and A and C on the other, a composition may be obtained from each pair, which will be represented by the point of intersection of the lines DB and AC. The reaction is in this case of the form:

$$D + B \rightleftarrows A + C.$$

In a four-component system we use a tetrahedron for depicting compositions. The shape of this tetrahedron may be arbitrary — both regular and irregular. By applying at its apices weights which are proportional to the quantities of the components, we shall find the point (M), which represents the corresponding composition, i. e. , the center of gravity of the mentioned weights.

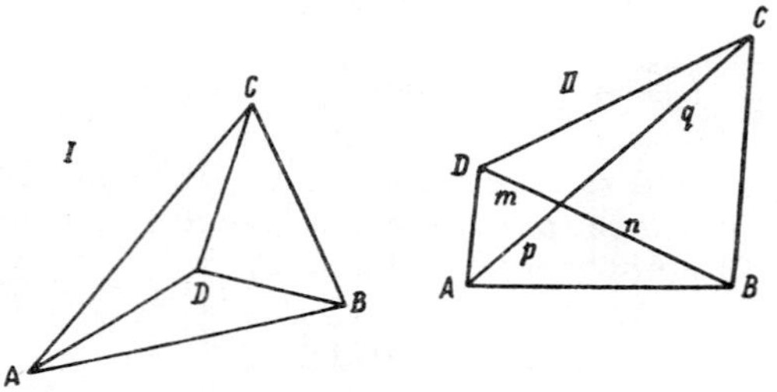

FIGURE 10. Two possible cases for the location of points which describe the compositions of phases in a ternary system, consisting of four phases:

I — the composition of one of the phases is within the triangle, the apices of which represent the compositions of three other phases; II — by connecting the composition points of the phases, a quadrilateral is formed.

34 If a straight line is drawn through M and one of the apices of the tetrahedron, e. g. , A, and produced till it meets the opposite face (Figure 11), one may imagine a weight which is applied at point M, its composition being formed by the four components as a resultant of the weight a of the component A and the sum of the other weights $b+c+d$, applied at point K. Hence $a:(b+c+d)=MK:AM$ and further on $a:(a+b+c+d)=MK:AK$, i. e. , in order to find the relative content of component A, we may apply the same construction as we did for the determination of the compositions of ternary systems.

The location of four points in a ternary system determines, as we have seen, two types of reactions which may occur at the interaction of these four compositions (phases); in a system of four components an interaction and reaction between five phases is possible, and the various positions of the corresponding five points determine different types of reactions. Figure 12 shows two possible types of location of five points according to the reactions $A+B+C+D \rightleftarrows E$ and $A+E \rightleftarrows B+C+D$. In the first case, the point E is located within the tetrahedron $ACBD$, in the second case — the line AE intersects the plane of the triangle BCD.

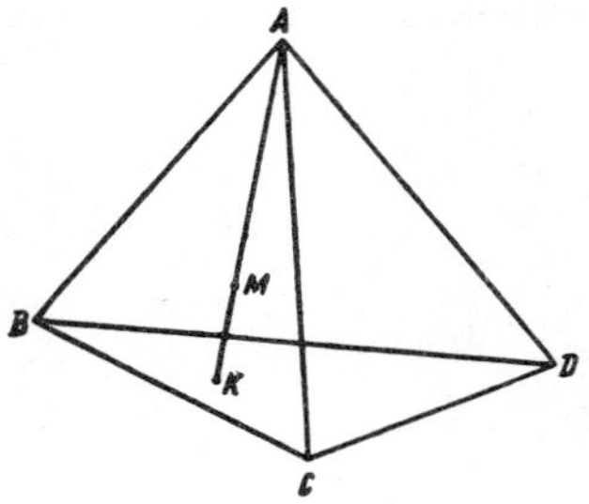

FIGURE 11. Representation of a four-component system with barycentric coordinates

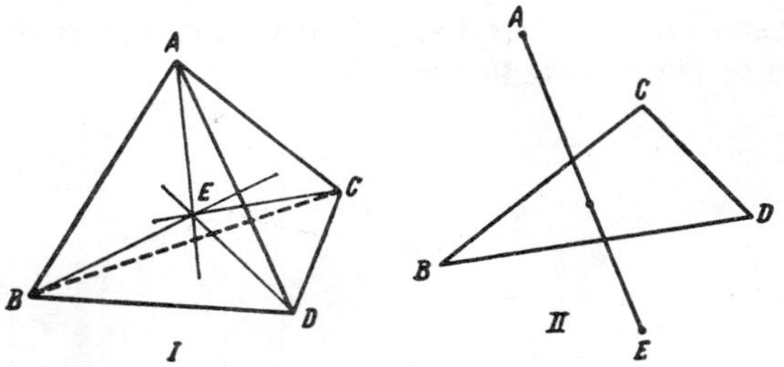

FIGURE 12. Two possible cases of location of composition points of phases in a four-component system, consisting of five phases

LIMITATIONS OF VISUAL PHYSICOCHEMICAL DIAGRAMS IN THREE DIMENSIONS

Physicochemical diagrams represent relations between the composition and the properties of systems. They are as sometimes termed "composition-property" diagrams. In order to depict on them not only compositions but also properties, dimensions of space must also be used for the demonstration of properties. In such a case the three dimensions of space are no longer at our disposal for the description of compositions; we then are forced to limit ourselves to fewer dimensions, and to simpler systems.

If both temperature and pressure are to be represented on the diagram, the representation of these two independent variables already call for two dimensions. Upon a flat, two-dimensional diagram, only the relations between temperature and pressure may be described, in this case repre-
35 senting a system which consists of one component. The depicting of relations between the composition of phases, temperature and pressure in a system of two components calls for a three-dimensional model, which may be represented upon a draft only by means of projections or sections. In the case of a ternary system, the relations between the composition and temperature may be represented only on a spatial diagram or model, which can be shown on a draft by means of one of the mentioned methods. Thus, we note that the possibilities of representing the states of various physicochemical systems in the form of visual diagrams is quite limited in comparison with the complexity of phenomena which are observed in natural processes. Nevertheless, the study of such simple systems is of great importance, as it provides us with a solid basis for the analysis of complex natural phenomena, which stems from concepts on simpler constituents of these phenomena. Thus, at the end of the previous chapter, the well known example of a physicochemical diagram was produced, demonstrating the

relations between temperature and pressure. This is a so-called diagram of state of a substance. We met with another example in the diagrams which expressed the dependence of the thermodynamic potential on temperature. In both types of diagrams, the change of composition is not included. The most simple and visual diagrams of binary systems, which help to explain the processes of changes on transformations from liquid phases into solid and vice versa, are of great interest for the interpretation of geological processes.

CHAPTER III

DERIVATION AND GENERAL REVIEW OF CRYSTALLIZATION DIAGRAMS OF BINARY SYSTEMS NOT CONTAINING VOLATILES

Of the two possible states of a system with a given composition, the more stable is the one with the lower thermodynamic potential. This enables the prediction of the direction of processes, as the system tends to pass from an unstable to a stable state. Hence, it is initially desirous to determine how the thermodynamic potential varies in phases of various composition, corresponding to the various states of the system.

Phases which consist of two components may present the following cases:

1. These components are pure substances. Each phase consists of only one component. The phases are immiscible, or insoluble in each other. They may only form mechanical mixtures.

2. Solutions showing unlimited ratios of components. The solutions may be solid (crystalline), liquid or gaseous. For liquid and solid phases, unlimited mutual solubility is only one of the possible cases. The gases are always mixed in all ratios, forming a homogeneous gas.

3. Solutions of components with limited mutual miscibility. These solid or liquid solutions cannot contain components in all possible ratios. Their ability to dissolve one another mutually is limited, and, as is said, a miscibility gap exists.

4. One, or several chemical compounds, consisting of two components. These compounds may be stable, both in the solid and liquid states; they may dissociate on melting, decompose below the melting point or pass into solid solutions of variable composition on increasing the temperature.

THERMODYNAMIC POTENTIALS OF BINARY MECHANICAL MIXTURES AND SOLUTIONS

1. Mechanical mixture of two pure components. Two pure components do not form solutions in all proportions. Let the thermodynamic potential of one of them equal ζ_1, and that of the other ζ_2; if the amount of the first, per unit mass, equals x, then the amount of the other will be $(1-x)$. It is known that the thermodynamic potential of a heterogeneous system such as a mechanical mixture is in general determined by the following equation:

$$\zeta = \frac{\Sigma m_i \zeta_i}{\Sigma m_i};$$

37 in our case

$$\zeta = \zeta_1 x + \zeta_2 (1 - x),$$

or, after transformation of this equation:

$$\zeta = (\zeta_1 - \zeta_2) x + \zeta_2,$$

where ζ_1 and ζ_2 are constant values, independent of x. Hence, the equation is that of a straight line (Figure 13).

2. **Solution of two components in unlimited proportions.** If the components form homogeneous solutions, in which they may participate in all proportions, the shape of the thermodynamic potential versus composition curve may be clarified from the following considerations. Let us take two close compositions, each of which, according to the conditions, is a homogeneous solution. If a mechanical mixture of an intermediate composition c is formed from the two, then, according to the above, the thermodynamic potential of this mixture will be the ordinate of point c (Figure 14) on a straight line, drawn through points a and b, which represent the two close compositions under consideration. Such a mechanical mixture is, however, unstable; according to the condition of complete miscibility, it should become a homogeneous solution. Hence, the thermodynamic potential of this solution, being the more stable phase, should be lower than that of the mechanical mixture. Thus, point c on the thermodynamic potential curve for the solution of this composition should lie on the same ordinate as c', but below it.

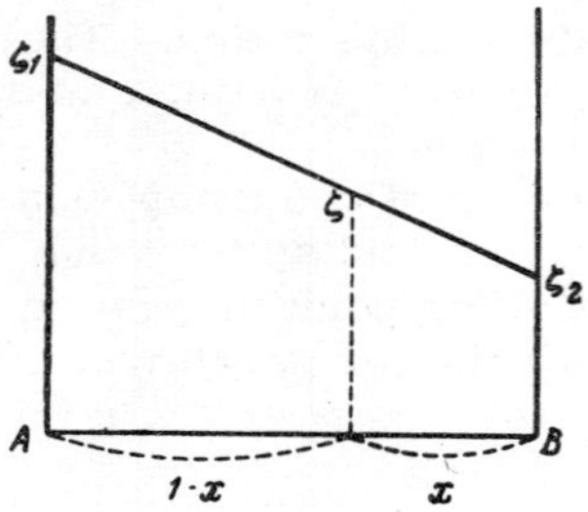

FIGURE 13. Thermodynamic potential of a mixture of two phases

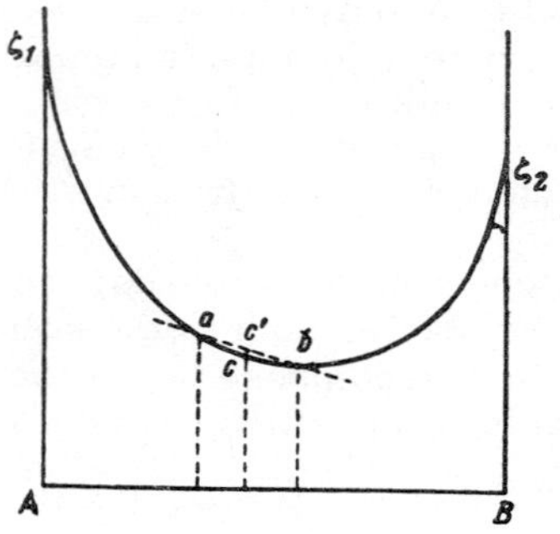

FIGURE 14. Thermodynamic potential of unlimited solutions

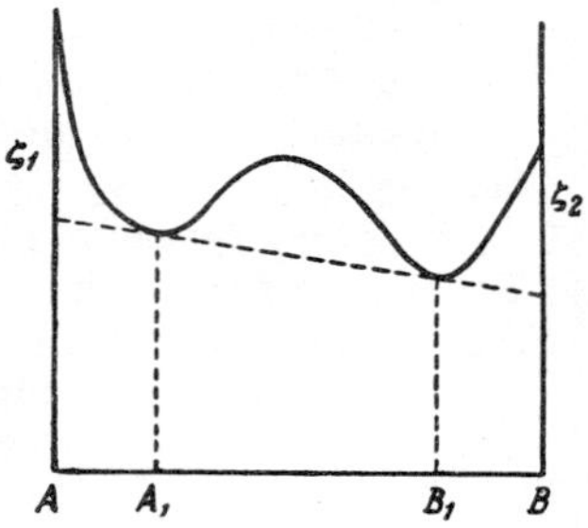

FIGURE 15. Thermodynamic potential of partially miscible solutions, with a solubility gap between points $A_1 - B_1$

Such a location of points should arise for any two points a and b, however close together, along the entire length of the curve. This may occur only if the entire length of the thermodynamic potential curve for homogeneous solutions is convex downward (see Figure 14).

3. **Solutions with limited miscibility.** In this case (Figure 15), compo-
38 nent A may form a homogeneous solution with component B only up to a certain limit, which is indicated on the diagram by point A_1, whereas component B may dissolve A to a certain limit, indicated by the point B_1.

A GENERAL CHARACTERISTIC OF EQUILIBRIUM DIAGRAMS OF UNARY SYSTEMS**

Before passing on to considering the compositional representation of multicomponent systems, let us review in brief the properties of the unary diagrams, although some of them have already been mentioned.

24 As the Phase Rule shows, the following cases of equilibrium are possible in a unary system: 1) the system consists of one phase and has two degrees of freedom (i. e., variance of both temperature and pressure); it is depicted on the diagram as an area (see Figure 2); 2) a system of two phases, possessing one degree of freedom, is represented on the diagram as a line which separates the fields of stability of single-phase systems, and 3) a system of three phases (i. e., the maximum number of phases) which has no degrees of freedom; its equilibrium is represented by a point where three lines, each representing two phases, intersect.

On Figure 3, two types of one-component systems are shown, both of them consisting of three phases: 1) gas + liquid + solid coexisting at the point D_1 and 2) gas + solid phase S_1 + solid phase S_2 coexisting at point D_{1-2}. There is one more type which is possible, but more rare — a case of coexistence of three solid phases S_1, S_2 and S_3 (e. g., in the system H_2O at high pressures). Two gaseous phases are impossible in a system, and two liquid phases are only possible if the system consists of more than one component, as the two liquid phases must differ from each other in their composition.

The liquid-gas equilibrium curve terminates in the so-called critical point. *Critical points on equilibrium diagrams are those points where two phases become identical both in their properties and composition (in polycomponent systems), i. e., they become one phase.* Such is the case in the example cited above where the difference between the liquid and gaseous phases disappears. It is obvious from the diagram that, according to the termination of the boundary curve, the transition from the area of a typical liquid phase to the area of a typical gaseous phase may occur in two ways: 1) on boiling of the liquid, i. e., by crossing the boundary curve, and 2) in an indirect way, i. e., by gradual conversion. The first case occurs for instance if water, at a pressure of 1 atm., is heated up to 100° and boiled. The temperature of the system will not change on boiling, as there is only one degree of freedom, and the pressure is constant. The water will gradually change into vapor at the expense of the so-called latent heat of vaporization. However, the same result, i. e., water vapor at 100°, may be obtained from water in a different way; first, the pressure may be raised above the critical (e. g., up to 300 atm.); secondly, the temperature may be raised above the critical (e. g., up to 400°); thirdly, the pressure may be lowered to 1 atm., and lastly, the temperature may be lowered correspondingly. All conversions will occur gradually without intersecting the two-phase lines. No clear boundary or discontinuous change of properties are observed on passing the critical pressure or critical temperature when taken separately. We shall confine ourselves to these remarks and shall return to this interesting problem further on in a special chapter (XIV).

In one-component systems the coexistence of two or even three solid phases of the same composition is possible. These are the so-called

Within the range $A_1 - B_1$, homogeneous phases are unstable and separate into two phases.

In such a case, the thermodynamic potential curve cannot be convex downward between the points A_1 and B_1, as this would indicate complete miscibility of A and B. On the contrary, the curve is convex upward over the range of the considered section. Thus, all intermediate solutions are unstable, and separate into A_1 and B_1, the composition of which may be determined by drawing a tangent to the curve, as shown on Figure 15. Within the ranges AA_1 and BB_1, the condition of complete miscibility is fulfilled.

These curves, which reflect the dependence of ζ on the composition in various cases of solubility, are geometrical expressions of the necessary conditions to be satisfied by this dependence and which correspond to the laws of thermodynamics, as summarized in the concept of the thermodynamic potential.

A certain type of curve is preserved as long as the conditions of mutual miscibility are maintained; when temperature is changed, the form of the curve changes, and its type may sometimes also change. When considering the processes of transformations in systems, it is the difference in curve types which is of interest to us; other properties of these curves may here be dispensed with. The mentioned properties are sufficient for the drawing of very important conclusions.

The differences in the type of relationship between the thermodynamic potential and the composition in the three considered cases are geometrically expressed as figures. The analysis of these figures also shows up those characteristics which precisely express the differences in the course of physical phenomena in various cases.

Let us consider the process of crystallization in a binary system, i.e., the conversion process of a given system from a homogeneous liquid phase to a heterogeneous system, which finally passes into a system of solid phases only. Depending on mutual miscibility of the components in various phases, different types of reactions may occur. We may anticipate their special features and review all the possible principal types.

THE CASE WITHOUT SOLID SOLUTIONS

In melts the components exhibit unlimited mutual miscibility; in solid phases they form no mutual solutions at all.

From the fact that crystals do melt at high temperatures, it should be concluded that, at these temperatures, the ζ of the solid phases, i. e., of the phases which are composed of atoms arranged in regular crystal lattices, is higher than the thermodynamic potentials of the liquid phases, which are comprised of the same atoms and atom groupings, but in the state of disorder, and are capable of shifting their positions.

If the temperature of the system is above the melting points of both the components, the thermodynamic potential of any mixture formed from the crystalline phases (conjecturing that these phases do possess crystal lattices at such a temperature), should be higher than the thermodynamic potential of a homogeneous phase of corresponding composition in the liquid

state. This is due to the fact that the entire thermodynamic potential curve of the liquid phases, which is convex downwards, should lie below the straight line representing the thermodynamic potentials of a mixture of
39 solid phases. On the other hand, if we consider a system of two completely crystallized components at any composition (i. e., at any ratio of the components), we should conclude that the straight line, representing the thermodynamic potentials of the mixture of solid phases, lies below the thermodynamic potential curve of the liquid phases at this temperature.

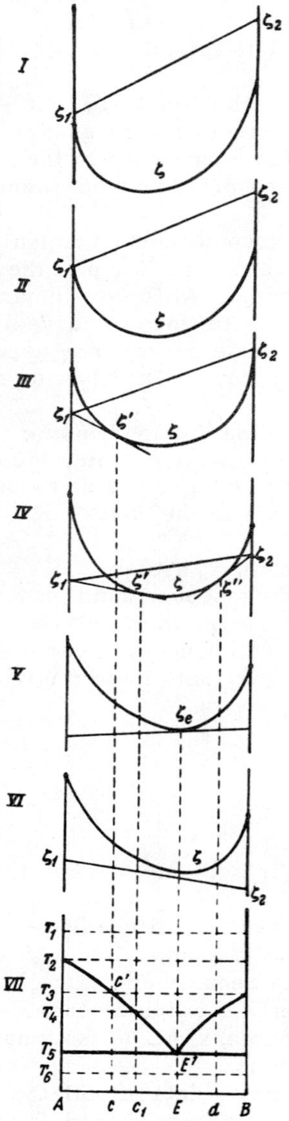

40
FIGURE 16. The derivation of a binary diagram with the absence of solid solutions and compounds

During the crystallization process potentials of both the solid and liquid phases change. This change, however, occurs in such a way that the curve of the potentials of the liquid phase is shifted from its position below the straight line of the potentials of mixtures of solids to a position completely above this line. On considering this process in continuity, we shall note some important geometrical relations.

Let points ζ_1 and ζ_2 be the ordinates of the thermodynamic potential for the corresponding solid phases and liquids. As long as points ζ_1 and ζ_2 are above the ζ curve, crystallization has not begun. As the temperature is lowered, the curve of ζ rises, and its terminals reach points ζ_1 and ζ_2. When, for example, the left end of the curve reaches point ζ_1 we obtain an equilibrium between the liquid and the solid phase of the composition of the pure component *A* (Figure 16, II). Further on ζ rises in relation to ζ_1 and ζ_2, leaving ζ_1 below the left end of the curve. A tangent to the ζ curve can be drawn from the point ζ_1. The point of contact ζ' represents a composition *c*. On examining the obtained figure (Figure 16, III), it may be concluded that, between *A* and *c*, the straight line connecting ζ_1 with the point of contact ζ' is lower than the ζ curve. This means that within the range of the mentioned compositions, the mechanical mixture of crystals *A* and liquid *c* is more stable than the homogeneous liquid of the same composition. This liquid will separate into crystals of *A* and liquid *c*. This liquid is poorer in *A* and richer in *B* than the initial liquid. The liquid phase of composition between *c* and *B* will be the more stable at this temperature.

If a temperature-composition diagram is constructed (Figure 16, VII), then, in order to represent the above-mentioned features of the system, a line should be drawn parallel to the abscissa with T_3 as its ordinate. This line is divided into two parts by the point *c'*, which corresponds to the composition of c: to the left of this point the system will consist of a liquid of composition c and crystals of A; to the right of this point it will be one homogeneous liquid.

At a temperature lower than the melting point of the solid phase A, but infinitely close to it, the point ζ_1 will be infinitely close to the point ζ'.

As the temperature is lowered, point ζ_1 will also be increasingly lowered in relation to the left end of the ζ curve. Correspondingly, the point of contact ζ' will be shifted increasingly to the right of composition A. This follows directly from the fact that the ζ curve is convex downwards throughout its length. Such a change indicates that the field of stability of systems consisting of crystals A and a liquid will broaden, and the liquid will be enriched in the component B.

At a temperature corresponding to the lower melting point of the component B, the thermodynamic potentials of its liquid and solid phases become equal, and, with a further drop in temperature, the point ζ_2 will drop below the ζ curve; now a tangent to the curve ζ may also be drawn from the right side of the diagram, from $\zeta_2\,\zeta''$; a field of stability appears consisting of mixtures of component B with liquids. On the thermodynamic potential diagram, their lowest values, which correspond to the stability of systems of various compositions at temperature T_4, will be represented by the ordinates of the following segments: the straight line $\zeta_1\zeta''$ (solid phase A + liquid), the curve $\zeta'\zeta''$ (liquid) and the straight line $\zeta''\zeta_2$ (solid phase B + liquid). On referring to the composition-temperature diagram, we determine the areas of coexistence of two phases at temperature T_4 (see Figure 16, VII); solid phase A plus liquid c_1 (a liquid of intermediate composition between c and B) and liquid d + solid phase B.

On lowering the temperature, points ζ' and ζ'' approach one another, and eventually coincide at a certain point ζ_e (see Figure 16, V), which corresponds to a definite composition E. This will be a case of invariant equilibrium, as the number of phases and their composition remain constant, and a transformation of the liquid E into a mixture $A + B$ occurs.

On the composition-temperature diagram, the melting point curves meet at a point E'. At lower temperatures the ζ curve lies entirely above the straight line $\zeta_1\zeta_2$. In this case the mixture of the solid phases A and B is stable, and our binary system will yield crystalline aggregates of A and B at all compositions.

Point E' is the eutectic point, and E represents the eutectic composition. The field below point E' is limited from above by a straight line passing through this point and parallel to the abscissa; it represents the area of existence of mixtures of solids A and B only.

Thus, four fields are distinguished on the temperature-composition diagram, each differing from the others in the number and composition of the phases which are stable in them.

Let us take a random composition (c on Figure 16, VII) and trace the course of its crystallization. Above temperature T_3 a single phase is observed; at temperature T_3 the crystallization of A commences; as the temperature drops, these crystals continue to form, and the liquid is enriched in component B. The change in composition of the liquid and in the temperature is represented by the section of the curve $c'E'$. At temperature T_4 the system will consist of a liquid of composition c_1 and crystals of A;

the relative amounts of liquid and crystals are determined, according to the rule of barycentric coordinates, by the ratio of segments $Ac:cc_1$. On cooling to the temperature T_5, the liquid will attain the composition E, and the
41 amount of crystals of A within it will be $\frac{cE}{AE}$. On further cooling of the liquid, A and B will crystallize out simultaneously, but the temperature will remain constant, owing to the liberation of latent heat of fusion, as long as liquid still remains. The point E' is an invariant point of stability of three phases: A, B and the eutectic solution E. The relative amounts of A and B, which simultaneously separate from the eutectic solution, may be determined from the ratio of the segments $BE:AE$, and the total amount of crystals of A in the entire mixture is represented by the ratio $cB:AB$.

Thus, the concept of thermodynamic potentials provides us with a means of geometrically portraying the eutectic scheme of crystallization of binary systems. It is evident that crystallization with the formation of eutectics is a necessary outcome of the first and second laws of thermodynamics, which are combined in the concept of the thermodynamic potential. This arises on condition that the solid phases of the components are completely immiscible, whereas the liquid phases are completely miscible. The eutectic scheme of crystallization is not an empirical generalization of isolated experiments but the outcome of general laws. Therefore, this scheme is applicable in cases of any natural phenomena, provided that the two above-mentioned conditions are fulfilled.

CRYSTALLIZATION WITH FORMATION OF CONTINUOUS SOLID SOLUTIONS

In this case the variations of thermodynamic potentials of both liquid and solid phases versus their composition will also be represented by curves which are convex downwards throughout their lengths.

At high temperature, when a system of any composition is in the molten liquid state, the thermodynamic potential curve of the solid phases will lie entirely above the potential curve of the liquid phases. The exact opposite holds true at sufficiently low temperatures. When the system is crystalline at any composition, the potential curve of liquid phases will lie above that of the solid phases. Thus, on lowering the temperature of crystallization, a shift of the potential curves occurs. The curve of the liquid phases, starting from below the curve of the solid, then rises above it. An intersection of the curves is thus unavoidable.

From the geometric configuration of the curves, which are convex downwards throughout, it follows that three cases of intersection are possible during such shifting.

1. The curves, on approaching one another, meet at one set of their end points; their intersection point is obviously the melting point of the more refractory, pure component. On further shifting of the thermodynamic potential curves, they always intersect at one point only. This point is shifted along the curves from the more refractory to the less refractory component. The last common point of the thermodynamic potential curves is their other set of end points — representing the melting point of the more fusible, pure component.

2. On cooling the system, the thermodynamic potential curves of the liquid and solid phases approach one another and intersect at a point representing a certain intermediate composition. A further shifting of the curves gives rise to two points of intersection. Both tend towards the
42 end points of the curves which will be reached, one after the other.

3. The thermodynamic potential curves approach one another and meet at one set of end points. Further on the curves will intersect in the vicinity of this point, and the intersection point will shift away from this end, but later on the curves will intersect at the other set of end points. A second intersection point thus arises which likewise shifts away from the last-mentioned set of end points. On further shifting of the curves, their intersection points approach one another and eventually coincide at the point where the curves separate from one another.

According to Roozeboom, there are three different types of crystallization of solid solutions, which correspond to these three possible geometric cases of intersection of the thermodynamic potential curves.

Type I

On cooling, the thermodynamic potentials of the solid and liquid phases of one of the components become equal at the melting point of this component.

Figure 17 shows the successive shifting of the intersection of the curves ζ_l and ζ_S and the change in the position of their common tangent. First the curves meet at one set of their end points (see Figure 17, II). On further cooling of the system, the curves intersect, and a common tangent may be drawn (see Figure 17, III). At this moment the stable phases are: 1) solid phases of composition from *A* to the phase with the potential ζ_{S_1}; 2) a mixture of the solid phase with potential ζ_{S_1} and the liquid with the potential ζ_{l_1} over the range which corresponds with the composition of these potentials; 3) liquid phases ranging from the composition corresponding to the potential ζ_{l_1} to the composition *B*.

On further cooling, the intersection point of the curves ζ_l and ζ_S will be shifted from left to right; the segment corresponding to the tangent to ζ_{l_1} and ζ_{S_1} on Figure 17, III. The length of this segment increases from zero, corresponding to the melting point of the more refractory, pure component (see Figure 17, II), to a certain value (see Figure 17, IV) and then decreases again to zero (at the melting point of the more fusible component). On
43 cooling below the melting point of this component, the curves ζ_l and ζ_S diverge.

If we construct a composition-temperature diagram, using the same methods as employed in the construction of the eutectic curve, the following fields may be distinguished upon it: liquid, liquid + solid solution, solid solution. These fields are limited by two curves, which represent the compositions of the coexistent liquid solution (the liquidus curve) and solid solution (solidus curve).

Let us trace the course of crystallization. We shall consider a composition (Figure 18) which is completely liquid at high temperatures (down to the temperature T_1). At the temperature T_1 a solid phase of composition

S_1 starts to crystallize out. At the temperature T_2 the system consists of liquid L_2 and the solid phase S_2 in the ratio $MS_2:ML_2$. As the temperature is lowered the points L_2 and S_2 are shifted to the right, i. e., the composition of both the liquid and the solid solution change. The ratio $L_2M:S_2M$ increases steadily, i. e., the amount of the solid phase increases, and that of the liquid decreases. At the temperature T_3 the solid solution reaches the composition M, and all the liquid is used up. Crystallization is terminated.

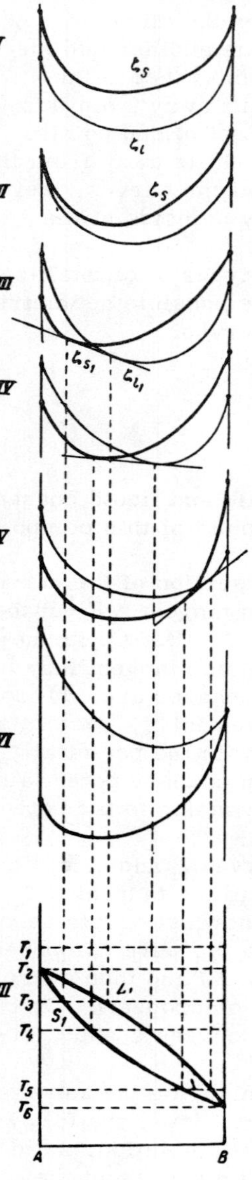

FIGURE 17. Derivation of a binary system diagram with a continuous series of solid solutions (Type I Roozeboom). I-VI — thermodynamic potential curves: ζ_S — of solid phases (bold lines); ζ_l — of liquid phases (thin lines);

It is evident from the considered process that the composition of the solid solution changes during crystallization, i. e., an exchange of matter occurs between the solid solution and the liquid. The solid solution is always richer in the more refractory component than is the liquid, and they both change in composition in the same direction: the change in the composition of the liquid appears to follow the change in composition of the solid solution.

It should be noted that the less the difference in melting temperature between the pure components A and B, the less is the vertical divergence of the solidus and liquidus curves. If these components have equal melting temperatures, these curves are theoretically fused into a single, horizontal, straight line.

Type II

The changes in thermodynamic potentials with decrease in temperature are in this case represented on Figure 19. At a certain moment the potentials of the liquid and solid phases are equal at a certain intermediate composition. The corresponding curves touch each other (see Figure 19, I). Later, the curves ζ_S and ζ_l intersect, as shown on Figure 19, II.

On drawing the common tangents to both curves, it will be noted that there are two such tangents. As cooling proceeds the curve of the solid phase ζ_S will overtake the curve of the liquid phase ζ_l on its downward path. The intersection points of the curve will diverge, in similarity to the contact points of the two tangents, approaching the margins of the diagram. When the melting point of the more refractory component is reached, Figure 19 will achieve the form III; after reaching the melting point of the more fusible component, Figure 19 will have the form IV. The melting point curve (see Figure 19, V) will have a maximum for a certain intermediate composition *C*. Its branches to the right and left of this maximum are similar to the melting point diagram of Type I. In this case, too, the solid solutions are richer than the liquid in that constituent, the addition of which raises the melting point of the melt. At the maximum point the compositions of the liquid and solid phases are identical, and the melt fuses and solidifies at the same temperature as an individual substance.

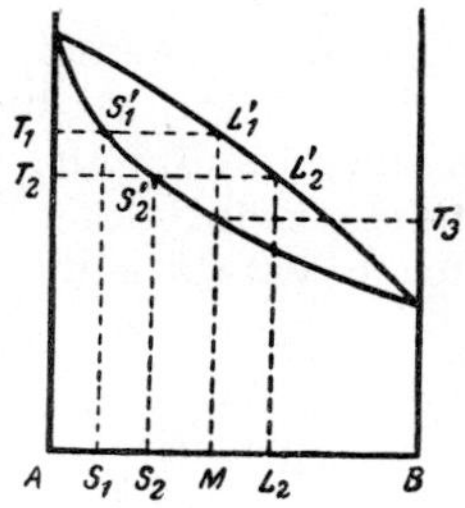

FIGURE 18. Crystallization of solid solutions of Type I

44 [This type of melting point diagram is very rare and has not been noted for systems of any importance in the consideration of mineral formation processes to the present day].

Type III

In the case when the first notable change on decreasing the temperature is that the potentials of the liquid and one of the pure components become equal, we have already seen (crystallization of Type I) that on the addition
45 of the other component the melting curve is shifted downward, and the composition of the liquid phase differs from that of the solid, which contains a greater amount of the more refractory component. But a case may arise, differing from Type I, where on further reduction of temperature the thermodynamic potentials of the other component and of the liquid will become equal not at the last moment, but earlier than occurs for some middle part of the curves. In this case, two points of intersection of curves ζ_s and ζ_l will arise, as seen from Figure 20, II and III. Correspondingly, in the part of the curve nearest to the more fusible component we will find: 1) a lowering of the point of crystallization on the addition of the other component, and a different composition of the solid and liquid phases; 2) the composition of the solid phase will be richer in that component, the addition of which raises the melting point.

It is obvious that on further lowering of temperature a situation is reached where the curves of the solid and liquid phases, ζ_s and ζ_l, will meet before the curve of the solid phase, throughout its length, drops below the liquid curve. This moment corresponds to the minimum upon the melting point curve (Figure 20, IV).

[The diagrams of Type I are characteristic of isomorphous mixtures, in which the relative difference in radius between the ions which replace one another is small (usually up to 15 %), whereas the diagrams of Type III are characteristic of mixed crystals with a greater difference in ionic radii (up to 40 %). This latter type covers, in particular, the diagrams of most systems composed of the analogous salts of potassium and sodium, which also have the same structure, e. g., the system KCl—NaCl, which was studied by Kurnakov and Zhemchuzhnyi].

EXSOLUTION OF SOLID SOLUTIONS BELOW THE MELTING POINT CURVE**

On lowering the temperature the stability of solid solutions decreases. Therefore, in several systems, in which a continuous series of solid solutions is formed on crystallization, their exsolution is observed below the crystallization curve. This phenomenon is especially characteristic of systems with melting point diagrams of Type III, i. e., for crystals with

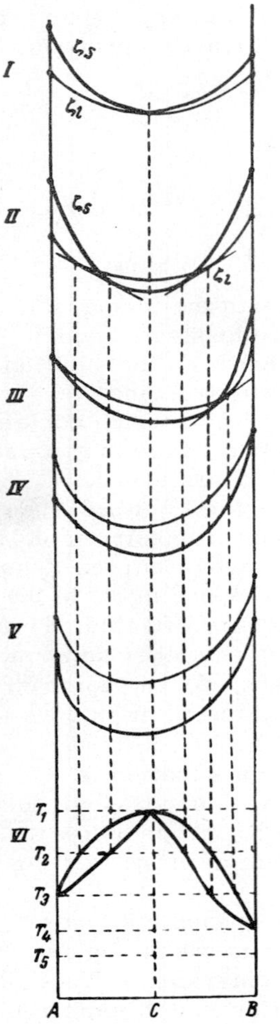

FIGURE 19. Derivation of a binary diagram with a continuous series of solid solutions (Type II of Roozeboom — a curve with a maximum)

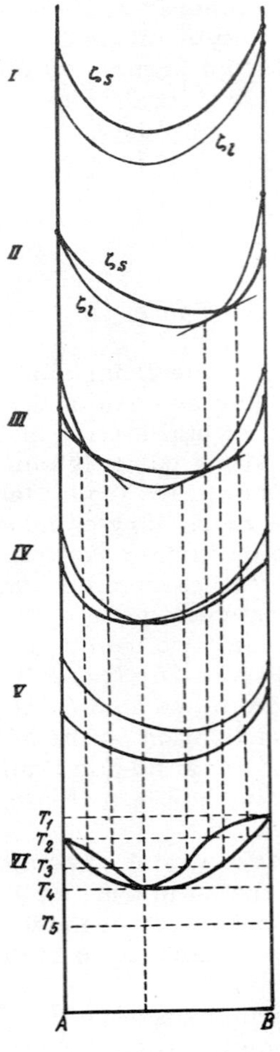

FIGURE 20. Derivation of a binary diagram with a continuous series of solid solutions (Type III of Roozeboom - a curve with a minimum). I-V - Thermodynamic potential of the solid (bold line) and liquid (thin line) phases.

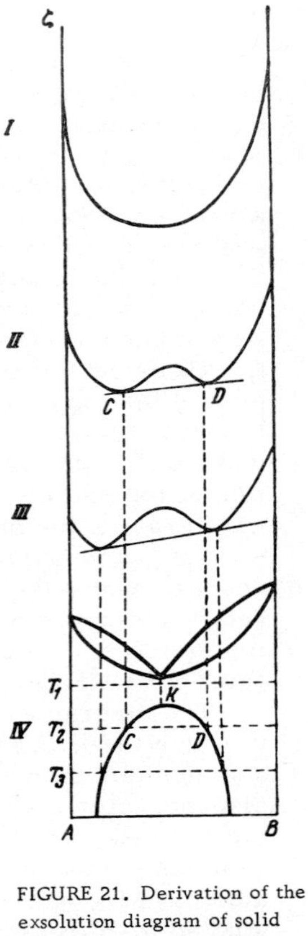

FIGURE 21. Derivation of the exsolution diagram of solid solutions below the melting point curve

considerable differences in the radii of the mutually replacing ions. In systems with melting point diagrams of Type I, the exsolution of solid solutions is much rarer.

At high temperatures, but already below the melting point curve, i.e., in the field of the solid phase, the curve of potentials of solid solutions will be convex downward (Figure 21, I). At lower temperatures a flexure which is convex upward appears on this curve, thus being similar to the case mentioned in relation to exsolution of liquid solution (Figure 21, II). A tangent may be drawn to such a curve and will contact it at two points ***C*** and ***D***. The location of these points of contact determines the limiting compositions of the solid solutions as the potential of the mixture of two solid phases is lower on this segment than is the potential of solid solutions. A further drop in temperature will not bring about any change in the type of the curve, but the distance between the points of contact, which determine the compositions of the limiting solid solutions, increases (Figure 21, III).

Correspondingly, one may draw two diverging curves on the composition-temperature diagram which represent the compositions of the limiting solid solutions in equilibrium with each other (Figure 21, IV). At a certain temperature between T_1 and T_2 (Figure 21, I and II), the configuration of the potential curve changes, i.e., a flexure appears, but the distance between the points of contant is extremely small. These two curves will meet at point ***k***, and the diagram will have the form of a curve, convex upward, with a maximum below the melting point curve. The maximum point is actually the critical point of exsolution, as here two coexisting solid phases have the same composition and properties, i.e., they are converted into a single phase.

CRYSTALLIZATION OF SOLID SOLUTIONS WITH LIMITED MISCIBILITY*

46

In the case of limited miscibility in the solid phase, we assume the presence of a flexure, convex upward, in the corresponding ζ_s curve. There may be two possible cases of intersection of this curve with ζ_l curve. The first arises when the potential curves intersect at only one point throughout the process of cooling and corresponds to Type IV of Roozeboom. This intersection point appears at the melting point of the more refractory component when the potential curves touch, and, being shifted, it disappears at the temperature of melting of the less refractory component, below which temperature the potential curves separate. The process occurs as shown on Figure 22. On comparing the potential diagrams, we obtain the following form of melting point curve. On the right side, beginning with the melting point of component ***B***, there are two curves: one for the liquid and the other for the solid, just as at the start of the Type I process. But the situation shown on Figure 22, IV, is quite specific to this curve. At this moment the common tangent to the curves contacts them at three points — two points on the solid solutions curve and one on the liquid curve. The location of these points is shown on the diagram. Three phases coexist at this moment: the liquid l and two solids S_1 and S_2. On the melting point diagram the composition of the liquid corresponding to this temperature is ***C*** and that of the solids A_1 and B_1.

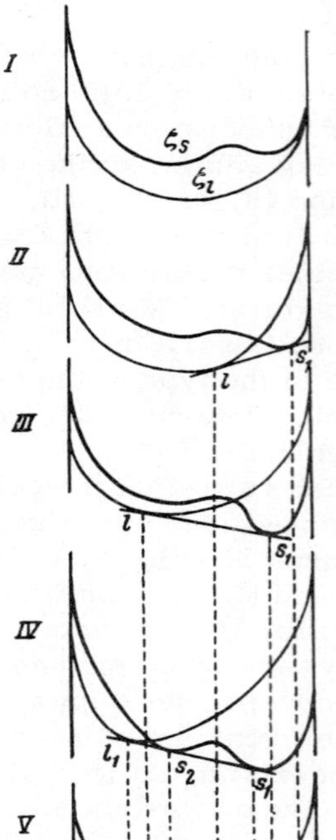

FIGURE 22. Derivation of a diagram of a binary system showing limited solid solutions (Roozeboom's Type IV)

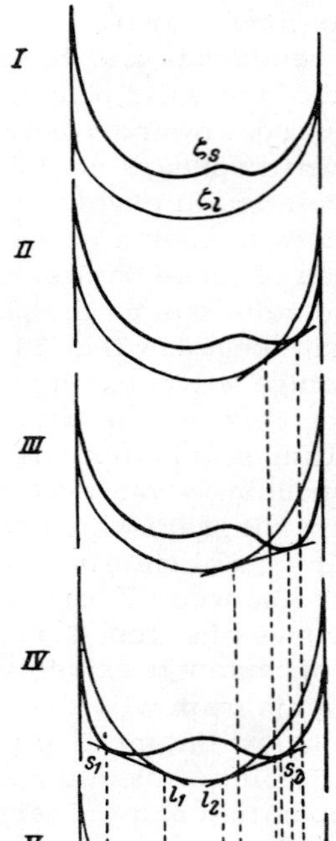

FIGURE 23. Derivation of a diagram of a binary system showing limited solid solutions (Roozeboom's Type V)

On further cooling we obtain two tangents (Figure 22, V). The tangent ls_3 corresponds to the mixture of two solid solutions, the tangent s_1s_2— to a mixture of the solid solution (richer in the component A than the preceding solutions) and the liquid. Melting point curves of this solid solution and liquid will intersect at the melting point of component A, in a manner similar to Type I.

The diagram of Type IV shows only one intersection point between the potential curves of the liquid and solid solution. Type V of Roozeboom (Figure 23) shows two intersection points. In this case the concentration of the liquid melt in equilibrium with two limiting solid solutions will be within the concentration boundaries of these solid phases (Figure 23, V).

On the temperature-composition diagram the course of crystallization is represented in the same way as was done for other types of crystallization. Bearing in mind what has already been said about the construction of melting point curves, the diagram hardly requires any explanation. At temperatures T_2 and T_3, one liquid and one solid phase may coexist in equilibrium; at the temperature T_4, depending on the initial composition, combinations of liquid l_1 and solid s_1 or liquid l_2 and solid s_2 are possible. At temperature T_5 the liquid l and two solid solutions s_1 and s_2 are in equilibrium. This moment corresponds to the eutectic point.

The general trend of crystallization is the same as in the case of a eutectic system of immiscible components, but the separated solutions will consist of A with some content of B to one side of the eutectic, and of B containing A on the other side. The composition of the solid phase will differ from that of the liquids, both of which change as crystallization proceeds.

[It is evident that a diagram of Type V may be constructed on the basis of that of Type III if the temperature of exsolution of solid solutions rises above the melting point curve. Such a transition occurs in several systems showing a gradual increase in the difference between the radii of the mutually
48 replacing ions in crystals of varying compositions]. Much more rarely — only in cases of isomorphous exchange of ions of various types — is a direct transition from Type I systems to Type IV systems observed.

On comparing Figure 16 with Figure 23, it becomes evident that the usual scheme of eutectic crystallization is the limiting case of this type of diagram (Roozeboom's Type V), on the decrease of the miscibility range in solid phases. It is self-evident that the derivation of diagrams of the former type (see Figure 22) cannot be applied directly in the absence of solid solutions, as, in the latter case, there is no convex part to the thermodynamic potential curve in the solid phases. The only possible diagram here is the one shown on Figure 16.

On decreasing the miscibility limits of solid phases for diagrams of Type IV (see Figure 22), we shall arrive at a diagram of the type shown on Figure 16, where the eutectic point is extremely close to the melting point of the more fusible component. The above-considered types of melting point diagrams of solid solutions were derived by the Dutch chemist Roozeboom, and are known in literature as the five Roozeboom types. It should be mentioned that Roozeboom arrived at his conclusions on the basis of the laws of N. N. Konovalov, which apply to equilibrium of gaseous and liquid phases in binary systems and show completely analogous relations. Originally, Roozeboom derived these types for a more complicated case — for three-component systems where water is one of the components and the other two are salts forming solid solutions. These diagrams merely reflected the relations between the salts of the solid and liquid phases, not taking the water into account. These are the so-called distribution curves (of V. S. Sobolev). Roozeboom stressed that the diagrams were constructed directly on the basis of Konovalov's laws. Somewhat later (1899), Roozeboom formulated a general derivation of the sited diagrams of melting of solid solutions.

CASE OF ISODIMORPHISM OF MIXED SOLUTIONS

[In constructing the melting point diagrams of Types IV and V nothing was said of the crystalline structure of the separated phases; however, in several cases, the crystalline structure of the pure components is identical, and the miscibility gap is merely due to considerable differences in the parameters of the lattice, this being connected, in ionic crystals, with large differences between ionic radii. If the stable modifications of both components in the solid state are of different crystalline structures, a gap in solid solubility is necessary, irrespective of how similar the ionic radii are. It can be shown that in this case of so-called isodimorphism we shall arrive at diagrams of Types IV and V].

If substances A and B are dimorphous, and yield two series of isomorphous solutions, A_1B_1 and A_2B_2, which differ in their crystalline structures, then each of the series has its own thermodynamic potential curve — the curves A_1B_1 and A_2B_2.

If A_1 is more stable than A_2, and, on the contrary— B_2 is more stable than B_1, the curves will be of the form shown on Figure 24, I. Their solid solutions within the limits S_1 to S_2 will decompose. To the right of S_2 there will be a solution of A in B_2, to the left of S_1— a solution of B in A_1. In deriving melting point curves we are confronted with a phenomenon similar to the one we encountered in the case of partial miscibility, but the part S_1S_2 of the diagram (Figure 24, I) corresponds to the concave part of the curve for cases of incomplete miscibility of the solid solution (Figure 24, II). Thus, in the case of isodimorphism, there are already two familiar types of
49 melting curves; Type IV and Type V. We may also arrive at these curves in a different manner — by constructing the melting point curve for each ζ_S curve (for A_1B_1 and A_2B_2) and ζ_l curve, omitting the unstable parts as shown on Figure 24, III and IV. The construction is so simple that further explanations are unnecessary.

THE COMPONENTS FORM A COMPOUND IMMISCIBLE IN THE SOLID COMPONENTS

Let the compound be of composition C. At high temperatures, the potential of the solid phase of compound C is higher than that of the liquid phase; at low temperatures it is lower. Thus, the point ζ_C for the compound C shifts from an area above the ζ_l curve to the area below it, but at the same time the potential curves ζ_A and ζ_B of the components A and

B also pass from the region above the curve ζ_l to the region below it. By constructing a tangent the composition of the liquid, coexistent with the solid phase at the given temperature, might be calculated. After carrying out these constructions, we shall see that two cases are possible: 1) when the point ζ_C of compound C crosses the ζ_l curve, the point C lies outside of the segment of the tangent drawn from point ζ_B to the curve ζ_l; 2) at the mentioned moment point C lies above the corresponding segment. Let us consider both cases.

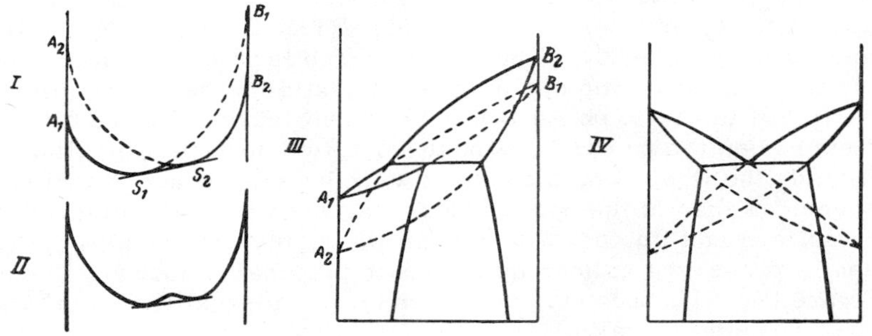

FIGURE 24. Thermodynamic potential curves and binary diagram showing isodimorphism

First case (Figure 25). Initially ζ of the solid phases of both components and their compounds is higher than that of the liquids. The system is a homogeneous liquid. As the temperature drops there arises a certain moment when the potential of the compound becomes equal to the potential of the liquid (Figure 25, III), i. e., ζ_C lies above the ζ_l curve. This may occur either before the potential of one of the components drops below the ζ_l curve, after the crystallization of one of the components has already begun (as shown on Figure 25, II), or even below the temperature of melting of both pure components. However, according to the stated condition, the point ζ_C will reach the potential curve of the liquid at the moment when the point of contact determining the liquid composition in equilibrium with the solid phase has not yet reached the composition C on the curve. Thus, at this moment, the potential of the compound becomes equal to the potential of the stable liquid of composition C, i. e., both phases will attain equilibrium; crystallization of a solid phase C from a liquid of the same composition will start.

At lower temperatures, the point of the compound C will lie below the potential curve of the liquid, i. e., it is the solid phase which is stable at this composition. Obviously, from this point (Figure 25, IV) not one, but two tangents may be drawn to the curve. The points of contact, as in the cases for pure components considered above, determine the composition of the liquids in equilibrium with the solid phase C. Instead of liquids with compositions l_1l_2, the combinations Cl_1 and Cl_2 appear, being more stable; i. e., the system passes into a mixture of compound C and the liquid.

On reviewing the diagram we note that the part of the curve to the right of composition C shows the same sequence of changes regarding the curve and the tangents to it, as we have seen in the general case of two immiscible components. In similarity, the point of contact (E_1) of the liquid curve and the straight line of the mixture $C + B$ (Figure 25, V) will correspond to the crystallization of the eutectic.

All that was said about the right side of the diagram (to the right of composition C) also holds true for its left side. In the case shown on our diagram, the temperature of crystallization of component A is lower than the temperature of crystallization of compound C. This is not reflected in the general form of the diagram, on the left side of which there is also a eutectic point (E). Therefore, the melting curve has a maximum which corresponds to C, and two eutectic points E and E_1 (Figure 25, VII).

Second case (Figure 26). The liquid, which is in equilibrium with one of the solid components (on our diagram B), attains the composition C before the potential of the solid phase C reaches the ζ_l curve (Figure 26, I). At the moment when the point of composition C lies upon the potential curve of the liquid, the liquid which coexists with the solid phase B will be richer in component A than is the compound C. As seen from the diagram (Figure 26, II), solid B and liquid l_1 will remain the stable phases, as $\zeta_{l_1+B} < \zeta_C$. This condition is preserved as long as the point ζ_C remains above the tangent $\zeta_B l_2$ (Figure 26, III), because, in this case, the inequalities $\zeta_{l_2+B} < \zeta_{C+B}$ and $\zeta_{l_2+B} < \zeta_{l_2+C}$ will also be valid.

Only at the moment when ζ_C of compound C falls on the tangent $l_3\zeta_B$, i.e., $\zeta_C = \zeta_{l_3+B}$, will all three phases be in equilibrium. On the melting point diagram (Figure 26, VII) this moment will correspond to the discontinuity point P. On further lowering of temperature (Figure 26, V) $\zeta_{C+l_4} < \zeta_{B+l_4}$, i. e., solid C crystallizes out from liquid C, the compositions to the right of C having already solidified completely (Figure 26, V). Finally, Figure 26,VI, corresponds to the eutectic crystallization C and A.

In similarity to the case of the melting point diagram where one of the components was polymorphous, here too the transition of the composition point C from its position above the straight line ζ_{l+B} (Figure 26, III), through this line (Figure 26, IV), to a position below it (Figure 26, VI), causes a discontinuity in the melting point curve. At this temperature, in melts of composition between P and B (Figure 26, VII), resorption of already formed crystals occurs. Below T_4 the melt must consist of two phases C and B, B constituting the $\frac{CD}{CB}$ fraction of the system (Figure 26, V). Above the transition point the system consists of liquid P and B, B constituting the $\frac{PD}{PB} = \frac{PC + CD}{PC + CB}$ fraction of the system.

Obviously $\frac{PD}{PB} > \frac{CD}{CB}$, i. e., at the transition point, a part of the solid B should dissolve.

POLYMORPHISM IN BINARY SYSTEMS**

Figures 27 and 28 represent the case when two modifications are solid
51 solutions of components, the relations of components in these solutions being different.

At high temperatures, a double curve solidus-liquidus will be obtained, corresponding to the solid modification of A (Figures 27 and 28). At a

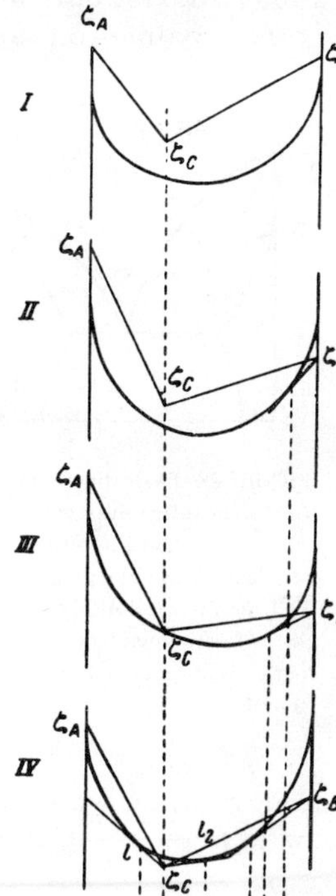

FIGURE 25. Derivation of the phase diagram of a binary system with a compound having a congruent melting point.

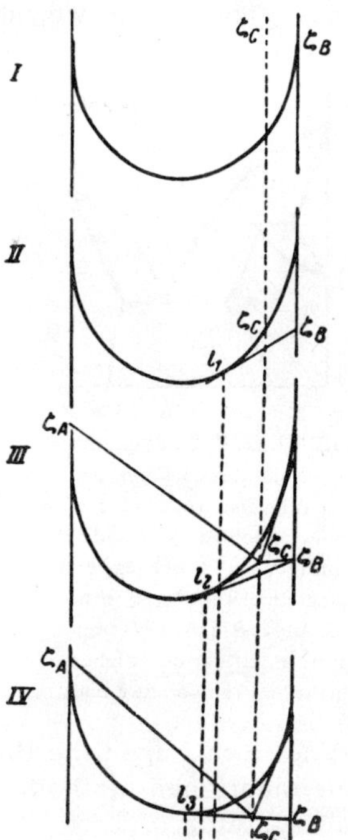

FIGURE 26. Derivation of the phase diagram of a binary system with a compound having an incongruent melting point.

52 certain temperature, a new modification A_1 will appear. According to the Phase Rule, two solid solutions and one liquid may coexist only at a single temperature. Therefore we shall obtain three points on a single horizontal straight line.

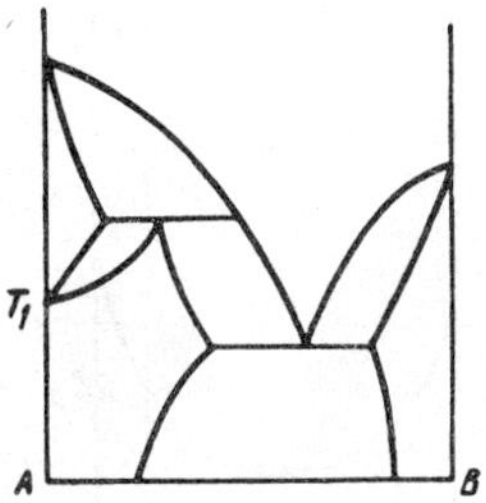

FIGURE 27. Phase diagram for one of the cases of polymorphous inversion of solid solutions (miscibility of A and B is limited, with a melting point diagram of Type V; the miscibility of B in the low temperature modification is higher than in the high temperature one).

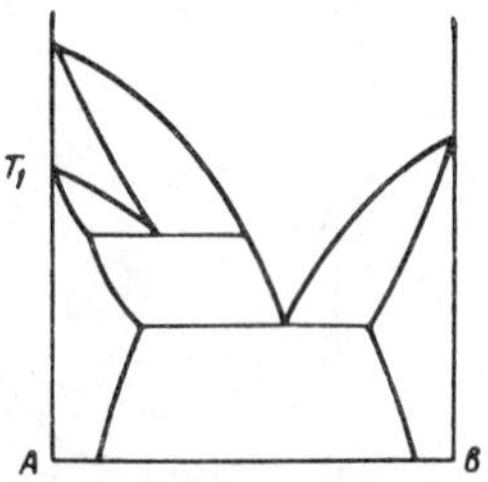

FIGURE 28. Phase diagram for one of the solid solutions (miscibility of A and B is limited, with a melting point diagram of Type V; miscibility of B in the low temperature modification is lower than the high temperature one).

We shall not dwell upon the thermodynamic derivation which is possible by the above-mentioned method. We shall now in brief mention the properties of these systems.

In Figures 27 and 28 the points T_1 mark the temperatures of polymorphous inversions of pure components. Depending on the solubility of the second component in these two modifications, two types of configuration of the curves of polymorphous inversion are possible. If the miscibility of the second component in the high temperature modification of *A* is higher than that of the low-temperature one, then the addition of component *B* lowers the temperature of polymorphous inversion (see Figure 28). If, however, the miscibility of component *B* in the high temperature modification is lower than in the low temperature one, then the addition of component *B* raises the temperature of the inversion point (see Figure 27). Examples of both types of diagrams are known in nature.

In the particular case when the miscibility of component *B* is the same in both phases, its addition should not influence the temperature of the inversion point, i. e., the inversion curves will fuse into one horizontal line. An even more particular case arises when component *B* is immiscible in solid *A*, or in other words, its miscibility in both polymorphous modifications is zero. Here again the addition of component *B* should have no influence on the inversion point of polymorphous modifications of component *A*. The diagram achieves the form of Figure 29. In practice, however, such a conclusion should be drawn with extreme caution, as a very slight solubility may be mistaken for the total absence of solid solutions. Quite a small admixture, and quite an insignificant difference in solubility may, however, exert a great influence on the variation of the thermodynamic potential, and shift the inversion point markedly. An example is the considerable
53 change in the inversion temperature of α- and β- quartz, as dependent on the

content of lithium salts and other admixtures in the solution (see p. 169).

These considerations go to demonstrate how careful one must be in using polymorphous inversions as temperature indicators in natural processes.

On the above-cited diagrams (see Figures 27 and 28) cases of equilibrium with polymorphous inversions of two solid phases and one liquid are shown. In other words, cases of intersection of the polymorphous inversion curve and the melting point curve. There are also cases when the inversion curve may lie below the melting point curve, and the invariant point of the mentioned type is absent. In cases where the solid solutions also remain continuous at lower temperatures, and both components undergo polymorphous inversion, the inversion curves are continuous too. In this case, two types of curves are possible, according to three Roozeboom types of melting point curves of solid solutions. If below the melting point curve there are exsolution curves of solid solutions (intersecting the melting point curve, or closed, with an upper critical point), the inversion curves intersect correspondingly with the curves of exsolution, and an invariant point appears at which three solid phases are in equilibrium: two polymorphous modifica-
54 tions of phase A_1 with different contents of component B, and phase B_1 with dissolved A. The combinations of curves are the same as for the case of intersection of inversion and melting point curves. The above-mentioned considerations of the influence of an addition of the second component on the inversion temperature and on its relative solubility in the two polymorphous modifications remain valid.

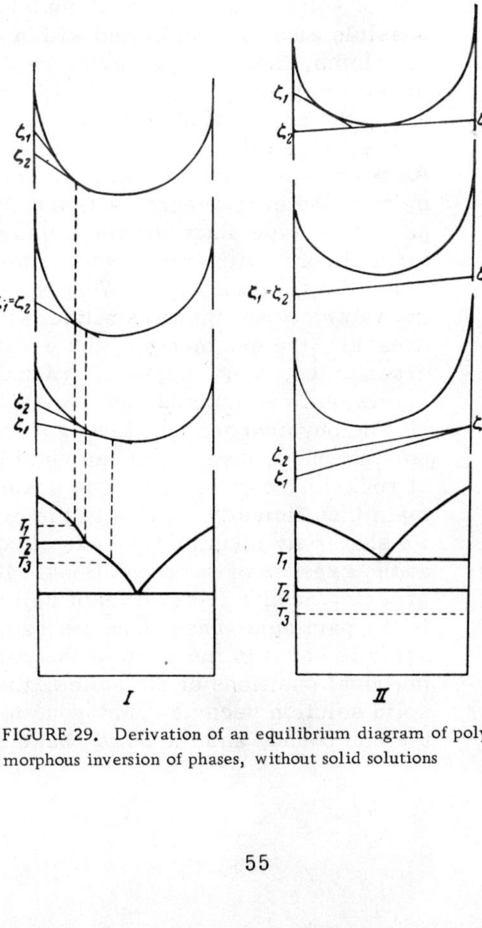

FIGURE 29. Derivation of an equilibrium diagram of polymorphous inversion of phases, without solid solutions

EXSOLUTION OF A COMPOUND BELOW THE TEMPERATURE OF MELTING, AND INCREASE OF ORDERING IN SOLID SOLUTIONS**

Hitherto we have considered cases of compounds melting with and without exsolution. However, there is a third, possible case, where the compound exsolves on heating below the temperature of melting. A phase diagram with such compounds, lacking solid solutions, is shown on Figure 30. At high temperatures, the thermodynamic potential of the compound is higher than the potential of the mixture of two phases, composed of pure components (Figure 30, I). At lower temperatures, the potential of the compound is lower than that of the mixture (Figure 30, III) and the stable phases will be *A* and *C* or *C* and *B* (Figure 30, IV). Such a temperature must exist, at which the thermodynamic potential of *C* will lie on the straight line which connects the potentials of *A* and *B* (Figure 30, II). At this temperature all three phases will be in equilibrium; at a constant pressure this equilibrium will be invariant.

The problem of formation of compounds at the expense of solid solutions is strictly connected with the problem of ordering in a crystal lattice. It has been established by X-ray analysis that, in several cases, changes in temperature cause gradual changes in the lattice, i. e., gradual polymorphous inversion at the expense of increased ordering in the location of the atoms. At higher temperatures, atoms or ions, which are similar in their radii and valency, may often replace one another freely, causing the formation of solid solutions. Such mutual replacement is possible also in a compound with a constant ratio of the atoms, these atoms having random distribution in two or more geometrically different positions, e. g., the atoms of aluminum and silicon in high-temperature feldspars and other aluminosilicates. As the temperature is lowered, the orderly arrangement of the atoms increases, i. e., atoms of one particular type start showing a marked preference for a certain lattice site, while atoms of another type occupy other sites. When such order is complete, the symmetry of the crystalline lattice usually increases; the symmetry of the crystal is often increased too, i. e., a type of gradual polymorphous inversion is completed, as it were.

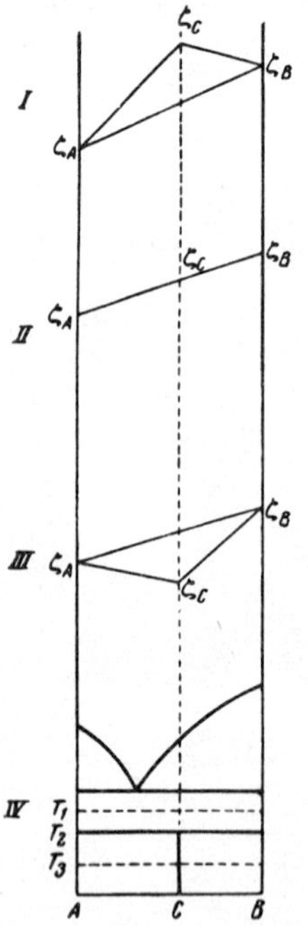

FIGURE 30. Derivation of a phase diagram for systems with compounds exsolving below the eutectic temperature.

The physicochemical laws governing such gradual polymorphous inversions, as established for a series of rock-forming minerals and assumed for others, remain insufficiently studied to the present date. Here we shall only mention the case of compounds forming at the expense of solid solutions. If an increase of
55 ordering occurs in crystals of solid solutions, then, in the particular case when the ratio of both components is equal to the ratio of the corresponding geometrical positions of the atoms, the structure of the solid solution becomes analogous to the structure of the compound, and the solid phase attains new

properties. During the process, the continuity of the solid solution may still be preserved, i. e., a continuous series will be observed between ***A*** and ***C*** as well as between ***B*** and ***C*** (Figure 31). Upon the phase diagram, the formation of such a compound will not be reflected. However, this case is characterized by the appearance of a specific point on the composition-property diagram (Kurnakov, 1940). Such a point of discontinuity should appear also on the thermodynamic potential curve (Figure 31, II).

On further lowering of the temperature two alternatives arise; either a decomposition of the solid solutions, or polymorphous inversion of the formed compound, which should be reflected on the phase diagram. As distinctive from the cases considered above, the exsolution will occur here between the phases ***A*** and ***C***, ***C*** and ***B***. Upon the thermodynamic potential curve, this will be manifest by the appearance of two concavities (Figure 31, III and IV) and the corresponding, four exsolution curves, with two critical points.

On the polymorphous inversion of a new compound, we shall be confronted with phenomena analogous to those considered above (see p. 52), the only difference being that the inversion curves will diverge from point ***C*** to both sides, i. e., they will have a maximum at point ***C***. These inversion curves will intersect the exsolution curves as shown on Figure 32.

CRYSTALLIZATION FROM TWO LIQUID PHASES

We shall consider only the most simple case of a binary system, i. e., the components do not form mixed crystals, and their liquid melt, which is homogeneous at high temperatures, separates into two immiscible liquid phases when temperature decreases.

We have already seen that the thermodynamic potential curve of the liquid is convex downwards throughout its length at high temperatures, and that when temperature is lowered it acquires a concavity, thus determining the area of separation into two liquid phases. Each of these liquid phases is a solution of one component in the other, but a discontinuity in composition exists between them (no intermediate compositions of solutions are possible). As temperature is lowered, the concavity usually increases (Figure 33), but in certain cases it may decrease or even disappear. In this case, a closed area of exsolution appears on the composition-temperature diagram, as is the case, for instance, with water-chloroform mixtures.

A crystallization diagram under conditions of exsolution of the liquid phase may be constructed, using the already familiar method of considering the relations between the liquid and solid phases.

At high temperatures (T_1) the liquid is homogeneous, and the curve is convex downward (Figure 33, I); when temperature is lowered, this curve attains a concavity which determines the field of discontinuity (Figure 33, II). In the considered example, which is shown on Figure 34, the crystallization of the more refractory component has already began at the same temperature, from a liquid even richer in it (Figure 34, II). At the temperature T_2, the following fields appear on the composition-temperature diagram (from right to left): solid + liquid; one liquid, relatively rich in the refractory component; two liquids; one liquid, relatively rich in the more fusible component.

56 As cooling proceeds, the point of contact on the curve ζ_l, determining the composition of the liquid phase (l_1), which is in equilibrium with the crystallizing solid phase, is shifted along the curve to the left, and eventually coincides (Figure 34, III) with the point of contact which determines the composition of the liquid which is richer in the refractory component. At this moment a common tangent may be drawn from the point of the solid phase to both sections of the liquid phase curves, ζ_l, which are convex downward. Three phases are here in equilibrium: two liquid and one solid. Here we are confronted with an invariant system, in which conversions occur at constant temperature, resulting in increases

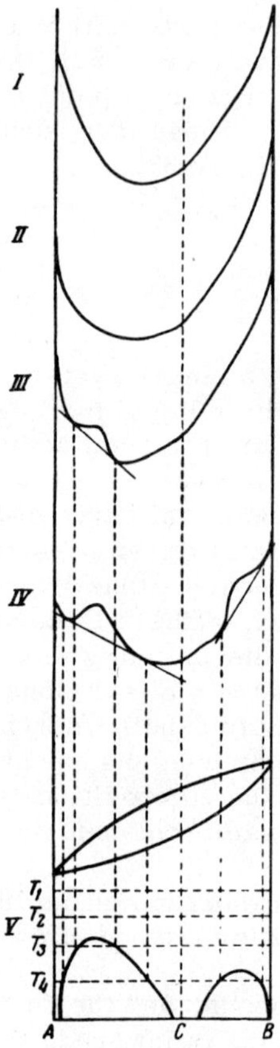

FIGURE 31. Derivation of a phase diagram in the case of the gradual formation of a compound at the expense of solid solutions

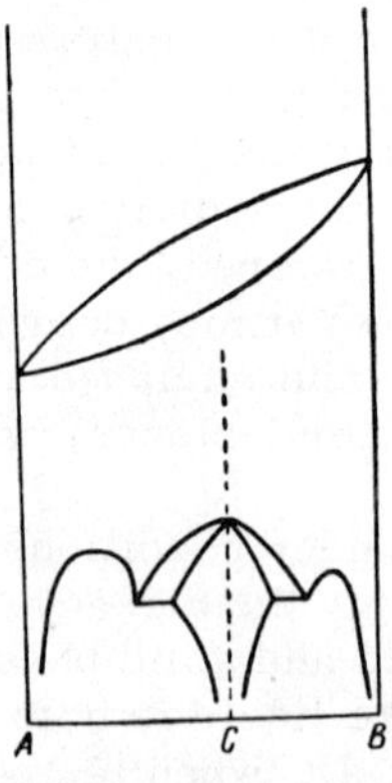

FIGURE 32. Phase diagram showing one of the possible cases of formation of compounds at the expense of solid solutions. The compounds undergo polymorphous inversions.

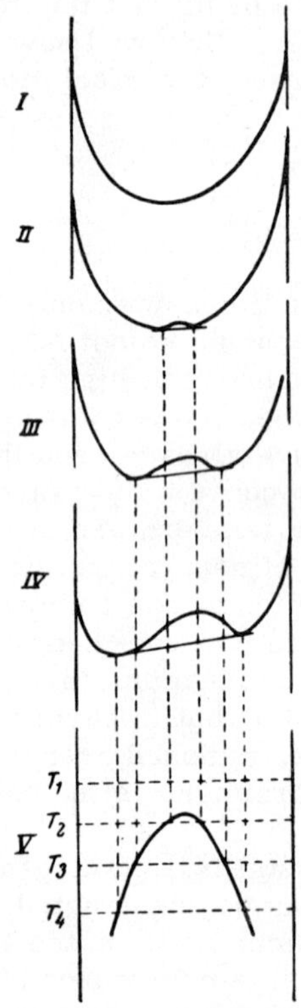

FIGURE 33. Derivation of a phase diagram of two immiscible liquids.

in the amounts of the solid phase and the liquid phase richer in the fusible component, at the expense of the liquid phase which is richer in the re-
57 fractory component. When the latter disappears, the crystallization of the more refractory component proceeds from the residual liquid, rich in the fusible component. At temperature T_4, the following fields are present (from right to left): the solid refractory component + liquid, rich in the fusible component; liquid; liquid + solid fusible component. Further cooling results in the eutectic crystallization of both solid phases from a liquid of the eutectic composition, rich in the fusible component (Figure 34, V); crystallization will continue at constant temperature (Figure 34, VI) and will terminate with the formation of a system consisting of two solid phases which represent the pure components. [More complicated cases, involving the formation of two liquids in the presence of solid solutions, are not considered here, but might be analyzed readily by applying the same methods].

GENERAL FEATURES OF BINARY DIAGRAMS

The above-cited examples of temperature-composition diagrams reveal certain regularities which merit noting as they are of use in considering more complicated systems.

We have considered such conversions in systems as the appearance of new phases and the disappearance of existent ones, and the change in composition of phases and their quantitative relations within a system. Let us now see how these phenomena are expressed in the binary systems under consideration.

The complete set of objects under investigation, i. e., all possible compositions of a system and the phases which comprise it, are uniquely represented by points on a straight line, the ends of which represent the components constituting the system and its phases.

According to the Phase Rule, in the above-considered cases of inversions, a system which consists of one phase will be termed bivariant, of two phases — univariant, of three — invariant. In the last case, the composition of phases remains constant in all inversions, and a reaction occurs between them, which changes only the relative amount of phases. In a univariant

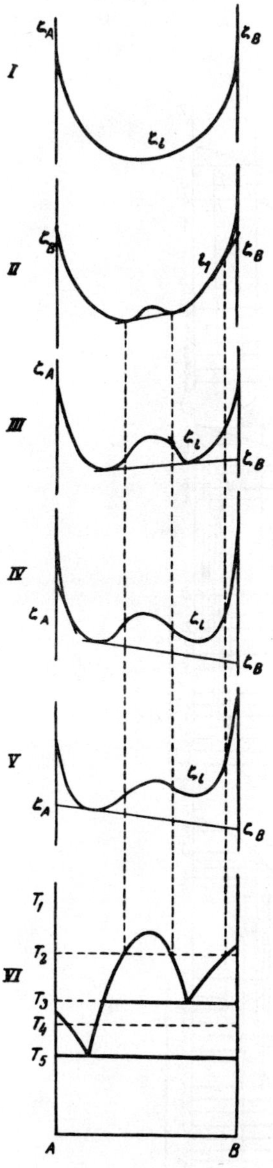

FIGURE 34. Derivation of a phase diagram for two immiscible liquids with the crystallization of a solid phase, in the simplest case.

system, the composition of phases may change, their relative amounts
58 changing at the same time (as these are dependent on the composition of the phases). When the entire system consists of one phase only, it is clear that the composition of this phase remains constant [if the composition of the entire system does not change] and only temperature and pressure vary independently.

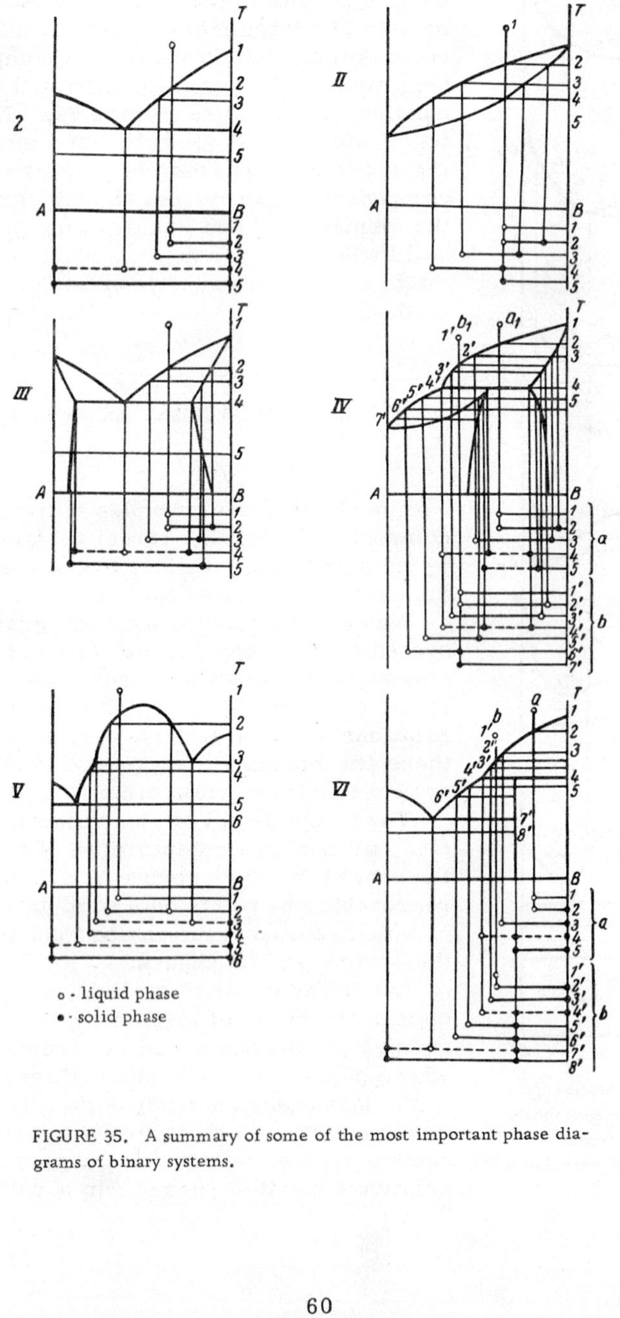

FIGURE 35. A summary of some of the most important phase diagrams of binary systems.

In studying systems, we deal with solid and liquid phases. Further on
59 we shall designate them correspondingly by black (solid) and white (liquid) circles (Figure 35). Various cases of phase conditions of binary systems will be represented on the diagrams by points upon the concentration line, the ends of which represent the components. A system consisting of one phase will be represented by one point which shows its composition; a system of two phases — by two points. These points determine a segment of the concentration line, which is termed the *phase segment*. The infinite number of points on this segment represent the infinite number of various compositions which may form these two phases. Obviously, the two-phase system chosen by us may also be represented by a point on the segment.

Thus, the composition of a two-phase system is represented by a point, which must lie on the phase segment. This point cannot lie outside the phase segment, even if the composition of the phases changes, and the points which describe them are shifted. Phase segments may be of three kinds: both phases may be solids; one phase solid, the other liquid, and both phases liquid. In our designation these will be represented by combinations of circles: two black, one black, one white, two white.

An invariant (three phase) system will be represented by three points on the concentration line. If one of the phases of such a system is a liquid, then there are two possibilities for the location of points: 1) the point which represents the composition of the liquid phase lies between the composition points of the solid phases; 2) the point which represents the composition of the liquid phase lies to one side of the two composition points of the solid phases. In our designation this will be reflected either by the location of the white circle between the two black ones, or to one side of them. In the case of two immiscible liquids, i. e., when there are two liquid phases and one solid, the location of the latter is always to one side of the liquid phase points.

In an invariant system, the inversion consists in quantitative changes only. The amount of certain phases decreases, of others — increases. In an invariant, binary system, which is represented by three points, two cases of inversion are possible: 1) the amount of the phase, represented by the middle point, decreases, while the amounts of the phases represented by the marginal points increase; 2) the amounts of both phases, represented by the marginal points, should decrease; this follows directly from the accepted method of representing compositions. When any one of the phases disappears, the remaining two phases determine the phase segment within which the composition point of the system must necessarily lie.

Let us consider the above-mentioned general peculiarities in some examples of melting point diagrams (see Figure 35).

Eutectics (see Figure 35, I). At the temperature T_1, the system consists of one liquid phase, as shown below the melting point diagram by a white circle on an auxiliary line marked *1*. At the temperature T_2, the crystallization of the solid phase starts. Upon the auxiliary line *2* a phase segment appears, the ends of which are marked by a white and a black circle; at incipient crystallization, the point of the liquid phase keeps its position, and the quantity of the solid phase is yet extremely small. As the quantity of the solid phase increases, the point of the liquid phase is shifted to the left, as shown on auxiliary line *3*; the composition of the solid phase remains constant, and its point remains fixed. At the eutectic temperature (T_4) a second solid phase appears. Further on the changes occur in an invariant system, the phase line of which is shown as a dotted line on our figure. Eventually, the liquid phase disappears, and the system is transformed into a two-phase system (at T_5).

Roozeboom's first type of mixed crystals (see Figure 35, II). Here we
60 find the following stages. At T_1 there is one liquid phase; at T_2 a solid phase appears, and a phase segment may be constructed; at T_3 crystallization of the solid solution proceeds, the phase segment is shifted to the left, and both solid and liquid points are shifted. Further cooling results in a position of the phase segment where all the liquid phase (at T_4) is used up; at T_5 only the solid phase remains.

Roozeboom's fifth type of mixed crystals (see Figure 35, III). At T_1, a single liquid phase exists; at T_2, a solid phase appears, and a phase segment may be constructed; at T_3 crystallization of mixed crystals occurs, and the phase segment is shifted to the left; T_4 corresponds to the eutectic point. At this moment a new solid solution appears, and further inversion occurs in an invariant system. It consists of a simultaneous crystallization of solid solutions, and is terminated by the disappearance of the liquid phase; at T_5, two solid phases remain, which vary somewhat in their composition on cooling, and exsolution of solid solutions.

Roozeboom's fourth type of mixed solutions (see Figure 35, IV). Here we shall consider two cases: points a_1 and b_1.

1. At T_1, there is one liquid phase; at T_2, an additional solid phase $T_{3'}$ appears; at T_3 crystallization of solid solution proceeds; at T_4 a third phase appears, which is also a solid solution. Further inversions occur in an invariant system — the marginal solid and liquid phases decrease, until all the liquid is used up at T_5.

2. At $T_{1'}$, one liquid phase exists; at $T_{2'}$, a solid phase appears; at $T_{3'}$ crystallization of the solid solution proceeds; at $T_{4'}$, a second solid phase appears; in the three-phase system invariant inversion occurs, the solid phase, represented by the middle point, crystallizes out, while the liquid and the other solid phases decrease until their solid phase disappears; at $T_{5'}$, after the marginal solid phase has disappeared, further crystallization of the remaining solid phase occurs; at $T_{6'}$, the liquid phase disappears; at $T_{7'}$, only the solid phase remains.

Formation of two immiscible liquids (liquation) (see Figure 35, V). At T_1, there is one liquid phase; at T_2 the system consists of two liquid phases; at T_3 a solid phase appears, consisting of the more refractory component; an invariant inversion in the three-phase system occurs, the liquid of the composition represented by the middle point decreases; at T_4 the system again consists of two phases—from the liquid, which is rich in the volatile component, the refractory component crystallizes out; at T_5-eutectic crystallization (invariant inversion): the liquid decreasing gradually, disappears; at T_6 the system consists of two solid phases.

A system with a compound having an incongruent melting point (see Figure 35, VI). We shall consider two cases. The first is characterized by the point *a* .

At T_1, there is one liquid phase; at T_2, the separation of the more refractory component as the solid phase begins; at T_3, this component proceeds to crystallize out — the system has two phases; at T_4, a compound appears, the system has three phases — invariant inversion; the liquid decreases together with the already formed component: the liquid is the first to disappear; at T_5 the system consists of two solid phases: the more refractory component and a compound.

The other case is characterized by point ***b***.

At $T_{1'}$ there is one liquid phase; at $T_{2'}$, the separation of the refractory
61 component begins, in the form of a solid phase; at $T_{3'}$, this solid phase continues its crystallization, the liquid decreases (two-phase system); at $T_{4'}$, a compound appears — a three-phase system; when an invariant inversion occurs, the liquid and the earlier crystallized component decrease, this component being the first to disappear; the solid phase of the compound and the liquid remain, and the crystallization of the compound proceeds (at $T_{5'}$ and $T_{6'}$), and at $T_{7'}$ the eutectic point is reached, the system consisting of three phases; on invariant inversion the liquid disappears; at $T_{8'}$ two phases remain: the fusible component, and the compound.

General conclusions. The consideration of the various examples allows one to draw some general conclusions.

Geometrically, the composition of each phase and the composition of the system is described by means of points. A system which consists of two or three phases, is described by two or three points respectively. It is logically necessary that the point of composition of the entire system should lie within the limits of the phase segment which represents a two-phase system. During inversions and changes in composition of phases, one or both end points of the phase segment are shifted, but the point of the system never leaves the segment. When the point coincides with the end point of the segment, one of the phases appears, or disappears.

The phenomenon of crystallization is expressed by the transition of the quantitative changes in a system, which are geometrically expressed by the shifting of the point in the phase segment, to qualitative changes; a one-phase system passes into a two-phase system, and may pass further on into a three-phase system and back into a two-phase system. The types of conversions and the order in which they occur are determined by the relations between the components in each particular case (ability of mutual miscibility of the components in the liquid and solid phase, the presence of a compound with congruent or incongruent melting point, etc). When these relations are given, the entire path of the process can be followed, by considering the above-mentioned geometrical features of the diagram which represents it.

THE REACTION PRINCIPLE

When any substance crystallizes from a solution two cases may possibly arise.

1. The crystallized solid phase remains unchanged from the start to the end of the crystallization process; all that happens is that crystals either continue their growth or increase in number.

2. The crystallized solid phase becomes unstable at a later stage, and it changes its composition by reacting with the liquid. We meet with

this case in all examples of crystallization of solid solutions and in systems in which compounds with incongruent melting points are formed. Let us once again dwell on this case.

All diagrams of crystallization of solid solutions show two pairs of curves: the liquidus curve (L) and the solidus (S). When temperature drops, the composition point of a system is shifted along the vertical line $M'M$ (Figure 36). At the temperature T_1 the crystallization of the solid phase of composition S_1 starts. At this moment its mass is negligibly small, and the composition of the system M coincides with the composition of the
62 liquid phase L_1, representing the other end point of the phase segment. At a lower temperature T_2 the liquid phase L_2 is in equilibrium with the solid phase S_2; the phase segment is L_2S_2. The ratio of the liquid phase to the solid is $MS_2 : ML_2$. In other words, the solid phase comprises the fraction ML_2/L_2S_2 of the system. The composition of the liquid will change from L_1 to L_2, and that of the solid from S_1 to S_2. At an even lower temperature, T_3, the solid solution S_3 will be in equilibrium with the liquid L_3; the phase segment will be L_3S_3; the ratio of solid to liquid is $ML_3 : MS_3$, i.e., the solid phase comprises the fraction ML_3/L_3S_3 of the system, which fraction is larger than at the temperature T_2. The compositions of the liquid and solid phases were both changed increasingly in the same direction, i.e., both were enriched in the more fusible component (the composition points were shifted to the left). Crystallization terminates at temperature T_4, when the liquid phase reaches the composition of L_4, and the composition of the solid phase is S_4. At this moment the liquid phase almost disappears, and the composition of the system coincides with the end of the phase segment representing the solid phase.

Thus, as temperature is lowered, the composition of both solid and liquid phases change continuously. The amount of the liquid decreases and that of the solid increases. The solid phase is enriched in the more fusible component, which passes into the solid solution as if replacing the more refractory component. A continuous interaction and interchange of matter occurs between the liquid and solid solution, or, in other words, a continuous reaction takes place between them.

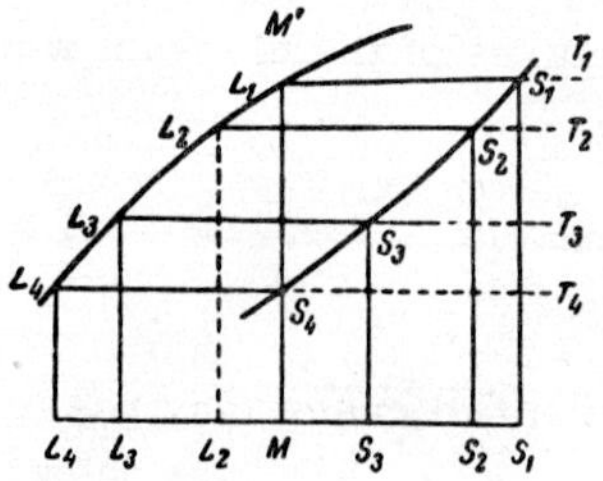

FIGURE 36. Continuous reaction between liquid and solid phases in a binary system with unlimited solid solutions

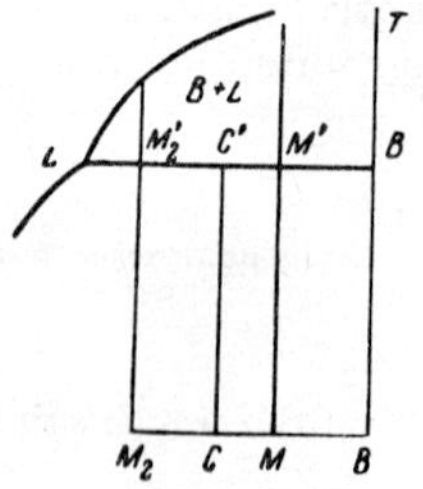

FIGURE 37. Reaction between the liquid and solid phases in a system with a compound having an incongruent melting point.

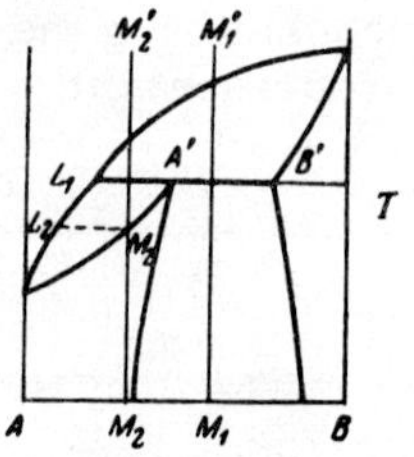

FIGURE 38. Reaction between the liquid and solid phases in a system with limited solid solutions of Type IV.

A different type of reaction is observed in the case of a system with a compound having an incongruent melting point. The part of the diagram of interest to us has the form shown on Figure 37. At a temperature corresponding to the line LB', an invariant inversion occurs in the three-phase system. Above this temperature, the solid phase B is in equilibrium with the liquid in the field $B+L$; below this temperature, either the compound C and the liquid are in equilibrium, provided that the total composition of the system is represented by a point which lies to the left of C, or a mixture of two solid phases results — the compound C and component B, provided that the total composition of the system is represented by a point to the right of C. In both cases the compound C is formed at the expense of the liquid and of the earlier separating solid phase B, which thus diminishes due to the action of the residual liquid, i.e., it interacts with the liquid and is dissolved as it were. The changes in the relative amounts of phases may easily be traced at this inversion. Let us consider two cases: 1) the system
63 is of composition M'; above the temperature T, it consists of the solid phase B and the liquid L; the ratio of the liquid L to the solid B is $M'B : LM'$. Below the temperature T, the system consists of the solid phases C and B, where $C : B = B'M' : C'M'$; above the temperature T, the system of the composition M_2' consists of a liquid and a solid phase B in the ratio $M_2'B : M_2'L$; below this temperature the system consists of liquid and C in the ratio of $M_2'C : M_2'L$.

In the considered case, if a compound having an incongruent melting point is present, the reaction between the liquid and the separated solid phase will not be continuous as is the case of solid solutions. A gap appears between the compositions of the solid phases B and C.

In a certain sense we are confronted with an intermediate type of crystallization reaction of solid solutions, which correspond to Roozeboom's melting curve of Type IV (Figure 38). At the temperature T reaction occurs. The solid solution B', with B as its dominant component, reacts with the liquid L_1; a solid solution of composition A' is formed during this reaction. Depending on the composition of the system, crystallization either terminates in the formation of the mixture $A'+B'$ (in the case of M_1'), or, in the case of A', in the formation of a mixture of L_1 and liquid M_2', being subsequently followed by a continuous reaction between the liquid and solid solution, until all the liquid is used up (point L_2), and a solid solution of the composition M_2 is formed. More complex cases are also possible, i.e., when compounds and limited solid solutions are present.

Thus, after considering the various cases of interaction between the liquid and solid phases, two principal types of interaction may be distinguished: a continuous type of reaction and a discontinuous one.

INTERACTION BETWEEN THE CRYSTALLIZING LIQUID AND FOREIGN INCLUSIONS*

Let us consider cases of addition of the solid phase of one of the components or of its solid compounds or solid solution with the other component to the crystallizing liquid. We shall confine ourselves to the most simple examples only.

Let us consider a system which crystallizes with the formation of binary eutectics; at the temperature T the melt of initial composition M consists of crystals A and liquid L, which is saturated with respect to this component A (Figure 39). Obviously, the addition of A at the same temperature will not cause any change in the equilibrium of the system. However, if temperature is kept constant and the solid phase B is added (point M_1), the liquid will dissolve this component as it is undersaturated with respect to it. The solution will require latent heat of fusion; the heat will be withdrawn from the liquid and the cooling should cause the crystallization of the phase A, with respect to which the liquid is saturated. This, however, does not occur in reality, as the passing into solution of the component B lowers the relative concentration of A, and the liquid is no longer saturated with respect to it. The result is that the liquid, being enriched in the component B, changes in composition and temperature (down to T_1) along the solubility curve of A from L to L_1. The amount of A decreases, until at L_1 the composition of the liquid reaches the overall composition of the new mixture; solids A and B disappear, and further on crystallization of A recommences up to the eutectic point E, where A and B crystallize out simultaneously. The component B, on addition to the mixture of liquid and crystals, will be dissolved provided that the change in the overall composition of the mixture caused by this addition does not go beyond the eutectic. The component B will be dissolved even if it is less refractory than A, as is seen from the example shown on our diagram (see Figure 39).

Let us consider another case: that of a continuous series of solid solu-
64 tions of Roozeboom's first type. Let L be a saturated liquid in equilibrium with the solid solution S. At the same temperature a solid solution S_1 is introduced (Figure 40). Let us assume that the quantity of S is negligible — practically zero. On considering the thermodynamic potential diagram we note that mixtures of S_1 and L are unstable, and will either turn into a mixture of liquids L and S or a solid solution intermediate between S and S_1.

[As the liquid L is in equilibrium with the solid solution S, the reaction $L+S_1 \rightarrow S$ will start here. The solid phase will increase in quantity, and, providing the system is not cooled, the temperature rises somewhat on account of the liberated latent heat of crystallization. The composition of the liquid should change together with that of the solid phase (which will be enriched in the more refractory component). The reaction proceeds until either all of the liquid or the added solid phase S_1 disappears].

The final result depends on the total composition of the initial liquid plus the solid phase which is added to it. Obviously, the composition of the resulting solid solution (after crystallization) will be more refractory than that obtained without the addition of the solid solution S_1, but more fusible than S_1.

Naturally, it is not possible to add to the liquid a more fusible or equally fusible solid solution than the one with respect to which it is saturated, as in this case it will fuse and will not be solid. Let us add a solid solution at its melting temperature (lower than the temperature of the liquid). The addition will cause the cooling of the liquid and crystallization from it of that solid solution, with respect to which it is saturated. The added fusible solution will pass into the liquid. This gives rise to a solid solution of an intermediate composition between the liquid and the added solid solution as the final product of crystallization.

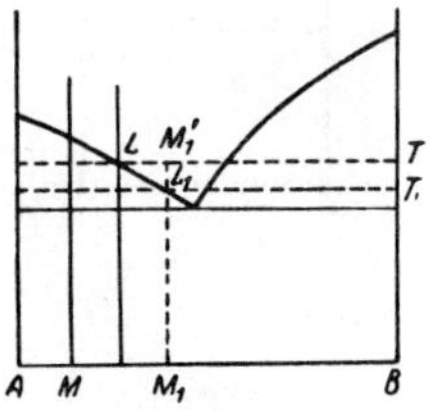

FIGURE 39. Scheme of reaction between the melt and an added solid phase for the simplest system of the eutectic type

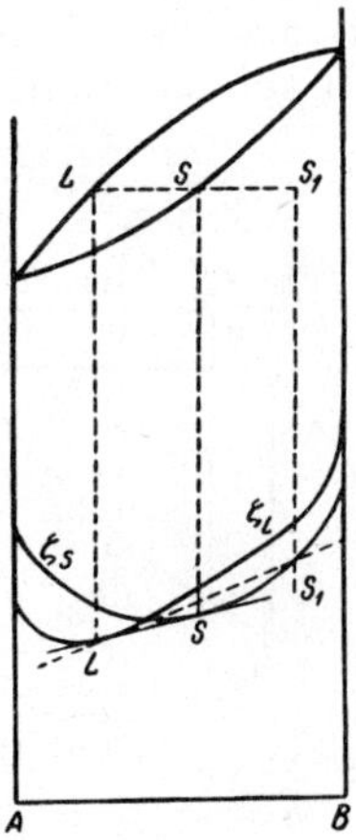

FIGURE 40. Scheme of interaction between the melt and an added solid phase for a diagram with solid solutions

ZONED CRYSTALS

Interaction between the crystals of a solid solution and the liquid, consisting of an interchange of matter between them during slow crystallization (provided there is enough time for such interchange) while maintaining the
65 equilibrium, results in a homogeneous crystal of solid solution. If cooling proceeds too rapidly, zoned crystals may be formed.

Let us imagine that the crystallization of the solid solution has proceeded from T_A to T_1 while continuously maintaining equilibrium (Figure 41, I). At T_1 crystals of the composition S_1 and the liquid L_1 coexist. The crystals will constitute a fraction $\frac{L_1 M_1}{L_1 S_1}$ of the whole system. Let the interaction between S_1 and the liquid now cease. The liquid will solidify from below T_1 down to T_2, reacting with the solid crystals formed, but not with the formerly separated solid phase S_1. The solid solution now crystallizing out will continue to grow around nuclei of the earlier formed crystals of composition S_1, without showing any change in its composition. At T_2 the overgrowth zone will be of composition S_2. It will constitute a fraction $\frac{L_2 M_2}{L_2 S_2}$ of the residual liquid of the composition L_1. Let us assume that the interaction between S_2 and the liquid again ceases further on. The zone of composition S_2 will be overgrown by crystals, and, assuming that these crystals will interact with the liquid up to the end, a third zone will form, which eventually will attain the composition S_3. Thus, instead of homogeneous crystals, we shall obtain crystals consisting of three zones with the compositions S_1, S_2 and S_3. The overall composition of the crystal will of course be S, but the inner zone will be richer in the more refractory component, and the outer zone — in the more fusible component than is this overall composition.

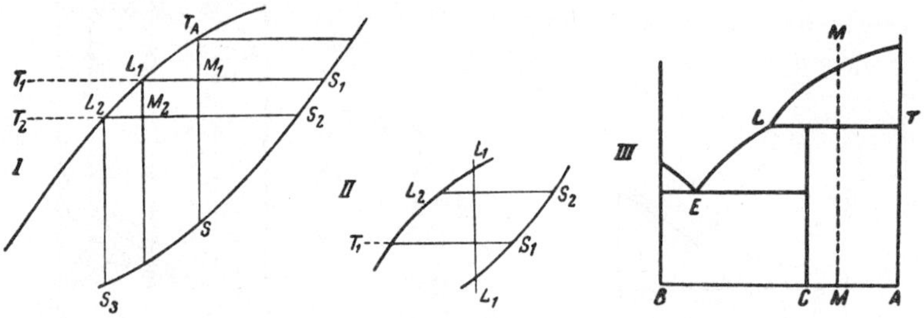

FIGURE 41. Scheme of formation of zoned crystals

Any number of such gaps and terminations in the interaction between the crystals and the liquid is possible; the gaps in the process may be arbitrarily small, and when they become infinitely small and very numerous, a continuous change of composition from the center to the peripheries occurs.

In case of supercooling the described phenomenon may yet change. If the liquid on cooling reaches the temperature T_1 without any formation of crystals, being supercooled, it starts its crystallization here (Figure 41, II) with the separation of solid solution S_1. During such crystallization the composition point of the liquid L_1 should shift to the left, and on liberation of the latent heat of fusion the supercooled melt will rise in temperature. The point of the liquid will therefore be shifted along the line L_1L_2 and, as crystallization proceeds, a solid solution of composition S_2 will be formed from the liquid of composition L_2. This solid solution will be richer than S_1 in the more refractory component. Thus supercooling may cause a reversed location of zones in relation to the normal case of a slow or interrupted reaction between the crystals and the liquid.

66 When compounds with incongruent melting points are present in a system, the interaction between the liquid and the crystals may terminate. This happens, for instance, when the compound *C*, on crystallization, forms a crust on earlier formed crystals of *A*, thus isolating them from the liquid and preventing their interaction. This may lead to important results. Consider, for instance, the composition represented by point *M* (Figure 41, III). During normal crystallization and interaction of the initial solid phase *A* with the liquid we obtain a mixture of *C* and *A* in the ratio $AM:CM$. The component *B* cannot appear in this case. If, during reaction between the solid phase *A* and the liquid *L* at temperature *T*, the crystals of *A* are isolated from the liquid and stop their reaction, the solution *L* will remain saturated with respect to *C*, which will continue to crystallize on cooling. The composition of the liquid changes until it reaches the eutectic composition, when the solid phase *B* will join *C* in crystallization. The appearance

of the latter is not anticipated, judging from the composition of the initial melt *M*.

The interruption of reaction between the crystals and the liquid, both in the case of solid solutions showing continuous interreaction, and in the case where a break occurs in the reaction, may also occur on removing the crystals from the liquid, e. g., as a result of their sinking to deeper parts of the reservoir or vice versa, due to their floating upward. This possibility is considered as the basis of so-called gravitation differentiation, which is the outcome of fractionation. This will be considered further on.

OVERSATURATION

Figure 42, I, shows a thermodynamic potential curve ζ_l for a homogeneous liquid consisting of *A* and *B*. Let us assume that a certain quantity of crystals *A* are formed from this solution; the thermodynamic potential of this solid phase is *AK*. As we have noted previously, the stable phases will be the mixture *A* and the liquids from *C* to *B*. The composition *C* corresponds to a saturated solution of component *A* in the liquid, and compositions from *C* to *B* — to unsaturated solutions with respect to this component. Liquids with compositions between *A* and *C*, for which the potential curve *PS* lies above *KS*, and which, hence, may exist only in unstable equilibrium, are oversaturated with respect to component *A*. If the temperature drops below the crystallization point of component *B*, we shall obtain a similar field for solutions oversaturated with respect to this component at the right-hand side of our diagram. In the case of liquids below the eutectic temperature, the thermodynamic potential curve lies entirely above the line *KL* (Figure 42, II), which represents the potentials of the mixture of solid phases; all liquids will be oversaturated with respect to both *A* and *B*.

Let us now consider the composition-temperature diagram (Figure 42, III) and note the location of the areas of undersaturation, saturation and oversaturation, in the simplest case of crystallization with formation of eutectics. One should bear in mind the construction of Figure 16.

Point *S* and the corresponding point *C* (see Figure 16) represent a composition which is also designated by the same letters in Figure 42, III. At the given temperature, solutions which are oversaturated with respect
67 to component *A* will have compositions lying to the left of *C*. The area of oversaturation lies below the melting point curve *PE*, and extends into the area of stability of the mixture *A*+*L*, but the solid phase *A* is not exsolved from the oversaturated solution. Similarly, solutions will be oversaturated with respect to *B* if they occur in the field below curve *QRE*. The solutions which are simultaneously oversaturated with respect to both *A* and *B* are those which lie in the field *MEN*, being limited upward by the extensions of curves *QE* and *PE*.

[As in the case of solid solutions, the liquid will be oversaturated with respect to the solid phase lying below the melting point curve — phase S_2 on Figure 42, IV and V.]

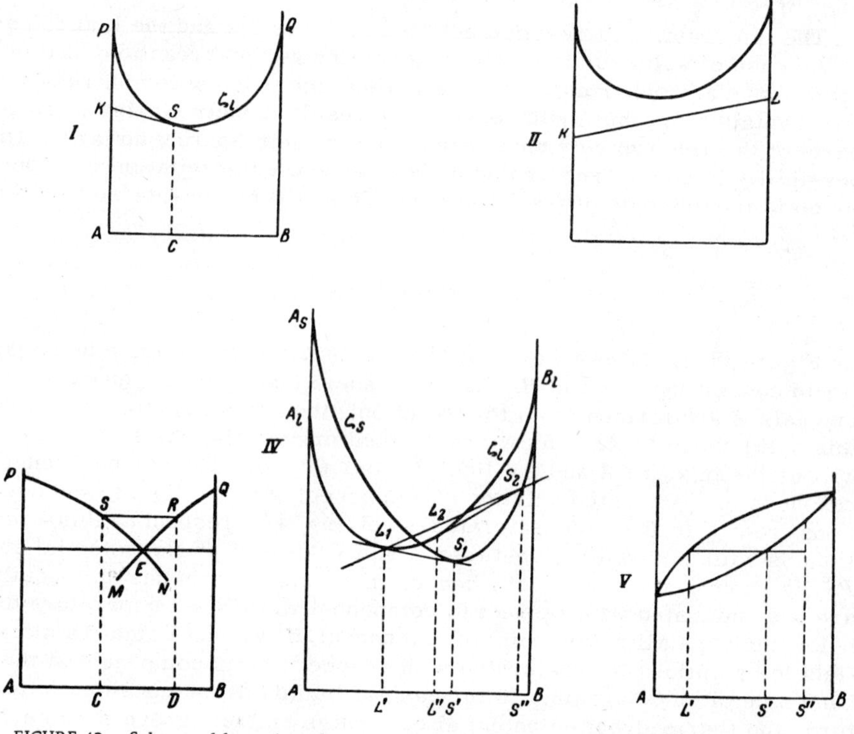

FIGURE 42. Scheme of formation of oversaturated solutions in a diagram of the simplest type

DEPENDENCE OF MELTING POINT DIAGRAMS OF BINARY SYSTEMS ON PRESSURE**

All above-considered diagrams illustrate equilibrium in binary systems as dependent on temperature with pressure constant. These diagrams are usually quite sufficient for a complete characterization of a system, as the influence of pressure on the change of melting temperatures both in unary and in binary systems is very small. However, in several cases it becomes necessary to construct a complete equilibrium diagram of a binary system, taking into account the influence of pressure.

68 The influence of pressure on the change of melting temperatures is relatively small, as this reaction proceeds with small volume changes. The melting point curve of diopside (Yoder, 1952) can be expressed by the following equation: $T = 1391.5 + 0.013P$. This equation was derived experimentally, and it corresponds to a 65° increase in the melting point when pressure is raised to 5,000 atm.

The general view of the simplest diagram of a binary system with a eutectic scheme of melting is shown on Figure 43. Here one may clearly discern that our usual diagrams are constant pressure sections of a three-dimensional diagram — the so-called isobaric sections. In three-dimensional representation the melting point curves correspond to melting surfaces, and the eutectic points to eutectic curves. The eutectic curve, in similarity to the melting point curve of a unary system, is actually a straight line with a gentle slope with respect to the pressure axis. However, because the increase in melting points of the components is variously influenced by pressure, the eutectic line is not parallel to the *PT* plane; i. e., in various sections not only the eutectic temperature will change, but also the composition of the liquid at the eutectic point. This is evident from Figure 44, where two isobaric sections at widely differing pressures are combined.

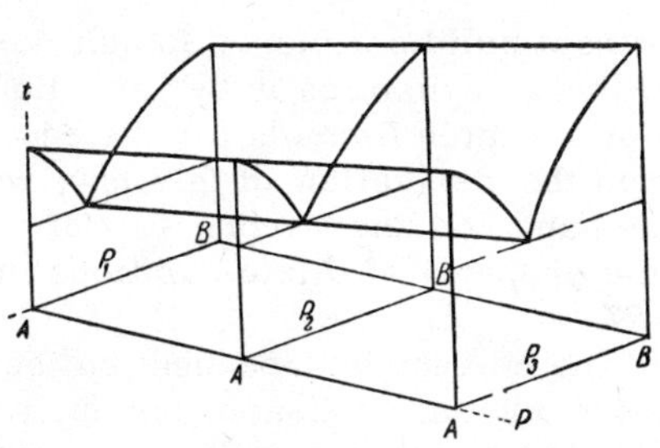

FIGURE 43. General view of the phase diagram of a binary system in the simplest case (eutectic) showing variance of both temperature and pressure

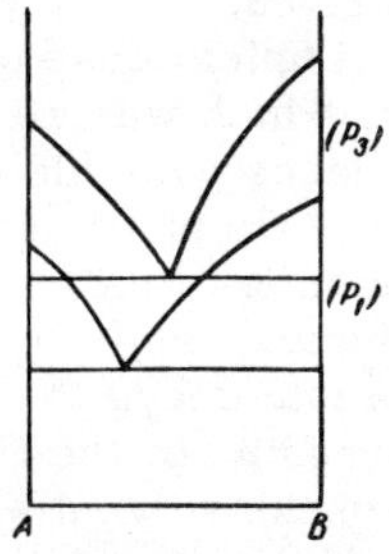

FIGURE 44. Two superimposed isobaric sections of a binary system of the simplest type

The cited data are often used in various theoretical discussions on the possibility of a partial melting of sections of the Earth's crust on a sudden decrease of pressure. One should, however, regard these considerations with extreme caution, as in nature the increase of pressure always causes an increase in the amount of water dissolved in the melt, thus lowering the melting temperature to a much greater extent.

On more complicated phase diagrams, other lines of three-phase equilibrium will appear — in particular the peritectic line. Cases of qualitative change of the section with changes of pressure are also possible, e. g., the transition of a compound from incongruent melting to congruent melting. Among the examples of qualitative change of the diagram which are of importance in petrology, one should mention, in addition to the appearance of new polymorphous modifications, the possibility of formation of new compounds of small specific volume, which are stable only at high pressures. Such a compound appears in the system albite-nepheline and is a mineral of the pyroxene group — jadeite. The condensation of the crystal
69 lattice in this case arises from the transition of aluminum from a fourfold coordination to a sixfold coordination, which results in a difference of specific volume of more than 15 % (see p. 186).

The construction of phase diagrams in general, and of binary systems in particular, is usually a direct outcome of experimental investigations. A description of the technique of such investigations is beyond the limits of the present book; only some of the most general principles will be considered further on in Chapter VI.

On the other hand, there are simple mathematical laws which make it possible, by means of equations, to construct the equilibrium lines of binary systems with a greater or lesser accuracy. These equations, in fact, do contain some constants of pure components which are experimentally derived.

For the simplest case of a diagram without solid solutions, Raoult derived a law, which was afterwards based thermodynamically by van't Hoff, and which makes possible the derivation of a simple formula for the equilibrium line. We shall not dwell here upon the derivation of this law; we shall only mention that it is based, on one hand, on the second law of thermodynamics, and on the other, on the property of dilute solutions in which the solute obeys the equation $PV=RT$.

The depression of the freezing point of the solvent is dependent on the properties of the solvent — its melting point and latent heat of fusion, and the amount of the dissolved compound (taken in molecular units, independent of the composition of this compound). The depression of temperature when 1 g-mol of a substance is dissolved in 100 g of a solvent is:

$$\Delta T=\frac{0{,}2T_1^2}{\Delta Q},$$

where T_1 — is the melting temperature of the solvent (absolute);
ΔQ — latent heat of fusion of the solvent.

In the case of silicates, the molecular structure of the substance in the melt is usually unknown, thus making the calculation of the equilibrium line impossible according to this equation. Here, however, this equation is used for the opposite case — it enables us to get an idea of the molecular weight, and thus of the character of the particles in solution, on the basis of experimental determinations of the equilibrium line, i. e., of T, and latent heat of fusion. It should also be kept in mind that the mentioned equation is applicable only to dilute solutions, provided that the solution is not accompanied by either evolution or absorption of heat.

For equilibrium diagrams of solid solutions, van Laar derived equations which permit the construction of equilibrium curves for continuous solid solutions of Roozeboom's Type I:

$$\ln\frac{x'_A}{x_A}=\frac{\Delta Q_A}{R}\left(\frac{1}{T}-\frac{1}{T_A}\right);\ \ln\frac{x'_B}{x_B}=\frac{\Delta Q_B}{R}\left(\frac{1}{T}-\frac{1}{T_B}\right),$$

where x' and x — are the mole fractions of the components in the solid and the liquid phase;
ΔQ_A and ΔQ_B — are the molar heats of fusion of both components;

T_A and T_B — are their melting temperatures (absolute);
T — is the temperature of equilibrium between the liquid and solid phases;
R — is the universal gas constant.

70 The mentioned equations also present only approximate figures, but in some particular cases there was good conformity with experimental data, e. g., for the system albite-anorthite. If equilibrium curves are constructed for a given system, the equation may be used for calculating the latent heats of fusion, the experimental determination of which is difficult in silicates.

It has already been mentioned above that data of modern crystallochemistry usually enable us to predict the character of a diagram, not only including the presence of solid solutions, but also the type of the equilibrium diagram (see p. 45). This possibility is not yet sufficiently appreciated by physical chemists, who often consider equilibrium without taking into account the crystallochemical properties of solid phases.

In concluding this review, it should be mentioned that the considerations cited on p. 29 are perfectly applicable to the calculation of equilibrium lines in *PT*-diagrams of binary systems. Such calculations are especially used in the construction of equilibrium lines of solid phases, the experimental investigation of which is difficult, e. g., the formation of jadeite in the system $NaAlSiO_4 - NaAlSi_3O_8$.

Calculations are also often applied to equilibrium lines related to the exsolution of the solid phase with the liberation of gas and the formation of another solid phase. Examples of such calculations will be cited further on (see pp. 186 and 345).

71 CHAPTER IV

DERIVATION AND GENERAL REVIEW OF CRYSTALLIZATION DIAGRAMS OF TERNARY SYSTEMS

GENERAL REMARKS ON TERNARY SYSTEMS*

The derivation of properties of ternary systems may be performed in the same way as in the case of binary systems, using the method of geometrical representation of the system's properties on diagrams. However, these composition-property diagrams will no longer be planar as was the case with binary systems. In order to represent the composition and phases of ternary systems, we must start from three points which represent the compositions of the components and which determine the plane and triangle of concentrations. The composition formed by three components is represented by a point on the diagram (composition point), which is located at the center of gravity of weights applied at the component points, and which are proportional to the quantities of these components.

In order to describe properties represented on the composition diagram (thermodynamic potential, temperature, etc) we must resort to a third dimension in space. Hence the composition-property diagram for a ternary system will be three-dimensional or, as it is sometimes termed, a model-diagram, i.e., a geometrical figure of higher dimension than the planar diagrams of binary systems.

According to the Phase Rule, on transition from a binary to a ternary system, either the number of degrees of freedom increases by one, the number of phases remaining constant, or the number of phases increases by one, the number of degrees of freedom remaining constant. Correspondingly, the geometrical forms representing the composition-property relationships in spatial diagrams also become changed: points are replaced by lines and lines by planes. Areas of stability of systems, which are represented in a binary system diagram by two-dimensional fields, are represented here by three-dimensional fields.

In order to clarify the relationship between these more complex models and the simpler ones, we shall consider the relations between the compositions of phases in a ternary system in a manner similar to that used for binary systems. Let us consider possible changes of some phase compositions, forming a ternary system of a given constant composition.

If the system represents only one phase, the composition of this phase will be represented by the same point as the overall composition of the system *P*.

Let another phase of composition *A* appear in this homogeneous phase. The appearance of *A* will determine the phase segment *PA*. On the initial

72 appearance of phase A (when its quantity is yet infinitesimal), point P, the composition point of the entire system, will be one of the end points of the phase segment (Figure 45, I). Should the amount of A increase and its composition remain constant, the composition of the other phase P can no longer remain constant. It will change and at a certain moment be represented by point L. The composition of the system formed by phases A and L and represented by the point P should obviously lie on the line AL, and the ratio of A to L is $PL:PA$ (Figure 45, II). Thus, with an increase in the amount of phase with composition remaining constant, the composition point of the other phase will move along the line PL, which is an extension of line AP.

FIGURE 45. Two-phase equilibrium in a ternary system with the composition of one of the phases remaining constant

If the composition varies, and the composition point moves from A to A_1 and hence to A_2, the composition point of the other phase will move from L to L_1 and L_2, and the point of the overall composition of the system P will always lie on the straight lines AL, A_1L_1, A_2L_2 (Figure 46, I), i.e., the phase segment will rotate around point P. As the amount of A increases, points L, L_1 and L_2 are shifted away from P. The variation of the first phase is represented by the line AA_1A_2, and that of the second phase by the curve LL_1L_2.

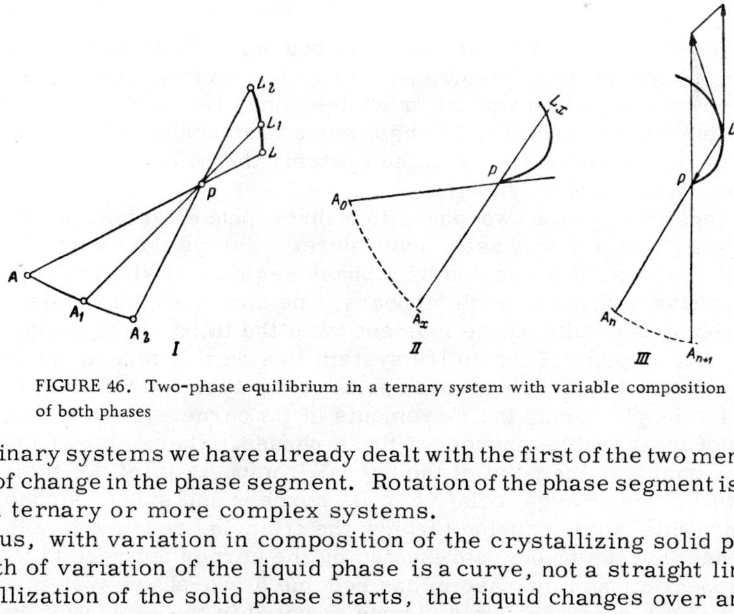

FIGURE 46. Two-phase equilibrium in a ternary system with variable composition of both phases

In binary systems we have already dealt with the first of the two mentioned cases of change in the phase segment. Rotation of the phase segment is possible only in ternary or more complex systems.

[Thus, with variation in composition of the crystallizing solid phases, the path of variation of the liquid phase is a curve, not a straight line. When crystallization of the solid phase starts, the liquid changes over an elementary segment along the corresponding straight line connecting the composition points of the liquid and solid phases (Figure 46, II). This straight line, which is the initial phase segment, is tangential to the

composition variation curve of the liquid at the point P. As is evident from Figure 46, II, the further phase segments are no longer tangential; they become secants. Complications arise due to the fact that at every point, there are, as it were, two simultaneous processes into which the complex
73 process of crystallization with a continuous variation in the composition of the solid phase may be resolved: the solution of the earlier formed phase and the formation of a phase of new composition, this time necessarily in a great quantity. The first process (Figure 46, III) shifts the composition point of the liquid nearer to point A_n; the second moves it correspondingly along a straight line away from point A_{n+1}. The total change of the liquid phase will correspond to the diagonal of a parallelogram which at its limit will coincide (when the distance between points A_n and A_{n+1} is decreased) with the tangent to the variation path of the liquid at this point. The interesting peculiarities of crystallization connected with this will be considered in greater detail below.]

A system of three phases with three components is univariant, provided that the pressure remains constant. Three phases determine a phase triangle just as two phases determine a phase segment. The total composition of the system is represented by a point, which must necessarily lie within the phase triangle, the corners of which represent the compositions of these three phases. The amount of each phase will be found by means of the rule of barycentric coordinates, being the weight components applied at the phase points, i. e., at the corners of the phase triangle, the total weight being applied at the overall composition point of the whole system. If this point lies on one of the sides of the triangle, this means that the amount of the phase represented by the opposite corner equals zero. This phase has either not yet made its appearance, or has already disappeared during the process which changed the composition and amount of the three phases of the system.

The changes in the composition of the phase are expressed in the shifting of the point representing this composition. In a system consisting of three phases — represented as a phase triangle — one, two or even all three of the corners may be shifted. The triangle moves and changes its shape, but as long as there are three phases in the system, the point of overall composition must lie within this triangle.

The transition from a two-phase to a three-phase system is marked by the appearance of a new phase; prior thereto, the system, which consisted of two phases, was represented by a phase segment. With the appearance of a new phase a phase triangle appears, one side of which consists of the former phase segment; at the moment when the third phase begins to form the composition point of the entire system lies on this side of the triangle. Further on, the overall composition point of the system is shifted to within the phase triangle during the movements of its corners; now the system consists of measurable amounts of three phases. On moving and changing the phase triangle, the point of the overall composition of the system is shifted within the triangle relative to its corners and sides; a case may arise when this point, passing through the triangle, reaches its other side. This means that the phase represented by the corner opposite to this side has disappeared, and the system has become a two-phase system.

Thus, in a ternary system a change is noted in the geometric relations observed in binary systems in the case when the entire phase segment was entirely shifted (on the crystallization of solid solutions). In binary systems we have already dealt with "areas of phase formation". These were segments of concentration lines which contained all composition points from which the considered phase may crystallize on cooling. In binary systems we have noted that two cases of crystallization are possible: 1) the particular composition point of the phase coincides with one of the points of the field of its formation — congruent melting, and 2) the composition point of the phase does not coincide with the composition points of its field of formation and it lies in an "alien" field — incongruent melting.

In the case of ternary systems, the fields of formation are areas — parts of the concentration triangle. The lines which divide them are lines of
74 coexistence of the phases formed in these fields. The composition of the separating phase may also be described by a point, which may either lie within the field of formation of this phase (congruent melting) or is located in an "alien" field (incongruent melting). The bold line on Figure 47 shows the boundary between two fields of release of phases *A* and *B*; these fields are marked by corresponding letters in circles. The points *A* and *B* represent the compositions of phases *A* and *B*. The composition of the system, which consists of the three phases, is marked by point *P*. Figure 47, I, shows the case of congruent melting of *A* and *B*, and Figure 47, II — a case of congruent melting of *B* and incongruent melting of *A*.

On cooling, a crystalline solid phase *B* begins to form in both cases, point *P* being located within its field. The composition of the liquid phase changes from *P* to L_0, where it reaches the boundary of the fields *A* and *B*. We obtain the phase triangle from *A*, *B* and L_0. The composition of phases *A* and *B* are assumed to be constant; the composition of *L* varies from L_0 to *L* along the boundary line between *A* and *B*. Point *P* remains within the phase triangle.

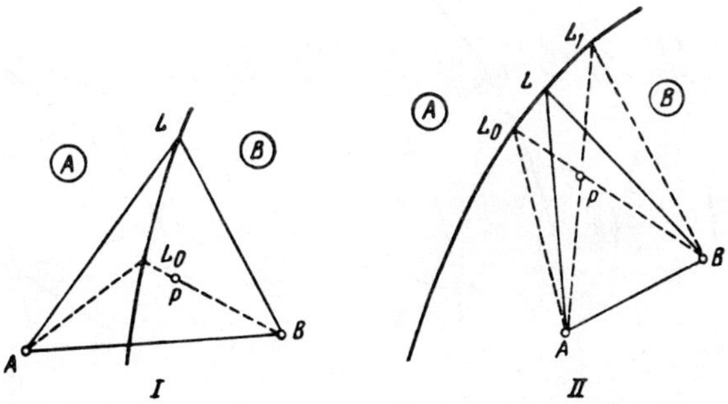

FIGURE 47. Three-phase equilibrium in a ternary system with two of the phases maintaining constant composition (solid phases)

In the case of congruent melting (Figure 47, I) as point L_0 moves, point *P* moves away from sides *BL* and *AL*; the quantity of both phases increases. In the case of incongruent melting (Figure 47, II), as point L_0 moves, point *P* moves away from side *BL*, and approaches side *AL*. Thus the amount of phase A increases, and that of the earlier separated phase B decreases as the liquid changes from P to L_0.

When the liquid reaches L_1, point P reaches the side of the triangle AL_1B; the amount of phase B is now zero, having disappeared, and the system consists of two phases A, and L_1. Further on the liquid phase L changes its composition from L_1 along the extension of AL_1.

[Thus, at every moment the liquid releases a certain total of solid phases, which may be represented by a point on the line AB (Figure 47, I and II); the elementary change in the composition of the liquid will occur along the line which connects the composition of the liquid with this point; in the limit, this straight line will be tangential to the path of change of the liquid composition. If both phases crystallize simultaneously, their sum will always lie between the points AB, and the tangent to the crystallization path passes within the phase triangle. If the intersection point of the tangent with the line AB lies beyond the limits of the segment AB, i. e., the tangent
75 to the crystallization path lies outside the phase triangle, we are no longer dealing with a sum of, but with a difference of phases A and B, i. e., one phase crystallizes while the other dissolves. Thus, the location of the tangent to the crystallization path in its relation to the phase triangle allows one readily to distinguish between the case of so-called cotectic (simultanous) crystallization and peritectic crystallization, i. e., crystallization with solution of one of the phases due to reaction.]

We have assumed that compositions A and B remain constant, but in other cases they may change, moving along lines (drawn thinner) on Figure 48. In these cases, obviously, congruent as well as incongruent melting is possible, as shown respectively on Figure 48, I and II.

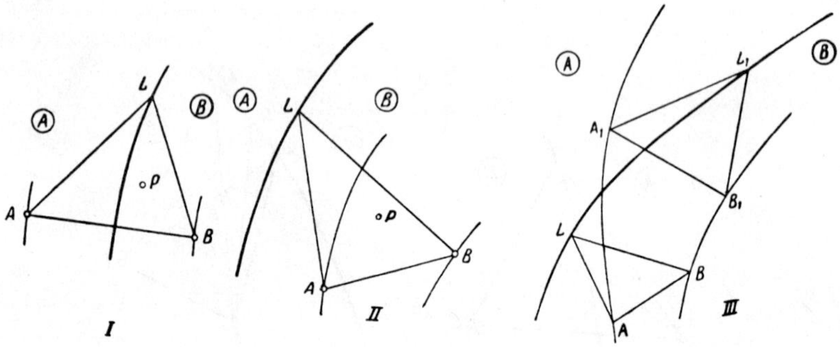

FIGURE 48. General case of three-phase equilibrium in a ternary system with variable compositions of all phases

[In some systems the character of crystallization changes from congruent to incongruent and vice versa. The composition line of one of the solid phases intersects the composition line of the liquid phase (Figure 48, III).

The above-mentioned rule of the tangent is fully applicable in this case (i. e., with a variable composition of solid phases) only at the initial moments of crystallization. Further on the process is complicated by "solution" of earlier formed crystals; therefore it will depend on their quantity. Interesting peculiarities of crystallization in such systems and some general principles will be considered further on in concrete examples (see pp. 230 and 239).

In a binary system at constant pressure we dealt with invariant inversions — the systems consisted of three phases (eutectics and peritectics). In a system of three components we shall deal with invariant inversions — the system consists of four phases. If all these phases are of different composition, i. e., if there are no polymorphous inversions in one of the phases, then various cases may arise, depending on the location of composition points. If crystallization is to be considered, i. e., a case when one of the phases (L) is liquid (represented by a blank circle on Figure 49) and three phases (A, B and C) are solid (black circles), then the number of cases is reduced to the three, as shown on Figure 49.

At invariant inversions occurring at constant temperature, the amounts of certain phases increase and those of others decrease. During crystallization, the amount of the liquid phase always decreases. This is represented by the minus sign on Figure 49. From the location of the composition points, the conjugate changes in the amounts of other phases and the
76 type of the occurring reaction becomes obvious. An increase in the quantity of phases on Figure 49 is marked by a plus sign, a decrease — by a minus sign. The following inversions are invariant:

$L \rightarrow A + B + C$ (Figure 49, I);
$L + C \rightarrow A + B$ (Figure 49, II);
$L + A + B \rightarrow C$ (Figure 49, III).

One of the phases marked by a minus sign will disappear before the others. The system will again revert to a three-phase system. Knowing the point of overall composition of the system, one may predict which of the phases, (which take part in the invariant inversion and which are marked by a minus), will disappear before the others, as the remaining three phases form a phase triangle within which is located the point of overall composition.

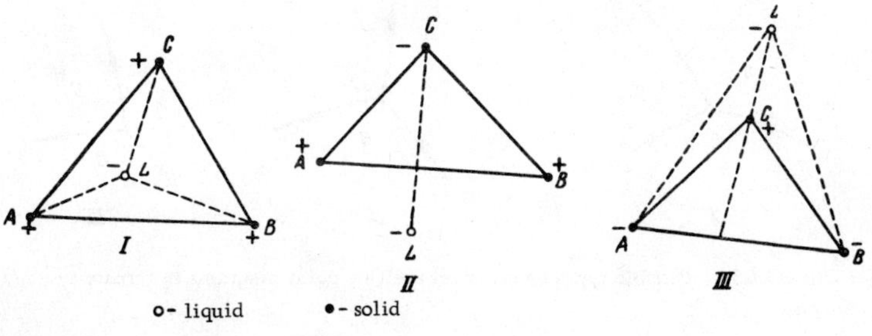

FIGURE 49. Possible cases of interrelation of composition of three solid phases and one liquid phase in four-phase equilibrium in a ternary system

In a binary system an invariant inversion corresponds to eutectics and peritectics, i.e., to singular points upon the melting curve or on the boundaries of the fields of separation of phase formation. In a ternary system there may also be cases where three phase fields, of *A*,*B* and *C*, meet in one point. These three solid phases coexist simultaneously with the liquid phase *L*, which now reaches a composition represented by the intersection point of the boundary lines.

On composition-temperature diagrams of binary systems, points of eutectics and peritectics corresponding to an invariant inversion differ from one another in that in the first case the melting curves meet in the eutectics, being directed down towards this point, whereas at the peritectic point there is only a break; the melting curves on both sides of this point slope in the same direction. In a ternary system the slopes of the boundary lines delimiting the fields may present three cases, as shown by arrows on Figure 50. It is evident which of the cases described on Figure 49 correspond with the combinations of arrows on Figure 50, at an invariant inversion.

1. The first case is obviously the case of eutectics; the composition of the liquid is represented by a point where the boundary lines meet, and it is the lowest point; if temperature drops the liquid can no longer exist. The invariant system becomes a system of three solid phases.

2. The composition of the liquid may vary along the line dividing *A* from *B* after the invariant four-phase system has inverted to a three-phase system; thus phase *C* disappears, and *A*, *B* and *L* remain. This corresponds to the second case on Figures 49 and 50.

77 3. The composition of the liquid, after the transition of an invariant four-phase system to a three-phase one, may change down the lines dividing either of *A* from *C* or *B* from *C*, depending on the overall composition of the system. This corresponds to the third of the cases represented on Figures 49 and 50.

In the second and third cases a sort of branching or forking of the curves is observed. In the second case the forking is termed as elevated (its branches are above the line dividing *A* from *B*), in the third case — as lowered forking (its branches are lower than the dividing line between *A* and *B*).

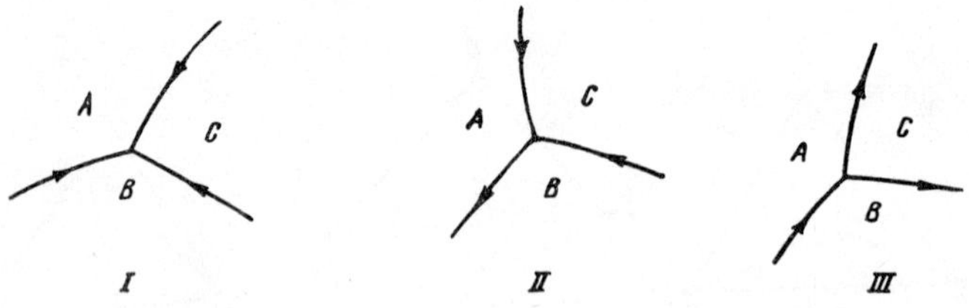

FIGURE 50. Possible types of invariant melting point diagrams in ternary systems

Let us confine ourselves to these comparisons of general features of diagrams of binary and ternary systems. We shall return to this problem when considering other features of these diagrams.

Using the same reasoning applied in the case of binary systems, we may arrive at conclusions on the form of the surface of the thermodynamic potential of a system which is a homogeneous phase of three components. Let us take any two sufficiently close compositions. For their mechanical mixture, the value of the potential should be one of the ordinates on the straight line connecting the ordinates representing the thermodynamic potentials of these compositions. As in the case of a homogeneous phase, these compositions do not form a mechanical mixture, and obviously their potentials will also be lower than the potential of a mechanical mixture. This will be true for any combination of the two compositions. Hence, a point on the potential surface of a homogeneous phase will lie below any section of a straight line intersecting this surface. This means that for a homogeneous phase the surface ζ is convex downward throughout its length.

It is possible to employ another line of reasoning. Let us consider compositions formed from the component *A* and all possible combinations of the components *B* and *C*, i.e., let us take all possible sections of the diagram produced by the plane *AE*, *E* shifting from *B* to *C* (Figure 51). In all these sections the ζ curve for compositions, shown by points of the line *EA*, should be the curve of a homogeneous phase of a binary system, i.e., it should be a curve which is convex downward throughout its length. The combination of an infinite number of such curves will yield the surface ζ of a homogeneous phase of a ternary system, which will obviously be convex downward throughout. This surface will contact the faces of the concentration prism along the ζ curves of the binary systems *A—B*, *B—C* and *A—C* which are included in the ternary system. The surface is suspended above the concentration triangle (the base of the prism).

78 On considering the mechanical mixture, we note that in similarity to the binary system, the thermodynamic potential of the mixture of two phases may be obtained by drawing a straight line through the ζ ordinates of these phases. The surface of thermodynamic potentials for the three phases will obviously be the plane drawn through the points of ζ and representing the potentials of the three phases. This is obvious, as any mechanical mixture of three phases may be formed by the addition to a corresponding mixture of two phases of an appropriate quantity of a third phase (Figure 52). The point ζ' lies upon the straight line $\zeta_1 \zeta_2$, and point ζ located upon
79 the straight line $\zeta'\zeta_3$ will probably fall within the plane $\zeta_1\zeta_2\zeta_3$. It is assumed that the given three phases are of constant composition.

If there are several phases of constant composition, then, at constant temperature and pressure, there are certain combinations of three phases. The concentration triangle *ABC* (Figure 53) may be divided by various methods into triangles, the corners of which are the composition points of the phases. The thermodynamic potential surface consists of several planes, each of which corresponds with one of these triangles, which are the projections of such planes upon the concentration triangle. Thus is obtained a polyhedron composed of triangular faces. For the given equilibrium conditions (temperature and pressure) this division may only be carried out in one definite way, its characteristics being as follows: if an arbitrary vertical plane *xy* is drawn, then the intersection line of the polyhedral surface of thermodynamic potentials with this plane will be a polygonal line convex downward throughout its length. This is only possible if the entire polyhedral ζ surface is convex downward.

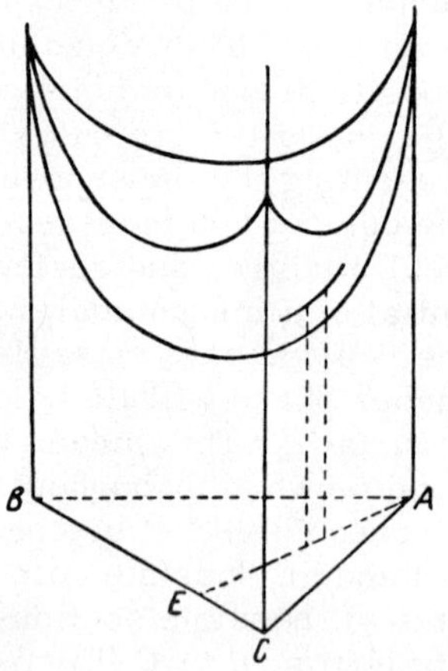

FIGURE 51. Thermodynamic potential of unlimited solutions in a ternary system

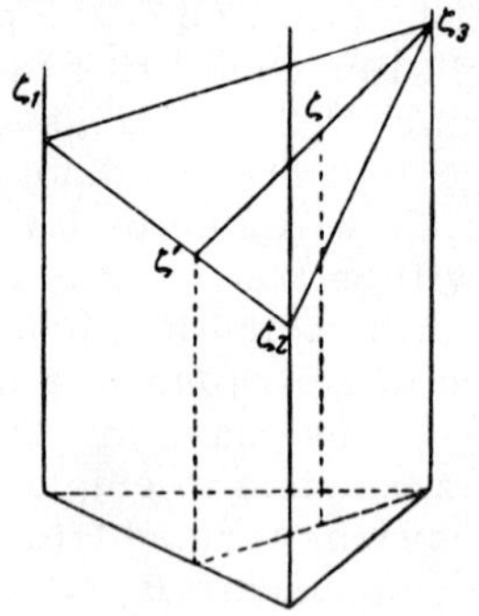

FIGURE 52. Thermodynamic potential of a mixture of three solid phases of constant composition in a ternary system

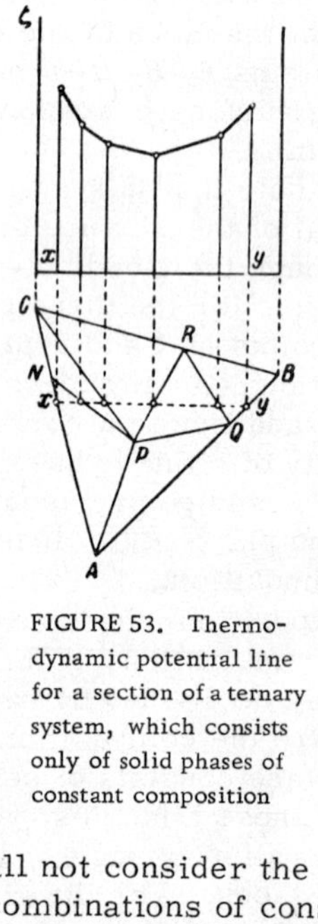

FIGURE 53. Thermodynamic potential line for a section of a ternary system, which consists only of solid phases of constant composition

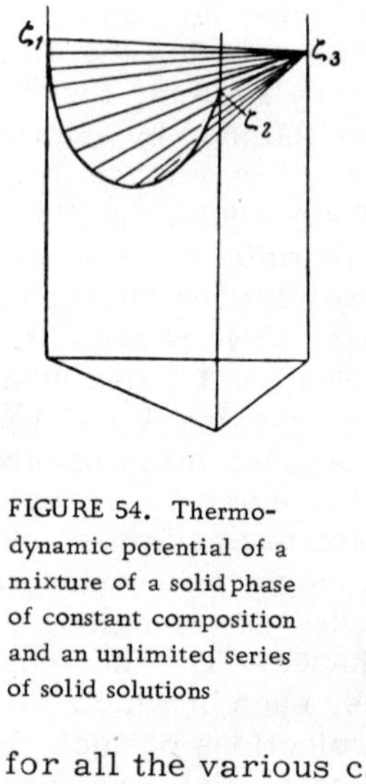

FIGURE 54. Thermodynamic potential of a mixture of a solid phase of constant composition and an unlimited series of solid solutions

We shall not consider the forms of ζ surfaces for all the various cases of phase combinations of constant and variable composition possible in ternary systems. We shall confine ourselves to the case where two of the components form a series of continuous solid solutions, and the third forms a pure crystalline phase. Obviously, in this case (Figure 54) the surface of thermodynamic potentials will be in the form of a cone, with its apex at ζ_3 and its generatrices passing through ζ_3 and through each of the points of the thermodynamic potential curve of the solid solutions ζ_1 ζ_2.

THERMODYNAMIC POTENTIAL OF A SYSTEM CONSISTING OF A HOMOGENEOUS LIQUID AND CRYSTALS OF CONSTANT COMPOSITION

Let us consider a system at constant temperature and pressure, consisting of a homogeneous liquid phase, in which crystals of constant composition have already formed. The thermodynamic potential of these crystals is therefore lower than that of the liquid phases with the same composition, and points ζ_1, ζ_2 and ζ_3 will lie below the ζ surface. Figure 55 shows three cases: 1) composition of the crystals coincides with that of one of the pure components — ζ_1; 2) the crystals consist of two components — ζ_2; and 3) crystals consist of three components — ζ_3. From points ζ_1, ζ_2 and ζ_3 cones may be drawn which envelope the potential surface ζ. The surface of these cones lies below the saccular surface of ζ. Hence, systems with compositions located upon these conical surfaces and represented by points within the projection upon the concentration triangle, will be more stable as a mixture of two phases. One of these phases which corresponds to the apex, is crystalline, the other — a liquid. Its composition corresponds to the point of tangency to the surface of that generatrix of the cone which passes through the total composition of the system.

In the described system, at a certain temperature, the solid phase of a constant composition is not in equilibrium with a liquid of constant composition, as was the case in binary systems, but with an infinite number of liquids. Their compositions are represented by points on the curve *L*, which is a projection of the line of tangency of the saccular surface of potentials and the conical surface, constructed as shown above. The composition of liquid *L* is completely determined by the total composition of the system, since in a system consisting of two phases, the composition of the liquid *L* must be the end of a phase segment, which passes through the point of total composition of the system. The line *L′L″* (Figure 56) shows
80 all possible compositions of the liquid which are stable at a certain temperature together with the solid phase *A*. Thus, this line is the isotherm of saturation of the liquid with respect to phase *A*. The field of coexistence of crystals *A* and the liquid at a given temperature is the area *AL′L″*. Points within this area represent the compositions of liquids saturated with respect to *A*; outside this area the solutions are undersaturated with respect to *A*, as (see Figure 56) the mixture *A+L* is stable between the points *A* and *L′L″*, whereas on the other side of the line *L′L″* only the liquid is stable. In a binary system, the place of the line *L′L″* is taken by a point.

When temperature falls, the points ζ_1 and ζ_2 are increasingly lowered in relation to the ζ surface. The conical surfaces which envelope the surface of thermodynamic potentials of the liquid phase widen increasingly; the line *L′L″* moves away from the composition of the solid phase *A*.

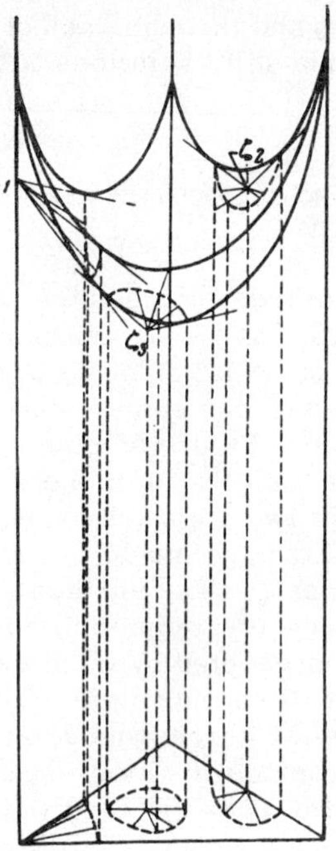

FIGURE 55. Possible cases of relations of thermodynamic potentials for the equilibrium of two phases in a ternary system at a constant composition of the solid phases

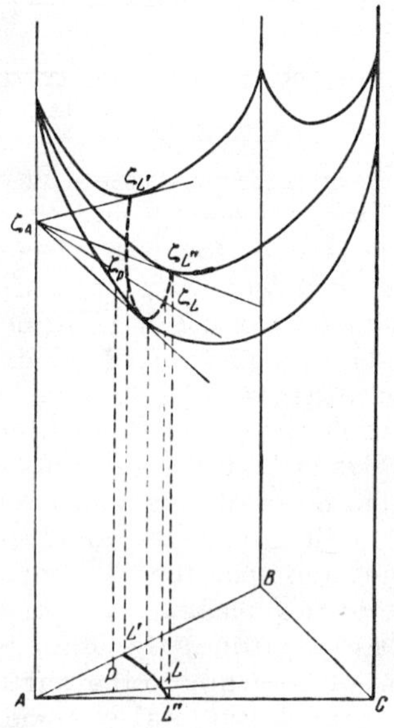

FIGURE 56. Construction of the equilibrium area of the liquid phase with a solid phase of constant composition in a ternary system

TWO SOLID PHASES OF CONSTANT COMPOSITION AND A LIQUID

On the cooling of a ternary melt, let us assume that the formation of one solid phase is followed by the formation of another. At first, the fields of liquid saturation with respect to the two phases do not touch. In a series of solutions of various composition, some will be saturated with respect to one of these phases, others by the other. No more than two phases may coexist simultaneously, one of them being a liquid. On cooling, the areas of saturation of the liquid with respect to both phases are broad-
81 ened until eventually they become contiguous. As the surface of thermodynamic potentials is convex downward throughout, it may be proved analytically that the point of contact between the isotherms lies upon the straight line connecting the composition points of the solid phases. The generatrices of the cones *AK* and *BK*, being tangential to the ζ surface at the same point *K*, are fused into one straight line *AB*, which is tangential to the ζ surface at this point *K* (Figure 57, I).

On further lowering of temperature, the line *AB*, which connects the ends of the ordinates ζ_A and ζ_B, will detach itself from the ζ surface. In this case a plane may be drawn through the straight line $\zeta_A \zeta_B$, which is tangential to the surface ζ at the point ζ_{l_1} (see Figure 60, IV)†. This plane

† [If points *A* and *B* lie within the concentration triangle, two tangential surfaces may be drawn, and correspondingly two phase triangles may be obtained, their apices being l_1 and l_2 respectively, as shown on Figure 57, II].

lies below the surface of the thermodynamic potentials, and its three points, ζ_A, ζ_B and ζl_1, determine the phase triangle, the apices of which will be the projections of these points upon the plane of the concentration triangle. Point ζ_l is easily located by constructing the isotherms of saturation for both solid phases at a given temperature; the intersection point of the isotherms is the sought point ζ_l (Figure 57, II). Indeed, l_1A (as well as l_2A) is the generatrix of a cone, the apex of which is ζ_A. l_1 (as well as l_2) will be the point of contact of the line l_1A with the surface ζ, being at the same time the point of contact of the line l_1B with the same surface. Two straight lines, l_1A and l_1B, both tangential to the surface ζ, at the same point, determine a plane, which is tangential to the same surface at the same point and which passes through ζ_A and ζ_B.

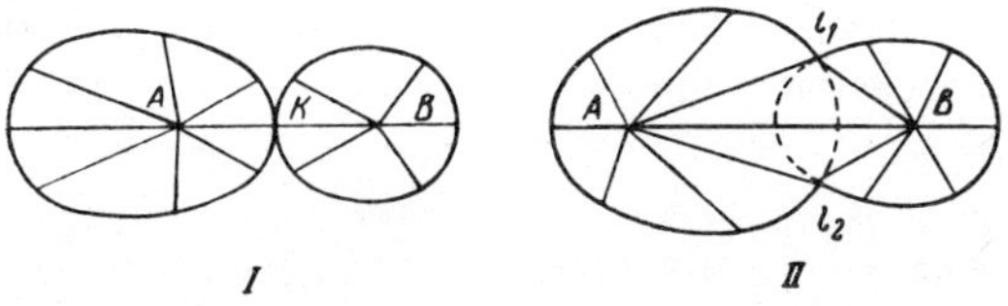

FIGURE 57. Transition from an equilibrium of a liquid phase with one solid phase to an equilibrium of a liquid with two solid phases

FORMATION OF A BINARY SOLID SOLUTION IN A THREE-COMPONENT SYSTEM

We shall limit ourselves to the case when *A* and *B* form a solid solution (Roozeboom's Type I) which does not contain the component *C*.

Let us consider the relationships of thermodynamic potentials at temperatures lower than the crystallization temperature of the more refractory, pure component *B*. From the various possible combinations of temperatures, we shall choose the case when $T_B > T_C > T_A$.

As shown on Figure 58, tangential lines D_1D and *BE* may be drawn on the faces of the thermodynamic potential prism. As in the case of binary systems, these lines determine the composition of the liquid (*D*) and the solid solutions *BA* (D_1), which are stable at a given temperature together with the liquid *D* (for the face *AB*) and the solid phase *B* coexistent with the liquid *E* upon the face *BC*. Through each point of the segment of the
82 curve $\zeta_{L_{AB}}$ of a solid solution, which lies below the liquid curve $\zeta_{S_{AB}}$, one may draw tangents upon the face *AB* from *B* to point D_1, to the surface ζ of three-component solutions (*1—1′*, *2—2′*, *3—3′*, *4—4′*). All cones thus derived will possess one enveloping linear surface (i. e., one formed by the movement of a straight line); this surface passes through the curve $\zeta_{S_{AB}}$ of the solid solutions *BA* and is tangential to the surface ζ of three-component solutions along a certain line. The technique of construction of such a surface will be considered in greater detail below (see Figure 60, III). The mentioned line of contact intersects the prism's faces at points *D* and *E*.

A part of such a linear surface between the segment of curve $\zeta_{S_{AB}}$ of the binary solid solution *BA* and the obtained line of contact will obviously lie below the ζ surface of ternary solutions; therefore the latter will split into two phases corresponding to the end points of the straight line which forms such a linear surface. These end points are: 1) a point on the curve $\zeta_{S_{AB}}$ of the solid solution *BA*, and 2) a point on the ζ surface of ternary solutions, which lies on the line of contact of this surface with the constructed linear surface (see Figure 58). The projection of the obtained line of contact is the isotherm of saturation of the ternary liquid solution with respect to the binary solid solution *BA*. For various overall compositions of the system, which lie within the limits of the exsolution fields (i. e., crystallization of the solid solution *BA*), the compositions of the liquid and of the solid solutions separating from it are different. Their points correspond to the ends of the generatrices of the linear surface. One of the ends (the composition of the liquid) will lie on the isotherm, the other (solid solution) upon the line *BA*. The phase segment is the projection of this line upon the plane of the concentration triangle.

SOLID SOLUTIONS OF THREE COMPONENTS WITH UNLIMITED MUTUAL SOLUBILITY AS SOLIDS

In this case, there are two saccular surfaces $\zeta_{L_{ABC}}$ and $\zeta_{S_{ABC}}$ (Figure 59). If the temperature is lower than the melting temperature of the pure component *A*, the surface $\zeta_{S_{ABC}}$ is located below the surface $\zeta_{L_{ABC}}$ in the corner *A* of the concentration triangle. Obviously, in such a case, the curves $\zeta_{S_{ABC}}$ and $\zeta_{L_{ABC}}$ intersect along some line *DE*, the ends of which are the intersection points of $\zeta_{S_{AB}}$ with $\zeta_{L_{AB}}$ and of $\zeta_{S_{AC}}$ and $\zeta_{L_{AC}}$ (points *D* and *E*) upon the faces of the prism of the considered diagram. Upon these same faces in binary systems, which are included in the ternary system, one may draw common tangents to $\zeta_{L_{AB}}$ and $\zeta_{S_{AB}}$ and to $\zeta_{L_{AC}}$ and $\zeta_{S_{AC}}$, while in ternary diagrams a linear surface may be constructed between segments of these tangents, which envelopes the surfaces $\zeta_{L_{ABC}}$ and $\zeta_{S_{ABC}}$ of the ternary system from below. This linear surface contacts both ζ saccular surfaces along certain lines, the projections of which are points *S* and *L* (Figure 59) upon the plane of the concentration triangle. In the area between *A* and the line *S* solid solutions are stable; between *S* and *L* a mixture of solid solution and liquid is stable, and to the right of *L* — the liquid phase only is stable. The composition of the solid solution and the liquid in the area between the lines *L* and *S* may be obtained as the projections of the end points of that generatrix of the linear surface which passes through the overall composition point of the system. The projection of this straight line is the phase segment.

83 We shall confine ourselves to the above-cited examples. From these examples we note the analogy existing in the construction of diagrams which represent, geometrically, the state of binary and ternary systems

as determined by their composition and thermodynamic potential, i. e., the magnitudes which are necessary for the derivation of composition-temperature diagrams.

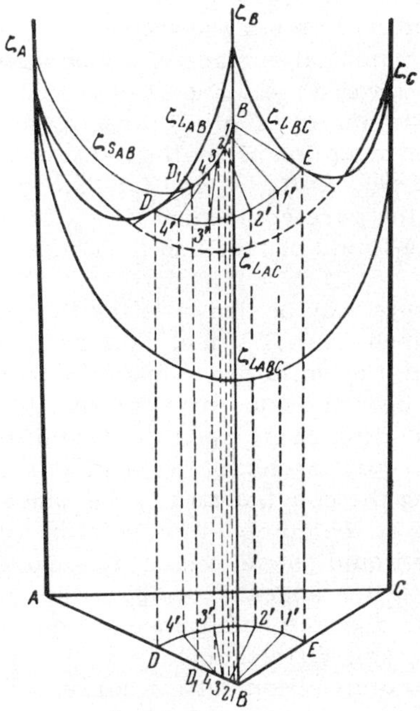

FIGURE 58. Relations of thermodynamic potentials in the case of equilibrium of a liquid with a solid phase of varying composition

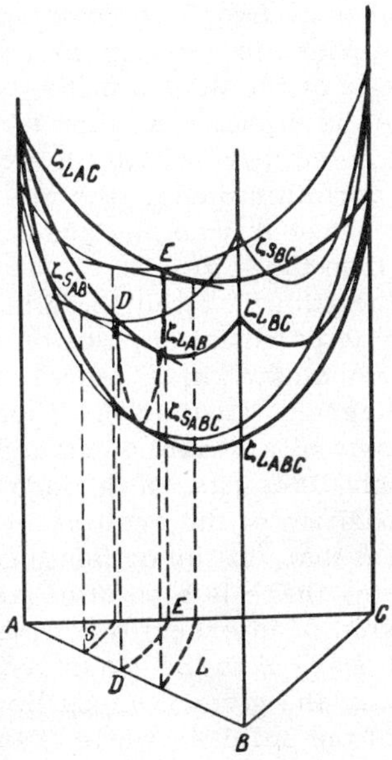

FIGURE 59. Relations of thermodynamic potentials of the liquid and solid phases with unlimited ternary solid solutions

Using such constructions of thermodynamic potentials, one is able to present a complete review of all possible cases of crystallization and to derive the composition-temperature diagrams for ternary systems, as was done for binary systems. This, however, would be too lengthy, as here we are confronted with many more possible combinations. We shall confine ourselves to two examples only, thus providing an idea of the method of construction of composition-temperature diagrams, which is completely analogous to the constructions used for binary systems. These two examples are of special importance to petrology.

DERIVATION OF A MELTING POINT DIAGRAM OF A TERNARY SYSTEM IN THE SIMPLEST CASE*

The simplest case of a ternary system is when three components are non-polymorphous, chemically non-interacting and miscible, in all weight ratios, into one homogeneous liquid phase, from which they crystallize in
84 the pure form on cooling. At constant pressure, such a system may

consist of one, two, three, and eventually four phases during the process of crystallization. In the last case the system is invariant, i. e., it may only exist at a definite temperature, pressure and composition of the liquid phase. The composition of the solid phases is always constant.

In order to obtain the curve of the melting surfaces, we shall use the same method as employed in the consideration of binary systems.

85 It has already been pointed out that the potential surface of a homogeneous liquid phase is entirely convex downward, toward the plane of the basal triangle of the diagram (the concentration triangle); ζ of all three solid phases is higher than this surface until the inception of cooling. On cooling, the ζ potentials of both the liquid and solid phases (ζ_A, ζ_B and ζ_C) change. At a certain moment (the melting point of the pure component *A*), ζ_A of the solid phase *A* becomes equal to the ζ of the liquid phase, and then becomes less than the latter.

When this is attained, a bundle of tangents may be drawn to the thermodynamic potential surface of the liquid phases from ζ_A, i. e., the end point of the ordinate which represents the potential of solid phase upon the potential diagram (Figure 60). Their projection upon the concentration triangle has the form of a bundle of straight lines originating at the apex *A*. One of these straight lines passes through the ordinate, corresponding to the given composition of the system. On considering the construction, it becomes evident that, at the ordinate corresponding to the composition point of the system, there is a point of potential of the liquid phase, which lies above the point of intersection of the ordinate with the tangent from ζ_A of the solid phase *A* to the ζ surface. This means that the homogeneous liquid phase of the given composition is unstable, and will decompose with separation of the solid *A*, while changing its own composition until it corresponds
86 to the point of contact. Upon the concentration triangle this change will be represented by the shifting of the point corresponding to the composition of the liquid along the line which connects the point of initial composition with the apex *A*, towards the side opposite to this apex. This will continue until the thermodynamic potential ζ of any other solid component, e. g., *B*, becomes lower than that of its liquid melt, i. e., until temperature drops below its melting point. From this moment on there are already two points (ζ_A and ζ_B of the solid components) upon the faces of the prism diagram (*A* and *B*), which will lie below the potential surface of the homogeneous liquid phase. [Hence two areas may be constructed in each of which the mixture of a solid phase with a liquid will be more stable than the liquid phase (see Figure 60, II). As temperature is lowered, points ζ_A and ζ_B are removed further and further away from curve $\zeta_{L_{AB}}$. At a certain moment (Figure 60, III), the straight line $\zeta_A\,\zeta_B$ will become a tangent to this curve; it will then pass below the curve. The moment of tangency corresponds to the eutectic temperature of components *A* and *B*. This temperature may either be higher than the melting temperature of the third component *C*, or lower. The latter case is represented on Figure 60, III.

Below the mentioned temperature of the two-component eutectic, a plane, tangential to the plane $\zeta_{L_{ABC}}$, may be drawn through points ζ_A and ζ_B. This plane will correspond with the thermodynamic potential of a mixture, consisting of a liquid phase, the composition of which is determined by the

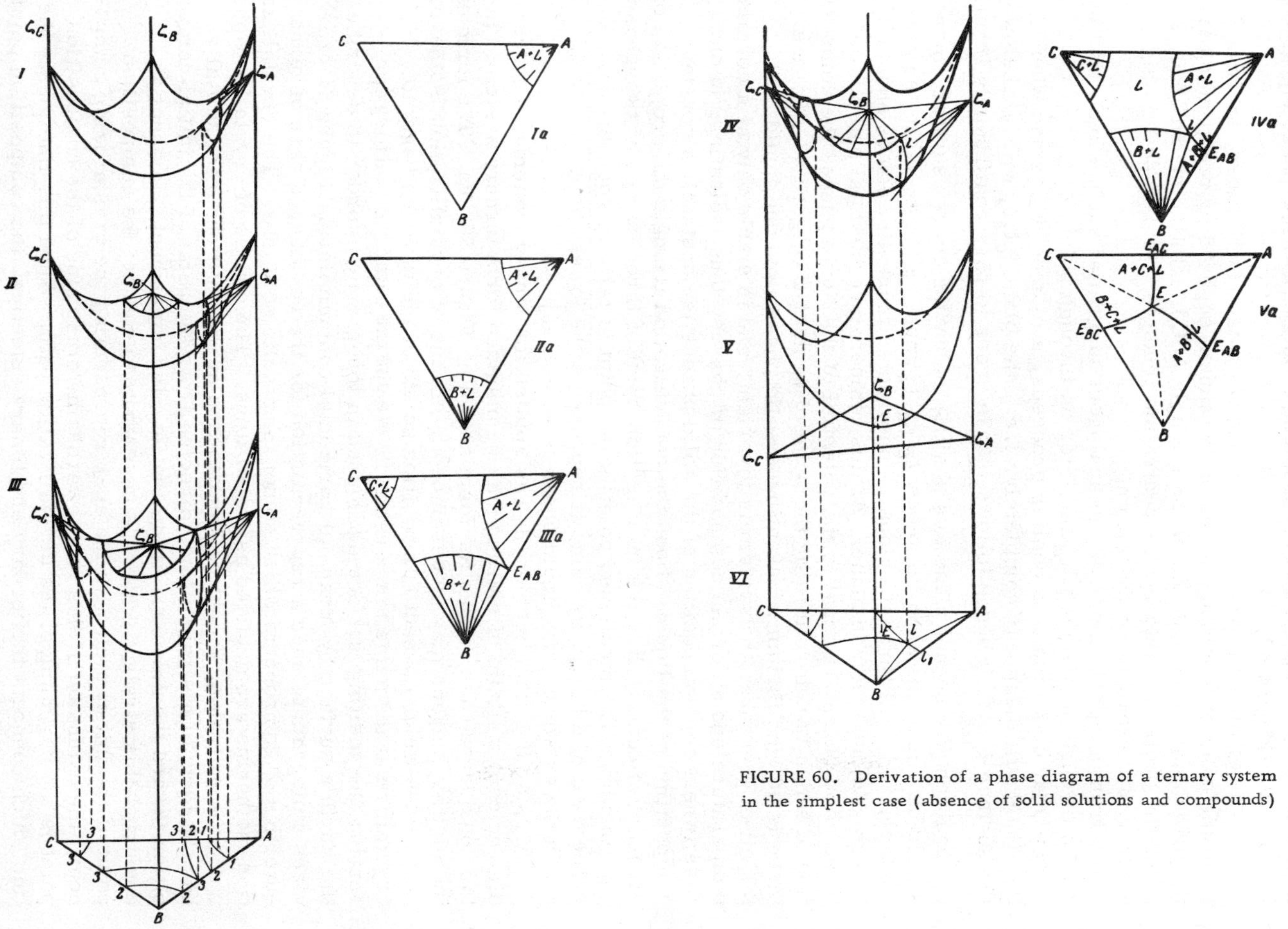

FIGURE 60. Derivation of a phase diagram of a ternary system in the simplest case (absence of solid solutions and compounds)

point of contact l (Figure 60, IV), and of the solid phases *A* and *B*. The potential of this mixture will be lower than the potential of the liquid on this segment; thus we obtain a corresponding triangle of three-phase equilibrium, *BlA* (Figure 60, VI).

As temperature is lowered, the point of contact l will move, starting from the point of binary eutectics on the prism edge *AB* (corresponding to the projection on the side *AB* of the triangle ABC), into the prism, i.e., along the projection into the concentration triangle, its path describing a curve of three-phase equilibrium.

As from a certain moment, an analogous tangential surface may also be drawn through the points $\zeta_C\ \zeta_B$ and later on through points $\zeta_A\ \zeta_C$.

One may claim that, as cooling proceeds, a moment will arise when the liquid will disappear completely, i. e., the plane $\zeta_C \zeta_B \zeta_A$ will fall below the surface $\zeta_{L_{ABC}}$ It hence follows that in order to pass continuously from the position of a secant plane to a position below the convex surface, there should be a moment when the plane $\zeta_C\ \zeta_B\ \zeta_A$ must be tangential to the given surface, i. e., all three previously constructed tangential planes should merge into one common plane. At this temperature the total potential of three solid phases is equal to the potential of the liquid phase, the composition of which is determined by the point of contact *E* (Figure 60, V); i. e., there is an equilibrium between three solid phases and one liquid. At the given pressure, the system is invariant, and is stable only at a certain temperature and a certain composition of phases, thus indicating, in our case (where the composition of the solid phases is constant), a constant composition of the liquid. Upon the concentration triangle the projection of the point of contact *E* is the intersection point of three curves of two-phase equilibrium. The mentioned point is the point of ternary equilibrium. As
87 in the case of a binary system, it represents the lowest temperature at which the existence of a liquid phase is possible.]

By using the same method as was applied for binary systems, a three-dimensional diagram of melting may be obtained for a ternary system. In the considered case, it has the form shown on Figure 60, Va. The form of the melting surface may be described by means of crystallization isotherms, i. e., lines which connect composition points of melts which begin to crystallize at the same temperature. We use the same method for representing the melting surface as is used in topography in order to describe the Earth's surface by means of horizontal contour lines. Figure 61 provides a diagram of such a representation for the considered case of ternary eutectics of components which do not form solid solutions. The crystallization isotherms are depicted as dotted lines. The course of crystallization, i. e., the path along which the composition of the melt changes, initially has the form of a straight line, directed from the point of the initial composition toward the side opposite to the forming component. In order to construct this straight line, it is necessary to connect the composition point of the initial melt with the apex representing the crystallizing component; the extension of this line towards the other side of the composition point is the beginning of the crystallization path. When this line meets the line which connects the ternary and binary eutectics, the composition point of the melt moves along this line (eutectic line) towards ternary eutectic.

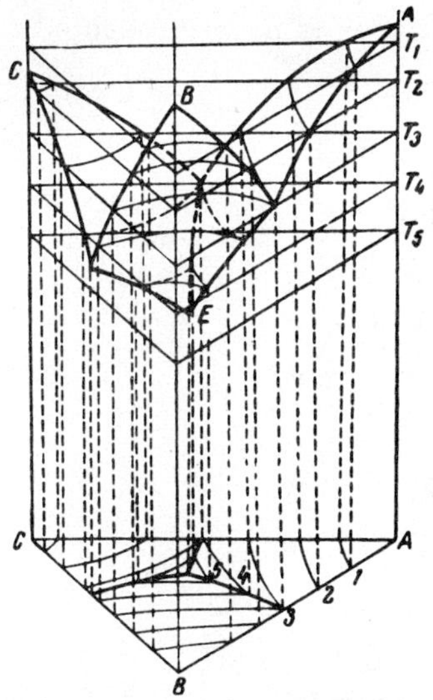

FIGURE 61. General view of the simplest phase diagram of a ternary system

The two components (from the binary eutectic of which this line originates) separate out. When the ternary eutectic is reached, all three components crystallize out simultaneously.

GENERAL REVIEW OF TERNARY SYSTEMS WITHOUT SOLID SOLUTIONS

[Before passing on to the second example of derivation of a ternary phase diagram of a system with solid solutions, we shall present a brief review of equilibrium diagrams of ternary systems without solid solutions. It should be mentioned that even with this reservation there are many possible variants, of which we shall only consider the most important.]

1. One of the components of the ternary system is dimorphous, with an enantiotropic inversion. The inversion point lies above the binary eutectic of *A* and *B* and below that of *A* and *C* (Figure 62, I). In this case the surface of crystallization of component *A* has a break at *K*. The line of this break (the inversion curve) is the isotherm *KK′* [provided that there is a complete absence of solubility in solid phases], i.e., it lies within the horizontal plane; according to the condition, this plane lies above **E** and below E_{AC}.

2. The other two components of a ternary system form a compound. As we have seen while considering binary systems, two cases may exist
88 here: 1) the case of an obvious maximum, or a congruent melting point curve and 2) the case of a concealed maximum, or incongruent melting point curve. Both cases may occur in a ternary system [various relations within the diagram being possible here].

Figure 62, II, shows a diagram with a congruently melting compound, where the addition of the third component does not change the character of this melting. Here, as in a binary system, the entire diagram may be divided into two independent diagrams by the line *CD*. Upon this line the curve E_1E_2 has a maximum *E′*, and in each of the obtained parts there is a ternary eutectic point. If the composition of the system is represented by a point within the triangle *ACD*, we must arrive at point E_1; if the composi-
89 tion is in the triangle *DCB* — at point E_2. Compositions which lie along the line *CD* will behave as a binary system, the crystallization of which terminates at point *E′*.

Figure 62, III, shows a diagram with an incongruently melting compound, the incongruency being also preserved within the triangle. The character of melting of the compounds may be determined easily from the location of point *D*, which lies in the crystallization field of a different solid phase — *A*. The line *CD* does not pass through the field of crystallization of ***D***. It will be shown that in this case the section ***CD*** is not a binary system, as phase A may be formed here, although its composition is not represented in this plane.

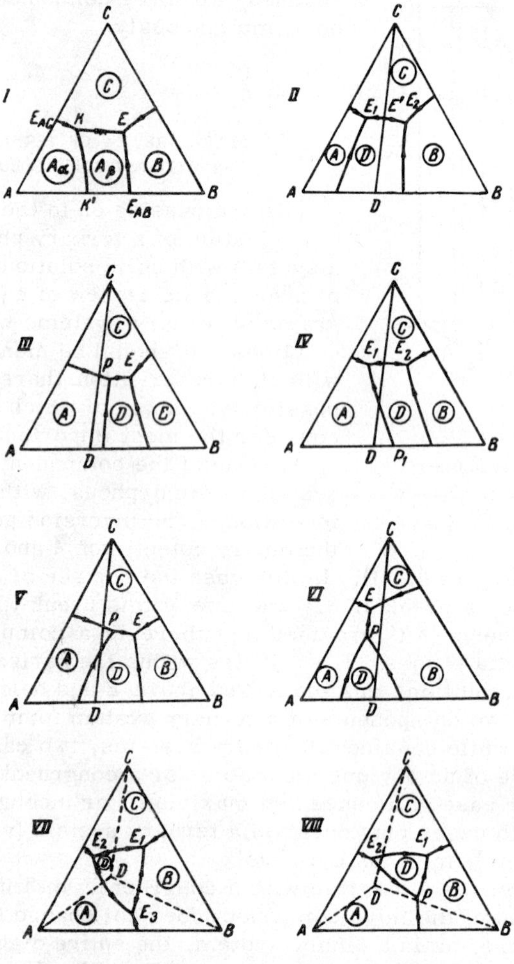

FIGURE 62. Examples of the most important cases of phase diagrams of ternary systems with compounds

Figure 62, IV, V and VI, represent more complex diagrams of cases when the character of melting of the compound changes within the concentration triangle, i. e., on the addition of the third component. Figure 62, IV, represents the case when an incongruently melting compound of a ternary system becomes congruent; here the equilibrium line E_1P_1 intersects the straight line CD. Whereas point P on Figure 62, III, is peritectic, corresponding to the reaction $A+L=C+D$, both points of four-phase equilibrium on Figure 62, IV, are eutectic (all arrows converge), i. e., at point E_1 there is a simultaneous crystallization of three solid phases: ***A, C*** and ***D***.

Figure 62, V and VI, show examples of exsolution of a congruently melting compound in a ternary system, which is again reflected on the diagram of intersection of the three-phase equilibrium lines, $A+D+L$ or $B+D+L$, by the straight line CD, causing the appearance on Figure 62, V, of a peritectic point similar to the point on Figure 62, III, with a reaction according to a scheme $A+L=C+D$. In the system described on Figure 62, VI, the addition of component C causes the eventual decomposition of compound D into its components according to $D \rightarrow A+B$. Thus, at the peritectic point P, the liquid does not participate in the reaction. Figure 62, VII and VIII, show two of the many possible variants of equilibrium diagrams with a compound consisting of three components; its composition point lies within the composition triangle. Figure 62, VII, describes the equilibrium when congruent melting of a compound occurs. In this simplest case, the sections AD, BD and CD are binary systems with a corresponding eutectic point, and the entire diagram may be divided into three independent ternary diagrams: ABD, ACD and BCD. Each of the mentioned diagrams has its own ternary eutectic point.

Figure 62, VIII, represents one of the cases of equilibrium with a compound which melts incongruently. Whereas on Figure 62, VII, the point which describes the composition of a compound was located within the field of its crystallization, point D here lies in the crystallization field of phase A. For systems with any number of components, this is a direct sign of incongruency of melting, as another phase (and not D) is the first to crystallize from a melt of composition D. E_1 and E_2 on Figure 62, VIII, are eutectic points (i. e., simultaneous crystallization of three solid phases occurs here), and P — a peritectic point (i. e., reaction according to the scheme $A+L \rightarrow B+D$).

EXAMPLES OF CRYSTALLIZATION OF TERNARY SYSTEMS WITH INCONGRUENTLY MELTING COMPOUNDS**

When deriving the diagram we briefly considered some examples of crystallization of the most simple ternary system with a eutectic point. Before passing on to systems with solid solutions, we shall cite some examples of crystallization in systems with incongruently melting compounds,
90 which are of special interest in petrography. Figure 63, I, shows such a diagram. Let the liquid be of composition M_1. The point from which crystallization will start lies within the crystallization field of phase A. Furthermore, the point lies within the triangle CBD, i. e., it may be anticipated that after complete cooling it should consist of these phases, i. e., that the crystals of phase A will disappear.

During crystallization of A, the liquid should change along the segment M_1a of the straight line passing through the apex A. When the liquid reaches the composition corresponding to the boundary curve, phase D will join A in crystallization. As seen from the relation of the tangent to the phase triangle, it is not a simultaneous crystallization of two solid phases that occurs here, but a reaction $A+L \rightarrow D$, the composition of the liquid changing toward point P. However, in the given system, the composition of the liquid will not reach this point, as the second side of the phase triangle, Db, passes through this point. This means that phase A has disappeared completely. Here again a degree of freedom appears, and the point representing the composition of the liquid will leave the boundary curve and intersect the crystallization field of *D* along the line *bf*, which passes through point *D*. After the boundary curve reaches the fields of *D* and ***B***, three phases are once again in equilibrium. In this case a simultaneous crystallization of phases *D* and *B* occurs, until the composition of the liquid reaches the ternary eutectic point *E*, where a third solid phase, *C* , joins them. The crystallization will terminate at a constant temperature and composition of the liquid.

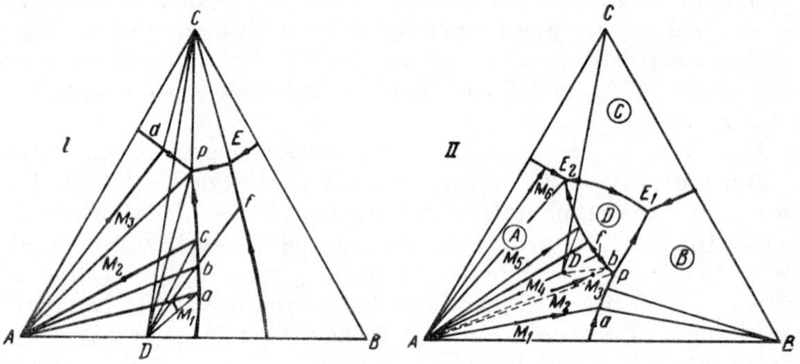

FIGURE 63. Course of crystallization in certain ternary systems with incongruently melting compounds

Point M_2, in similarity to point M_1, lies in the field of phase *A*, but is, however, located within the triangle ***ACD***. Hence crystallization terminates here with these three phases. After the liquid reaches the composition *c* upon the boundary curve, phase *D* joins phase *A*. As aforementioned, the amount of phase *A* then decreases, but in this case the liquid reaches point *P* before phase *A* disappears (as the system lies within the triangle *ADP*). Crystallization of phase *C* starts at this point, the reaction being $A+L\rightarrow C+D$.

The system will possess no degree of freedom (at constant pressure), until one of the phases disappears. In dissimilarity to point *E*, two cases are possible here: either phase *A* is used up and crystallization of the liquid proceeds along the boundary curve ***PE*** , terminating at point *E*, or the liquid is used up, and crystallization terminates at point *P*. A particular
91 case is the simultaneous exhaustion of both phases *A* and *L*. Naturally, the result of crystallization depends on the location of the given point in the system. The location of point M_2 within the triangle *ACD* corresponds to the second case, i. e. , crystallization ceases at the peritectic point.

The third point M_3 is similar to M_2 both with respect to the initial crystallization (in the field of *A*) and to the final result (termination of crystallization at a peritectic point). The only difference is the location of the segment M_3d, which reaches the boundary curve of fields *A* and *C* and not of *A* and *D*, as in previous cases. Thus, the second solid phase is *C*, *A* and *C* crystallizing out simultaneously. Phase *D* only joins these two at point *P*, and hence the amount of phase *A* will decrease, but the liquid is used up before *A*, and crystallization is completed at *P*.

An example of a ternary system with an incongruently melting compound is presented on Figure 63, II. Let us consider crystallization of systems which lie in the field of *A*, within which is also located the point *D*. This case is of greatest interest.

The location of the system M_1 within the crystallization field of *A*, as well as within the triangle *ABD*, determines the beginning and the end of crystallization. On crystallization of phase *A*, the composition of the liquid changes along a straight line to point *a* on the boundary curve. Here the simultaneous crystallization of phases *A* and *B* commences, the liquid changing along the curve to point *P*. At *P*, phase *D* joins them; the system becomes (at constant pressure) invariant, and a reaction occurs according to the scheme $A+L \rightarrow B+D$. From the position of the given point, it is evident that this reaction comes to an end when the liquid is completely used up.

The system of the given point M_2 is quite similar to that of M_1, and differs from it only in that on the crystallization of *A* the liquid, in this case, reaches the boundary line between *A* and *D*. This is followed by the crystallization of phase *D* according to the scheme $A+L \rightarrow D$ (this is evident from the location of the tangent to the crystallization path in relation to the phase triangle). During such crystallization, phase *A* may be used up before the liquid reaches the invariant point; in our case, however, this does not occur, and crystallization will again be terminated at point *P*.

In order that crystallization should not cease at the peritectic point, the initial composition should lie in the triangle *BCD*, e. g., point M_3. After reaching the boundary curve at the same place as in the former case and arriving at point *P*, the liquid is not used up until the solid phase *A* disappears completely. When one of the phases disappears, a degree of freedom appears, i. e., the temperature begins to fall, and the liquid will change along the boundary curve up to point E_1, where the process is terminated by eutectic crystallization.

The composition of point M_4 upon line *AD* will behave quite peculiarly. Here crystallization of *A* will also proceed up to the boundary curve, i. e., even when the liquid is exactly of composition *D*. At the boundary line, the usual reaction $A+L \rightarrow D$ begins, but in dissimilarity to the previous cases the compositions of all three reacting phases are located upon a single straight line. Therefore, the point M_4 will not change its location until the reaction is completed, i. e., over this section the system behaves as a binary system, and point *f* is a maximum point upon the curve PE_2 (this is the general rule for systems with the above-mentioned type of location of the three phase compositions). In our case, the crystallization of the liquid will cease before phase *A* is used up.

Thus, the maximum on the boundary curve between the fields of *A* and
92 *D* lies on the straight line *AD*. If we had selected point M_5 which, after crystallization of phase *A* had commenced, would reach the boundary curve at the other side of the maximum, then after phase *D* had joined in crystallization, the liquid would have changed in the direction of point E_2. As the amount of phase *A* decreases, the liquid does not necessarily attain the eutectic point. However, in the given case, the system lies within the triangle ADE_2. This means that the liquid will reach point E_2, where

simultaneous crystallization of all three phases will occur.

It should be mentioned that systems may exist in which the solution of phase A is followed by its crystallization together with D, even before the eutectic point is reached. This calls for a change in the location of the tangent to the crystallization path in relation to the phase triangle (see Figure 74).

Crystallization of a system represented by point M_6 will begin with separation of the solid phase A, which is afterward joined by phase C; the liquid will be used up after the third phase D joins the others at the eutectic point E_2.

DERIVATION OF A TERNARY DIAGRAM WITH SOLID SOLUTIONS*

[As a second example of the derivation of a ternary diagram by means of thermodynamic potentials, we shall consider the following case, which is of especially great importance in petrography.]

Two components of a ternary system form solid solutions with a melting curve of Roozeboom's Type I. The third component does not form solid solutions with them (Figure 64).

Let us construct diagrams of thermodynamic potentials ζ; for a homogeneous liquid we shall obtain the surface ζ_l, which is everywhere convex downward (see Figure 51); for solid solutions of components A and B — we shall obtain the curve ζ_{AB} on the face AB, and for component C — the point ζ_C upon the corresponding edge of the prism. At high temperatures, ζ_C and the curve ζ_{AB} lie above the surface ζ_l. At sufficiently low temperatures, ζ_{AB} and ζ_C lie so far below the surface ζ_l that all points of the cone (formed by straight lines drawn through ζ_C and through the points of curve ζ_{AB}) will likewise lie below this surface. The ordinates of the points of such a cone are obviously the potentials $(\zeta_C+\zeta_{AB})$ of mixture C with various solid solutions AB. When temperature is lowered, this conical surface moves from above surface ζ_l to a position below it. By geometrically examining the possible cases of their intersection, the melting point curves of ternary systems with solid solutions may be derived. This may be done in a manner similar to the examination of the intersection of surface ζ_l and plane ζ_{ABC} for the case of ternary eutectics.

Let the conical surface $\zeta_C\,\zeta_{AB}$ intersect with surface ζ_l as shown on Figure 64. A plane may be drawn which passes through ζ_C and is tangential to surface ζ_l and curve ζ_{AB}. For its construction we shall draw straight lines from ζ_C, tangential to the surface ζ_l; they will also form a conical surface, which envelopes ζ_l. The trace of this surface on face AB of the prism is shown by the dotted line $k'k''$. On the prism face AB we shall draw a common tangent $k\zeta_{AB}$ to the curves $k'k''$ and ζ_{AB}; its point of contact with $k'k''$ is k. Now, $k\zeta_C$ is one of the generatrices of the cone which envelopes surface $\zeta_{l\,ABC}$; therefore it has a common point of contact L with
93 the latter. This will be the required point. Points ζ_{AB}, ζ_L and ζ_C thus lie in a plane which is tangential to both surface $\zeta_{l\,ABC}$ and to the curve ζ_{AB}, forming a triangle whose projection on the concentration triangle will be $L(AB)C$ (Figure 64, I).

At the moment shown on the diagram the plane of this triangle lies below surface ζ_l of a homogeneous liquid phase. Therefore, the homogeneous liquid phase of composition d, the location of which is shown on the plane of the concentration triangle within $\Delta L(AB)C$, would decompose into three phases: C, (AB) and liquid L. The three points, which are marked by these letters on the concentration triangle, form the phase triangle. The relative quantities of phases C and (AB) are determined by the position of point d of the given composition, according to the rule of barycentric coordinates. Connecting d and C, and producing CL to intersect with AB, we note an important feature, i. e., that the composition of the solid phase (the solid solution of AB) is richer in the more refractory component B, whereas the composition of the liquid is richer in the more fusible component.

[To the right of the phase triangle, i. e., between lines $(AB)C$ and BC upon the concentration triangle, all points of the conical surface $\zeta_C \zeta_{AB}$ are lower than the points of the liquid potential surface, and at a given temperature two solid phases are stable: C, and mixed crystals of variable composition. To the left of the phase triangle, areas of two-phase equilibrium are noted: $C+L$ and $(AB)+L$ as well as the stability field of one liquid. The area of coexistence of the solid phase C with the liquid is determined, as in the above-considered case, by the location of the points of contact of the straight lines, drawn from point ζ_C to the surface $\zeta_{l_{ABC}}$ between the lines $\zeta_C L$ and $\zeta_C l_1$; the potential of the sum of the solid and liquid phase is here lower than the potential of the liquid.

It is more complicated to determine the curve which limits the field of coexistence of mixed crystals with the liquid phase. Such a curve should originate at point L, because the amount of the solid phase C on the AB side of the concentration triangle is zero, i. e., we deal with a two-phase equilibrium. The end of the curve on side AB may also be determined easily, by drawing a common tangent to the curves $\zeta_{l_{AB}}$ and ζ_{AB} on the face of the prism. Point 1 here corresponds with the composition of the liquid phase (projection upon triangle ABC), and point $1'$— with the composition of the coexisting solid phase. In order to construct a point upon the liquidus curve within the field, one should assume a certain composition of system, E (see Figure 64, III)*. Then, by drawing a bundle of vertical planes through the ordinate, one may obtain a series of sections of surface $\zeta_{l_{ABC}}$ and a series of intersection points of each of these planes with the curve $\zeta_{s_{AB}}$. In each of these planes we shall draw a tangent from its intersection point with $\zeta_{s_{AB}}$ to the line of intersection with plane $\zeta_{l_{ABC}}$. A series of straight
94 lines is obtained (forming a helical surface), which intersect the ordinate of E. By selecting an intersection point such that its ordinate is lower than all others, we shall find on it points $2'$ and 2, which correspond with the

* In order to simplify the diagram, this detail is given for the construction of an analogous curve, at higher temperature, for potential relations as on Figure 64, II.

In this case, the curve $K'\ K''$ lies below the curve ζ_{AB}, and no phase triangle will be obtained, but fields of two-phase equilibrium appear: liquid + solid C and liquid + solid solutions AB. The curve L_1L_2 for a liquid which is in equilibrium with solid C may be obtained by projecting the points of contact made by tangents drawn from point ζ_C to the surface $\zeta_{l_{ABC}}$. We shall use the method illustrated on Figure 64, III to construct the composition curve of the liquid which is in equilibrium with solid solutions.

potentials of the solid solution $\zeta_{s_{AB}}$ and of the liquid $\zeta_{l_{ABC}}$, a mixture of which, while having the same composition as E, has a lower potential. Therefore a decomposition occurs into (AB) of composition $2'$ and a liquid of composition 2. By changing the location of point E, one may construct the required curve, as shown on the concentration triangle (see Figure 64, I).

In the area of concentration between the curves Ll_1 and L 1 and point A at given temperature, only the liquid phase is stable, possessing here the lowest potential.

The above-constructed thermodynamic potential diagram applies to a certain temperature T. Using the same construction for different temperatures, one obtains phase triangles $C(AB)L_1$, which differ both in shape and position. [As temperature falls, the triangle is shifted so as to displace point L toward A (the point of the more fusible component). In the limiting case, when line $\zeta_C\zeta_A$ contacts the curve on the face of prism AC, the phase triangle becomes a straight line coinciding with AC, and crystallization terminates here (at the temperature of eutectic crystallization of C with the
95 more fusible component A). If the temperature rises above the initial value T, the triangle is shifted in the opposite direction, until it finally becomes a straight line which coincides with side BC at the moment when the straight line $\zeta_C\zeta_B$ contacts the curve $\zeta_{l_{CB}}$ on the prism's face BC. At higher temperatures, a simultaneous crystallization of two solid phases is impossible, and our system will only possess a liquid stability field and areas of two-phase equilibrium $L+AB$ and $L+C$.

Figure 64, II, shows the construction of an equilibrium diagram at a temperature higher than the eutectic temperature of BC, but lower than the melting temperature of the pure components B and C. By drawing a system of tangents from point ζ_C to the plane, we determine the boundary of coexistence between phase C and the liquid $(L'LL'')$. By constructing an auxiliary curve $K'K''$, which is the trace of intersection of the obtained conical sur-
96 face with the face AB of the prism (compare with curve $K'K''$ on Figure 64, I), one may readily see that it is impossible to draw a common tangential plane to surfaces $\zeta_C\zeta_{AB}$ and $\zeta_C\zeta_{l_{AB}}$, as the curve $K'K''$ lies entirely below the curve ζ_{AB}.

The composition curve of the liquid which is in equilibrium with the solid solution AB is constructed by the above-mentioned method (see Figure 64, III). Initially, two of its points are determined on the faces AB and BC of the prism: in the case of AB a common tangent is drawn to curves $\zeta_{l_{AB}}$ and ζ_{AB}; in the case of BC — a tangent to $\zeta_{l_{BC}}$ from point ζ_B.

If the temperature is raised, the melting point of one of the components is reached (on our diagram component C), i. e., above this point only the crystallization of one solid phase is possible (in our case of mixed crystals AB). Eventually, when temperature exceeds the melting temperatures of both components (i. e., in our case above the melting point of B), the entire conical surface $\zeta_C\zeta_{AB}$ will lie above plane ζ_l, and one liquid phase is stable at all concentrations.]

Now we understand the shape of the melting point diagram of a ternary system in the considered case. Its construction is completely analogous to that of a binary system. Figure 65, I shows a three-dimensional diagram of a

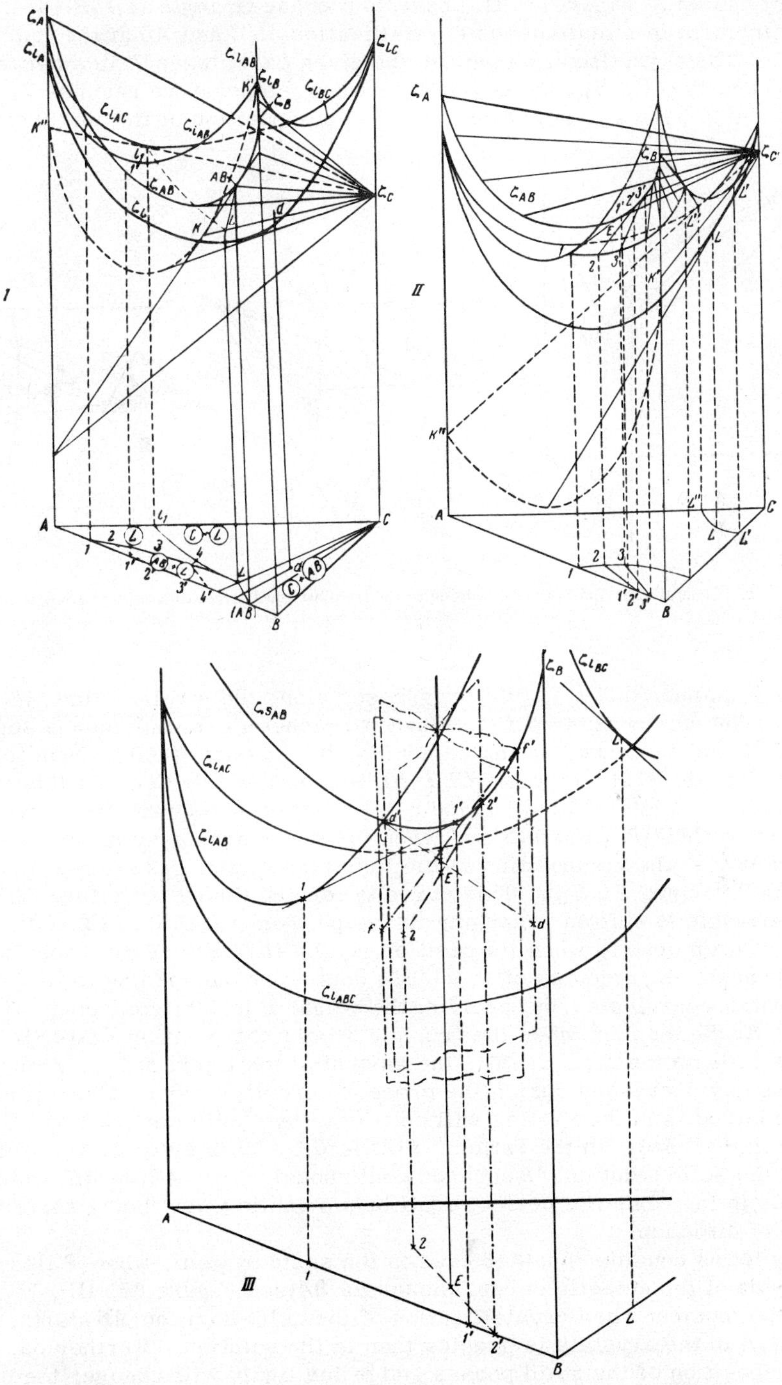

FIGURE 64. Derivation of one of the simplest ternary diagrams when solid solutions are present

ternary system. Figure 65, II, presents a phase triangle $CL_1(AB)_1$ at the initial moment of simultaneous crystallization of C and AB at the temperature T_1. The crystallization path in the given case, when P lies between point C and line $E_{AC}E_{BC}$, is as follows: when temperature reaches T_1, component C begins to separate out. The composition of the liquid changes

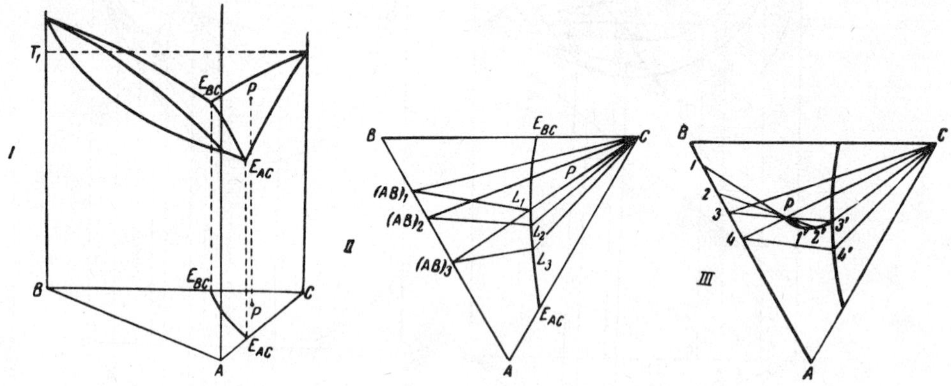

FIGURE 65. General view and projection of one of the simplest equilibrium diagrams of a ternary system with solid solutions

in such a manner that its point progresses along the straight line CL_1, which is the continuation of CP. When it reaches L_1, simultaneous separation of AB and C starts. In the beginning, the amount of AB is infinitesimal — point P in the phase triangle $CL_1(AB)_1$ lies on the side CL_1. At this moment, the system consists of liquid L and a solid phase, the quantities of which are in the ratio $CP:L_1P$. Further on, when the simultaneous separation of AB and C takes place, the composition of the liquid phase changes from L_1 to L_2 and
97 L_3 along the curve $E_{BC}E_{AC}$. This curve is termed the eutectic line. The phase triangle is shifted, changing its shape from $CL_1(AB)$ to $CL_2(AB)_2$, etc. At a certain moment, when its position is $CL_2(AB)_2$, the system consists of three phases, the amounts of which are determined according to the rule of barycentric coordinates, by resolving the weight P into the components L_2, C and $(AB)_2$. At the moment when the triangle attains the position $CL_3(AB)_3$, i. e., when P falls on the side $C(AB)_3$, the amount of the liquid phase, which had continuously decreased during the process, becomes zero. Crystallization is terminated, and the system will consist of two solid components, C and solid solution $(AB)_3$, in the ratio of $P(AB)_3:CP$. As is evident, the composition of the solid solution AB changed continuously from AB_1 to AB_3 which is possible in the case of a perfect equilibrium continuously being restored by means of diffusion.

Now let us consider another case in the same system, when P lies on the other side of the eutectic curve, closer to BA (see Figure 65, III).

At the moment when crystallization of the solid solution AB starts, the ratio $B:A$ in the crystals is greater than in the solution. Further on, both the composition of the solid phases and of the liquid will change; the phase sections 1—$1'$, 2—$2'$, 3—$3'$ continuously pass through point p. The construction of a curve corresponding to the change of the liquid phase is considered on page 72. When the composition of ***AB*** is ***3***, the liquid should be of composition ***3'***; at this moment, the composition of the liquid also reaches the eutectic curve. ***C*** begins to separate simultaneously with ***AB***; further on the change of composition of the liquid follows the line ***3'—4'***, and crystallization terminates when the amount of liquid becomes zero at point ***4'***.

CONODES (SIDES OF PHASE TRIANGLES) AND CRYSTALLIZATION PATHS

During the experimental study and description of ternary systems with solid solutions, it is not only necessary to obtain data on the surface of melting (i. e., on the composition of the liquid phase which is in equilibrium with the crystals), but also on the composition of the solid phases which are in equilibrium with liquids of various compositions. Thus, in order to represent a system upon a plain, the drawing of three-phase equilibrium lines and isotherms proves to be insufficient. In order to assess the composition of solid solutions (for the case of solid solutions consisting of two components only) the most important factors are the sides of the phase triangle, or the so-called conodal straight lines which connect compositions of liquid and solid phases (or any two phases) which are in equilibrium in the presence of a third phase, i. e., in our case — that connect the composition of the liquid on the boundary curve $E_{AC}\,E_{BC}$ with crystals ***AB*** in the presence of crystals ***C***. Such experimentally determined conodes should be included in the diagram (Figure 66, I) within certain intervals. If a point representing a system is located between two conodes, the conodes may easily be drawn by means of interpolation.

The knowledge of the location of the conodes enables one to follow the course of crystallization, beginning with the incipient crystallization of a third phase. In order to construct the path of change of the liquid within the field of crystallization of one solid phase of various composition, the above-mentioned conodes are insufficient. In order to arrive at the composition of any solid phase which is in equilibrium with a liquid, the so-called crystallization paths must be drawn. A crystallization path is a line so constructed that the tangent to it at any point, if produced up to the solid solutions line, shows the composition of the conjugate solid
98 phase. As we shall see below, these lines are really paths of crystallization, i. e., paths indicating the liquid phase at inequilibrium, i. e., when the separating crystals are continuously removed in some way, and no reaction between liquid and crystals occurs.

On Figure 66, II, a diagram of the above-considered system is shown with conodes and paths of crystallization. Lines which correspond with binary solid solutions (in this case the side ***AB*** of the phase triangle) will in future always be featured as bolder lines.

If data on crystallization paths are lacking, the composition of the solid solution in equilibrium with a given liquid may be approximately determined by means of a binary diagram, i. e., assuming that the third component does not influence the ratio of the first two, either in the solid or in the liquid phase. This is achieved (see Figure 66, II) by drawing a straight line through the given point of composition of the liquid and the opposite apex of the triangle, up to its intersection with side ***AB*** at point ***l***, i. e., by determining the component ratio in the liquid phase. We shall then transfer the obtained point to the liquidus curve of the binary diagram — ***l***, and determine the composition of the solid phase ***S*** in the usual manner. The obtained point is transferred to the side of the concentration triangle as **l'**.

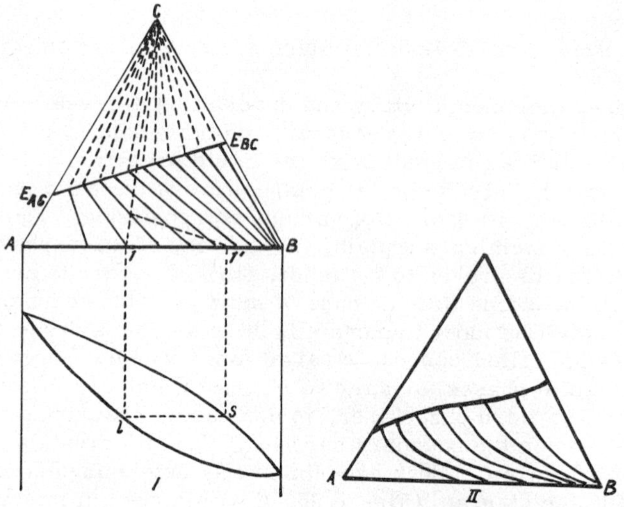

FIGURE 66. Conodes (1) and crystallization paths (II) for one of the simplest ternary diagrams with solid solutions

EXAMPLES OF TERNARY SYSTEMS WITH SOLID SOLUTIONS*

We considered a ternary system in which two components form a solid solution of Roozeboom's Type I. This system is of special importance in igneous petrology. Without dwelling upon other types of diagrams, we shall merely mention them as examples.

A case of a ternary system where two components form a mixed solution of Type III. The isotherms and crystallization paths are shown on Figure 67; the curve $E_{AC}E_{BC}$ (eutectic curve) has a minimum at point M_E in correspondence with the minimum in system ***AB*** at point **M**.

Two components of a ternary system are isodimorphous, or form a discontinuous series. In order to understand this case, we shall recall the diagram for the case of isodimorphism of a binary system (see Figure 24).

99 [Figure 68 provides a general view of such a diagram of a ternary system in the most simple case, where the binary system ***CB*** with limited solid solutions is related to Roozeboom's Type V (with an eutectic). Figure 69 shows a rarer case, where the system ***CB*** is related to Type IV. The

figure shows the projection of the diagram upon the concentration triangle together with the isotherms, and the development of the prism faces. In contrast to point *E* on Figure 68, the invariant point *P* is in this case a peritectic point — a point of reaction and not a eutectic point.]

One of the possible cases in a ternary system is when two pure components, forming a binary system of Type III, split into solid solutions of Type IV on the lowering of temperature after the addition of a third component. Here a critical point occurs on the surface of the melting point diagram, below which the isotherms have a turning point, as shown on Figure 70.

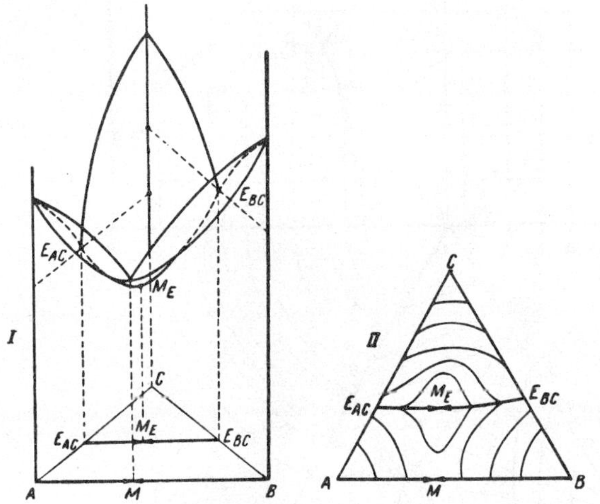

FIGURE 67. General view and projection of a ternary equilibrium diagram with solid solutions when the system *AB* is of Roozeboom's Type III

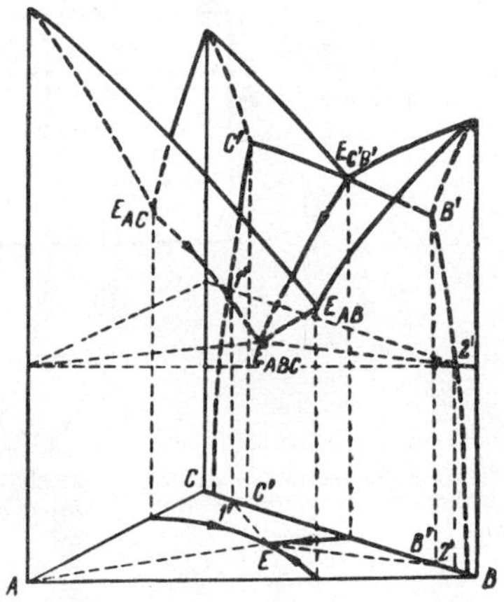

FIGURE 68. General view of a ternary equilibrium diagram when one of the binary systems has limited solid solutions (Type V)

Possible variants of diagrams with solid solutions are quite numerous. Even more numerous are diagrams of systems in which, in addition to solid solutions, compounds are present which themselves often form solid solutions. We shall cite two examples below.

Figure 71 shows a fairly simple case of a system in which components *A* and *B* form a continuous series of solid solutions of Type I, whereas component *C* forms with *A* and *B* the compounds D_1 and D_2, which also form a continuous series of solid solutions. Both compounds melt congruently. Within the diagram there are two cotectic lines, one of them corresponding to a simultaneous separation of phase *C* together with a solid solution of a binary compound (D_1D_2), the other — with a simultaneous separation of the same solid solutions of compound D_1D_2 with solid solutions (*AB*). In the latter case a new type of a phase triangle is observed, all three apices of which are mobile; when temperature falls, the apex 1 moves along the boundary curves, and the apices representing the solid phases — along the corresponding straight lines. Thus, in addition to the conodes which connect the

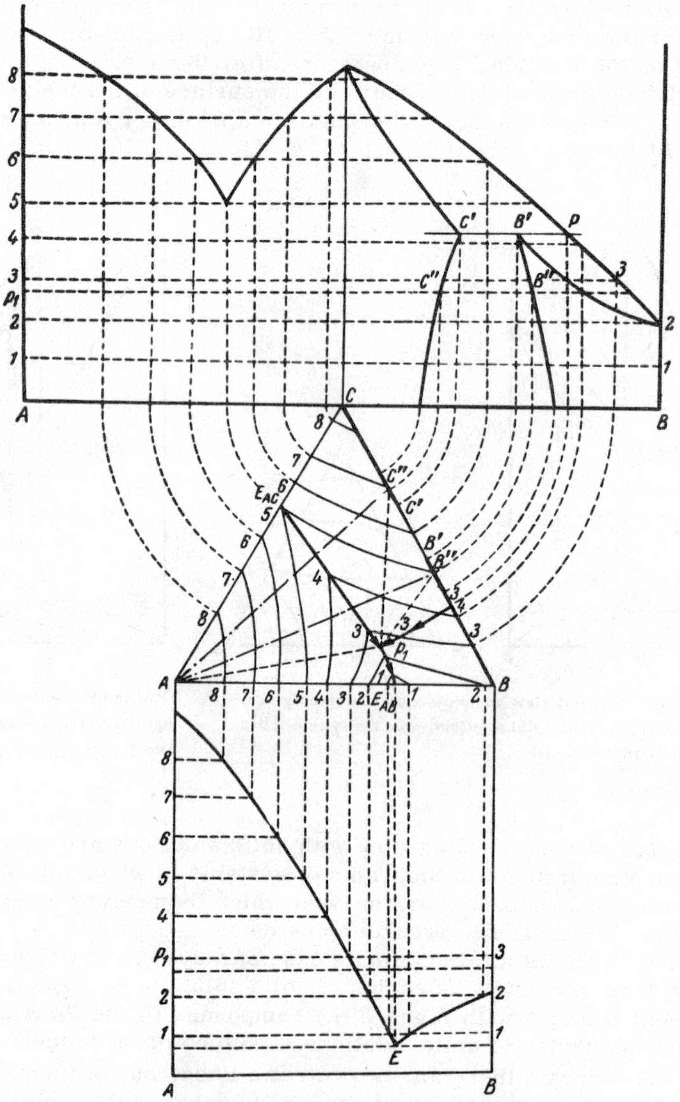

FIGURE 69. Equilibrium diagram (the prism faces unfolded) of a ternary system, in which the binary system *CB* belongs to Roozeboom's Type IV

101 conjugate compositions of the liquid and the solid phases, there are also conodes which correspond to the third side of the phase triangle, i. e., they connect conjugate compositions of solid phases which terminate crystallization. These conodes may also be drawn for diagrams at temperatures below melting, i. e., for paragenetic diagrams of rocks.

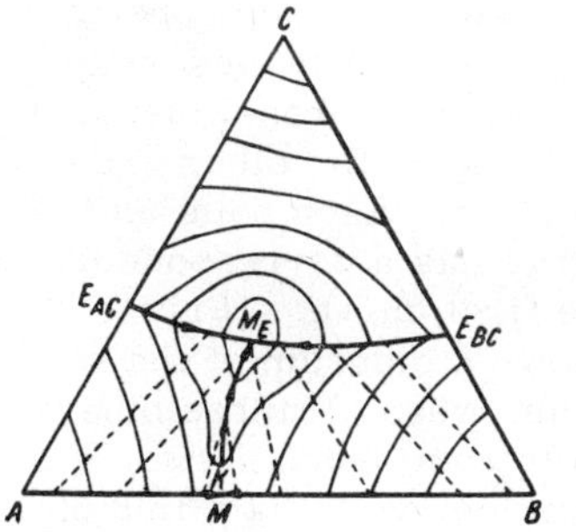

FIGURE 70. Equilibrium diagram of a ternary system, in which system *AB* is related to Roozenboom's Type III; however, the decomposition of solid solutions starts within this system.

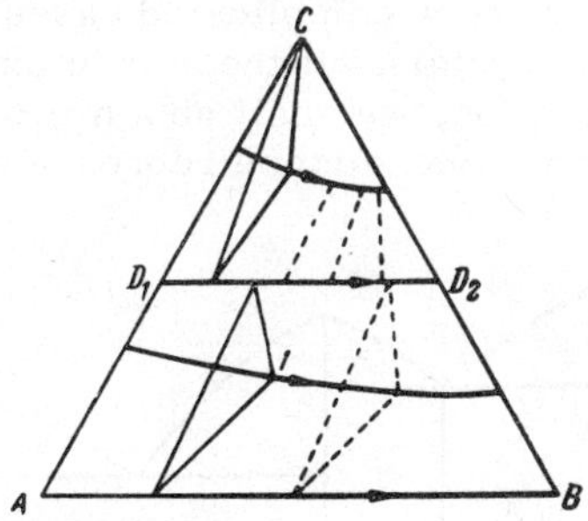

FIGURE 71. An example of an equilibrium diagram of a ternary system with a compound and binary solid solutions. Compounds D_1 and D_2 melt congruently, yielding a continuous series of solid solutions of Type I.

It is worth noting that the ratio ***A*:*B*** is generally not equal in both series of solid solutions. However, the conodes usually do slope in the same direction. This is of great importance in petrology, as in igneous rocks an equilibrium is found between two or more femic minerals, each of which represents a series of solid solutions of a magnesium and an iron component. In 1935–1936, D. S. Korzhinskii (during his study of parageneses of crystalline schists of Aldan, at the same time as, but independent of the work of N. Bowen and Schairer on the system $Mg_2SiO_4 - Fe_2SiO_4 - SiO_2$), showed that there is a variable relation between the magnesium component and the ferrous component in femic minerals in equilibrium; at the present, a great deal of importance is attached to the study of the regularities of conjugate compositions of these minerals.

Figure 72 shows an example, where components *A* and *B* form a compound with an incongruent melting point, this compound forming a continuous series of solid solutions with component *C*. In systems *AC* and *BC* there is a simple eutectic point. Whereas one of the boundary lines is in this case a simple cotectic curve, i. e., it represents simultaneous crystallization of two phases, the other line changes its character, as it connects the eutectic and peritectic points. This diagram shows two phase triangles which correspond to reaction and simultaneous crystallization respectively.

As already mentioned, in the case of crystallization of solid solutions, the rule of position of the tangent in relation to the phase triangle is only of limited importance; it is only fully applicable in the case when reaction between the liquid and the earlier separated solid solutions is impossible.

102 Complex phenomena occurring here at the change in nature of the curves (which are often neglected even in most comprehensive texts on physical chemistry) are dealt with in detail below, by considering a concrete example.

A TERNARY SYSTEM WITH SOLID SOLUTIONS OF THREE COMPONENTS*

Even more complicated cases arise when all three components form mixed crystals. If these solutions form a continuous series without decomposition, we shall obtain two surfaces upon the three-dimensional diagram; one surface represents the liquid phase, the other — the solid solution. Each point on the second curve has a corresponding point on the first curve. [Figure 73, I shows a diagram of the simplest case, when all three binary systems yield continuous series of solid solutions with diagrams of Type I. Figure 73, II shows the concentration triangle, upon which are projected the isotherms of the liquid surface (by solid lines) as well as those of the solid phases (dotted lines). Crystallization paths are depicted on Figure 73, III, and the method of their construction is shown, again. As in the case of systems with binary solid solutions, these crystallization paths are only drawn for cases of interrupted equilibrium, when reaction between the liquid and solid phases, resulting in the formation of zoned crystals, is impossible. The straight lines which connect conjugate compositions of the liquid and solid phases at each temperature are tangential to these curves; i. e., when crystallization paths are given, conodes for each isotherm may be readily constructed.

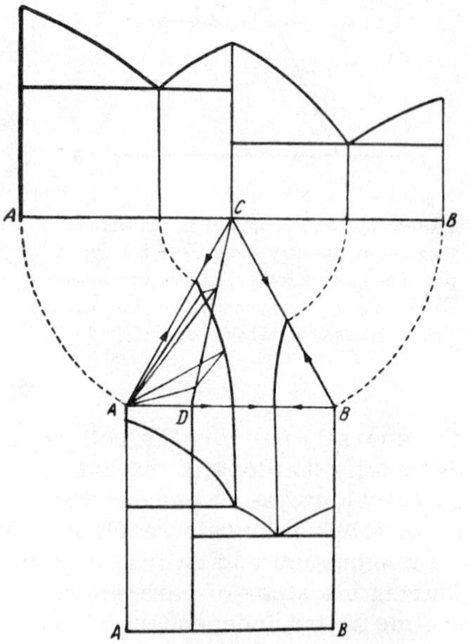

FIGURE 72. Example of a ternary equilibrium diagram with a compound and binary solid solutions. Compound **D**, which melts incongruently, forms a continuous series of solid solutions with component **C**.

With equilibrium crystallization, the location of the line of change of the liquid and solid phases depends on the location of the point. These lines cannot be drawn in advance, as a network of numerous intersecting lines is obtained. The accurate construction of these lines for each system is rather difficult, but in all cases their terminal points may easily be established, as with equilibrium crystallization the composition of the final crystals should correspond to the composition of the given liquid. After the point has been projected upon the liquidus surface (Figure 73, IV) and the conode for the corresponding isotherm is drawn, we shall obtain the initial points for the composition of the liquid phase — ***a*** and of the solid ***a'***, as well as the temperature of initial crystallization. By projecting the

same point onto the solidus surface (point b', which naturally coincides with point a on the projection of the concentration triangle), we obtain the temperature at which crystallization terminates and the final point for the curve of the solidus change ($a'b'$); by drawing a conode for this isotherm,
103 we determine the final point of change of liquid — b, i. e., the composition of the final fraction of the liquid phase. The conode $a'a$, which corresponds to the temperature of initial crystallization, is tangential to the curve of change of composition of the liquid phase at the initial point a; the conode for the temperature of termination of crystallization bb' — is tangential to the line of change of the solid phase at its final point (b'). For each intermediate isotherm, the conode cc', which connects the compositions of the liquid and solid phases, passes through the point of given composition (as the system consists of two phases throughout) and intersects both curves.]

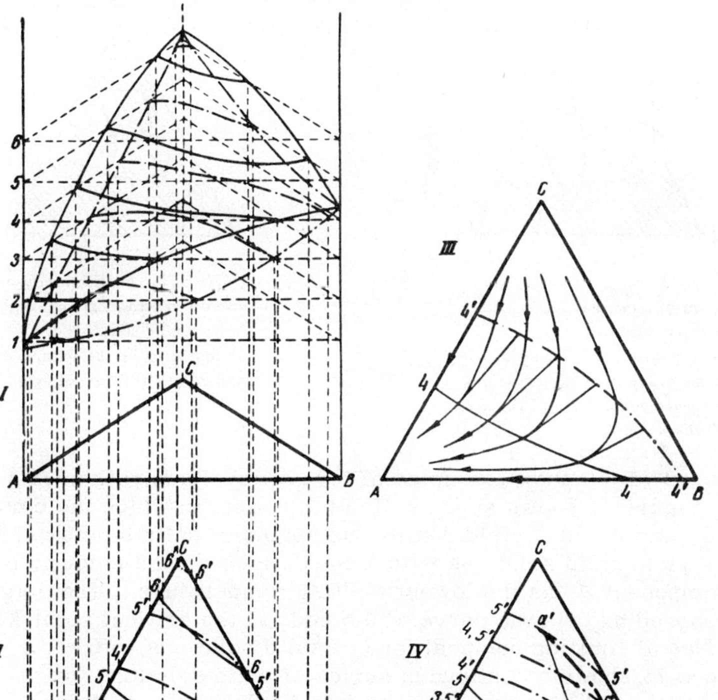

FIGURE 73. Example of an equilibrium diagram of a ternary system with continuous solid solutions of three components:

1 — general view of the equilibrium diagram. Solid lines here and further on indicate isotherms upon the liquidus surface; broken lines with dots — isotherms upon the surface of the solidus; II — concentration triangle with the corresponding isotherms; III — concentration triangle with the so-called crystallization paths; IV — crystallization path of a system with composition a

In the case when the three components of the system do not form a continuous series of solid solutions (limited isomorphism) a splitting of the system into three phases occurs within certain limits. Here, as in the above-considered cases, there is also a phase triangle, but its apices are not restricted to the apices or sides of the principal concentration tri-
104 angle, but also occur within this triangle. When temperature varies, the composition of phases change, the phase triangle is shifted, and its apices trace out certain curves. A combination of three such curves, corresponding to the three apices of the phase triangle, is termed the curve triade. One of them corresponds to the composition of the liquid, the other two — to the compositions of the solid phases (Figures 74—76). When the phase triangle is shifted, the apex corresponding to the liquid phase moves in advance of the apices of the solid phases.

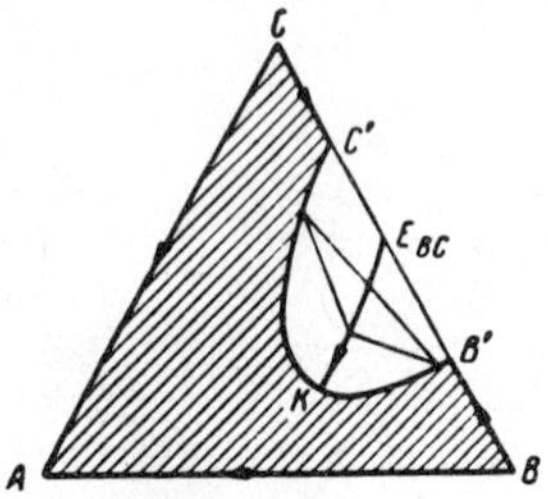

FIGURE 74. Equilibrium diagram of a ternary system with ternary solid solutions. The binary system ***CB*** has solid solutions with a discontinuity and a eutectic point of crystallization (Type V)

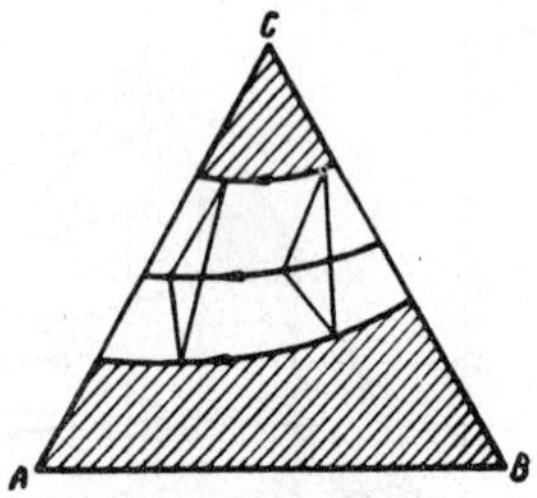

FIGURE 75. Equilibrium diagram of a ternary system with ternary solid solutions and a discontinuity in the systems ***AB*** and ***BC*** (both systems of Type V)

Figures 74—76 show examples of ternary systems with such solid solutions. Figure 74 shows an example of a system in which the components *A* and *C*, and *A* and *B* yield continuous series of solid solutions, whereas *C* and *B* yield solid solutions with a discontinuity and a eutectic point (Type V). Component *A* has the lowest melting temperature. The diagram has a corresponding cotectic curve which ends at the critical point *K*, at which two series of limiting compositions of solid solutions converge.

Figure 75 shows a continuous series of solid solutions for the system *A—B*, whereas *A—C* and *B—C* form solid solutions with discontinuities and eutectic points. Eutectic points are here connected by a cotectic line. On Figure 76 one of the binary systems is changed — a comparatively rare case is chosen, i. e., a diagram of Type IV. Here the curve of three-phase equilibrium connects the peritectic and eutectic points, intersecting the solid solutions curve, and undergoing a change in type.

105 We shall limit ourselves to the above-cited examples of ternary diagrams with solid solutions. Numerous concrete examples of silicate systems of the greatest importance in petrology will be presented further on. A more general review of ternary systems is given in the special work of D. A. Petrov.

FORMATION OF ZONED CRYSTALS IN TERNARY SYSTEMS*

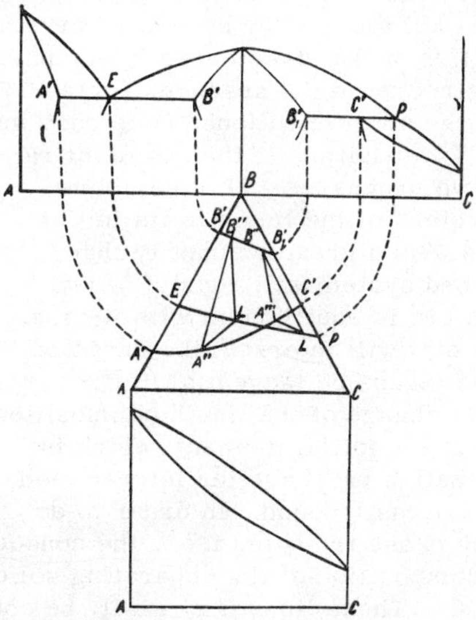

FIGURE 76. Equilibrium diagram of a ternary system with ternary solid solutions and with a discontinuity in the systems A—B and B—C (the first of Type V, the second of Type IV)

In ternary systems, just as in binary, when the system contains components which form solid solutions an exchange of matter occurs between the separated crystal and the liquid phase during crystallization. If this exchange is prevented for some reason, zoned crystals are formed as in the case of binary systems. Let us recall the path of the process in a binary system. With complete equilibrium between the crystals and the liquid, exchange of matter between the solid and liquid phases occurs throughout the course of crystallization. We shall consider a ternary system, the diagram of which is represented on Figure 77, I. Let P_1P_1 and P_2P_2 be the crystallization paths and E_1E_2 the eutectic line. Crystallization from a melt of composition *1* starts with the separation of a solid solution with the composition *1′*. The composition of the liquid varies along the curve (*1*—*4*).

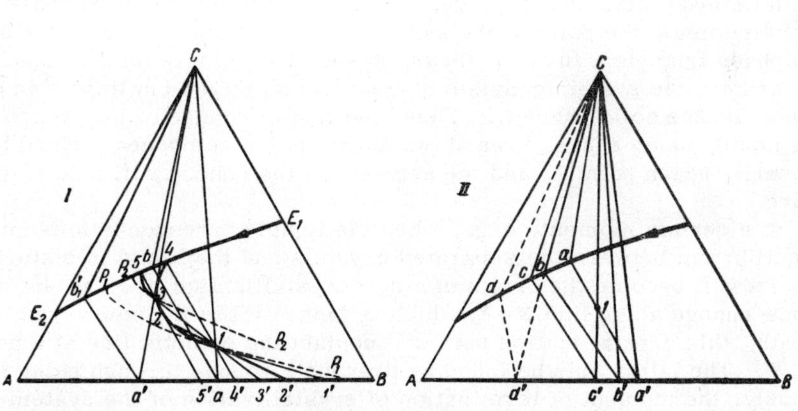

FIGURE 77. Formation of zoned crystals in a ternary system

1 — with crystallization of one phase in the field of solid solutions; II — with simultaneous crystallization of two phases along the cotectic boundary line

[If the process continued to be confined to separation of crystals with a continuously changing composition, the change of the liquid will correspond to the segment of the straight line connecting the points of the liquid

and solid phases which are in equilibrium, i. e., this straight line will always be tangential to the path of change of the liquid. However, as shown above, in the case of equilibrium crystallization the process is more complex: not only new crystals separate, but the earlier separated crystals change their composition. The peculiarities of the given curve in comparison to the course of crystallization, constructed in the absence of interaction, are evident even directly without the above-mentioned constructions. On shifting the liquid point from position *2* to position *3*, the resultant reaction may be considered as the sum of two processes: 1) a complete solution of crystals *2′*, which results in return of the liquid to its initial position *1*, and 2) separation of crystals *3′* in a greater quantity than crystals *2′*. At every moment the described system will consist of two
106 phases: the liquid and the crystals which are in equilibrium with it; i. e., the mentioned straight lines (**2—2′**; **3—3′**, etc) will intersect the curve of change of the liquid composition and by no means be tangential to it.

The accurate construction of the line of change of the liquid composition is difficult. However, both its direction at the initial moment (which is determined by the tangent to the crystallization path) and the intersection point of this line with the eutectic curve are easily found. In order to do this, one should construct the sides of the phase triangle, i. e., the conodes which show the equilibrium between the composition of the separating solid solutions, and the liquid with crystals of ***C***. These lines may easily be constructed either directly from experimental data, or by means of the already drawn paths of crystallization, by drawing tangents to these paths at the intersection points with the eutectic curve. Consider a conode passing through point *1* (see Figure 77, I) which represents the composition of the system. If such a conode is not given on the diagram, it may easily be constructed by interpolation. At the moment when the composition of the liquid falls upon the eutectic curve, and the quantity of the third phase is yet infinitesimal, the point of the system should lie upon one of the sides of the phase triangle. In other words, the point should lie on the conode ***4—4′***, as here the system consists of liquid *4* and mixed crystals ***4′*** in equilibrium with the solid phase ***C***. Thus, the line of change of the composition of the liquid, being convex toward the more fusible component, should necessarily reach point *4*, and the segment of the conode will be a chord of this arc.

If, at a certain moment, e. g., when the liquid corresponds to point ***3′***, the equilibrium between the separated crystals and the liquid is disturbed, this point will become the origin of a new crystallization line, as the liquid will now change along line ***3′—3***, which is tangential to the new crystallization path; this crystallization path will contact the eutectic line at a new point, *5*. The latter may be found by drawing a conode through point *3*. Obviously, the moment of termination of crystallization of the system will have changed. With complete equilibrium the final phase triangle is ***Cab***, but in our case it will be *Ca′b′*, and the composition of the last liquid phase (here ***b′***) is somewhat shifted toward point ***A***.

The composition of the mixed crystals will eventually correspond to point ***a′***, and a marginal growth of this composition will take place on cores of composition ***3′***.

With repeated interruptions, a series of zones may be obtained, and eventually a continuous change in the composition of the mixed crystals may occur in the case of a total absence of interchange of matter between the solid and liquid phases. In the latter case, the line connecting the compositions of the liquid and solid phases, will continuously be tangential to the line of composition change of the liquid phase. Thus, the crystallization paths which we mentioned correspond to such a course of change of the liquid, in the absence of equilibrium with the earlier separated crystals.

It is obvious from the aforementioned that, with continuous interchange, i. e., in an equilibrium system, the crystallization path should be constructed for each point separately. If one of the points on such a line is taken as a new initial point, then its crystallization path will be different. The construction of an entire system of such lines is meaningless, as these lines will intersect; there is a point in drawing them for several isotherms on different figures. However, in order to consider the path of crystallization, the absence of such a system is not essential, as only in rare cases is the knowledge of the exact position of such lines necessary for quantitative calculations.

It should be noted that if the exchange between the mixed crystals and the liquid phase is disturbed in a three-phase system, i. e., already at the
107 eutectic line, then the point of variation of the liquid cannot leave its path, and the change of the crystallization path will be expressed only in the dislocation of the final phase triangle ***Cd'd*** (see Figure 77, II). Obviously, this does not concern systems where the boundary line is of a reactional character (i. e., where it is a line of incongruent melting).

Finally, we shall compare phenomena which occur during the formation of zoned crystals or fractionation during the crystallization of solid solutions (i. e., separation from the system of previously precipitated crystals) in binary and ternary systems. In a binary system, the composition of the newly formed crystals as well as that of the last fraction of the liquid phase will be richer in the more fusible component ***A*** than these phases when in perfect equilibrium; this will occur at the expense of part of the crystals which are richer in the more refractory component ***B***. In a ternary system, besides the mentioned phenomena, a change occurs in the crystallization path of the liquid until the eutectic curve is reached, accompanied by a change in composition of the last portion of the liquid phase, affecting not only the ratio ***A*:*B*** but also the amount of component ***C***.

Cases of disturbed equilibrium in more complex ternary systems, especially in systems with incongruent melting of solid solutions, will be considered below together with actual diagrams.]

LIQUATION AND CRYSTALLIZATION OF IMMISCIBLE LIQUIDS IN TERNARY SYSTEMS**

If the components of a ternary system are not completely miscible in all proportions in the liquid phase, then here, as in binary systems, the coexistence of liquids of various composition is possible. There is, however, a series of variants, which is dependent on the relation of the phases in the limiting binary systems, and in the ternary system itself, up to the formation of three immiscible liquids coexisting with the solid phase at a constant temperature. Without dwelling upon the thermodynamic

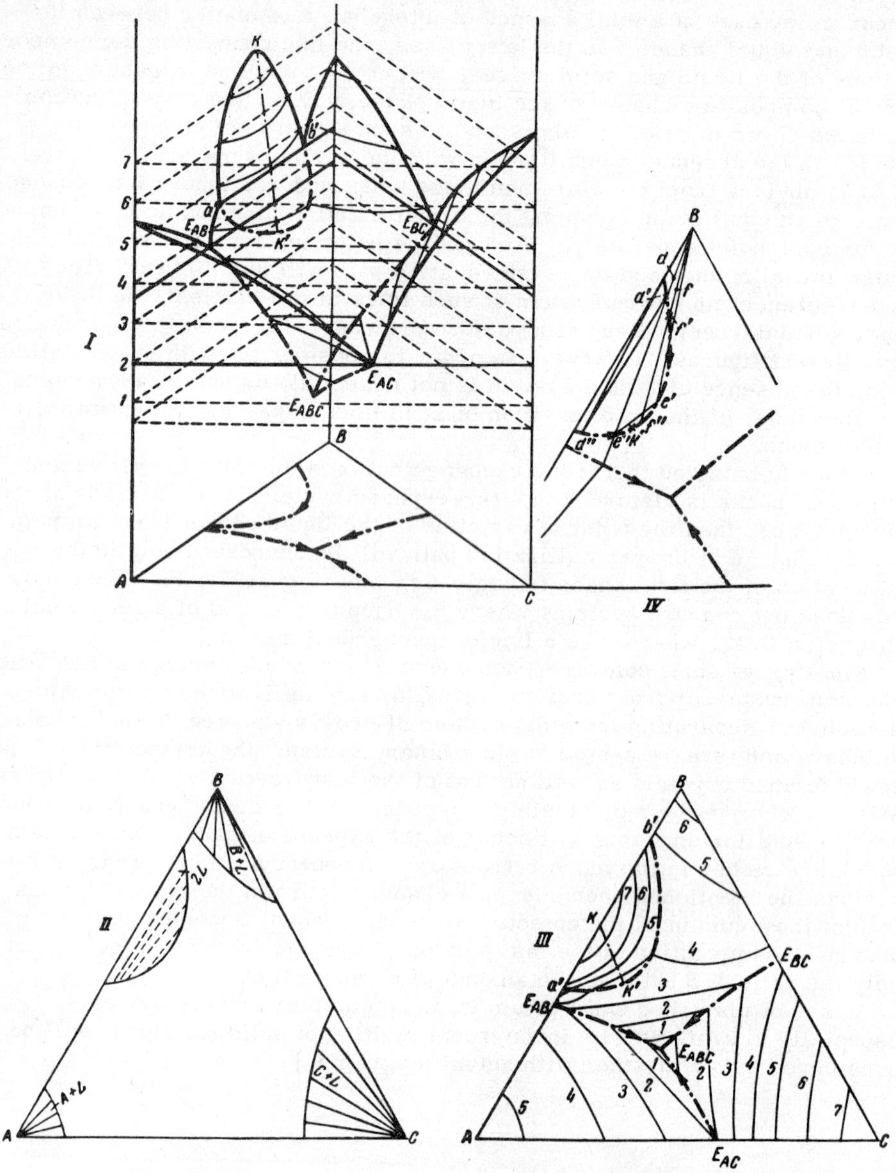

FIGURE 78. Equilibrium diagram of a ternary system with the formation of immiscible liquids in a binary system $A-B$:

1 — general view of the diagram in three dimensions; II — isothermal section at a temperature above the crystallization temperature of the immiscible liquids. Conodes are shown in the area of coexistence of two liquids; III — concentration triangle with isotherms; IV — decomposition into two liquid phases in the presence of a solid phase B

derivation of such diagrams, we shall consider some of the most simple examples, which correspond to the case of liquation in one of the two limiting binary systems.

Let us consider the ternary system ***A—B—C*** in which the system ***A—B*** splits into two immiscible liquids, where complete miscibility in the liquid phase takes place in the systems ***B—C*** and ***A—C***. For the sake of simplicity, we shall also assume that compounds and solid solutions are absent in the solid phases, i. e., eutectic crystallization is observed.

In similarity to the two curves in a binary system corresponding to the composition of the coexisting liquid phases, the ternary system should have two surfaces, which meet within the prism, forming a dome as shown on Figure 78, I. The line of mergence of these surfaces, starting from point *K* of the binary diagram, is a critical line and represents the moment of transformation of two liquid phases into one. Figure 78, II shows an isothermal section of a diagram at a temperature higher than the crystallization temperature of two liquids. Within the enclosed area the liquid will split into two phases — conjugate compositions connected by conodes. At the critical point *K* these conodes become points, i. e., the liquids will have the same composition.

The described dome-shaped surface is intersected by the surface of melting. Let us consider a case when this intersection occurs within the crystallization field of one solid phase only, i. e., there is no four-phase equilibrium of two liquids and two solids.

As in the case of three-phase equilibrium, only one degree of freedom remains, and temperature should vary with the change of composition of the liquid, i. e., the curve *a′K′b′* on Figure 78, III is no longer isothermal.
109 The critical point *K′* at the termination of the critical curve corresponds either to the highest or to the lowest temperature of coexistence of the solid phase *B* with two liquids (in our case it is the lowest temperature).

If we consider a composition corresponding to the point *d* (Figure 78, IV), the solid phase *B* will be the first to crystallize, and the liquid will change along the straight line *Bd′* until it reaches the boundary curve. At point *d′* a second liquid phase of composition *d″* appears. *Bd′d″* constitutes the phase triangle. On further crystallization, the composition of the liquid phases will change in the direction of the arrows, and point *d* will enter the phase triangle. For the given system, point *d* is again on the side of the phase triangle *Be′e″*, when the first liquid phase is completely exhausted, and crystallization proceeds again from one liquid, its composition varying along the straight line *Be″*. However, another case is possible on considering a given composition *f*. The continuation of line *Bf* will intersect the same liquid phase curve at two points before reaching the critical point. Here the amount of the other liquid phase will first increase, and then decrease until it disappears completely.

Another case worth mentioning is when a splitting into two immiscible liquids occurs in the two limiting systems *A—B* and *B—C*, the area of immiscibility passing through the entire ternary system. A corresponding diagram is shown on Figure 79, where curves of three-phase equilibrium of two liquids with the solid phase *B* are depicted. The path of crystallization and splitting is clear from the aforementioned.

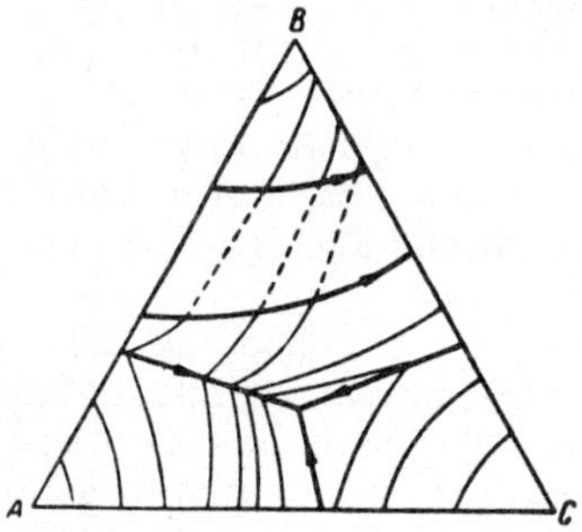

FIGURE 79. Equilibrium diagram of a ternary system with the formation of immiscible liquids in binary systems *A—B* and *B—C*

In order to derive ternary diagrams we used the construction of sections at constant temperature, i. e., the so-called isothermal sections. Such isothermal sections are, in general, of great importance both in experimental studies and in a more detailed consideration of the system's features. Further on we shall consider the features of such sections in general.

In melting point diagrams the isothermal section separates the stability field of the liquid phase from the field containing solid phases either with or without the liquid phase. It is necessary to draw on the concentration triangle a line of intersection of the melting surface with a horizontal plane at a certain temperature, i. e., at a certain level. Such lines are in fact drawn on every diagram in the form of isotherms, which are in actuality the basis for the construction of complete isothermal sections.

As the entire system is considered at constant pressure and the section is isothermal, the maximum number of coexisting phases should be three. Thus, a part of the diagram depicting the start of crystallization should be divided into separate areas: an area of equilibrium between the liquid and one solid phase; an area of equilibrium between the liquid and two solid
110 phases; an area of solid phases, the maximum number of which in each area should not exceed three.

It follows from the Phase Rule that with the presence of three phases in an isothermal section, there is no degree of freedom, i. e., all three phases are of constant composition. Thus, the field of coexistence of one liquid and two solid phases will be described in the section as a phase triangle, the apices of which correspond to the compositions of these phases. The apex of the triangle corresponding to the liquid phase (L_1 and L_2 on Figure 80, I and II) is easily located at the intersection point of the given isotherm with the lines of three-phase equilibrium. The apices corresponding to the solid phases either have a constant position (Figure 80, I) or (with the presence of solid solutions — Figure 80, II) they are determined by the conodes, the locations of which are determined by the experimental study of the system.

Areas located between the side of the phase triangle and the section of the curve representing the composition of the liquid phase are fields of equilibrium of two phases: liquid + solid phase. If the latter is of variable composition, straight lines should be drawn in the isothermal section, which connect the conjugate compositions, the location of which is again determined experimentally (they are not shown on Figure 80, II). On the basis of the obtained experimental data the above-described "crystallization paths" are constructed. These permit one to find the conodes for each isotherm. In order to do this one should draw a tangent to the path of crystallization at the point of its intersection with the isotherm and produce this tangent up to the line of solid solutions.

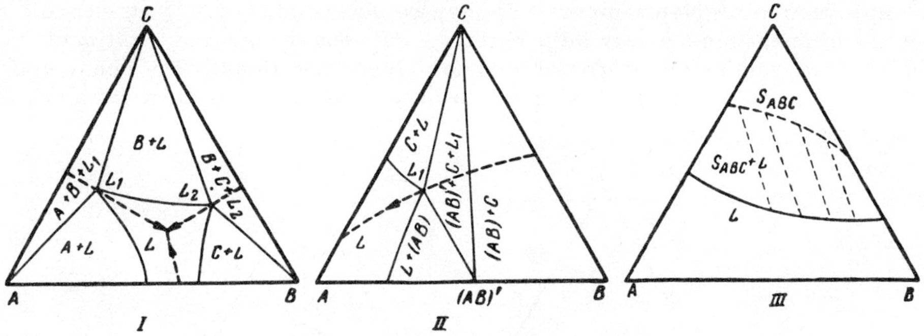

FIGURE 80. Examples of isothermal sections of ternary diagrams

I — a system without solid solutions and compounds; II — a system with continuous solid solutions; III — a system with ternary solid solutions

111 Figure 80, I and II shows typical isothermal sections for simplest systems without and with solid solutions (see also sections on Figures 60 and 64).

The presence of ternary solid solutions gives rise to two surfaces on the diagram — the liquidus and the solidus. Correspondingly, two lines of intersection, determining the composition of the liquid and solid phases, will appear upon the isothermal section. The compositions of the conjugate phases should be connected by experimentally predetermined straight lines (Figure 80, III).

To the right of the phase triangle, for the section on Figure 80, II, and an appreciable area of Figure 80, III, only solid phases exist (two and one respectively). The division of the area of solid phases in more complex cases is considered further on.

If the given temperature is lower than the lowest melting temperature in a given ternary system (in the simplest case — below the ternary eutectic point), then the plane of the isothermal section does not intersect with the surface of melting, and an equilibrium of solid phases only will be observed on the section. Once again the Phase Rule enables us to clarify the character of such sections.

In the case when all solid phases of the system are of a constant composition, i. e., there are no solid solutions, the concentration triangle of such an isothermal section should be divided into separate triangles — areas of coexistence of three phases. It is self evident that, if there are no com-
112 pounds present, the possible solid phases constitute the components of the system and there will be only one triangle, i. e., the concentration triangle.

When compounds of constant composition are present there are two possible cases: 1) triangulation (division into triangles) may be performed only in one way (Figure 81, I and II); 2) triangulation may be performed in several ways (Figure 81, III and IV).

In the other case triangulation should be performed directly from observation of equilibrium parageneses. A limited number of observed parageneses is usually sufficient in order to select the correct variant. After such a selection has been made, the paragenesis of a system of any composition at a given temperature may be determined, i. e., an entire series of parageneses may be predicted. Moreover, one may distinguish which parageneses are impossible under given conditions. The construction of such paragenetic diagrams is now widely used in the study of metamorphic rocks.

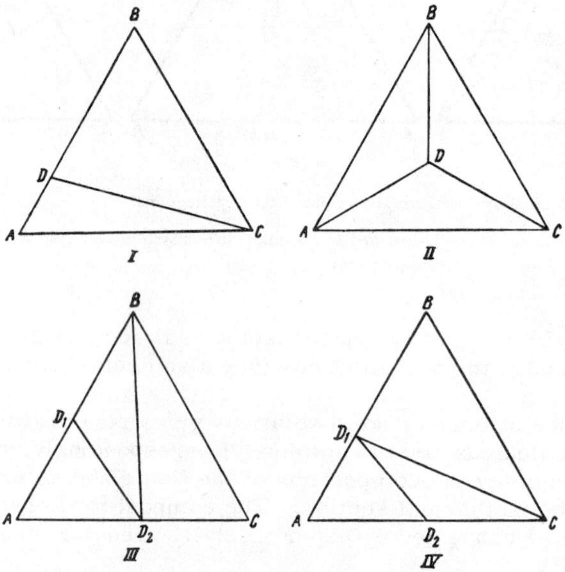

FIGURE 81. Isothermal sections of ternary systems at a temperature below crystallization temperature (paragenetic diagrams):

I — a system with one compound of two components; II — a system with one compound of three components; III and IV — a system with two binary compounds (two possible variants are shown)

Figure 82 shows a classical example of such a diagram for contact meta-
113 morphic hornfels of the Oslo region, as constructed by V. Goldschmidt in 1911. With the minerals present in hornfels, it is evident that there are seven possible variants of triangulation. There being only two observed two-mineral parageneses, diposide-anorthite, and hypersthene-anorthite, six variants are excluded, and we arrive at the one (Figure 82, IV) which completely determines all parageneses of the given system. It should be mentioned that such a diagram only holds for a certain pressure and temperature, which are in this case determined by the conditions prevailing near the contact with hypabyssal intrusions.

In cases when the solid phases are in the form of solid solutions, the character of the diagram becomes complicated. In the presence of solid solutions consisting of two components only, areas will appear, within which only two solid phases are stable, their composition being variable. These areas may be triangular in shape; the apex of such a triangle corresponds to a certain composition of the solid phase, and its base to the solid solution. The areas may also be quadrangular (usually a trapezium),

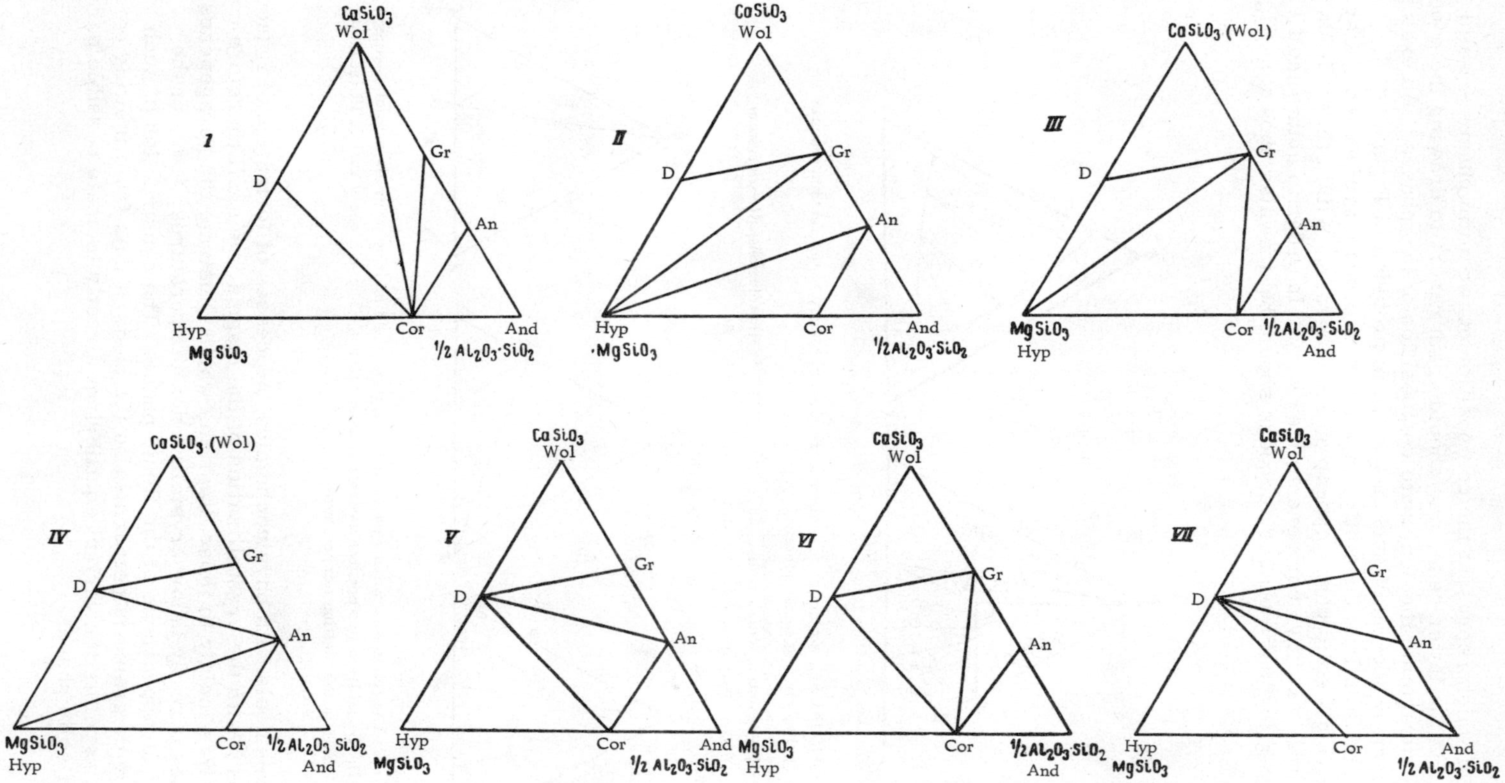

FIGURE 82. Paragenetic diagram (isothermal section below the temperature of crystallization) of a complex ternary system with several compounds

All seven geometrically possible variants are shown. Section IV is the real paragenetic diagram of hornfels after Goldschmidt. Wol — wollastonite; D — diopside; Hyp — hypersthene; Cor — cordierite; And — andalusite; An — anorthite; Gr — grossularite

two sides of which represent the composition of two solid phases — solid solutions. In the latter case, one should again resort to drawing the conodes
114 in order to determine the conjugate composition of two phases. An example of an isothermal section of such a system is provided in Figure 83, I.

When solid solutions consisting of three components are present, stability fields of a single solid phase appear on the diagram. Within these fields the concentration of two components may vary. In order to obtain the field of stability of two solid phases, conodes should be drawn as shown on Figure 83, II.

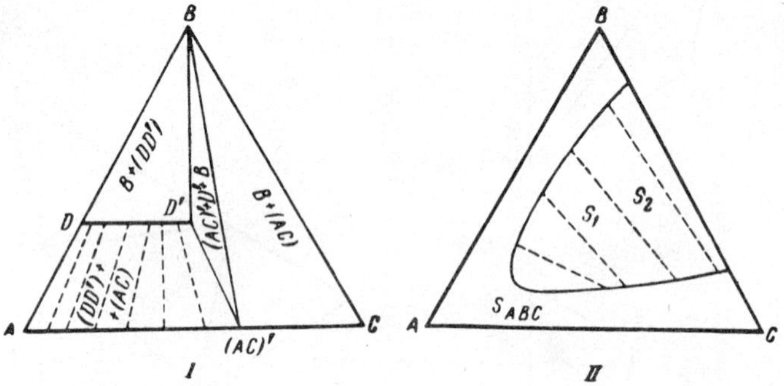

FIGURE 83. Examples of isothermal sections of ternary systems at temperatures below melting temperatures:

I — a system with a compound and two series of solid solutions; II — a system with ternary solid solutions, which are limited for *BC*

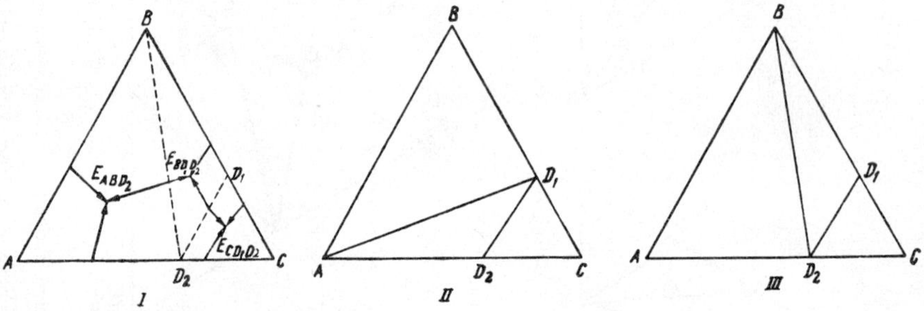

FIGURE 84. The relation between a paragenetic diagram and a melting point diagram of a ternary system (I). II and III — two possible variants of triangulation. The real one is variant III (at least in the vicinity of the melting temperatures)

This book deals with physicochemical processes of formation of igneous rocks, i. e., with the crystallization of the magma, we shall therefore confine ourselves mainly to those diagrams which illustrate the parageneses of solid phases directly below crystallization temperatures, i. e., at the
115 moment of disappearance of the liquid phase. The construction of such paragenetic diagrams is performed on the basis of data from melting point diagrams, as points of invariant equilibrium directly indicate possible parageneses of solid phases.

Figure 84, I, shows a diagram with two compounds; here two variants of triangulation are possible (Figure 84, II and III). The equilibrium between solid phases and the liquid at the eutectic points shows that the real variant, determining the parageneses in the vicinity of the temperatures of melting, is the diagram shown on Figure 84, III. On considering the diagrams of existing silicate systems, one often uses such isothermal constructions, even in the case of solid solutions.

The division of the concentration triangle into the mentioned parageneses is usually (when melting point diagrams are available) not difficult (with the obvious proviso that experimental data for the construction of conodes are at hand). Such a division is quite useful, as it permits the prediction of the final result of crystallization irrespective of its path; i. e., it allows one to determine the composition of the rock for any initial composition of the system. This is widely used in the consideration of diagrams of silicate systems.

VERTICAL SECTIONS OF TERNARY DIAGRAMS**

Together with isothermal sections, i. e., with horizontal sections (planes parallel to the base triangle), vertical sections are also used in considering ternary diagrams, i. e., sections perpendicular to the base triangle and parallel to the temperature axis. Such sections are usually drawn through points corresponding to the compositions of individual phases, and in some cases also through other parts of the system (e. g., a series of sections parallel to the prism face). These vertical sections are often the basic material for constructing the system during laboratory investigations. It is necessary to become familiar with their features.

At the first glance, sections of this type usually resemble diagrams of binary systems. In actuality, however, there are three possible cases.

1. The given section is truly a binary system, i. e., the compositions of all separating phases lie within the plane of the considered section, and may be represented in the form of a chemical equation, if the compositions which correspond with the extreme points of the sections are taken as the initial components.

2. The given section is binary only over a certain range, where it obeys the above-mentioned regularities. Elsewhere it does not correspond to a binary system.

3. The given section is pseudobinary throughout, i. e., the phase compositions do not lie within the plane of the section and they cannot be expressed as a sum of the above-mentioned components.

It is evident that the number of phases correspond to a binary system only in sections of the first type or in part of the sections of the second type. In a pseudobinary section, the number of phases corresponds to a ternary system, and all one can do is to trace, for each composition, the temperature limits of appearance and disappearance of new phases, but the composition of these phases cannot be plotted.

In order to construct any vertical section, the first thing to do is to draw the line of intersection of a given plane with the surface of melting, i. e.,
116 the boundary which limits the field of stability of the liquid phase only.

Below this curve the liquid is joined by one, then two and finally (in the general case) three solid phases. On the other hand, numerous changes are also possible in the solid system after the disappearance of the liquid.

The section may be constructed in two possible ways. The first makes use of a series of isothermal sections, upon each of which may be found points which divide the horizontal (i. e., isothermal) lines of the given vertical section into parts with different number of phases. By joining these points with a smooth curve the areas themselves may be constructed directly.

The second method is the more usual and will be used by us here. It consists in the direct construction of a vertical section from the diagram. In order to do this it is sufficient to follow the phase change of a series of compositions taken in the section plane, with change in temperature, and then connect the points obtained by means of curves, or in particular cases by means of straight lines. It is very important to find the terminal points of these lines on the margins of the diagram (or on lines which lie in a plane dividing the complex diagram into several simple ones), and on boundary lines of the melting surface.

In order to clarify this method of constructing sections, we shall consider some examples below.

Figure 85, I, shows the position of three sections on the most simple diagram with a eutectic point. Figure 85, II, III and IV show the corresponding vertical sections. In this case, for each composition within the triangle, crystallization will terminate at the ternary eutectic point, i. e., on the diagram the field of solid phases is always separated from the field where solid phases are in equilibrium with a liquid by a straight line parallel to the base and corresponding to the eutectic temperature.

In the left part of the section *B—1* (Figure 85, II) the system is binary, since during crystallization of phase ***B***, the point representing the composition of the liquid lies in the section plane. For all compositions the third phase appears at a temperature corresponding to point *f*, i. e., the fields of two-phase and three-phase equilibrium are here separated from each other by a horizontal line. Crystallization of compositions in the area ***f—1*** starts with the solid phase *A*. In order to understand the further path of crystallization, we shall draw a straight line ***AE*** (Figure 85, I) which will cut the section line at point *g*. For a composition corresponding to ***g*** the crystallization of the solid phase ***A*** will continue up to the temperature of the ternary eutectic, where two other phases will join it simultaneously. Thus, at the section point ***g*** on the curve dividing two-phase fields from three-phase fields, will lie on a straight line corresponding to an equilibrium of four phases (three solids and a liquid). The field of three-phase equilibrium wedges out. The corresponding curve will obviously have its source at point *f*. Over the *1—g* portion of the section the second solid phase to appear will be ***C***. For point *1* it joins at the temperature of binary eutectics, the corresponding curve terminating again at point ***g***, which lies upon a horizontal straight line, at the temperature of ternary eutectics.

A section (Figure 85, III) along line *2—3* does not pass through any of the apices. The line which delineates the area of the liquid phase has a rather complex configuration, and is comprised of three parts, corresponding to the composition of the first crystallizing phases. In the left part of

the section where *A* is the first to crystallize, the second solid phase to appear is ***B***. The curve which outlines the area of three-phase equilibrium, has its origin at the prism face at the temperature of the binary eutectic, and proceeds to point ***a*** on the corresponding boundary curve. The curve for the right part of the section is constructed in the same way. Here, phase *C* is the first to crystallize, followed by phase ***B***. In the middle part of the
117 section, the nature of the second crystallizing phase depends on the location of the point. In this case we again draw line *BE* which intersects the section line at *c*. For composition *c*, the crystallization of *B* continues up to the point of ternary eutectics, where the boundary curves of three-phase equilibrium fields meet. These curves originate on the given section at points *a* and *b* respectively.

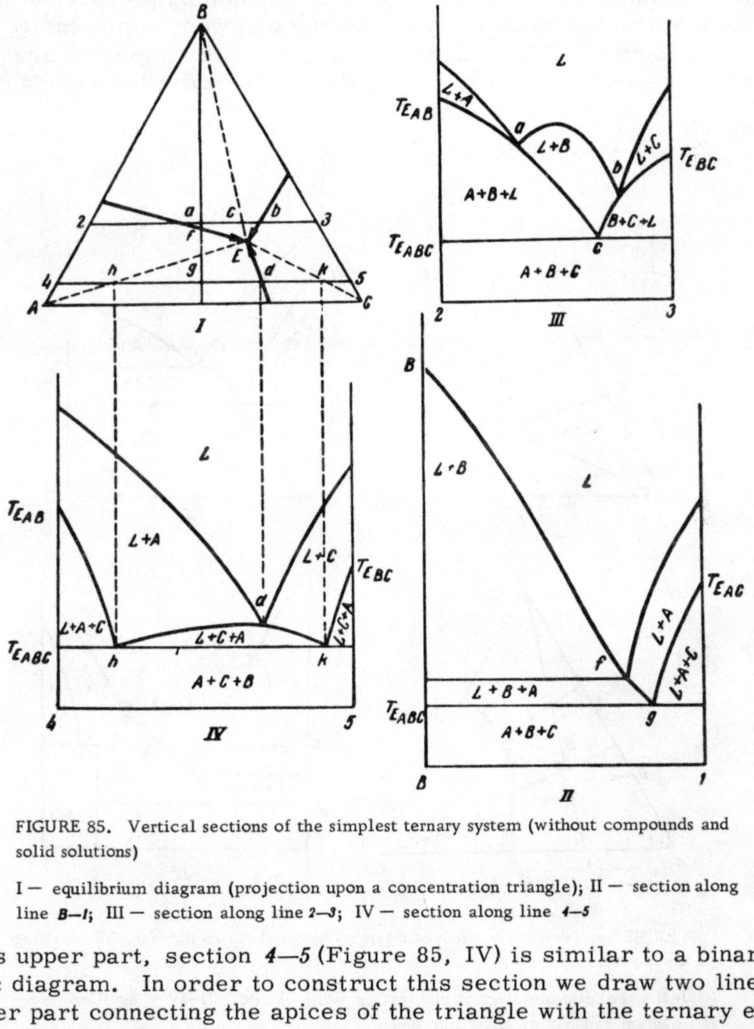

FIGURE 85. Vertical sections of the simplest ternary system (without compounds and solid solutions)

I — equilibrium diagram (projection upon a concentration triangle); II — section along line *B—1*; III — section along line *2—3*; IV — section along line *4—5*

In its upper part, section *4—5* (Figure 85, IV) is similar to a binary eutectic diagram. In order to construct this section we draw two lines in the lower part connecting the apices of the triangle with the ternary eutectics — *AE* and *CE*; two points are obtained — ***h*** and ***k***, which correspond to compositions for which the crystallization of one solid phase (*A* and *C* respectively) continues up to the temperature of ternary eutectics. On the prism faces, the curves separating fields of two-phase equilibrium from those of three-phase equilibrium, initiate at the temperatures of the corresponding binary eutectics. These curves terminate at *h* and *k*. In the middle part of the section these boundary curves start at a point situated on a boundary curve of the diagram, and again terminate at points *h* and *k*.

118 It is obvious that all curves of the sections may be constructed with absolute accuracy if isotherms are present on the diagram. The curves which separate the liquid field from that of crystallization, are constructed directly according to the intersection points of the section line with the isotherm. In order to obtain any point on the curves separating the two-phase area from the three-phase area, the section point should be connected with the corresponding apex of the triangle, and should be produced up to the intersection with the boundary curve on the diagram; temperature is read from the isotherm. These constructions are quite simple but are not shown as the insertion of the isotherms would have overcrowded our diagram.

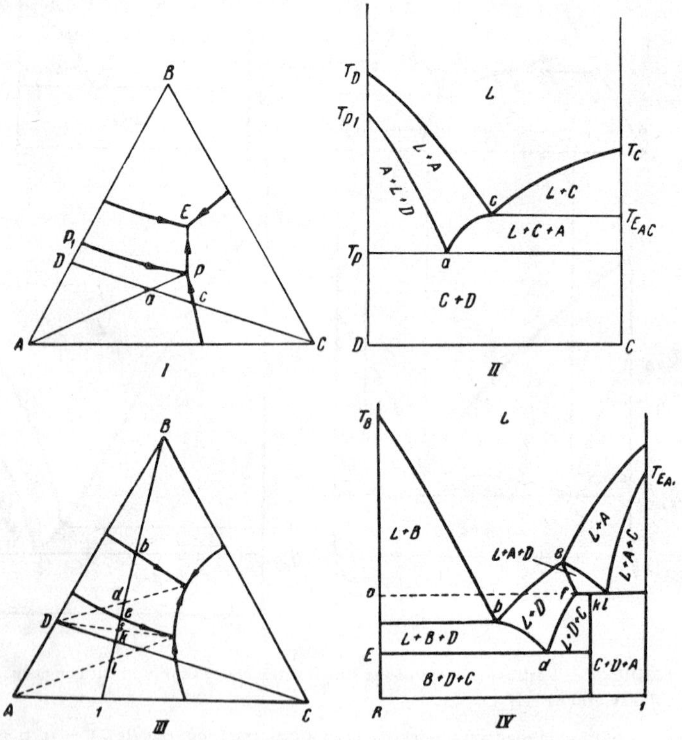

FIGURE 86. Vertical sections of a ternary system with an incongruently melting compound:

I–II — equilibrium diagram and section along line *DC*; III–IV — equilibrium diagram and section along line *B–1*

Figure 86, I, shows a more complicated diagram with a binary, incongruently melting compound. The section plane is drawn through one of the apices of the triangle and the compound (Figure 86, II). This section is pseudobinary throughout, as for any composition phases appear during crystallization, which cannot be contained within the section plane. In the left part of the section, crystallization starts with separation of phase *A*, while the three-phase equilibrium curve starts from a point corresponding to the temperature of the binary peritectic point T_{p_1}. Further on this curve subsides in correspondence with temperatures of the boundary line P_1P, up to the temperature of a ternary peritectic point T_p corresponding to
119 point *a* of the section. At this point, the second crystallizing phase will already be *C* and not *D*. The boundary curve again descends from point *C* on this curve to the ternary peritectic point. The right part of the section represents a composition for which crystallization starts with the separation of phase *C*. Initially, the system behaves as a binary one, the composition of the liquid remaining within the section plane. At point *c* a second solid phase *A* appears for all compositions over this range; i. e., the boundary line is here a horizontal straight line. The third phase appears at the temperature of the ternary peritectic point. As all our compositions lie upon the line *CD*, the reaction at point *a* will be terminated by the simultaneous disappearance of phases *A* and *L*, i. e., two solid phases *C* and *D* will be in equilibrium below the horizontal line.

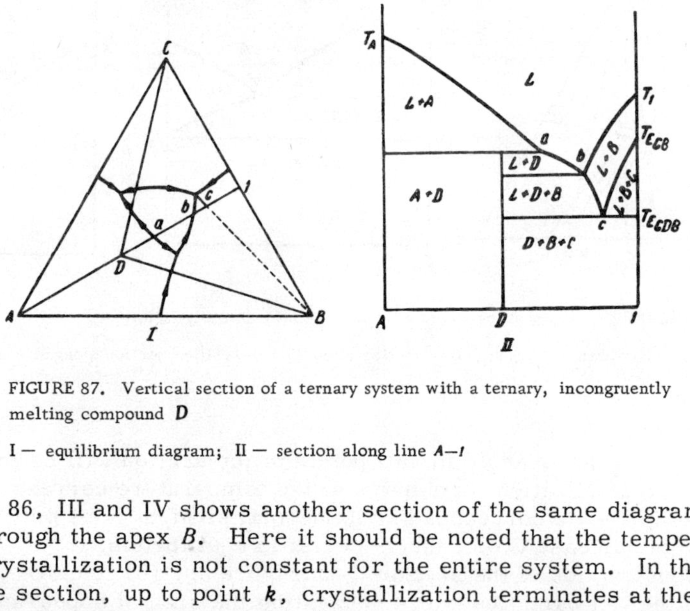

FIGURE 87. Vertical section of a ternary system with a ternary, incongruently melting compound **D**

I — equilibrium diagram; II — section along line *A—1*

Figure 86, III and IV shows another section of the same diagram, which passes through the apex *B*. Here it should be noted that the temperature of final crystallization is not constant for the entire system. In the right part of the section, up to point *k*, crystallization terminates at the peritectic temperature, whereas in the left part it terminates at eutectic temperature. Between points *k* and *f* there is an area where crystallization reaches the peritectic point, but continues below this temperature after phase *A* is completely exhausted. For compositions intermediate between points *e* and *f* phase *A* eventually terminates on the boundary curve. Thus, the three-phase equilibrium is again replaced by a two-phase equilibrium. Other relations are clear from the aforementioned.

Figure 87 shows an example of sections of the most complex system without solid solutions — with a ternary incongruent melting compound. The section is drawn through one of the apices of the triangle and the compound. As seen on Figure 87, II, the section, in its left part, is binary and corresponds to a binary system with an incongruently melting compound. Crystallization ceases at the temperature of the point *a*, where there is a maximum on the boundary curve. The right part of the section is not binary and here crystallization ceases at a temperature corresponding to the invariant point.

Figure 88, I, presents one of the simplest systems with binary solid solutions. Section *C—1* (Figure 88, II) is drawn through an apex representing a component which does not form solid solutions. The start of crystallization of the second solid phase on the left side of this section is determined by the temperature of point *a'*, i. e., we obtain a horizontal line. On the right
120 side of section, where crystallization starts from solid solutions, the second phase starts crystallizing at varying temperature, depending on the location of the point. This temperature may easily be determined by drawing a conode through the given point of the section and producing it up to the boundary curve. As previously, corresponding points on the diagram and on the sections are denoted by the same letter.

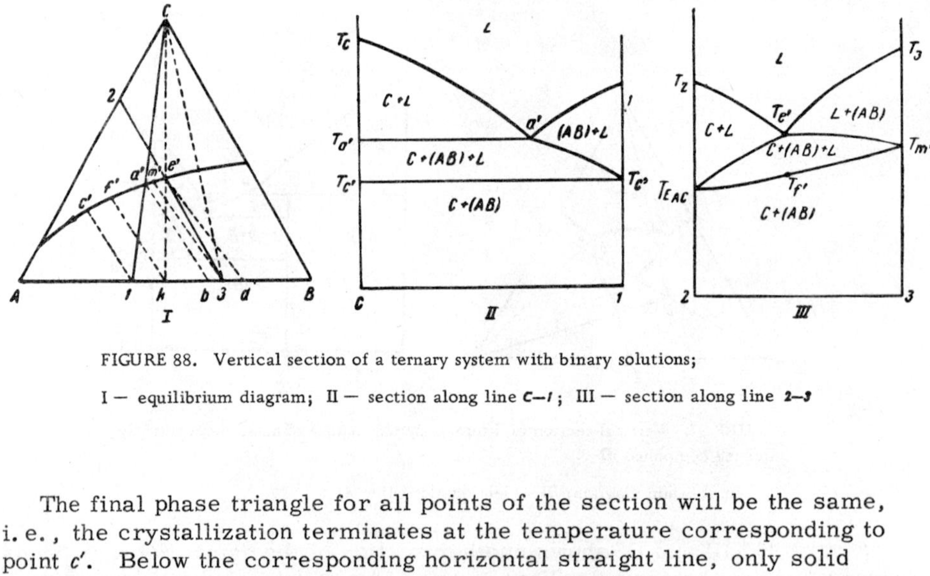

FIGURE 88. Vertical section of a ternary system with binary solutions;

I — equilibrium diagram; II — section along line *C—1*; III — section along line *2—3*

The final phase triangle for all points of the section will be the same, i. e., the crystallization terminates at the temperature corresponding to point *c'*. Below the corresponding horizontal straight line, only solid phases — in our case two of them — exist in equilibrium.

The construction of the section *2—3* (Figure 88, III) is more complicated. Here, in the left part, the curve separating the area of appearance of the second solid phase starts at a binary eutectic point $T_{E_{AC}}$ and terminates at the temperature $T_{e'}$ on the boundary curve. In the right part of the section the appearance of a second phase may be determined for every point of the section upon the conode. As distinct from the previous case, the temperature at the end of crystallization is in this case different. For point *2*

(Figure 88, I), it corresponds to the binary eutectic, for point *3* it coincides with the temperature of appearance of a second phase, and for point *e′* (as for any other point of the section) it is determined by constructing the final phase triangle ***Ckf′***.

Figure 89, II, shows a vertical section for the simplest case of a ternary system with ternary solid solutions. For each point, the start of crystallization is determined by the intersection point of the section with the liquidus surface, and its termination — by a point upon the solidus surface. We
121 construct curves which delineate various areas of the section. The section is not binary, as on varying the compositions of both phases do not lie within the section plane, as is seen on Figure 73 (see previously).

All above-mentioned sections are only considered between the moment of incipient crystallization and the disappearance of the liquid. Here, however, as in binary systems, several changes also are possible in the fields of the solid phases. In the absence of solid solutions these changes may amount to polymorphous inversions and to the formation (or disintegration) of compounds, as well as to interaction of solid phases. The number of phases will always attain the possible maximum, and inversions occur at constant temperature. On the diagram, a horizontal straight line will correspond to this moment. This line borders different areas of the section; it is the so-called "temperature stage" of inversion.

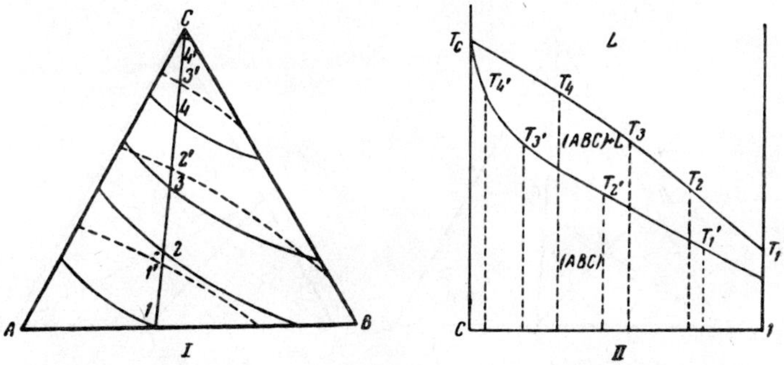

FIGURE 89. Vertical section of a ternary system with ternary solid solutions:

I — equilibrium diagram; II — section along line *C*–*1* (solid lines on the liquidus surface, dotted lines on the solidus surface)

In order to construct such a section (if the case is not merely one of polymorphous inversion) it is expedient to take as a basis isothermal sections above and below the temperature of inversion; these sections provide a means of accurately constructing vertical sections. From the general concepts on ternary systems it is clear that two types of reactions may occur:

$$C + F \rightleftarrows D + B \text{ and } D + B + C \rightleftarrows H.$$

Figure 90, I, II and III present three isothermal sections of an arbitrary system, which correspond to both types of reactions. All sections are drawn at temperatures lower than those at which the liquid phase can exist. Section I passes into II at t_1 and II into III at t_2. Figure 90, IV, shows a vertical section along line *A—1*. In this case, the change in parageneses occurs along horizontal lines corresponding to inversion temperatures. On some parts of these lines four solid phases are in equilibrium.

When solid phases of variable composition are present, the change in character of isothermal lines may be gradual. Thus, on considering the isothermal section shown on Figure 83, I, we note that as the temperature falls the area of solid solutions increases (DD_1), and the triangle of three-phase parageneses is shifted from left to right. On the vertical section (Figure 183, I and II), along the line representing the solid solutions, there are corresponding curves which demonstrate the shifting of points *D′a*. In
122 the case of a concrete system similar to the one described below (see Figures 131, 154, 155) a polymorphous inversion of the phase (DD_1) with variable composition also takes place. Polymorphous inversion curves intersect this boundary curve as shown on Figure 90, II.

In such systems a great number of variants is possible. There are no difficulties in the construction of these systems as long as the above-mentioned simple rules are obeyed and experimental data for the considered system are available.

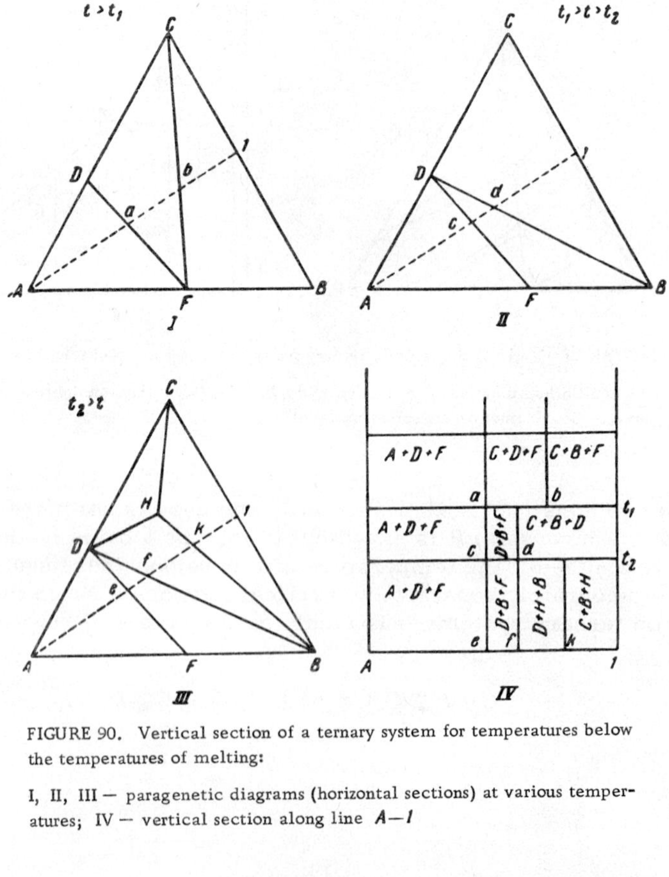

FIGURE 90. Vertical section of a ternary system for temperatures below the temperatures of melting:

I, II, III — paragenetic diagrams (horizontal sections) at various temperatures; IV — vertical section along line *A—1*

TERNARY MUTUAL SYSTEMS**

A peculiar position is occupied by the so-called ternary mutual systems, which at a first glance are liable to be classed as quaternary systems. One is usually referring to such systems as consist of two pairs of salts with common cations and anions, i. e., salts of the type *AC*, *AD*, *BC*, and ***BD***, which may form more complex compounds and solid solutions. At a first glance one may consider *A*, *B*, *C* and *D* as the components of such a system. However, it may readily be shown that not all these components are independent variables, as for each phase they should obey the equation (in atomic percentages)

$$A + B = C + D$$

123 Hence, only three components may be given, and the amount of the fourth is determined from the above-mentioned equation. On the other hand, it is impossible in such a system to select the components in such a way that each phase is expressed as a sum of components. One must resort to chemical equations where the minus sign also appears. If any three of the above-mentioned salts are selected as components, the fourth salt may be expressed as follows:

$$BD = AD + BC - AC.$$

In describing such a system, we are confronted by a new obstacle. The solution to this problem should be sought in the general properties of barycentric coordinates. Let the three mentioned components be located (as usual) at the apices of a triangle, which in our case should be considered as a right-angled isosceles triangle rather than as an equilateral triangle (Figure 92). The point representing the sum of $AD+BC$ should lie on the side of the triangles, the ratio should be 1 : 1 at the midpoint of this side, i. e., at point ***a***. If phase *BD* was the algebraic sum of component *AC* and the above-mentioned composition represented by *a*, the point representing *BD* would lie on the straight line connecting the corresponding points.
124 However, as we are here dealing with a difference, and not a sum, the mentioned point will not lie between points ***a*** and *AC*.

Assume that the required position of point *BD* has been found. According to the properties of barycentric coordinates one may write:

$$(AC)a : [(AC)(BD)] = 1 : 2,$$

as in this case the molecular quantities of the considered phases are in the ratio 2 : 1. Hence, it is clear that, in order to obtain the unknown point, line ***(AC)a*** should be produced for a distance equal in length to ***(AC)a***. Obviously, in the case of a right-angled triangle, we shall obtain a square, the corners of which will represent the composition of the initial "four components".

The above-mentioned systems are widely used in mineralogy, but are not often utilized in physicochemical petrology. The same relations may be observed as in normal ternary systems. In most cases, such a quaternary system may be divided into two normal ternary systems by drawing the so-called stable diagonal. It follows from the Phase Rule that after the system has completely crystallized no more than three phases can exist in equilibrium. Hence, triangulation should also be carried in this case, and if salts ***AD*** and ***BC*** are capable of mutual coexistence in equilibrium, then ***AC*** and ***BD*** (under the given conditions and also in most other cases) are not in equilibrium. Usually, on the basis of crystallochemical data, one may even predict which of the salt pairs will be in equilibrium (V. S. Sobolev, 1949). In the case where the direction of the diagonal connecting the coexisting salts corresponds to the binary section, one may carry out the above-mentioned division and consider each system independently. However, when compounds are present, more complex cases may arise, in which case the system should be investigated in its entirety. Such a system is depicted as a rectangular prism.

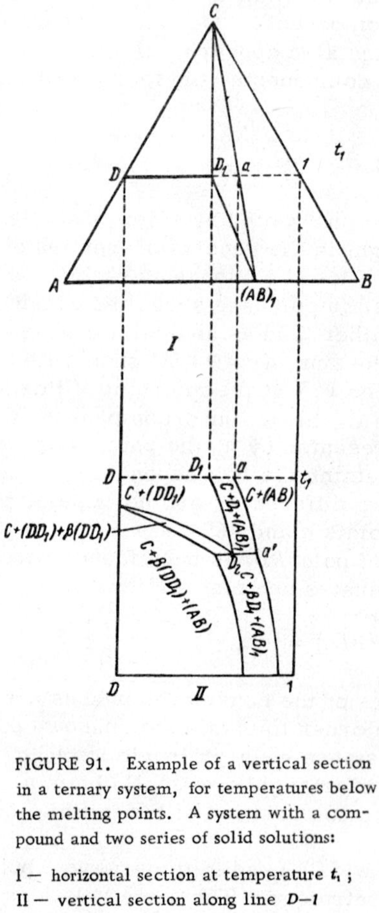

FIGURE 91. Example of a vertical section in a ternary system, for temperatures below the melting points. A system with a compound and two series of solid solutions:

I — horizontal section at temperature t_1;
II — vertical section along line *D–1*

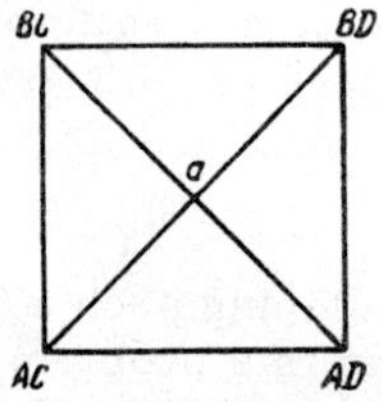

FIGURE 92. Schematic representation of a ternary mutual system

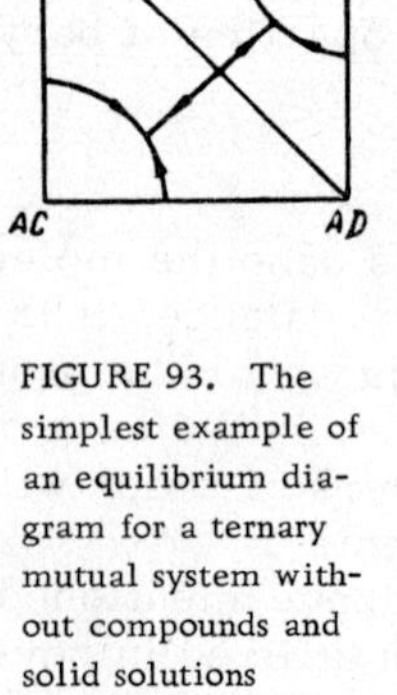

FIGURE 93. The simplest example of an equilibrium diagram for a ternary mutual system without compounds and solid solutions

Figure 93 shows the most simple case of a ternary mutual system without solid solutions and compounds. Phases ***AD*** and ***BC*** exhibit eutectic crystallization (in the same manner as for all sides of the square), and the system may be divided in two along the corresponding diagonal. Phases ***AC*** and ***BD*** are unstable when together.

125 CHAPTER V

SYSTEMS OF FOUR OR MORE COMPONENTS*

Previously it was shown how barycentric coordinates enable one to describe the composition of a phase by a point within a tetrahedron (Figure 94) or in general by a point in space. According to the Phase Rule, a quaternary system, consisting of a melt of a solid at constant temperature and pressure, possesses two degrees of freedom. Melts which are saturated with respect to one of the solid phases, are represented by a two dimensional geometric form — a surface. The intersection of two such surfaces yields the line of saturation in two solid phases at constant temperature and pressure. The intersection of three surfaces, i. e., a point in space, represents a system — melt and three solid phases. Isotherms of quaternary systems may be described by means of systems of intersecting surfaces.

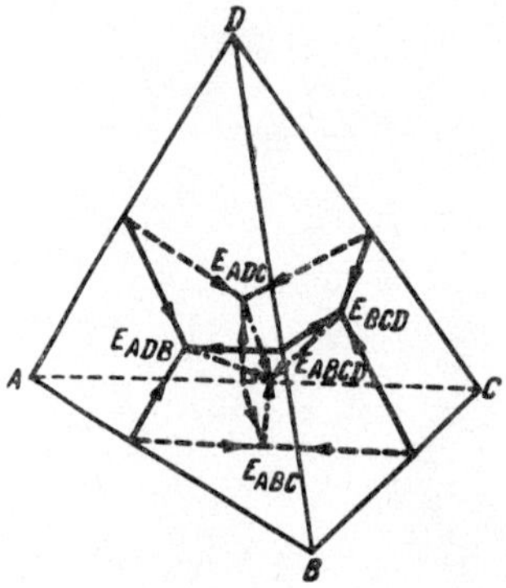

FIGURE 94. The simplest diagram of equilibrium of a four-component system without compounds and solid solutions (lines with dots and dashes show lines of equilibrium within the tetrahedron)

As the temperature varies the isotherm is shifted. If several such isotherms are drawn, corresponding surfaces may be drawn through the lines representing the composition of the liquid in equilibrium with two solid phases, and curves which are obtained as a result of intersection of the constructed surfaces, may be drawn through the points representing the composition of the liquid in equilibrium with three solid phases at the given temperature (Figure 94). The tetrahedron is divided into fields, each of which represents the area of saturation of a particular component, which begins to separate on the cooling of any composition located within the field in question. From the moment when this component begins to separate, the composition of the liquid changes and its composition point moves along the crystallization path. When it reaches the surface separating the areas of saturation of two solid phases, both phases begin to separate simultaneously, and the composition of the liquid (the third phase) changes, its composition point moving along the boundary surface. On

further movement along this surface, the composition point meets the curve of intersection of this surface with other surfaces (the curved edge of the saturation field of the first among the separating components). Three
126 crystalline compounds will now separate simultaneously from the liquid, and the composition of the liquid changes along this curve, either until a fourth solid phase appears, or until one of the three crystallizing phases is exhausted. The fourth point, at which the curve terminates, may be either a eutectic point or a reaction point.

It is evident from the above that the increase in the number of components leads to increasingly complex geometric forms for studying the properties of such systems. Already in the case of quaternary systems we are only able to depict the composition of the system and of phases by points in space. In order to describe either temperature or thermodynamic potentials, we would require a fourth dimension. In the case of a five-component system, even the composition cannot be represented in three dimensions.

On passing from simpler systems to more complex ones, we may observe general properties, which are repeated in a correspondingly modified form in various systems: e. g., the properties of a phase triangle, which are analogous to the properties of a segment of a straight line in a binary system and to those of a sphenohedron in a quaternary system. Further on, we shall see how the properties of a binary system included within a ternary are reflected in the properties of the latter, and vice versa — from a ternary system a binary can be isolated as a component of lesser complexity. A decrease in the number of components by one, while maintaining the number of phases, results in a unit decrease in the number of degrees of freedom. Therefore, the surface of two-phase equilibrium in a ternary system passes into the curve of a binary system, and a line of three-phase equilibrium — into a corresponding point. By a similar method, systems of lower order may in general be derived from multicomponent systems. From the properties of the former one may arrive at conclusions on properties of the latter.

[In similarity to the vertical sections of ternary systems one may draw analogous sections for quaternary systems. By drawing a planar section through the tetrahedron we obtain a figure analogous to the concentration triangle, and in some positions the sections are truly triangular. Here we are able to represent temperature as a separate coordinate, making the diagram an ordinary prism, and projecting the isotherms upon the base. As in sections of a binary system, the following cases also arise here:

1. The given section is really a ternary system, i. e., all compositions of phases participating in the equilibrium are described by points within the plane of the section.

2. Some part of the given section meets the above-mentioned requirements of ternary equilibrium, whereas in other parts the compositions of the phases are not described by points in the section plane, and the system is quaternary.

3. The compositions of phases participating in equilibrium do not lie in the section plane, i. e., the section is quaternary, and the diagram is pseudoternary at any composition.

It is evident that in order to obtain ternary or pseudoternary diagrams one may construct binary sections, which correspond in the general case to the conditions of a quaternary system, and in particular cases — to a ternary or even a binary system. The subsequent application of this method enables one to obtain binary sections from any number of components. However, the greater the number of components, the more complicated such sections are, and their general investigation calls for a complex mathematical analysis.]

127 CHAPTER VI

GENERAL REMARKS ON THE METHODS OF INVESTIGATING SYSTEMS*

[We have noted previously that as the number of components increases, the systems become more and more complex, and their representation less visual. On the other hand, equilibrium diagrams of simpler systems allow one to assess, to a certain degree, the properties of systems of higher orders, especially in those parts where the composition is fairly similar to that of the considered simple system. The most important problem of investigation is the correct choice of such simple systems the study of which] may lead us to important conclusions as to the properties of the magma, which is a complex multicomponent melt — solution.

For the sake of simplification, several methods may be used.

1. Magma usually contains volatile components in solution, which are more abundant in acid magmas. We may investigate the corresponding dry melts which lack these volatiles. A similar case occurs in nature during the cooling of effusive magmas — lavas. [However, this does not eliminate the study of silicate systems with volatiles, particularly with water; recently this research yielded numerous data on the formation of minerals in igneous rocks.]

2. Further simplification may be achieved by excluding some minor constituents. [First and foremost to be disregarded are those components which isomorphously replace the major components and dissolve in the phases of the system. In such a case the solution can never reach saturation and no new specific phase can appear. These components are termed disseminated. One may also disregard such minor components as form separate phases either alone or together with other components of the system, as their content is very low. Thus they do not influence the relations of other phases in the system. As an example we may mention ZrO_2 in zircon or P_2O_5 in apatite. These components, in analogy to the corresponding minerals, may be termed as accessories.]

3. We know that oxides, similar in their chemical character, may be classified in groups in rocks: e.g., K_2O and Na_2O, MgO and FeO. Instead of considering rocks which contain systems of all these oxides we may study similar systems of oxides of the various groups, considering each of them separately. Thus, a system may be considered where Na_2O is the sole alkali, instead of a system with both Na_2O and K_2O; in a similar manner MgO may be considered instead of systems with MgO and FeO, etc.

[However, the investigation of such partial systems does not provide full knowledge of the systems containing both mentioned components, especially if a discontinuity in the series of solid solutions is possible.]

4. During processes of formation of chemical compounds of some components, the equilibrium in the system is influenced by the concentrations of these components. However, when compounds are formed, they act as compo-
128 nents if they are not dissociated. Thus the number of equilibrium factors decreases. Thus, instead of a quaternary system SiO_2—Al_2O_3—MgO, a binary system CaO · Al_2O_3 · $2SiO_2$—CaOMgO · $2SiO_2$ is formed.

5. In selecting such simplified systems one may be guided by the parageneses of minerals in rocks: thus, quartz is not found together with corundum and nepheline, and enstatite — together with wollastonite, etc. All such simplified systems may be termed haplomagmatic systems. Bowen termed the system $mNaAlSi_3O_8$—$CaAl_2Si_2O_8$—SiO_2—$MgSiO_4$ as haplobasaltic and haplodioritic† (the latter is preferably termed haploandesitic). Similarly, the system mMg_2SiO_4—nFe_2SiO_4—$pSiO_2$ may be termed haploperidotitic, the system $NaAlSiO_4$—$KAlSiO_4$—SiO_2 below line $NaAlSi_3O_8$—$KAlSi_3O_8$ — haplophonolitic, etc.

The derivation of the above-considered melting point curves allows one to determine their types with exactitude; but the form of the curves is considered only from the qualitative point of view. The applied method cannot yield more than this; its advantage, however, is in the fact that we were not obliged to introduce any hypothetical assumptions on the state of the solution into our considerations.

After a theoretical consideration of various cases of crystallization systems, we shall turn to investigated systems of rock-forming minerals, which are the source from which our concepts on the crystallization of magma have stemmed.

The investigation of mineral systems may be carried out by several methods:

1. Nowadays, a large amount of experimental data on the study of most important silicate systems is forthcoming from investigations of physicochemical petrology. As a result of these investigations melting point curves of the studied systems are available for known experimental conditions. Conclusions from the obtained results are most reliable.

2. Conclusions on the relations of minerals and the character of their melting point curves may be drawn by studying the structure and composition of rocks. This is the analytical method.

[This method was widely used by petrographers during the first stage of development of physicochemical petrology, when experimental data for the most important systems were as yet unavailable, and technical knowledge was insufficient for the study of the most simple systems. However, even today this method is used if the corresponding system has not yet been investigated, and is especially applicable for the elucidation of the complexities of multicomponent systems as compared to simplified systems actually studied in the laboratory. This method is most important in the investigation of reactions between solid phases, i. e., equilibrium conditions in metamorphic rocks, as these reactions have not as yet been reproduced in the laboratory.]

3. Finally, an approximate idea on the positions of the eutectic points and other features of the mineral melts may be obtained from the equation for the lowering of the melting point. Vogt made partial use of this method, but the results are unreliable. The main setback of this method is that all

† **απλος** simple, simplified (Greek).

calculations of this sort call for exact data on the molecular ratios of the dissolved substances, and such data are very difficult to obtain for silicate melts. In many cases, there are also insufficient data on the latent heat of melting of the minerals, which data is necessary for such calculations.

129 [At present such calculations are used mainly for the construction of equilibrium diagrams, both in the case of change of pressure on the release of gaseous phase, and (mainly) in the case of reactions between solid phases.]

Synthetic experimental investigation of melting point curves has chiefly been carried out by two methods: a) the method of determining temperatures of crystallization during the more or less slow cooling of the melt and b) the quenching method. [Here, optical and, more recently, X-ray determination of the solid phases are widely used, enabling the determination of both constant and variable compositions. Conventional mineralogical diagrams are used.]

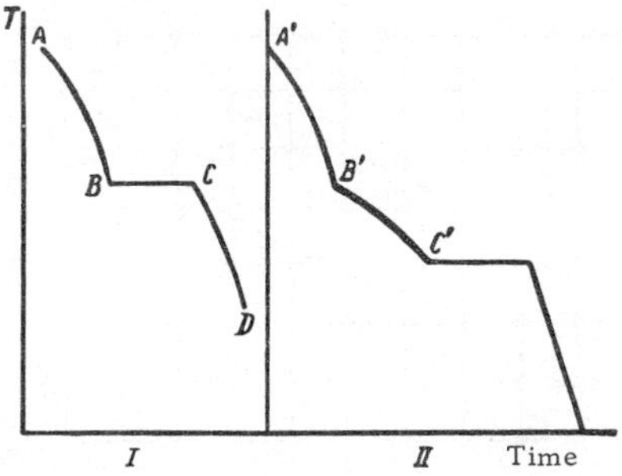

FIGURE 95. Cooling curves of melts

I — for a unary system; II — for a binary system with a eutectic point

In the first case the investigated melt is heated above the temperature of incipient crystallization, a thermocouple connected with a galvanometer, and a recorder, is introduced into the melt, and the melt is allowed to cool. As the temperature falls, a curve depicting this drop is obtained on the recording paper (Figure 95). This curve shows the mutual relationship between two variables — temperature and time.

When the temperature has dropped to the point of incipient crystallization, the liberated latent heat of fusion will arrest the fall in temperature. In this case, if the melt consists of only one pure substance (one component), the temperature will remain constant during its crystallization. On the cooling curve the horizontal portion *BC* corresponds with this halt in temperature drop. A similar horizontal break in the curve will occur in the case of eutectics. If the melt consists of more than one component, and its composition is not that of eutectics, the cooling curve will show a break, but *B′C′* will be inclined and not horizontal, as the fall in temperature, although partially arrested, continues up to the eutectic temperature (see Figure 95).

Figure 96 shows an example of relations between cooling curves (temperature—time) and melting point diagrams (composition—temperature). Knowing
130 the compositions of the investigated melts, and having the cooling curves at hand, one may construct a melting point diagram. This method of construction is evident from Figure 96.

A series of samples is taken the compositions of which are the same as that of the investigated melt. For a prolonged period each sample is kept at constant temperature. On rapid cooling (quenching) the attained equilibrium is "frozen", and the phases forming the system at this state of equilibrium are determined by a microscopic study of the melt. This series of experiments is repeated at various temperatures. Knowing these compositions and temperatures, and the state of the systems which were fixed by rapid cooling, one may construct curves of melting point diagrams as limits of various states of the systems (Figure 97, I, II).

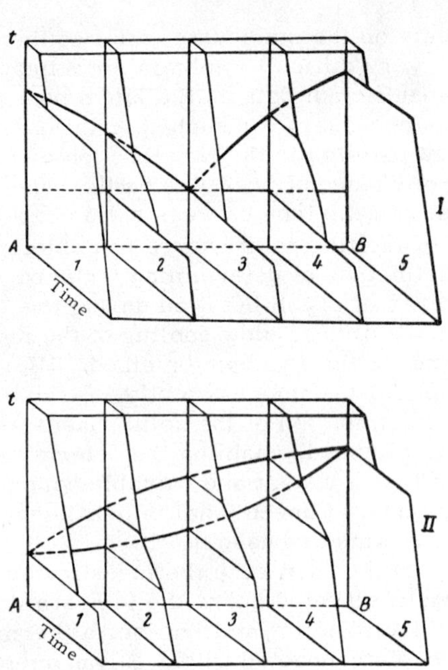

FIGURE 96. Relation of cooling curves to melting point diagrams:

I — a binary system with a eutectic point; II — binary system with solid solutions of Type I

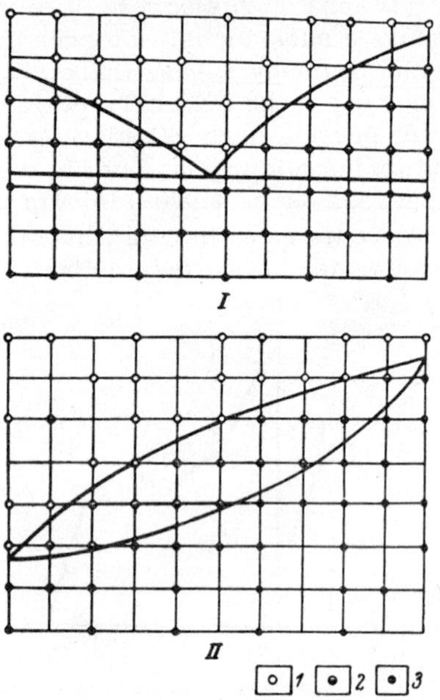

FIGURE 97. A scheme for constructing binary equilibrium diagrams by the quenching method

I— binary system with a eutectic point; II— binary system with solid solutions. 1 — supercooled liquid (glass); 2— glass and crystals; 3— crystals

The method of analysis of microstructures and chemical compositions of the minerals and of the rocks which they form, is the only method applied directly by many investigators engaged in the study of rocks. By comparing the chemical composition of minerals it is possible to determine whether they form solid solutions, and if so, whether these solutions constitute a continuous series or not. The study of physical and especially optical properties of these isomorphous mixtures (solid solutions) shows that the continuous variation of chemical composition is accompanied by a continuous variation of the properties.

If the components do not form solid solutions, eutectics should perforce be present. In this case, the eutectics determine the crystallization of the melt. Therefore, by observing the order of separation of two minerals in rocks consisting of these two minerals only, and by comparing the chemical composition of these rocks or the relative quantities of the minerals that form them, the composition of the eutectics may be obtained. An interesting example of crystallization is presented by the anorthite-olivine rocks (allivalites) from the Isle of Rhum (Hebrides). In samples rich in olivine, olivine is the first mineral to separate, whereas in leucocratic samples anorthite crystallizes first. In places where the minerals are present in their eutectic ratios they crystallize out simultaneously. In some cases of porphyries, and especially granite-porphyries, the matrix approaches the eutectic composition. This probably causes the peculiar texture of melanocratic gabbros of the Urals which were termed tilaites. Duparc termed this texture "cryptic".

131 In some cases, the simultaneous crystallization of both components from a mixture of eutectic composition may be recognized directly from the characteristic pegmatitic texture. [However, similar textures may also be of different origin — especially as the result of metasomatic recrystallization of the rock in a solid state.] The consistency in composition of quartz-feldspar micropegmatites is explained by many by their eutectic composition. [The most recent views of the authors on the genesis of granitic pegmatites are cited further on.]

Incongruently melting compounds may be detected by reaction rims of one mineral around another (if there is no doubt as to the primary origin of these rims), e.g. the abundant rims of orthorhombic pyroxene which surround olivine.

When components form a solid solution, as is known, the composition of the separated solid phase differs from that of the liquid. The solid phase will be enriched in that component, the addition of which raises the melting point. Owing to an incomplete equilibrium in rocks between the solution and the separating crystals, zonal textures are often found in isomorphous crystals. The composition of the inner zones corresponds to that of the solid phase separating at higher temperatures. After the presence of a continuous series of solid solutions has been established on the basis of chemical analysis and the continuous change of physical (e.g. optical) properties, one may distinguish Types I, II, and III from the peculiarities of zoned crystals. In the case of Type I, the inner zones are always richer than the outer in the same component. In Type III, in members of isomorphous series approaching the composition of component B, the inner parts of zoned crystals are richer in component B; in cases approaching the composition of A, the inner zones are richer in this component, i. e., the inner zones are enriched in the predominating component. In Type II the situation is reversed.

If components do not form a continuous series of solid solutions, this discontinuity may be easily detected after a sufficient number of analyses have been made. The classification according to Types IV or V is thus determined. The latter two types are distinguished from one another on the basis of analytical indications.

In Type IV (see Figure 38) a solution with an intermediate composition between L_1 and A' will separate mixed crystals of A with a comparatively small content of B. No crystals of B will separate after the formation of crystals of A. Solutions with a composition ranging between B and B' will separate crystals of B with a small amount of A. The whole quantity of A is included in these mixed crystals. A solution with a composition between B' and L_1 will first separate mixed crystals of B with a certain content of A, and the composition of the solution changes, approaching L_1; here crystallization of B terminates and A crystallizes out. At a temperature corresponding to point L_1 resorption is observed. Thus, in Type IV the crystallization of B after that of A is excluded, and in several cases resorption of B is observed. B always crystallizes before A. Eutectics do not exist. As is evident from the scheme, A may contain more B in solid solution that the latter may contain A. In zoned crystals, the core is always richer in component B. In Type V, A is the first to crystallize from solutions rich in this component, whereas B is the first to crystallize from solutions rich in B. An eutectic is present. In zoned crystals of A, the core is enriched in this component; in zoned crystals of B the core is richer in B than the margins. Both A and B may contain comparatively smaller amounts of each other as compared to Type IV.

Thus, by analysing the relations of minerals and their chemical composition, one may arrive at an idea of the character of the melting curves.

Part II

THE MOST IMPORTANT SILICATE SYSTEMS LACKING VOLATILE COMPONENTS

132 CHAPTER VII

SOME FEATURES OF SILICATE MELTS

In previous chapters, we discussed crystallization processes which take place during the cooling of melts under ideal conditions of equilibrium. We assumed that disturbed equilibrium was reestablished without meeting with any obstacles, thus adapting the system to changing conditions. This idealization of the process is necessary when considering these phenomena. Investigators in both the exact and the natural sciences use this method to isolate the basic elements of natural phenomena from other coexisting elements, which introduce the enormous complexity characterizing processes in nature. Only by means of resorting to such an isolation is it possible to refer investigated phenomena to more general principles, and thus to explain them.

The next step is the evaluation of all those conditions which make the process more complicated in nature. In our consideration of silicate melts we shall now discuss this part of the problem. First of all it is necessary to consider some features of silicate melts.

VISCOSITY

When substances such as metals and salts pass from the solid to the liquid state on heating, or are crystallized from a melt on cooling, the velocity of inversion in most cases prevents superheating or supercooling. In the case of silicates this situation is different. Glasses, which are supercooled, metastable systems, are evidence of the silicate's ability to undergo supercooling. These glasses may remain in this state for several geological periods, as is evident from the volcanic glasses of ancient eruptions. Silicates may, to a great extent, also undergo superheating; in these systems, too, equilibrium between the solid and liquid phases is not attained. In contrast to salts and metals, the melting point of silicates cannot be determined in most cases. In the case of silicates one preferably refers to the melting interval. A long time interval is required in order to attain equilibrium between solid and liquid phases. Therefore, a special statistic method is employed for the determination of melting points.

The above-mentioned properties of silicate melts are associated with a
133 considerable amount of inner friction in these melts. In all transformations which occur in a liquid, e. g., formation of a solid phase, the movement of constituents due to diffusion, and even in chemical reactions, a relative movement of parts of the system occurs, e. g., the transfer of matter from the liquid to solid in the case of crystallization, and diffusion which supplies the material for the growing crystal. Every shifting of particles within a

liquid meets with resistance, which is similar to the friction which resists the motion of all bodies. This internal friction, which is termed viscosity, is a characteristic feature of melts, and should be given prime consideration.

The simplest case of internal friction in a liquid is observed when one part of a liquid moves with constant velocity relative to another part, and resistance originates at the interface between the two parts. The inner frictional force also involves particles of adjacent layers, and the velocity of their motion decreases gradually in successive layers further removed from the interface.

The inner frictional force between two infinitely close layers, separated by the distance dx and with a difference of velocities equal to dv, is

$$F=\eta S\frac{dv}{dx},$$

where S is the area of the interface.

This force acts upon both the contiguous layers. It decreases the velocity of the faster layer, and increases that of the slower one.

For a given liquid, the coefficient of internal friction or viscosity η is numerically equal to the force acting upon 1 cm^2 of a liquid as it moves in parallel layers with uniformly increasing velocities, such that the velocities of layers 1 cm apart will differ by 1 cm/sec. If the force is measured in dynes, the viscosity η will be expressed in poise. The reciprocal value $\left(\frac{1}{\eta}\right)$ is sometimes termed "fluidity".

The viscosity of fluids has a direct influence on the velocity of motion of bodies in the fluid and on the velocity of a jet or flow of this liquid. Both these phenomena are used for the measurement of vis-
134 cosity values. The motion of bodies within liquids is used to determine viscosity experimentally; observations of velocities of flow may provide a direct indication of the viscosity of lavas during volcanic eruptions.

TABLE 1

Substance	t, °C	η, poise
Water	20	$1\cdot10^{-2}$
Glycerine	20	$1\cdot10$
Gelatine	20	$1\cdot10^{6}-1\cdot10^{8}$
Diopside	1400	$3.8\cdot10^{0}$
Anorthite	1400	$3.8\cdot10$
Albite	1400	$4\cdot10^{4}$
Orthoclase	1400	$1\cdot10^{8}$
Diabase	1400	$6\cdot10 - 9\cdot10$
Basalt	1400	$5.5\cdot10$
Nepheline basalt	1400	$8\cdot10$
Olivine basalt	1400	$1.2\cdot10^{2}-1.36\cdot10^{2}$
Andesite basalt	1400	$1.38\cdot10^{2}$
Vesuvius lava	1400	$2.5\cdot10^{2}$
Andesite	1400	$1.17\cdot10^{3}$
Obsidian	1400	$1.7\cdot10^{5}$

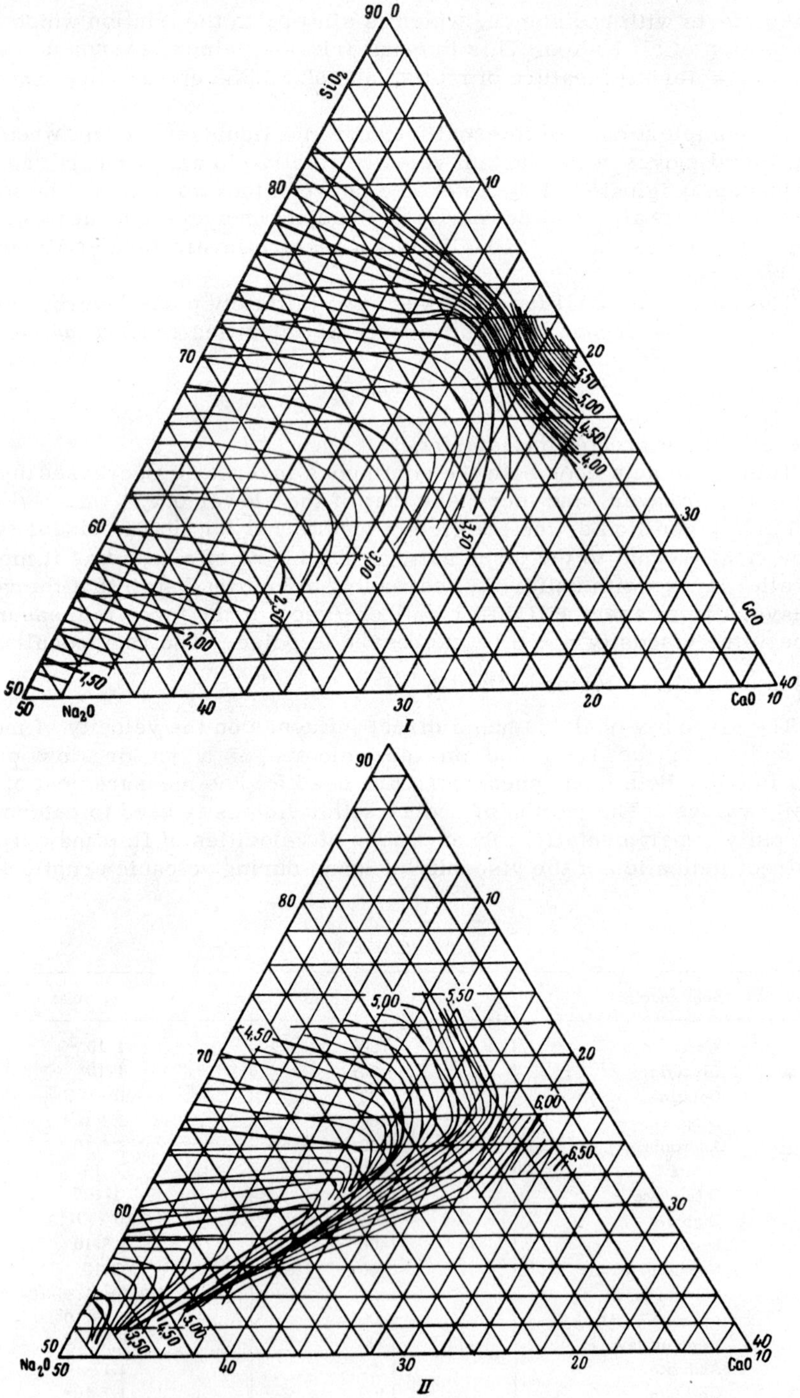

FIGURE 98. Isokom diagram for soda-lime silicate glasses (after Washburn, Shelton and Libman):

I — for t = 1,100°; II — for t = 900°

Without entering into the details of the methods of experimental determination of viscosity values, we shall only mention that the following observations are used here: 1) determination of the velocity of sinking of a ball or other body in a liquid, or on the contrary — the determination of the velocity of rising of bubbles; 2) determination of the damping of a pendulum's oscillations in a liquid; 3) determination of the resistance to rotation of a cylinder in a liquid. The determination of viscosity of silicate melts at high temperatures is mainly carried out by the last-mentioned method.

Viscosities of some familiar liquids and of some rocks and minerals are presented in Table 1.

On considering Table 1, it is evident that at normal temperatures (20° C) the viscosity of glycerine is thousands of times greater than that of water. The viscosity of gelatine is 10^5—10^7 times that of glycerine. The viscosity of an anorthite melt at 1,400°C is ten times that of a diopside melt, and that of albite is a thousand times that of anorthite; a melt of orthoclase is
135 ten thousand times more viscous than a melt of albite. We may also note that the viscosity of rocks varies, increasing in acidic rocks rich in silica up to 1,000—10,000 times that of basic rocks. Thus, the viscosity of silicate melts is a function of their composition. General trends in the influence of various constituents of silicate melts on their viscosity have been known for some time from the study of slags and other artificial silicate melts (glasses). The addition of iron (especially of ferrous) oxides and of manganese lowers viscosity. Slags with high contents of magnesium are less viscous than the corresponding calcium silicates; magnesium, however, lowers viscosity to a lesser extent than does iron. Alumina and silica, as well as oxides of titanium and chromium, increase viscosity. Nevertheless, the knowledge of these general peculiarities in influence of various oxides on viscosity is insufficient for considering the viscosity of complex silicate melts. To a certain degree this influence is also dependent on temperature, which generally has a great influence on viscosity. As an example we shall cite artificial glasses of the following composition: SiO_2, 50—80%, CaO, 0-30% and Na_2O, 10-50%. The diagrams on Figure 98 show lines connecting points of compositions with equal viscosities. These lines are termed "isokoms". As is seen from these diagrams, the form of the isokoms also changes to a certain degree at different temperatures. During the investigation of the viscosity of some binary systems, which form
137 eutectics or compounds during crystallization, no relationship between the viscosity and the location of such singular points in these systems was detected (Figure 99)†.

As is evident from the above-cited values, the viscosity of rock melts varies considerably. Melts of acid rocks are much more viscous than melts of basic rocks. On the basis of the investigation of approximately 35 rock samples, carried out by M. P. Volarovich and the Japanese scientist Kani, one may arrive at the conclusion that the viscosity of rock melts depends firstly on the relations between the femic and salic constituents of the rock, which is expressed as the parameter *b* of the numerical characteristics

† Preston disputed the correctness of these measurements, and presented curves upon which breaks are most evident (with a minimum) on the viscosity curves at a point corresponding to the compound $Na_2O \cdot SiO_2$. This, however, was not confirmed by Lillie's investigations.

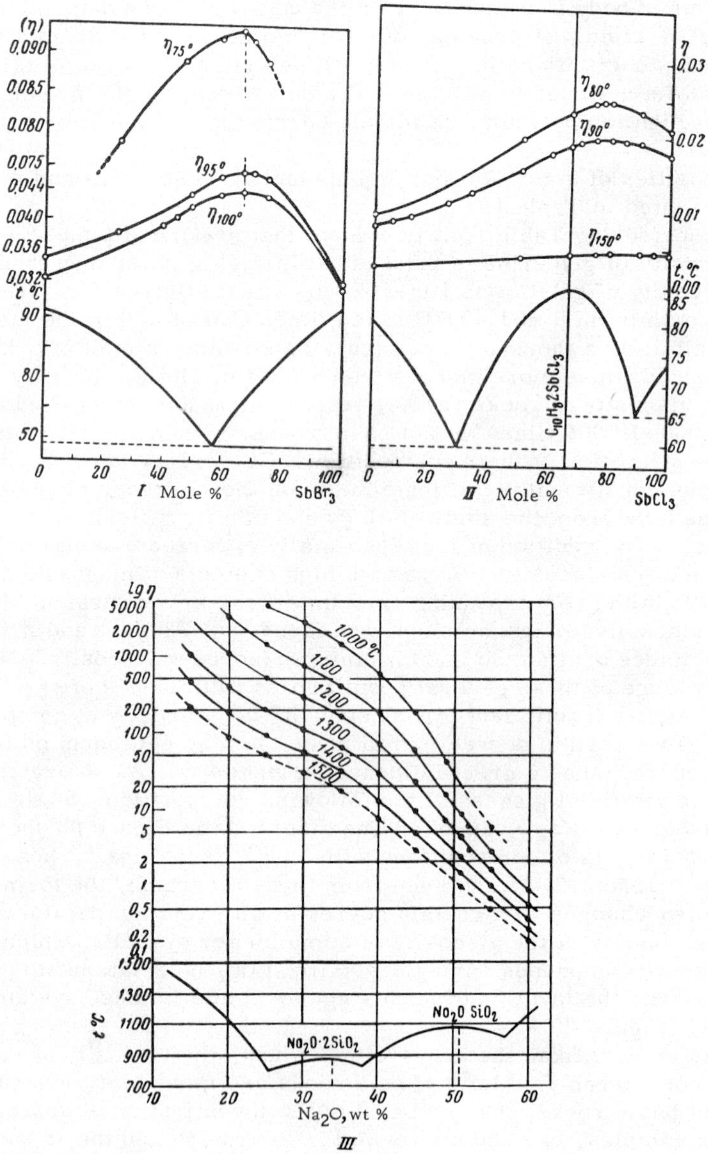

FIGURE 99. Some binary systems and their viscosity isotherms:

I — the system triphenylmethane — antimony tribromide; II — the system naphthalene — antimony trichloride (after N. S. Kurnakov); III — the system SiO_2 — Na_2O (after Heidtkamp and Endell) of chemical composition†. For equal values of *b*, viscosity also depends on the parameter *Q*, which expresses the surplus or deficiency of silica, as well as on other factors of the chemical composition. Thus, at the present moment we are only able to evaluate the approximate viscosity of a rock melt from the petrographical properties of the rock.

As magmatic melts also contain volatile constituents, the presence of which considerably lowers the viscosity, experimental data on the viscosity of rock melts may serve only as one of the bases in the analysis of natural phenomena connected with the viscosity of magmatic melts. During the consideration of these phenomena, the presence of volatiles should be taken into account. The water dissolved in the magma is of special importance. It is held within the magmatic solution by pressure. No sufficient quantitative data on the influence of magmatic water on viscosity are available. Isolated experimental data show that after dissolving water under pressure, the silicate melts become liquid at relatively low temperatures. Some time ago Barus showed that under such conditions water causes glass to melt below 200°. Important data on the solubility of water in granite melts were obtained by Goranson (see Part III).

Geological observations show that at depth granites are injected into the most minute joints during intrusion. One can scarcely believe that such intrusions are possible if the granitic magma has a high viscosity. On the other hand, when extruded at the surface after loosing the dissolved water and other volatiles, the granitic (rhyolitic) magma is very viscous, is usually restricted to domes, and does not form surface flows of great lateral extent. Obsidians are acidic volcanic glasses which contain a part of the magmatic water. On melting they liberate this water and other gases, and bubble up. After cooling, the glass of the remelted obsidian is much less fusible, and its melt is much more viscous.

The viscosity of silicate melts is influenced by admixtures of such "fluxes" as fluorite. M. P. Volarovich investigated the viscosity of diorite melts with and without admixtures of this substance at temperatures 1,400° and 1,230°. The results are shown on Table 2.

We shall also mention the experiments of English on the influence of B_2O_3 admixtures on sodium glasses. At 800° the admixture of boric acid at first sharply lowers the viscosity of the glass, and when a content of 25 % is reached, the further addition of boric acid lowers the viscosity considerably slower. At 700° a content of up to 15 % boric acid causes an increase of
138 viscosity. At a content above 15 %, a decrease of viscosity takes place. English associates this phenomenon with the formation of certain compounds in the melt.

† [T.N. The numerical characteristics are parameters introduced by A.N. Zavaritskii to express the chemical composition of rocks. Four parameters are used — a : c : b : s, where a + b + c + s = 100.

a = the relative amount of alkalis combined with Al atoms in aluminosilicates,

c = the relative amount of Ca atoms bound as $CaAl_2$,

b = the relative amount of atoms of other metals, excluding Ti and Si, which do not enter into Na aluminosilicates,

s = the relative amount of Si atoms.

For fuller details on Zavaritskii's numerical characteristics and their derivation see Translator's Appendix at end of book].

TABLE 2

t, °C	Viscosity η, poise		
	without addition	7% CaF_2	13% CaF_2
1400	708	206	107
1230	5990	1925	677

The viscosity of silicate melts, as in the case of any other liquid, depends on the temperature. Viscosity of water at 20° is 0.01 poise; at 0° — 0.018 poise, i.e., almost twice as large. The dependence of viscosity of silicate melts (glasses) on temperature has been expressed in various empirical formulas by different scholars. We shall cite the formula of Le-Chatelier.

$$\lg \lg \eta = P - Mt,$$

where η — is viscosity;
t — temperature;
P and M — coefficients.

Certain theoretical considerations on the structure of liquids, which will not be mentioned here, resulted in the formula

$$\eta = Ae^{\frac{B}{RT}},$$

where η — is viscosity;
T — absolute temperature;
R — the gas content;
A and B — constants.

It is worth mentioning that in this formula the coefficient B is proportional to the heat of fusion. It is evident from the cited formulas that it is more convenient to plot the value of $\lg\eta$ instead of η in order to demonstrate the dependence of viscosity on temperature (Figure 100).

The viscosity of magmatic melts is of great importance in the process of rock formation. One may discern the direct influence of viscosity as reflected in the shape of igneous bodies, in textural features, in the character of volcanic activity, in the velocity of lava flow, etc, and the indirect influence which is expressed in the varying influence of viscosity on crystallization processes, diffusion in melts, etc. We shall only mention the most important of these phenomena.

The difference in the viscosity of magmatic melts is most evident from the shape of both intrusive and extrusive rocks. Obviously, the less viscous melts, when injected under pressure between layers of rocks, will attain a far greater lateral extent than the more viscous melts. Hence, intrusive bodies of basaltic composition, i.e., formed from relatively non-viscous magma, are distributed over large areas. The famous intrusive body of Whin sill, with an average thickness of 25—30 m, extends
139 over more than 130 km. Examples of similar relatively thin, diabasic or

doleritic intrusive bodies may be found in different countries. Acid rocks, with higher viscosity, are on the contrary characterized under similar conditions by laccoliths — lenticular bodies which have been forced between strata in such a manner as to raise the overlying beds.

However, as already mentioned, at greater depth the granitic magma contains volatile components which are confined there by the greater pressure. This magma is more readily intruded between strata of rocks or into joints, thus indicating the low viscosity of these melts. Pressure in itself increases the viscosity of the liquids, but this influence is negligible if one takes into account the opposing effect of the volatiles, which are maintained in solution by this pressure.

The differences in the viscosities of lavas are no less manifest than these and may be observed in the shapes of extrusive bodies formed by the outpouring of lavas. Less viscous lavas are spread over many kilometers, their flows often being quite thin. Many of the basaltic flows serve as such an example. When the amounts of outpoured lava are large, they overflow areas in a manner similar to water during a flood. Viscous rhyolitic, dacitic, and sometimes andesitic lavas form domes; sometimes an eruption is accompanied by the formation of obelisks, as was observed during the famous eruption of Mont Pelée.

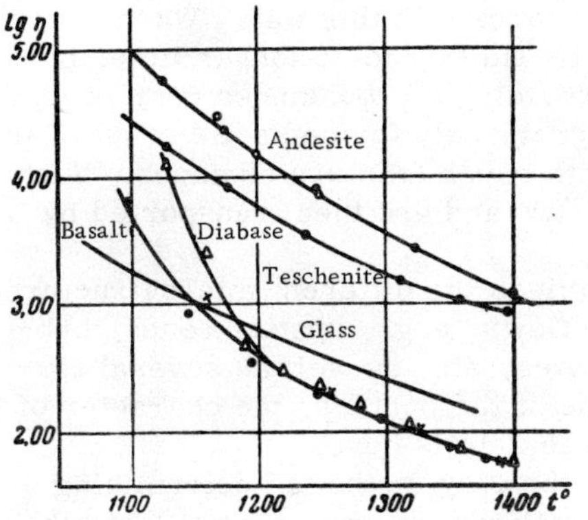

FIGURE 100. Dependence of viscosity on temperature for various melts of igneous rocks (after Volarovich)

The type of eruption depends firstly on the viscosity of the lava, which influences the character of liberation of gases from the magma, which in its turn determines the type of eruption. If these gases are evolved from near-surface parts of the lava masses, which are of low viscosity, the free evolution of these gases sometimes forms lava fountains characteristic of the Hawaiian type of eruption; in the case of a somewhat more viscous lava the emanation of gases causes the blowing out of molten lava — characteristic of the Strombolian type eruption; when the lava is even more viscous, the eruption becomes Vulcanian. The most viscous lavas form domes and even eruptions of the Baidansan type†.

In the case when gases are emitted from lavas which are located at great depths, and which are of relatively high viscosity, the eruptions resemble those of Mount Etna and Vesuvius; when the lavas are even more viscous, eruptions of the Mont Pelée type are characteristic.

The shape of volcanic bombs is very characteristic. In the case of lavas of low viscosity they are twisted, pear-shaped, or ellipsoidal; viscous lavas are characterized by "bread crust bombs".

† [T.N. This is apparently identical to what is usually termed the Plinian type of eruption in Western literature].

As is known, the upper surface of lava flows consolidates in two different forms: the block lava (aa) and ropy lava (pahoehoe). The surface of block lava somewhat resembles the accumulation of ice blocks on the surface of an irregularly frozen water body. The surface of ropy lava forms smooth flows resembling those formed when molten viscous liquids such as pitch, metallurgic slags, etc, are poured out and allowed to cool. Microscopic investigation shows that ropy lavas have a glassy groundmass, whereas block lavas have a microlitic groundmass, which is poor in glass. Both ropy and block lavas may sometimes be of identical composition; Washington explained the difference between them as being due to the varying increase in viscosity during the cooling of lava, and the associated emanation of gases. As already mentioned, the content of dissolved gases greatly lowers the viscosity of lavas. If the gases emanate at the initial stage of consolidation of a lava, the magma rapidly becomes viscous, the crystallization of the groundmass is hampered, and a glassy groundmass is obtained on cooling. Pahoehoe lavas are formed in this way. When a lava liberates its gases slowly, it maintains its fluidity for a longer time, but will subsequently crystallize rapidly, providing a groundmass poor in glass; the lava loses its fluidity abruptly; the previously formed solid crust of the flow is broken up and torn into blocks with rough fracture surfaces. Such blocks accumulate on the surface of the flow and are then transported by the current.

140

The viscosity of the lavas also determines the different microstructures which are characteristic of various lava flows, e. g., spatter cones, bubbles and vesicules in lavas, lava-stalactites, etc, as well as several textural features such as the form and shape of the bubbles, the character of jointing, etc. We shall not deal with these details here.

The change of rate of flow of lavas is the only method of determining the viscosity under natural conditions, when the lava has not yet lost all the dissolved volatiles.

Almost fifty years ago an attempt was made by Becker† to determine the viscosity by such a method. "The viscosity of the lava is manifest by its slow rate of flow. Thus, during the eruption of Kilauea in 1840, the lava flowed for 17.7 km down the slope from a height of 379.5 m in two days. According to Wilkes and Dana, this flow was fed by several fissures along its path, and was not erupted from one single vent. Under such conditions the heat of the flow was well preserved. The average velocity of flow of this lava down a slope of 2° was approximately 400 m per hour, or 6.7 m per minute. A stream of water of the same section on the same slope would have flowed with a velocity of about 9.66 km/hour, or approximately 24 times faster. As the density of lava is 2.5 times that of water, the kinetic viscosity of lava proves to be 2.5 × 24 or 65 times that of water".

Becker based his calculation of the mentioned flow on the data, published by Brigham, of T. Kohn, who was an eye witness to this phenomenon. According to Palmer ††, Becker used too low a value for the viscosity of lava flow, as this velocity applied to the frontal part of the flow, where the motion was

† [T.N. No reference is given. It seems to refer to G. F. Becker, "Some Queries on Rock Differentiation", Amer. Jour. Sci., Vol III, 4th ser. 1897].

†† [T.N. Reference not quoted; probably — H. S. Palmer, "A study of the viscosity of lava", Bull. Hawaiian Observ., Vol. XV, pp. 1-4. 1927].

already hampered. The frontal part of the flow moves more slowly than the flow as a whole, as the lava is hampered, due to accumulation and filling in of irregularities of the surface. Palmer made another attempt at calculating the viscosity of lava using data of Jaggar, who witnessed that the velocity of the Alika flow (pahoehoe) in its main stream was 17.7 km/hr, or 4.88 m/sec. He assumed the channel to be 6 or 12 m wide and 4.6—9.2 m deep. According to a later topographical survey, the gradient
141 of the channel is 152.4 m per 1.2 km, i. e., about 12.6°†. Palmer assumed the specific gravity of the lava to be 1.4. This is the value given by Jagger for the lava of Kilauea which is saturated with gases.

In order to determine the viscosity of lavas, Palmer, like Becker, calculates the flow velocity of a stream of water of the same dimensions, down the same gradient and meeting with the same resistance. These calculations of Palmer and Becker were based on the formulas of Chezy and Kutter, which are usually used in hydraulics. These formulas are empirical and apply to cases of turbulent flow, when eddies are formed within the flow. The application of these formulas in this case led Palmer to the conclusion that the Alika flow was 15 times as viscous as water.

However, Nichols †† showed that the lava flow is laminar and not turbulent, i. e., the liquid layers flow without eddies. Therefore we should not compare the velocities of flow of lava and water, as was done by Becker and Palmer.

Because of their viscosity, lavas move by laminar and not turbulent flow. In less viscous liquids, when the mean velocity exceeds a certain limit, the character of flow changes, eddies appear and the flow becomes turbulent. In water streams we deal with such flows. The velocity of turbulent flows is almost independent of the viscosity of the liquid. It is dependent on its density, the gradient, and the dimensions and surface irregularities of the channel (i. e., the resistance to motion).

Reynolds proposed the following rule for the flow of liquids in pipes when v is average velocity, r the radius of the pipe, ρ the density of the liquid, and η its viscosity; the flow is laminar if $\frac{vr\rho}{\eta} < 1000$ and is turbulent if $\frac{vr\rho}{\eta} > 1000$.

In order to indicate the difference in the character of motion of lava and water flows, Nichols applied the Jeffreys formula in determining the viscosity of lava of the same Alika flow, using the data of Jaggar. The Jeffreys formula for laminar flow is

$$v = \frac{g \sin \alpha \cdot d^2 \rho}{3\eta},$$

where v is the mean velocity;

g is the acceleration due to gravity;

α is the angle of inclination;

d is the depth of the flow;

η is the coefficient of viscosity (or simply — viscosity);

ρ is the specific gravity.

† [T.N. The value in the English original is given as 12.6 %].

†† [T.N. Reference not quoted. R. L. Nichols, Jour. Geol. 47, 290. 1939].

Nichols obtained a value for lava viscosity of $\eta = 4.3 \times 10^4$ poise, i.e., approximately 290,000 times the value calculated by Palmer for this flow.

Later, Nichols calculated the viscosity of the 1887 flow in the Kau district, Hawaii. According to Stearns, the velocity of this flow was 1.13 km/hour, the average thickness 2.2 m, the average gradient 65.8 m per km and the average width 1,220 m. Accepting Jaggar's value of 1.4 for the specific gravity of lava, Nichols in this case obtained a viscosity of $4.77 \cdot 10^4$ poise. Using these data and the formulas of Chezy and Kutter, it is found that under the same conditions, water would flow 40 times as fast as the lava of the 1887 flow.

142 During the eruption of the parasitic crater of the Klyuchevskaya volcano in Kamchatka, the viscosity of the lava flow was observed by S. I. Naboko. The lava, on issuing from the vent, flowed down a slope with a gradient of 10° with a velocity of 30 m/min, the depth of the flow being 1 m. Assuming that the specific gravity of a lava when saturated with gases is 1.4, as was determined by Jaggar in Kilauea, Naboko calculated the viscosity coefficient as being $1.6 \cdot 10^4$ poise. If the specific gravity of lava is accepted as 2.0, then $\eta = 2.6 \times 10^4$ poise.

According to Kani and Volarovich, the obtained values for basaltic lavas at temperatures of approximately 1,100—1,200° are approximately of the same order as for the melts. As we shall see further on, the basaltic lavas sometimes have a temperature lower than 870°. Many of these lavas are below 1,000° and only rarely do they attain a temperature of 1,200—1,250°. Thus, at lower temperatures they may possibly possess a fluidity equal to that of melts which originated from the same rocks at higher temperatures. This may be explained by the influence of gases dissolved in the lava. According to experiments of A. A. Leontieva on obsidian, which was quenched preliminarily in order to remove its water content, and then with the same obsidian unquenched, with 0.12 % water content, the viscosity of the quenched obsidian proved to be higher. The mentioned water content (0.12 %) lowers the viscosity 4—5 times.

In all events, the magnitude of viscosities in lavas and the corresponding melts of basic rocks (basalts) are of the same order.

Nichols, after obtaining the value of viscosity of a basaltic lava from data on the velocity of lava flows during eruptions, used this value to calculate the possible rate of flow of the McCarthy flow, Valencia County, New Mexico. This flow consists of pahoehoe lava and is over 48 km long. The last 9—10 km of the flow were studied in detail by Nichols. The gradient here is 4.9 m per km. The average thickness of the flow is somewhat above 9 m; the width of the flow is approximately 450 m. If a viscosity equal to that of the 1887 Hawaii flow is accepted for this flow, the velocity of laminar flow is approximately 7.7 km/hour. In this and other similar cases lava flows are considered as a single entity.

Of great importance is the influence of the lava's viscosity on the sinking or rising of separated crystals or on the rising of gas bubbles in it.

These phenomena are the basis for hypotheses of gravitational differentiation of melts and the so-called "two-phase convection" in lava and magmatic basins. Only few investigators consider that viscosity constitutes a factor sufficient to overcome the action of gravity in heterogeneous systems such as magmas with the separated crystals.

The velocity of a sinking sphere in a viscous medium is determined by Stoke's equation as

$$v = \frac{2gr^2(d_1 - d_2)}{9\eta},$$

where r is the radius of the sphere;
d_1 and d_2 are the respective densities of the sphere and the medium;
η is the viscosity of the medium.

M. P. Volarovich, using Stoke's formula, calculated that in a "dry" melt of diabase composition at a temperature of 1,220°, crystals of pyroxene with a diameter of 0.5 mm will sink with a velocity of 12 mm/hour, whereas similar crystals of labradorite will float upward at a rate of 18 mm/hour.

143 Bowen observed the sinking of olivine crystals after having separated from a melt in a crucible. On a larger scale this phenomenon occurs during the melting of basalt in ovens. Numerous examples of accumulation of heavy minerals in the lower parts of intrusions or even extrusions are known in nature. The accumulation of olivine crystals in the lower part of the basalt flow of Tertiary age in Oregon serves as an excellent example of this phenomenon. This part cooled rapidly due to the impeding of the flow at the bottom. Numerous examples are known of rising of lighter crystals, which enrich the upper parts of intrusions or lava flows (e. g., leucite in the Vesuvius lavas). The phenomenon of gravitational differentiation has been convincingly described several times. One of the first people to indicate its importance was Charles Darwin.

In some cases one may assume the formation of immiscible liquids in nature, e. g., immiscibility of silicate and sulfide liquids, similar to that encountered in metallurgical processes. In such an emulsion, drops of the heavier liquid will naturally sink. When drops of a viscous liquid sink within another liquid, the viscosity of which approaches that of the first, Stoke's formula should be replaced by the formula of Adamar-Rabchinski

$$v = \frac{gr^2(d_1 - d_2)}{3\eta}.$$

Thus, for instance the rate of sinking of a drop of iron sulfide with a diameter of 0.5 mm and a density of 3.95 in a silicate melt of a feldspar-pyroxene mixture with a viscosity of $2 \cdot 10^4$ poise and a density of 2.89 at a temperature of 1,300° is approximately 1.6 mm/hour.

Although the above-cited figures are all approximate, they are sufficiently exact to give one an idea of the order of velocities of the sinking and rising of crystals and drops.

Viscosity also determines the liberation of gas bubbles from magma and lava. As is known, Daly ascribes great importance to this process as an active factor during the differentiation of magma. He introduced the concept of two-phase convection, when currents arise in the lava mass due to movement of entire bodies of the liquid phase (lava) and gaseous phase (gas bubbles).

Using Stoke's formula at a constant viscosity, Daly calculated the velocity of rising of small bubbles. He showed that they rise very slowly, and therefore, in his opinion, the bubbles which form (e. g., during retrograde boiling) tend to accumulate, and such enriched areas may float up as entities. Such masses, the dimensions of which are quite large, will rise relatively rapidly. On rising they carry heat upward; this process of heat transport is somewhat opposed by the expansion of the bubbles due to the decrease of pressure on the rising; this expansion is associated with a certain cooling of gases.

Such an ascent of the foaming lava, accompanied by the expansion of the bubbles, occurs more readily in a lava with a low viscosity. In craters which are filled by more viscous lava, gases escape to the surface under much higher pressure, and through more constant channels; they may act in a manner similar to a giant blow pipe.

144 Slow diffusion in silicate melts also depends on the viscosity. As is known, the motivating force of diffusion is osmotic pressure, i. e., the pressure caused by the velocity of the dissolved particles moving in the solvent. This motion, as any other, meets with the resistance of internal friction, i.e., viscosity of the melt. This explains why the rate of diffusion is much lower in silica melts than in aqueous solutions.

The quantitative relations during diffusion are given according to Fick's law, which only holds true for dilute solutions. If diffusion occurs only in the direction of the x axis, and the concentration at a point with a coordinate x is c, then, according to Fick's law, the amount of dissolved material which passes through a section q during a time dt will be

$$dm = -K\frac{dc}{dx}dt,$$

where K is the diffusion coefficient.

This value is inversely proportional to the viscosity of the melt (directly proportional to the absolute temperature, and inversely proportional to the radius of the diffusing molecules). Apparently it represents the number of grams passing through 1 cm^2 in a unit of time at a concentration gradient accepted as unity.

The difference in amounts of material passing through two infinitely close sections is:

$$-K\frac{\partial^2 c}{\partial x^2}dxdt.$$

The amount of dissolved substance in the layer between the two sections will decrease by this value. Hence,

$$\frac{\partial c}{\partial t} = K\frac{\partial^2 c}{\partial x^2}.$$

This equation with partial derivatives determines the diffusion at each point of the melt. On integrating it, one is able to calculate the distribution of concentrations resulting from diffusion after a certain time.

Diffusion is visually manifest in glass if a coloring agent is introduced and spreads in it.

Of special interest are the experiments of Bowen, who investigated mutual diffusion in melts of diopside and plagioclase.

Using a small cylindrical crucible (5—6 mm in diameter and 10—20 mm high), Bowen filled the lower part with fused diopside, and placed above it a layer of molten plagioclase. The sample was heated to approximately 1,500° C, and was maintained at this temperature for a long time. Diffusion took place between the heavier layer of molten diopside and the lighter layer of molten plagioclase. After a certain time, the melt was rapidly cooled, being converted to glass. A prism was cut from the cylinder in order to enable the determination of the refractive indices in its various parts. The refractive index assisted in determining the composition of the glass and in tracing the mutual diffusion of diopside and plagioclase.

Figure 101 diagrammatically represents the results of certain experiments. The results lead one to the conclusion that diffusion is facilitated in the case of a more basic plagioclase. This is probably due to the viscosity of the plagioclase melt. The diagrams also show that the plagioclase diffuses more readily into the diopside melt than does the diopside into the plagioclase melt. As is seen from the results, diffusion under the experimental conditions is much slower than in salt solutions or molten metals.
145 It may be compared to the diffusion of relatively viscous organic compounds (e.g., glycerine in propyl alcohol).

Bowen assumes that the ability to diffuse is not much greater in magmas than in the investigated melts of diopside and plagioclase; in many viscous magmas it is even less. Using this assumption as a basis, and accepting the diffusion coefficient as $K = 0.25\ cm^2/day$ (the highest value observed in his experiments), he calculated the time required for an igneous body cooling from the margins to attain a state when the concentration due to diffusion becomes equal to the reciprocal absolute temperature, i.e., an equilibrium of osmotic pressure could have been achieved, causing the so-called Soret effect which was very widely used in petrology. It assumes that the distribution of temperatures in such a cooling mass depends on thermal conductivity. Bowen took as an example a layered body 20 m thick. Calculation shows that the rate of diffusion is 10^5—10^6 times the rate of temperature decrease due to thermal conductivity. Hence, the cooling of such a body will occur so rapidly that the resulting temperature gradient cannot be reflected in differences in composition due to diffusion, caused by a gradient in osmotic pressure. According to Bowen, the Soret effect cannot manifest itself during the

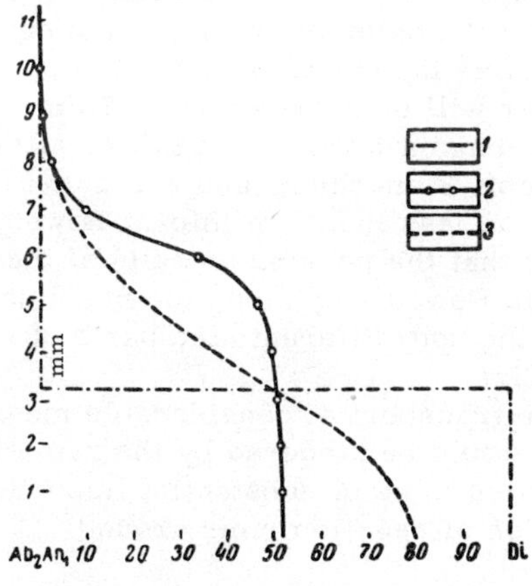

FIGURE 101. Diagram of the results of certain experiments on diffusion in melts of diopside and plagioclase (after Bowen)

1 — the initial state; 2 — the final state after 48 hours; 3 — the curve as calculated from Fick's formula with K = 0.015; Ab — albite; An — anorthite; Di — diopside

cooling of magmatic masses. The increase in dimensions is insignificant, as when the thickness of the body increases n times, the time required for diffusion and realization of the Soret effect as well as for cooling will increase n^2 times.

Such considerations are justified for the conditions assumed by the author; but one cannot regard as proven the fact that the viscosity η of the magma is not lowered considerably by the presence of volatile components, and the possibility of prolonged preservation of temperature differences due to a supply of magma in the central parts, or due to other causes, cannot be denied.

Bowen also discusses the role played by diffusion in the formation of marginal zones, according to Harker's hypothesis. The latter assumed that as a magmatic body cooled, crystallization of some component started when the magma became saturated with respect to it. This crystallization caused a diffusion flow of the crystallizing substance; in this way a marginal zone may grow. Bowen calculated that for the diffusion coefficient of 0.25 cm^2 per day (as assumed by him), the effect of such separation of the crystalline substance at the contact will have spread only 0.33 m into the depth of the magmatic body after 240 days, and 3.3 m after 64 years. The
146 thickness of the crystalline layer growing on the walls is independent of the reservoir dimensions (provided the inflow of substance is solely the result of diffusion) and depends only upon the thickness of the adjacent part of the magma from which the substance is derived by diffusion. For instance, if a mineral being deposited on the walls constitutes up to 20 % of the magma volume, and if the marginal crystalline layer continues to form until a quarter of this amount (i. e. , up to 5 % of the layer) has been deposited from it on the walls, it is seen from the calculation that in the course of nine months the forming marginal crystalline layer will be only 0.75 cm thick. After 16 years, the crystalline layer will be approximately 2 cm thick and will have formed entirely from a magma layer 1.5 m thick; after 260 years, the crystalline deposit will be only 8 cm thick, and will have been entirely derived from a marginal region less than 7 m thick. Bowen, on the basis of his assumptions, concludes that the possible results of the diffusion process would be insignificant. In Bowen's opinion, the marginal zones are to be explained (as by Daly) as the non-differentiated parts of the magma which were the first to crystallize.

Bowen denied the role of diffusion in the transport of considerable masses of matter over large distances, since this would be hindered by the rate of magma cooling; however, he admits diffusion to be of substantial importance over small distances, e. g. , in the formation of reaction rims around foreign bodies.

CRYSTALLIZATION

When considering the systems we dealt only with the boundaries of the various states, not touching upon how the transition from one state to another occurs. Thus, we have up to now omitted the process of crystallization, which is one of the main processes occurring during inversions. We shall now consider this process.

The process of crystallization may be considered from two aspects. Firstly, crystallization starts in certain places in the melt, i.e., embryonic crystals are formed; then the separated crystals grow at lesser or greater rate. Thus, two values characterize crystallization: 1) the number of crystallization centers (embryos); 2) the rate of growth of the formed crystals. Crystallization occurs due to the cooling of the system, i.e., as a result of a decrease in the energy of thermal motion in it. Let us recall the concepts of state of matter from the point of view of the molecular-kinetic theory.

The crystal is a spatial lattice, comprised of regularly orientated atoms. They may vibrate within certain limits due to thermal motion, but the preservation of shape and crystalline properties proves that the regular orientation of the atoms, which is typical of the crystalline lattice, is not disturbed. On heating, the thermal motion of the particles exceeds a certain barrier, and the lattice of the crystal is destroyed. In liquids (and gases), the particles are capable of free oscillatory motion. This is illustrated by the fluidity of liquids which do not preserve their shape and change it readily when subjected to external forces. Not all bonds between the atoms of a crystalline substance are destroyed by melting. This is demonstrated, for instance, by the fact that the latent heat of fusion usually constitutes only a small part of the total energy necessary to destroy the crystalline structure (heat of sublimation). However, the fluidity of liquids does show that the
147 particles which retained the regular crystalline arrangement now become capable of movement, enabling the entire mass of liquid to change its shape under the influence of small forces. On melting, the stable bonds holding the atoms of the crystal in a regular pattern are destroyed, and the anisotropy, which is a property of the crystalline state, is lost. In order to start the incipient crystallization of a liquid on cooling, the velocity of thermal motion of the particles should be decreased to the limit, below which the formed crystalline bonds between these particles cannot be destroyed. Thus, when atoms or groups of atoms have aggregated, embryonic crystals may be formed in various places. Hence, there are certain limits of thermal motion, i.e., temperature, within which the appearance of crystallization centers is optimal. The velocities of the particles should be sufficiently great to allow for their mobility, so that sufficient collisions necessary for the growth of a crystal may occur, but these velocities should not be so high as to destroy the crystalline bonds which are formed at such collisions.

After the appearance of crystallization centers, the growth of crystals starts. The atoms comprising these crystals are arranged in a regular pattern, and an ever increasing number of atoms join the first ones, being orientated in continuity with the spatial lattice. Obviously, this process is also governed by limiting temperature values, which correspond to the limiting velocities of thermal motion; the atoms, or more generally the particles from which the crystal is constructed, should be sufficiently mobile to acquire the required orientation. Their velocities, however, should not be too high, so that the atoms may maintain the positions which they acquire during the growth of the crystal.

These elementary concepts show that both the rate of formation of crystallization nuclei and the rate of crystal growth are limited by optimal temperatures, and that these temperatures are not always constant, as the number of centers appearing during crystallization and the rate of growth of crystals are complex functions of temperature.

The dependence of a value on temperature may be represented by a curve on a diagram. Both above-mentioned functions are represented by curves with a maximum, which is somewhat lower than the melting point, as the latter corresponds to the inception of the crystallization process and is therefore a limiting point. The optimal temperatures, both for the formation of crystallization centers and for the rate of crystal growth, correspond to a certain supercooling, the amount of which depends on the properties of the liquid and varies for different liquids. Obviously, as these temperatures are conditioned by the degree of mobility of the particles necessary for the construction of the crystal, they are dependent on the internal friction, i. e., on the viscosity of the liquid.

Viscosity hampers both the motion enabling the orientation of particles (atoms), and diffusion supplying the material for the construction of the crystal. It is known that the viscosity of the liquid increases as an exponential function on cooling, being especially rapid in the area of supercooling. Therefore, the maximum rate of crystal growth is already attained with a small degree of supercooling of the liquid. The maximum number of crystallization centers usually appears with a somewhat greater supercooling, although in some cases both maxima almost coincide. Usually, the maxima are better defined on the curve showing the number of crystal-
148 lization centers than on the curve of the rate of crystal growth, which is smoother. Figure 102 shows two possible relations between the curves of rate of formation of crystallization centers and rate of crystal growth.

The texture of the obtained crystalline aggregate, i. e., its granularity, depends on the relation between these curves, and on the degree of supercooling. Figure 102, II, shows that at temperature T_1 the number of forming crystallization centers is small, and the crystals very rapidly attain considerable sizes. Should crystallization occur at this temperature (or should the temperature drop slowly), the result would be a coarse-grained aggregate. At temperature T_2, the number of forming centers of crystallization is large, but the crystals grow relatively slowly. Obviously, crystallization under these conditions results in a fine-grained aggregate of crystals. If crystallization is accompanied by continuously falling temperatures, which pass through the values T_1 and later T_2, then the corresponding rate of temperature decrease may initially result in the formation of only few centers of crystallization and the growth of large crystals, whereas on further decrease of temperature and increase of viscosity, this crystallization is replaced by slow crystal growth from many centers of crystallization. The final product is individual large crystals within a fine-grained groundmass. Thus, a porphyritic texture may be obtained without the sharp change of crystallization conditions, as is customarily accepted as being the case during the formation of porphyritic textures in volcanic rocks, where phenocrysts are considered as intratelluric formations formed prior to the eruption of the lava and the crystallization of its groundmass.

[In the considered case, the overall rate of crystallization, which is defined as the percentage of crystalline substance formed per unit time from the total mass of the system, attains its maximum somewhere

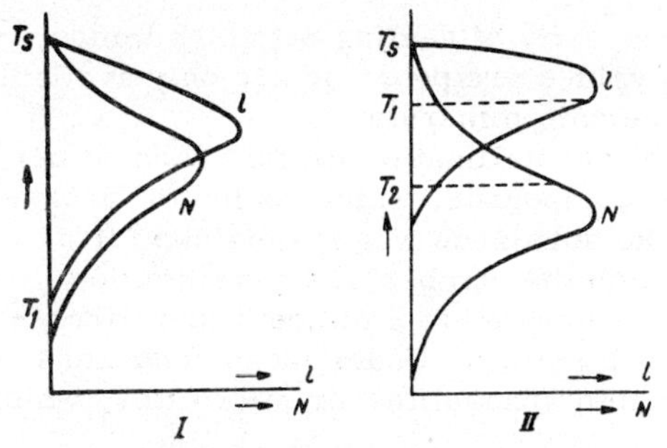

FIGURE 102. The relation between the curves of rate of formation of crystallization centers (N) and rate of crystal growth (l);

I — with maxima close to one another; II — the maxima do not coincide

between the maxima of the curves shown on the diagram. This rate will be considerably low, as under these conditions only a few centers of crystallization are formed, and the linear rate of their growth is small. For the case shown on Figure 102, I, the maximum overall rate is attained under the conditions of formation of maximum numbers of crystal centers, and at a maximum linear rate. Hence, the crystallization rate is greater here, and with equal cooling rates the probability of supercooling, with the formation of glass, decreases, as does the probability of formation of porphyritic and fine-grained textures.] As the crystal is an anisotropic substance, the velocity of its growth differs in various directions. This difference varies in dependence on the viscosity of the melt. The shape of the forming crystals depends on this.

One may state in general that the forms of crystals separating from
149 non-viscous melts approach isometrism, whereas at higher viscosity, the crystals expand at different velocities in various directions, forming crystals elongated in one crystallographic direction relative to the others. Feldspars in igneous rocks are a good example of the above-mentioned phenomenon. In plutonic rocks they form phenocrysts in porphyritic rocks, which have isometric outlines, or are thick tabulae. Such crystals are formed under conditions of relatively low viscosity. In the groundmass of porphyritic rocks, the microlites are lathlike, elongated in one direction or prismatic and spicular. A good example of the change in the form of microlites is found in the series gabbro—diabase—dolerite—basalt. The form of the microlites is so characteristic of the entire appearance of the microtexture of groundmasses in porphyritic rocks that during the initial stages of the development of the petrographical science, such a texture was distinguished by French workers as a specific type of "microlitic texture". This microtexture of the groundmass was accepted as a characteristic texture of trachytes, andesites and (in part) basalts.

If the melt is very viscous, the rapid crystal growth shows preference for one particular direction, forming thin, elongated needles or fibers. If these fibers grow from a common center, they form radial bundles — spherulites or spherocrystals. Spherulitic structures are especially characteristic of silica-rich volcanic rocks. Sometimes one may discern that the spherulites grew in so viscous a medium that it could even form cracks. In basic rocks spherulitic microtextures are only encountered in rapidly cooled masses, e. g., the variolitic textures in diabases and spilites. Well-developed spherulites are sometimes also formed in artificial glasses.

The measurement of the number of forming crystallization centers, and of the velocity of crystal growth, in silicate melts encounters certain difficulties, which arise due to the necessity of dealing with high temperatures. Therefore, the data on these values are poor and are only available for the rate of growth of crystals of certain minerals.

The determination of the number of crystallization centers formed per unit time was carried out for organic compounds, which cooled to form glasses, by the following method. The substance was maintained for a certain time at the temperature at which the number of crystallization centers were to be determined, and then heated to a temperature corresponding to the maximum rate of crystal growth. Under these conditions, the crystallization centers developed into spherulites or microlites, which were counted.

The rate of crystal growth may be determined by a direct observation of this process under a microscope, which is attached to a heating oven.

Using Delters microscope furnace, the following data were obtained on the number of forming crystallization centers *n* for various minerals with corresponding supercooling.

	t	*n*
Diopside	100	15
Spinel	174	47
Gehlenite	185	45
Melilite	100	43
Hedenbergite	200	34

There is no apparent relation between the number of forming crystallization centers and the chemical composition. Possibly the simplicity of the structure of molecules is of great importance here. In general it appears
150 that more crystallization centers are formed in orthosilicates than in metasilicates. The latter, however, have a greater rate of crystal growth.

Whereas a quantitative determination of the number of crystallization centers is difficult, simple experiments of melting are sufficient for detecting qualitative differences in the crystallization ability of various silicates.

For instance, the metasilicates of Ca, Mn and Fe crystallize readily, whereas such compounds as $CaMnSi_2O_6$ and $CaFeSi_2O_6$ will yield glass at the same rate of cooling. Admixtures of small quantities of Al_2O_3 and Fe_2O_3 in augites and basaltic hornblendes do not considerably influence the crystallization ability. A larger admixture of alumina does, however, lower this ability. Sillimanite Al_2SiO_5 (or mullite) has a considerable crystallization ability. Alkali silicates (e. g., nepheline and leucite) will crystallize much less readily than alkali-earth silicates (e. g., anorthite, gehlenite). Acid aluminosilicates — orthoclase and albite — have a low crystallization ability, and usually cool as glasses. Acid plagioclases crystallize less readily than basic ones.

On comparing natural minerals with artificial compounds, Delter noted the greater crystallization ability of the former, with the exception of natural diopside, which crystallizes less readily than the artificial one.

According to Delter, the crystallization ability may be expressed as the ratio of the crystalline phase to glass under equal cooling conditions. This ratio is the sum of the number of crystallization centers and the rate of crystal growth. In practice, one may determine this value by noting the time needed for the crystallization of a unit mass of the melt, or by measuring the quantity of crystalline substance formed during unit time.

The following values may give an idea of the relative crystallizing abilities as determined by the mentioned method. The values represent the percentage of the crystallized part of a melt, which was heated by Delter to 100° above the melting point, and then cooled to red heat relatively quickly (in 5 min). Spinel and magnetite crystallized entirely. Olivine also crystallized in the form of a fine aggregate; bronzite yielded 75—80 % crystals, hypersthene 70—80 %, hedenbergite 60—70 %, augite 60—65 %, anorthite and labradorite 40—45 %, nepheline and leucite 30—35 %; albite, orthoclase and quartz cooled to form glass.

Determination of the degree of supercooling corresponding to the maximum crystallization, or the optimal devitrification as it is sometimes termed, was carried out as follows. A supercooled melt (glass) was located in a pipe, one end of which was heated to melting point of the substance, whereas the other side was maintained at normal temperature; a gradual thermal gradient was set up in the pipe; at a certain point along the pipe, in the area of temperatures corresponding to the optimal devitrification, a characteristic turbidity appears due to the formation of embryonic crystals or even microscopic spherulites and crystallites. Crystallization attains a maximum in the area of optimum temperatures, and decreases towards both ends. At a temperature corresponding to the crystallization point (melting point), a sharp boundary is observed. On passing to the area of low temperatures, the turbidity disappears, either gradually or abruptly.

As already mentioned, optimal temperatures of supercooling also exist for the rate of crystal growth. At the melting point a kinetic equilibrium exists between the number of particles which enter the lattice of the crystal and the number which leave the crystal due to thermal motion. The rate of
151 crystal growth is zero. On further cooling it increases and reaches a maximum, which, as already mentioned, is less sharp than the maximum of the number of crystallization centers. The gentle summit of this maximum is determined by the action of two opposing factors: the maximum rate of crystallization characteristic for each substance, and the contracting influence of heat of crystallization, which depends on external conditions of thermal conductivity. Therefore, the rate of crystallization versus degree of supercooling curve may have different shapes under different conditions. In silicates it is in most cases a smooth, gentle curve (Figure 103).

During his investigations, Delter found that individual, larger crystals are formed at a supercooling of 20—40°, corresponding to point B on the diagram. At point C, where supercooling already hampers the growth of crystals considerably, its value may be accepted as 60—120°. Delter roughly assesses that the rate of formation of crystals under optimal conditions is 4—10 times as great as that of crystals which started to separate at insufficient supercooling. Augite (see Figure 103) and fayalite have the greatest rates of crystal growth among common rock-forming crystals. These minerals are difficult to obtain as glasses. The rates of crystallization of labradorite and anorthite are lower, although still fairly high; during crystallization of these minerals in such experiments, grains are rarely formed and spicules or needles are more common. The crystallization rate in dry melts is practically zero. Nepheline crystallizes less readily than plagioclase, and the rate of its crystal growth is slower; crystal

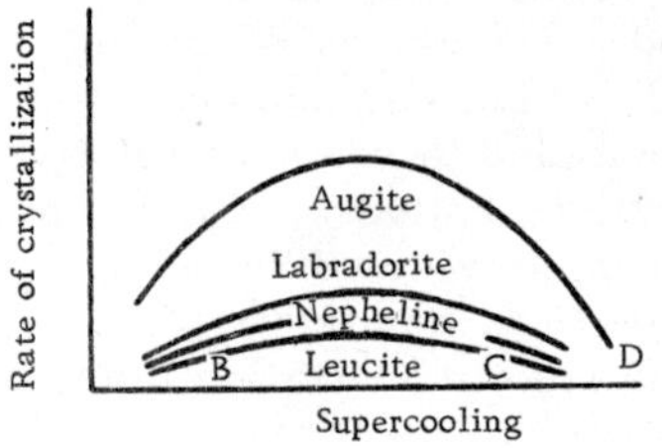

FIGURE 103. Crystallization rate of some minerals (after Delter)

skeletons are easily formed. The rate of crystallization is apparently even lower for leucite. Magnetite has a very low rate of crystallization and a large number of centers of crystallization.

In minerals which are solid solutions, the rate of crystallization changes continuously, depending on the composition. Thus, in plagioclases, it increases with the increase in anorthite content; in olivines — with the increase in fayalite content. The curve showing rate of crystal growth versus composition has various shapes in various cases: sometimes it has a maximum, sometimes it has a minimum, and sometimes it changes continuously in one direction. For chemical compounds, the rate of crystal growth is not an average of the rates of crystallization of the individual components. This value is not additive but constitutive. For diopside, for instance, which may be considered as a compound of $CaSiO_3$ and $MgSiO_3$, the rate of crystallization is lower than the mean rate of crystallization of its individual components.

The relative maximum values of rates of crystal growth for various readily crystallizing minerals are cited below (after Delter).

Augite	20	Nepheline . . .	5
Labradorite	7	Leucite	3–4
Olivine	3–10	Magnetite . . .	1–2

The absolute value of the rate of crystal growth under usual conditions is 1–2 μ/min. Hence, one may assume that the crystallization of an augite crystal 1 cm long takes about 7-9 days. Medium-grained gabbros (dolerites) are more or less crystallized after 10 days. Large crystals in the lavas of Mount Vesuvius could have formed over a period of 1–2 months.

152 Although Delter's data should be considered as only roughly approximate, they are of essential interest to the petrologist. Quite a considerable amount of data on the rate of crystallization was obtained by Kittl. He melted natural minerals and during their cooling measured the rate of crystallization as dependent on temperature by means of a microscope furnace. Despite the fact that crystallization rates are usually very small, they may in all events be measured by means of a micrometer ocular. In most cases the formation of thin spicular crystallites was observed. The increase of their length per unit time was accepted as the maximum rate of crystal growth in the direction of the major zone. The results of these investigations are shown on Figure 104, where curves of rates of crystallization for various minerals are depicted. It is interesting to note that in many cases these curves are not smooth but show a sharp maximum. In general Kittl's data correspond in their order of magnitude to those of Delter. During these experiments it was noted that orthosilicates yielded a larger number of crystallization centers than metasilicates.

Very interesting data were obtained by A. A. Leont'eva during the crystallization of remelted basalt. The melt was maintained at 1,500° and then poured out. The obtained black glass was afterwards crystallized in small crucibles. The first mineral to separate was magnetite at 1,300°; plagioclase started to separate at approximately 1,250—1,260°; later, pyroxene started to separate at 1,150° together with magnetite and plagioclase. As thin sections showed, the melt at 1,260° C contained only a few needles of plagioclase, approximately 7 μ in size; at 1,220° many needles (approximately 50 μ long) had formed, which on further crystallization became narrow plates. A. A. Leont'eva presents the following rates of crystal growth for plagioclases:

Temperature in °C	1,260	1,220	1,200	1,150	1,125	1,100
Rate of crystal growth in μ/min	0.2	1.7	2.3	5.6	1.9	1.0

The maximum number of centers of crystallization corresponds more or less to the maximum rate of crystal growth of 5.6 μ/min at 1,150° C.

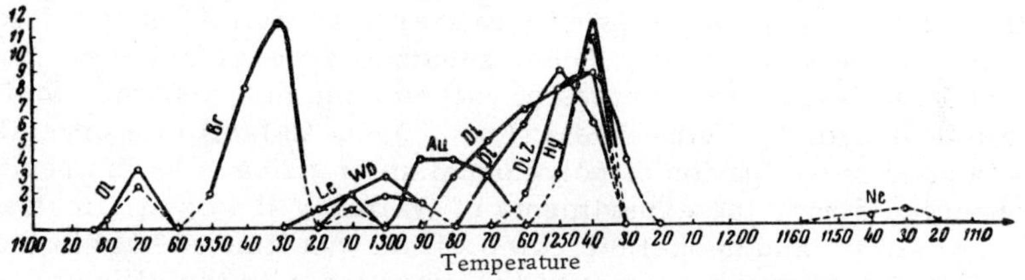

FIGURE 104. Rate of crystallization of some typical minerals (after Kittl):

Legend: Ol — olivine; Br — bronzite; Le — leucite; Wo — wollastonite; Au — augite; Di — diopside; Hy — hypersthene; Ne — nepheline

For magnetite the maximum rate of growth of approximately 1 μ/min was observed at 1,080°; the number of centers of crystallization is, however, always large. In diabase glass the maximum rate of crystallization of magnetite was observed at 1,160°, and its value was only 0.3 μ/min.

In addition to these data, Leontieva obtained data on the rates of crystallization of hematite from aegirine glass, and of sodium metasilicate from sodium glass. Their magnitudes are of the same order — 1—2 μ/min.

153 A. A. Leont'eva also investigated the dependence of the rate of growth of crystals (v) on the viscosity (η). She found a good agreement between experimental data and the proposed theoretical relation

$$v = \frac{K}{\eta} + K_0 \lg \eta,$$

where K and K_0 are constants.

As the maximum rate of crystal growth of different minerals is attained at different values of supercooling, i.e., at different temperatures and different viscosities, when various minerals are crystallizing from multicomponent melts, the relations of crystal growth may even be inversely proportional to those obtained during the crystallization of these minerals from pure melts. It is possible that a mixture of augite and plagioclase presents such a case.

The increase in viscosity of melts, which lowers both the number of crystallization centers and the rate of crystal growth, occurs very rapidly during cooling. A lowering of the crystallization temperature is observed in cases of formation of eutectics. Therefore, eutectic mixtures will often crystallize with greater difficulty than pure components. Thus, for instance, diopside and anorthite, if considered separately, will crystallize readily from their melts, whereas their eutectic mixtures usually yield glass under the same conditions of cooling. Vogt considered obsidians and the microfelsitic groundmass of porphyries as eutectics which are a product of devitrification.

Some time ago it was already mentioned that the addition of certain substances to silicate melts facilitates the crystallization of minerals on the cooling of these melts — hence the name of these substances — "mineralizers". Later this term acquired a broader and rather undefined meaning. Acids of tungsten and molybdenum are especially well known in this respect. By utilization of these acids as mineralizers, artificial quartz, feldspars and nepheline were obtained. These substances were, however, not always effective with respect to silicates of calcium and magnesium. Sodium tungstate was used by Fenner as a fluxing agent to facilitate crystallization, in his famous investigation of polymorphism of silica. An interesting fact was observed during the experiments of Wallace: the devitrification of glasses of albitic and nephelinic composition was limited only to those parts of the melts close to the contact of the tungstate with the silicate; crystallization occurred at a temperature of approximately 950° C.

The amount of mineralizers necessary for effecting crystallization varies in different melts. According to Morozewicz, the addition of 1% of tungstic acid is sufficient for the crystallization of quartz, but 10 % of this acid is needed to effect crystallization of a feldspar. The action of mineralizers is here considered as a catalytic one. Chlorides are good catalyzers. D. P. Grigor'ev obtained beautiful crystals of various minerals from silicate melts on the addition of fluorite (see Part III).

The results of Bren's experiments which were published some time ago, are very instructive. He managed with difficulty to crystallize albite and anorthoclase from their melts after addition of calcium chloride. In order to obtain albite it was sufficient to maintain a glass of this composition with an admixture of calcium chloride at approximately 750°; after 90 days of heating numerous small crystals were formed, part of which were fibrous, and the substance became entirely crystalline. Anorthoclase was obtained in a similar way. A glass of the necessary composition, to which 1 % calcium chloride was added, was maintained at a temperature of approxi-
154 mately 800—900° for 25 days. The chloride vaporized, and bubbles formed in the mass. At this moment only few small spicules and crystalline tabulae were present; their nature could not be determined. Then, as the temperature fell gradually down to 600° over a period of 30 days, the glass, after losing almost all of its chlorine, was completely devitrified, forming crystals of anorthoclase which were large enough for the identification of their optical properties and the measurement of $2V$. Bren mentions that the temperature of 600° is indicative of the fact that no gas or vapor is necessary for the crystallization of an anorthoclase melt, as at this temperature chlorides are non-volatile; in addition, these chlorides had already been removed during the preceding heating.

During the crystallization of natural magmatic melts, water undoubtedly plays a major role, being dissolved in the magma at lesser or greater pressures. One of its major influences is the lowering of viscosity of the melt, thus facilitating diffusion and crystallization.

THE GLASSY STATE

As is known, the cooling of silicate melts to form glass is a very characteristic phenomenon. These melts have a special affinity both to superheating and to supercooling. Therefore, in actuality, instead of the phase boundaries upon the diagrams there should be boundary zones, which extend on either side of the boundary. Within these boundaries the system is in a metastable state. The unstable equilibrium of the metastable area cannot be maintained when passing beyond the boundaries of the latter into the area of the labile state of the system. However, during supercooling, the thermal motion of the particles decreases more and more, the system becomes less and less subject to inversions calling for such motion, and the area of the metastable state increases greatly. On such cooling the metastable supercooled liquid passes continuously into glass.

In order to obtain the silicate melt in the glassy state, a sufficiently rapid cooling is necessary. Under such conditions, the velocity of the thermal movement of the particles (temperature) falls so rapidly that crystal lattices, which demand a certain time for their construction, as well as a sufficient mobility of particles, do not have time to form.

When cooling is accompanied by crystallization, and latent heat of fusion is liberated, a horizontal or more gently sloping section appears on the cooling curve (representing the change of temperature versus time). When the melt cools to form glass, the cooling curve is smooth, and does not exhibit such a marked change in the slope as appears during crystallization. The physical properties of a solid crystalline substance differ markedly from those of the same substance in the molten state. Therefore, all curves of physical properties versus temperature show a sharp change of slope. When passing from the liquid state to the glassy state, such changes are either entirely absent, or a particularly sharp change of properties occurs here, which will be considered below.

We know more about the structure of crystalline substances than about that of glass. Various hypotheses have been proposed concerning the structure of glass. In order to evaluate this structure we are obliged to
155 recall the main features of similarity and difference between crystalline substances, glasses and liquids.

Glass is usually considered as a solid, as its mechanical properties (preservation of form, resistance to change) are more close to those of a solid body than to a liquid. This state, however, is the result of continuous cooling of molten glass. As the melt is cooled, its viscosity increases, and when it reaches about 10^{13} poise, the glass becomes brittle and elastic, independent of its chemical composition. Already at a viscosity of 10^{9} poise, the substance will only undergo mechanical deformation with great difficulty, and in this respect resembles solid crystalline bodies. The preservation of form and resistance to its change show that particles of the substance do not possess the ability of translational and rotational movement which are characteristic of liquids.

On the other hand, a characteristic feature of glass is its isotropy, i. e., it does not possess the regular lattices causing the anisotropy of crystalline substances. This regularity cannot be detected even by means of rather detailed X-ray investigations. The absence of regular atomic lattices distinguishes glass from crystals, and in this respect glass resembles a very viscous liquid. The transition from the liquid state to the crystalline state calls for a certain mobility of particles, but despite this, crystallization is possible in any liquid — even in a very viscous one. In this case, however, its rate is very slow and a long period of time is required. When the cooling melt attains the brittle state, crystallization ceases in it; however, if a certain mobility is imparted to the particles by heating the glass to softening, crystals will form in it.

Thus, we may consider glass as being constructed of disordered particles, the bonds connecting which are sufficiently strong to prevent free dislocation. The mechanical properties of glass indicate that the stability of these bonds is of the same order as in solids.

By comparing other physical properties of silicates in the crystalline and glassy state, we may arrive at an even better concept of the peculiarities of the internal structure of glasses.

In most silicate minerals the specific gravity of the crystal is higher than that of the glass of a corresponding composition. This lowering of the specific gravity of minerals during remelting is measured in the following percents:

Quartz	17	Amphiboles	7–12
Orthoclase	7.5–8	Olivine	16
Anorthite	~ 2.7	Albite	~ 9
Leucite	~ 3	Labradorite	~ 5.5
Pyroxenes	12–14		

It is evident from the presented values that the change in specific gravity is not large, and its order of magnitude for glasses and for the corresponding minerals remains the same.

A similar relation between silicate minerals and their glasses is also observed on comparing other physical properties, e.g., the refractive index. As has been mentioned, the refractive indices of glasses are even used for identifying their composition.

All the features mentioned above lead one to the conclusion that a glass already possesses some elements of structure which are characteristic of the silicates in general, despite the absence of crystal lattices, which are typical of minerals and which cause their anisotropy.

156 The X-ray investigation of glasses, aimed at determining their diffraction lattice, yields one or several broad, diffuse rings. These differ essentially from the rings of the powder pattern of a crystalline substance, which are more numerous and sharply defined. From X-ray studies of glasses one may conclude that although their structure is similar to those of liquids (being amorphous), each atom in a glass has permanent neighbors over a fairly definite distance, whereas in liquids these neighboring atoms or molecules change continuously.

The crystalline structure of silicates is characterized by the presence of silica (SiO_4) tetrahedra. These tetrahedra and the atomic bonding within them determine the major physical properties of silicates. Naturally, it may be assumed that silica tetrahedra exist in glasses too, linked together by fixed bonds, but are located in a random and not in an ordered orientation. This assumption is the basis for the hypotheses of glass structure as proposed by Zachariasen and Warren. This point of view also explains the reaction of glass to X-rays.

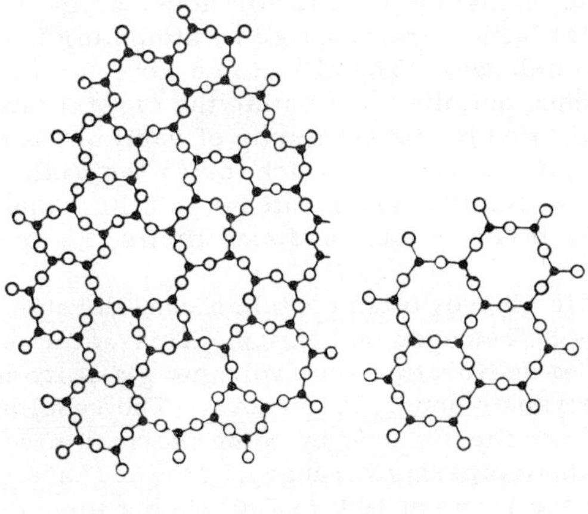

FIGURE 105. Schematic comparison of glass (left) and crystalline (right) structures

According to these concepts, glass is comprised of a three dimensional lattice of silica (SiO_4) tetrahedra, which form cells; but in contrast to the crystalline silicates these unit cells are irregular and disordered. Such a spatial cell is very difficult to represent graphically, but one may obtain a clear picture of the difference between the structure of glasses and crystals by replacing the three dimensional lattice by a two dimensional one — representing the tetradedra by triangles. From these planar schemes one may then draw an analogy to the true spatial lattices of silica (SiO_4) tetrahedra. Figure 105 compares schematically the structure of a glass and a crystal. Only the difference in the arrangement of those atoms which form the basic structure of the triangles are shown.

In silicate crystals, within regularly constructed lattices of SiO_4 tetrahedra, cations (or groups of atoms) are also regularly located, being bound to the SiO_4 tetrahedra within a common crystal lattice. In glasses,
157 the ions, atoms and groups of atoms are located in the loops of a network, thus constituting the basic structure of glass. Understandably, their arrangement is subject to that of the loops and lacks the regularity of arrangement of the sites of the crystal lattice. Thus, a uniformly random arrangement of particles is obtained, causing the isotropic properties of the entire structure.

The mentioned concept of the structure of glasses is in actuality only schematic, being similar to that of crystals as lattices of atoms. First of all, one cannot speak of the immobility of atoms in the sites of the SiO_4 tetrahedral network. Thermal oscillations should also take place here, as in the case of crystals. Therefore one may speak of the absence of translational and rotational movements of particles only in the same sense as when referring to crystals. Strictly speaking, even in crystals the possibility of translational motion cannot be denied completely, as is demonstrated by diffusion in crystals. Some similarities and differences may be noted between solid solutions (as mixed crystals are) and solutions in glass. During the formation of solid solutions, atoms (or ions) of an element in the crystal lattice are replaced by atoms (or ions) of another element (or by complexes). The new atoms (or ions) involved in the solid solution must occupy definite sites within the crystal lattice, and the replacement is facilitated by the similarity of radii of the two sets of ions. In the network of SiO_4 tetrahedra which forms the basic structure of glass, the dissolved atoms and ions are located within its loops and are thus not restricted to certain lattice sites and size limits, as are the atoms in crystals.

Natural volcanic glasses often contain a considerable amount of dissolved water, which may be removed on heating. This water is removed continuously. Attempts at dehydrating such volcanic glass from near Batum were made by D. S. Belyankin and V. P. Ivanova. The results yielded a smooth curve which was complicated only by some thermal effects caused by oxidation of iron in the temperature range of 300—350° and 600—650°. Dehydration occurs in the range of 100° to 700° C, but the bulk of the water is already liberated below 350° C. The dehydrated glass may reabsorb a certain part of the removed water. It is interesting that the refractive index of the glass initially decreases when the bulk of the water is lost, up to 350°, and then subsequently increases on further heating, thus indicating certain changes in the structure of the glass. Thus, on the one hand, the behavior of water in glass resembles that of zeolitic water, but on the other hand we note a difference in that the silicate solvent of this water undergoes structural changes after its removal. These changes should probably be considered as changes in the loops of the SiO_4 tetrahedra network.

Volcanic glasses also contain other volatile components in solution, which are liberated on heating within the temperature range 800—1,100°. The classical investigations of Bren are famous. He attached great importance to these gases as providing the explanation to the process of eruption. Table 3 presents examples of the composition of gases liberated on heating volcanic glasses.

As mentioned above, when a liquid is supercooled to form a glass, no sharp discontinuity in its properties occurs as is observed during crystallization. The supercooled, viscous liquid passes continuously into glass. However, if we construct a curve of change of viscosity versus temperature, it becomes evident that this smooth curve nevertheless shows some changes of slope. The theoretical empirical equations connecting these magnitudes are more or less in agreement with the experimental findings only over a certain range of temperatures: above the temperature which is termed the temperature of glass softening.

TABLE 3

Sample	Gas content	Composition of gas									
		in percent								in mg	
		Cl_2	HCl	SO_2	CO_2	CO	CH_4	H_2	N_2	NH_4Cl	Chlorides
1	167	—	61.11	0.55	14.30	10.55	—	12.36	11.13	5.4	755
2	397	11.76	5.88	12.16	49.58	7.64	0.5	8.00	4.00	Undetermined	
3	310	32.25	2.15	0.75	56.23	5.70	—	1.82	1.06	8	100
4	543	49.94	15.54	11.61	6.78	5.68	—	—	10.36	54	—
5	423	13.30	13.30	—	37.50	14.6	3.1	6.60	11.40	33	Undet.

Note. The gas content is given in cm at NTP. It represents the volume of gas which was liberated on the heating the glass up to explosion or melting of one kilogram of the rock. 1 — obsidian, San Andreas, Mexico; 2 — obsidian, Savan, Samoa; 3 — obsidian, Alagez, Armenia; 4 — andesitic obsidian, Krakatau 5 — pitchstone, Saxony.

The general trend of viscosity change as a function of supercooling is represented by a curve of the type shown on Figure 106. At temperature T_s the liquid and crystalline phases are in equilibrium. The viscosity of even slightly supercooled substances is not high — being of the magnitude of 10^2–10^3 poise.

On further supercooling, a rapid increase in viscosity occurs, being somewhat arrested when the viscosity attains values of the order of 10^9 poise. As already mentioned, at such a viscosity the substance has a great resistance to mechanical deformations, and in this respect resembles solid bodies. The temperature approximately corresponding to this viscosity can therefore be accepted as the temperature of incipient softening (T_f). On further cooling and increase of viscosity, when the latter attains magnitudes of about 10^{13} poise, the substance already possesses elastic properties similar to those of solid bodies. Further cooling has little influence. This occurs at temperature T_g (see Figure 106) which may be considered as the temperature of complete solidification. Thus, within the range $T_f - T_g$, the viscosity change versus temperature curve has a rather complex form. A more gentle curve is characteristic of this interval. It is termed the anomalous interval. This term is used as essential changes in the physicochemical properties of glass occur here. Such changes were carefully studied in artificial glasses. It is sufficient to mention only a few of them. Thus, a change in the heat capacity of the substance occurs within this interval. Figure 107 compares this change of heat capacity in the case of crystallization and of cooling to form glass, according to Tammann. Within the anomalous interval a change occurs in the volume of the glass, which change varies in dependence on the rate of heating. Above and below the anomalous interval the volume of glass increases almost linearily on heating. Within the anomalous interval a rather rapid increase of volume occurs, the coefficient of expansion rising from the value corresponding to low temperatures to a value which corresponds to softened glass. Above and below the anomalous interval the expansion coefficient remains quite constant. Sharp changes also occur in the value of the refractive index within the anomalous interval.

159 All these and other phenomena observed during the transition of glass from a brittle state to a viscous liquid show that some inner structural changes occur at this transition. Several hypotheses were proposed in order to explain these changes. Some investigators (A. A. Lebedev, A. Le Chatelier and others) compare such changes with polymorphous inversions in solid bodies.

During polymorphous inversions in solid bodies, a reorientation of atoms or groups of atoms occurs in the crystal. In silicates, a change in the mutual orientation of the SiO_4 tetrahedra takes place, e. g., at 575° a certain rotation of these tetrahedra occurs in the lattice of quartz, resulting in the inversion of the α-modification to β-modification, or vice versa. A. A. Lebedev indicated that in artificial glasses the above-mentioned change of inner structure occurs within the range of 500—600°, which corresponds to the α-β inversion of quartz. Although this connection was doubted by many authors, such an explanation is possible in view of the above-explained concept of the structure of glasses built of SiO_4 tetrahedra. Indeed, although the SiO_4 tetrahedra are randomly orientated in glasses, a certain mutual arrangement may, however, occur in some places in this metastable state, corresponding to an equilibrium arrangement at the given temperature. Such oriented silica tetrahedra do not form crystals; it may be stated with assurance that crystal lattices larger than 10 kX are not present in glass. This mutual orientation may, however, change to another one, as happens in true crystals. The total action of such dislocations of SiO_4 tetrahedra may result in an energy effect, which is similar to the effect of the corresponding polymorphous inversion.

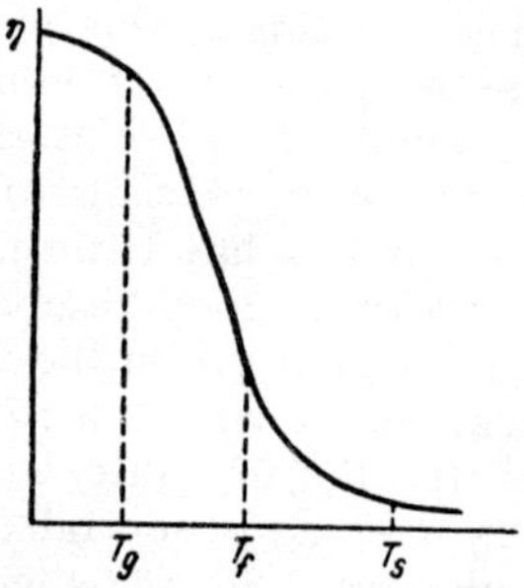

FIGURE 106. Viscosity change versus supercooling (after Botvinkin)

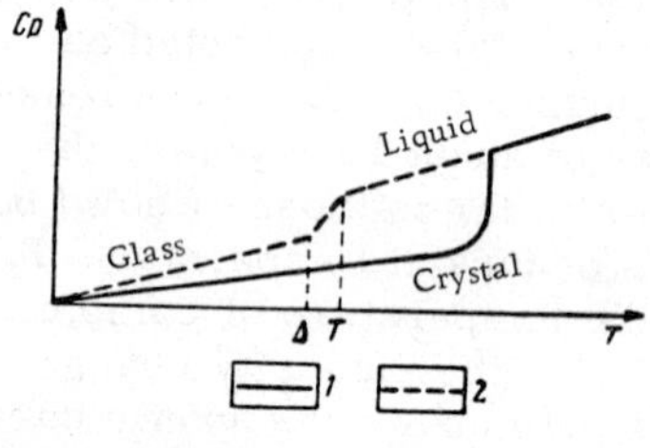

FIGURE 107. Comparison of change in specific heat of substances in the case of crystallization (1) and cooling to form glass (2) (after Tammann)

Thus, the presented concept of structure of glasses, as developed by Zachariasen and Warren, and accepted by many other authors, successfully explains several essential features of silicate glasses. From this point of view, the similarity in several physical properties of glasses and crystals of the same composition becomes more comprehensible. It also explains why glasses have a lower density, are isotropic and have a conchoidal fracture instead of cleavage planes as found in crystals. Finally, it also explains why glasses have no constant melting temperature, but exhibit a gradual transition from the hard, brittle state to a molten liquid. N. V. Belov

160 explains this by the dependence of the stability of bonds between silica tetrahedra on their mutual location. In a crystal this arrangement is equal for all tetrahedra, and when the kinetic energy of the particles' motion attains a value corresponding to the melting point, all bonds between the silica tetrahedra are broken, and the substance melts. In glass, the mutual location of silica tetrahedra is random. On heating, the bonds between them are broken gradually, starting with the weakest, and passing to the more stable ones. At the same time the glass begins to soften, and its fluidity increases until reaching the state of a true liquid. According to this point of view, the anomalous interval begins from the moment of incipient breaking of bonds between SiO_4 tetrahedra, and continues up to the moment when a sufficient number of bonds have been broken to make the glass soft; the glass then becomes able to change its shape comparatively easily when subjected to external stresses.

In all our considerations of the features of glasses in connection with their structure, we have continually spoken about SiO_4 tetrahedra. In aluminosilicate glasses, as in crystalline aluminosilicates, a part of the silicon atoms in the SiO_4 tetrahedra is probably replaced by atoms of aluminum. The fact that silicate melts rich in silica possess a pronounced ability to form glasses shows that in order to obtain aluminosilicate glasses, the replacement of silicon by aluminum should not exceed a certain limit. Feldspars illustrate this excellently. Orthoclase and albite, in which only one quarter of the silicon atoms is replaced by aluminum, usually yield glasses on the cooling of their melts, whereas anorthite, where a half of the silicon atoms is replaced by aluminum, crystallizes readily.

DEVITRIFICATION

Glasses are metastable systems, which tend to pass into a stable crystalline state. They owe their existence to the fact that the transition from the metastable to the stable state occurs at an almost "infinitesimal" speed. As already mentioned, natural volcanic glasses may be preserved for several geological periods. In the majority of cases, however, ancient glasses are inverted to fine crystalline aggregates. This inversion is facilitated by pressure. This is obvious, as the inversion of glass to crystalline substance is accompanied by a decrease in volume.

The inversion of glass to a crystalline aggregate is termed devitrification. This term is more used in industry than in geology. In industry one speaks of the devitrification of industrial glasses on heating them to softening.

Petrographers, when speaking of devitrification, usually understand this term to refer to changes which occur at a later stage, in already solidified and cooled glasses, and which are caused both by the prolonged existence of the glass, which is a metastable system, and by the action of other factors, especially of orogenic processes. We shall use the term in this connotation. In order to describe the inversion of even very soft glass to a crystalline aggregate at high temperature, during natural cooling, we shall preserve the term "crystallization of glass".

Various processes occur in natural volcanic glasses during their alteration and devitrification.

161 1. During the cooling of glass physical changes already occur, which are connected with a decrease in the volume, and cracks appear, which are often loci for the inception of a new process. Glasses attain a specific perlitic texture. The term "perlite" was used more than a century and a half ago to describe silica-rich volcanic glasses, which form rounded pieces on crushing — resembling pearls (perlstein; our term "perlite" was used by Bedan in 1822). Their form is due to the concentric tensional cracks in the glass. Marekanite from the Marekanka River on the Okhotsk coast is a well known example of perlite. The perlitic texture is specific to glasses rich in silica and is characteristic of them only.

2. Volcanic glasses undergo hydration — they absorb water. One cannot always determine whether the water is primary or secondary. Opinions differ as to the origin of the water in pitchstones. Whereas some consider it as being of primary origin, others consider it as secondary. Pitchstone is an acid glass with a 4—5 % water content, thus differing from obsidian, which usually contains less than 1 % water. One may speak with greater assurance about the secondary hydration of basic basalt glasses.

3. Crystallization proper of glasses consists in its transformation to fine crystal aggregates, which are often undetectable even under the microscope. Here two cases may arise.

A. The glass disintegrates, inverting to a crystalline aggregate of minerals, which are found as primary constituents in igneous rocks. One is not always able to distinguish at which stage of the rock formation such crystallization takes place, and whether it is a specific form of crystallization associated with the cooling of very viscous glasses, or whether it is a much later secondary process of devitrification.

B. The glass disintegrates into a fine aggregate of minerals, which are not characteristic of igneous rocks, but are rather products of metamorphic processes. This is probably a secondary crystallization. It may either involve the rocks regionally, or may be of a local extent, depending on specific conditions (e. g., zeolitization of glasses). The regional extent of such forms of glass disintegration is a characteristic indicator of anchimetamorphic paleotypal rocks.

We shall only consider the inversions of the extreme types of glasses: the rhyolotic-acid, and the basaltic-basic ones.

Microfelsite is a characteristic product of devitrification of acid glasses. Its composition may be represented as a mixture of 3/4 alkali feldspar and approximately 1/4 quartz; sometimes the content of quartz is even higher.

In certain pitchstones, e. g., in the famous pitchstone of Arran Island (Scotland), samples of which are found in almost every petrographic collection, a certain reaction to polarized light is observed. On examining a thin section of this pitchstone under crossed nicols, we may see that, unlike glass, only narrow areas surrounding crystallites disseminated in the rock remain completely dark, whereas the mass as a whole is not characterized by complete extinction. On removing the analyzer, a certain turbidity is observed in the rock.

Under crossed nicols, microfelsite shows a distinct reaction to polarized light, but individual crystalline particles, being several times smaller than the thin section, and having similar, low (quartz, feldspar) refractive indices, are indistinguishable. The larger these particles are, the clearer
162 becomes the texture of microfelsite. It becomes evident that we deal with

a fine aggregate of yet indeterminable minerals, the individual crystals of which are not yet distinguishable. Microfelsite passes into felsite, in which the smallest grains have no clearly defined borders; the feldspar is sometimes fibrous and forms bundles; quartz occurs as irregular grains. On further increase of grain size, felsite passes into a fine aggregate of irregular grains of minerals (already distinguishable as quartz and feldspar under the microscope), which differ sharply in their structure from microlitic textures formed during the primary crystallization of viscous rock melts. Spherulitic microstructures are often associated with microfelsitic texture, but the growth of well-developed spherulites seems, in all events, to be conditioned by a sufficiently viscous and not too brittle glass, enabling the fibers of the minerals to grow with a sufficient velocity, thus forming spherulites.

The inversion of glass to microfelsite, and the varying fineness of the microfelsite structure, are sometimes unequally distributed throughout the mass of the former glass. The inversion to microfelsite along perlitic cracks is rather typical. In older perlites such cracks are usually surrounded by microfelsitic rims. It has already been mentioned that perlitic cracks sometimes serve as points of initial growth of spherulites, but this is a relatively rare case, whereas the inversion of glass to microfelsite is more often observed. In all probability, this process is subsequent to the formation of spherulites. The irregularity in the texture of microfelsite often allows one to distinguish under crossed nicols the initial flow structure of the glass which passed into microfelsite, with its characteristic bends in streams of viscous liquid, etc.

Glass is, as mentioned, a metastable system, and therefore in ancient rocks rich in silica, glass passes into microfelsite and felsite. Although the quartz and feldspar crystals comprising the last mentioned are very small, this rock type consists of crystalline phases and is therefore a stable system. This stability is visually revealed by halleflinta, which are included in Pre-Cambrian complexes and have a fine felsitic texture.

The alteration of basaltic glasses follows a different course owing to their different composition.

Peacock and Fuller distinguish thus between basalt glasses. A completely fresh glass is termed a sideromelane. It is usually observed in quenched parts of the flows. Fuller, in particular, observed this type of glass near the Columbia River, at the end of a basaltic flow which poured into a lake. Sideromelane is a dark, homogeneous, fresh glass which resembles bottle glass, especially when powdered. Its hardness is approximately 5, and its specific gravity without the bubbles is 2.7. The refractive index of the examined sample was found to be 1.583. The term sideromelane was used by Sartorius von Waltershausen to describe fragments of glass found by him in palagonitic tuffs of Sicily and Iceland. Rosenbusch called such glass a hyalobasalt.

In chilled parts of dikes and in flows which cooled too slowly for sideromelane to be formed, the glass becomes turbid due to the separation of magnetite dust, or magnetite globulites, cumulites and crystals. This glass is termed by the authors as tachylyte. The term "tachylyte" was used by Breithaupt in 1826 without a microscopic description. Fuller (1932) cites an interesting example of inversion of sideromelane to tachylyte around crystals of labradorite and pyroxene, which very much resemble what the

author had observed in the "shikhlunite" glassy varieties of rocks: a dark
163 (in this case opaque) margin appears around each microlite in the transparent glass. The margin sometimes has offshoots 0.003 to 0.01 mm wide and up to 0.1 mm in length. These samples were taken from the intermediate layer between the completely glassy (sideromelanic) crust and the inner parts of pillow lava, 4 cm from the surface.

Palagonite is a hydration product of basaltic glass (sideromelane). It is a turbid transparent glass, ranging in color from brown to yellow. Sometimes one may succeed in observing the inversion of sideromelane to palagonite on minute fragments of tuffs. This is accompanied by the lowering of the refractive index, e.g., from 1.592 to 1.47. A palagonite gel is obtained which may subsequently become turbid, and signs of devitrification appear in it — "fibrous palagonite". Sometimes the refractive index drops to 1.45; this last-mentioned feature is dependent on the water content.

It is assumed that the water required for palagonitization is derived from external sources, namely the water from basins into which the lava was poured, or glaciers (in Iceland).

Chlorophaeite is an isotropic substance which fills amygdules in basalts. It becomes optically active on devitrification. Fresh chlorophaeite has a greenish color, but darkens with the passage of time, probably due to oxidation (hence the term: chloros — green, phaeos — brown); the fracture is conchoidal, the luster resinous or dull, the streak brown, the hardness — 1.5—3 and the specific gravity — 1.8—2.25; chlorophaeite does not melt, producing a magnetic mass; the composition is approximately $(Fe, Al)_2O_3 \cdot 2(Mg, Fe, Ca)O \cdot 4SiO_2 + 10H_2O$. Chlorophaeite is probably formed due to the presence of magmatic water in the lava itself. Under the microscope chlorophaeite is bright-green, olive-green, orange-brown (in most cases) and reddish; it is isotropic, or if birefringent, it becomes fibrous due to chloritic fibers. The refractive index is 1.50—1.54. It appears not only in amygdules but also in the groundmass.

MAGMA TEMPERATURE

We shall consider one additional feature of the natural silicate melts (the magmas) — the temperatures of magmatic processes. Firstly, we shall consider the direct measurements of these temperatures, as is possible in the case of lavas (magma poured out at the surface).

Direct measurements of the temperatures of lavas in volcanoes are available for basaltic and similar basic magmas, the eruptions of which are distinguished by a less violent character.

The older temperature measurements of lavas of Stromboli, Etna and Vesuvius yield values of 1,000—1,100°, and on Kilauea — approximately 1,200°. Perret (1924) determined the temperatures of the Vesuvius lavas as being within the range of 1,000° to slightly above 1,200°. According to his data, the temperature in the upper layers (30—50 cm) of the lava lake of Kilauea was 1,050° C.

Jaggar (1917), using Seger cones which he placed on an iron pipe and lowered into the lava of the Kulauea lake, found that the temperature of lava varies with depth. At the surface it is approximately 1,000°, at a depth

of 1 m it decreases to 900°, and at a depth of 13 m, i. e., at the lake's bottom, it increases to 1,170°. It is assumed that the increase in temperature at the surface is associated with reactions of oxidation by atmospheric oxygen. Jaggar assumes that the temperature in the depths of the lava lake is actually lower than in the investigated part; individual measurements give values of approximately 750°. The temperatures of the Kilauea lavas fluctuate within wide limits (down to 400°), depending on depth,
164 liberation of gases and reactions at the surface. Jaggar assumes the basaltic magma is already comparatively liquid at 750—850°. Apparently the presence of volatile components is of influence here.

An idea on the order of magnitude of temperatures of magmatic silicate melts may naturally be obtained from the data on melting temperatures of rock-forming minerals; these temperatures fluctuate within the limits of 950° to 1,800° — usually 1,000—1,500°. One should, however, bear in mind the fact that when considering magmas we deal with phenomena of solid solutions, eutectics and, lastly, the presence of volatiles, which lower the melting temperature and escape on crystallization. During the formation of igneous rocks from cooling magma, the crystallization process proceeds over a certain range of temperatures. This range does not and cannot coincide with the interval of melting of rocks, from which the volatile mineralizers were removed. Basalts melt within a range of 985—1,260°, granites —within a range of 1,215—1,260°.

The polymorphous inversions of certain rock-forming minerals and their decomposition constitute a convenient geological thermometer, as their corresponding temperatures have been determined experimentally. Thus, in particular, the inversions of silica modifications are widely used: cristobalite-tridymite-α-quartz-β-quartz. This system is considered further on. Here we shall only mention the application of the obtained data to the determination of magma temperatures.

As no crystalline modifications will separate from the melt at a temperature above that of the field of its stability, the presence of a certain modification may assist one in arriving at the temperature of crystallization of the rocks containing this modification.

Quartz may only separate directly from melts at atmospheric pressure at temperatures below 870°; on cooling it is slowly inverted at 573° from α-quartz to β-quartz. Tridymite may crystallize at temperatures below 1,470°.

However, the high-temperature modification may appear in the stability field of the low-temperature one — as a transitional unstable form. Thus, at a temperature below 870° tridymite may form instead of quartz, from a melt to which fluxing agents were added, and be slowly inverted to quartz. On the other hand, if quartz is heated above 870°, it may be preserved for a long time, not being inverted to tridymite and cristobalite. In nature tridymite and cristobalite are found in the groundmass of rocks or in pores of this groundmass — mostly in acid volcanic rocks which contain quartz phenocrysts. This indicates that cristobalite and tridymite were in this case formed below 870°, and possibly far below this temperature.

Quartz is found as phenocrysts of high-temperature α-quartz in rhyolites, dacites, etc. Hence, one may conclude that the rocks which contain these phenocrysts were crystallized at temperatures probably lower than 870° and higher than 575°. As these phenocrysts were formed at depth, under a certain pressure, the inversion temperatures should be somewhat higher, although their order remains close to the mentioned figures.

As already mentioned, the possibility does exist that even basaltic magma remains liquid at a temperature of about 870° due to the presence of mineralizers†.

165 In pegmatites β-quartz is found together with α-quartz. Therefore, it is possible that temperatures were lower than 573° in the final stages of pegmatite formation. The bulk of the quartz, which often forms graphic intergrowths with feldspars, has apparently crystallized at a temperature above the α - β quartz inversion. This is indicated by the inversion of quartz ichthyoglypts, the outlines of which are the result of interaction of the crystalline lattices of α-quartz and feldspar to form aggregates of β-quartz grains.

[As shown by experiments by Tuttle †† and Bowen, at a pressure of 400 atm, granite begins to melt at a temperature of 640° in the presence of water. A further increase of pressure has little influence on the change of the melting temperature; the mentioned value probably approaches the lower limit of the residual melt of a granitic magma. Granite may be completely fused at this temperature in the presence of 9 % of water. Under the mentioned conditions, a water content of 2 % will cause complete melting at 830°.]

In volcanic rocks hornblende is represented by the brown variety, in plutonic rocks — by the green one. Green hornblende passes into brown on heating in an inert gas (nitrogen) at approximately 750°. This may be considered as a proof that volcanic rocks were crystallized at a temperature above 750°. In certain volcanic rocks the brown hornblende undergoes dissociation and becomes opaque due to the separation of iron oxides. Such a dissociation (opacitization) is achieved artificially on heating above 1,050°.

Important indications on the magma's temperature may be obtained from changes in biotite, which dissociates on heating in an inert atmosphere to temperatures above 850°. Undissociated crystals of biotite in rhyolites and other acid rocks indicate that their crystallization probably occurred below this temperature.

Valuable indications on magma temperatures are presented by changes which may be observed in inclusions of foreign rocks within the magma. Sometimes it is observed that the material constituting the inclusions has been fused to form glass (so-called buchite, etc). The strongest alterations are observed in basalts. Here, the melting of shales and argillaceous sands is observed, and sometimes also of biotite, orthoclase and even quartz; inclusions of granite fused to a vesicular glass have been described. One

† [According to data of B. I. Piip, the temperature of basaltic lavas of the Klyuchevskaya volcano reached 1,200° during eruption. Indirect measurements yielded a temperature of the order of 1,100°, and the limit of plasticity fluctuated within the range of 980—1,040°. For the 1938 eruption he cites data of isolated measurements by Ivanov — a temperature of 870°, and even of 690° for marginal parts. B. I. Piip explains this by the increased content of dissolved gases in this flow. However, there are reasons for believing that such isolated measurements are erroneous, and that the temperature of crystallization of basaltic lavas (but not of intrusive rocks) does not fall below 1,000°].

†† See Tuttle in the list of references.

should, however, bear in mind that this is not the melting of individual minerals on being heated to their melting points, but is due to reaction between various minerals, or between minerals and the magma. The appearance of such minerals as mullite in sandstone inclusions constitute a spectacular example of this phenomenon.

In some cases, the inclusions of granite or arkose in basalts or diabases do not show any signs of melting, whereas in other cases they are more or less remelted. Experiments show that the melting of granite minerals occurs at temperatures of 1,025-1,100°, a fact which is used for assessing the temperature of magmas which contain altered or unaltered inclusions of these rocks†.

Sometimes tridymite or cristobalite appear in quartzose or quartz-rich
166 inclusions. Lacroix mentions that the inversion to tridymite and cristobalite occurs more often in trachytic rocks than in basalts. Sometimes the inversion to tridymite involves the quartz grains completely, and the aggregate has an appearance similar to that obtained in laboratories; sometimes glass appears between the crystals tridymite. This probably indicates a temperature higher than 870°, although tridymite and cristobalite may appear as unstable transitional forms also at lower (than 870°) temperatures. The inversion of wollastonite to pseudowollastonite at temperatures above 1,180° allows one to consider that the lavas containing limestone fragments altered to wollastonite had a temperature below 1,180°. Pseudowollastonite is not found in nature.

Marble inclusions may also serve as a certain indicator of magma temperature. Calcite at normal pressure undergoes dissociation at 910°, but with increased pressure dissociation occurs at higher temperature; a pressure of 20 atm raises this temperature to 1,110°.

In volcanic rocks oversaturated in silica (rhyolites, dacites) a micropoikilitic texture of the groundmass is often found. Geijer explained it as a result of tridymite inversion to quartz, which is accompanied by accumulative recrystallization. He found quartz pseudomorphs after tabular crystals — probably tridymite. The wide distribution of micropoikilitic textures leads one to the idea that the crystallization of the groundmass of these rocks probably occurs at temperatures above 870°, or, taking into account the possible appearance of tridymite as a transitional metastable form, somewhat below this temperature.

Owing to the higher pressure at depth, the magma contains a great amount of fusible or volatile components, which lower the crystallization temperature. As we shall see further on, direct attempts at melting granite with water in an autoclave showed that granite melts remain liquid at increasingly lower temperatures, as the water content is increased. Dry granite only begins to melt at 900—950°, whereas with a water content of 6 % it becomes liquid at 700°.

One may also judge the temperature of formation of plutonic rocks by making use of the temperatures of inversion of one modification to another. One should, however, bear in mind that these temperatures of inversion are considerably influenced by pressure; thus the inversion temperature of $\beta \rightleftarrows \alpha$-quartz at a depth of 10 km and with a hydrostatic pressure of the

† [Thus, in studying the partly fused xenoliths of granites in the leucite basalts of the Gauss mountains in the Antarctic, O. S. Vyalov and V. S. Sobolev (1958) arrived at the conclusion that the temperature of the lavas was within the range 1,150—1,200°].

overlying rocks increases up to 644°. At such depths the quartz-tridymite inversion temperature is raised by approximately 100° C.

After considering the available data, the following conclusion may be drawn.

Direct data on temperatures of lavas are rather scanty and indefinite. Temperature changes of basaltic lavas of Kilauea yield values ranging between 750 (?) and 1,200°; similar temperature values were recorded in other volcanoes (Vesuvius, Kluchevskaya). Rhyolitic lavas probably have lower temperatures.

The melting temperatures of minerals only allow us to assess the maximum temperatures of a magma. The inversion and decomposition of minerals, among which the most important is the transition of one form to another (in particular the inversion or decomposition of hornblendes and biotites), indicate that in acid volcanic rocks the highest temperature is
167 approximately 900°. A large number of these rocks crystallize below 750°. The presence of volatiles plays as great a role here as in basaltic lavas. The action of lavas on inclusions also shows that temperatures of lavas only rarely reach 1,250°, and are usually below 1,150°.

In general it may be considered that basaltic magmas only rarely attain a temperature of approximately 1,250°, very often — below 1,000°, sometimes — even below 850—900°. Rhyolitic magmas have lower temperatures, rarely exceeding 850—900°. Usually the temperature is lower than that of the decomposition of biotite (850°) and even lower than the temperature of inversion of normal hornblende to basaltic hornblende (750°). It is possible that in the majority of cases the acid magma has a temperature of 850° or even somewhat less.

After considering the problem of determining the temperature of magma, one can arrive at an essential conclusion concerning the phenomenon of lava crystallization.

Basic basaltic lavas remain liquid and relatively fluid down to temperatures at which acid lavas are already very viscous. Thus, on cooling, acid lavas lose their fluidity earlier than the basic ones, but the crystallization of the latter occurs at higher temperatures. Acid lavas crystallize at high viscosities. These features are clearly expressed in the different microstructures of acid and basic rocks.

UNARY SYSTEMS

Only two unary systems will be considered in this chapter, both of which are of utmost importance for the understanding of the processes of mineral formation in igneous rocks.

Both these systems have sufficient available experimental data or diagrams based on thermodynamic data. The polymorphous inversions of minerals, the phase relations of which are as yet insufficiently determined, are considered below in the description of binary systems.

THE SYSTEM SiO_2

The system SiO_2, being of exceptional importance for the understanding of processes of mineral formation, has been thoroughly investigated despite the technical difficulties involved. The first detailed investigation of this system was carried out by Fenner (1913), who constructed an equilibrium diagram for the system at low pressures. Later some non-essential amendments were made during experimentation at high pressures.

Figure 108 shows the system SiO_2 at low pressures. One should bear in mind that the pressures of the solid-gas and liquid-gas curves are not shown to scale as, in actuality, they are minute and cannot be measured over the range of low temperatures. Their construction is of purely methodical interest, as it enables us to consider the metastable equilibrium.

As is evident from the diagram, there are four polymorphous modifications within the field of stable equilibrium; α-cristobalite, α-tridymite, α-quartz and β-quartz. Inversion temperatures at low pressures are shown on the diagram.

According to the character of the inversions the mentioned points are of two types. The cristobalite-tridymite and tridymite-quartz inversions are accompanied by the complete destruction of the crystal lattice of the initial modifications. Inversion is extremely slow and is difficult to achieve under laboratory conditions without the addition of special admixtures. The inversions are sluggish, both on heating and on cooling. In practice, it is equally difficult to obtain crystals from the melt, which usually cools to form a glass.

The α-quartz (hexagonal modification) — β-quartz (trigonal modification) inversion is entirely different. No destruction of the crystal lattice is involved. An instantaneous rearrangement of the atoms is all that takes

169 place. The difference in inversion temperatures on heating and cooling does not usually exceed several tenths of a degree. Deviations from the inversion temperature of several degrees and more (up to 36°C according to Keith and Tuttle) are associated with the presence of considerable admixtures of lithium (0.1—0.01 %), i. e., the system is in this case binary.

The considerable delay in the inversions of the high temperature phases readily allows for their preservation in the field of unstable equilibrium. This only applies to cristobalite and tridymite. The high temperature quartz is never actually found and pseudomorphs after it are all that remain.

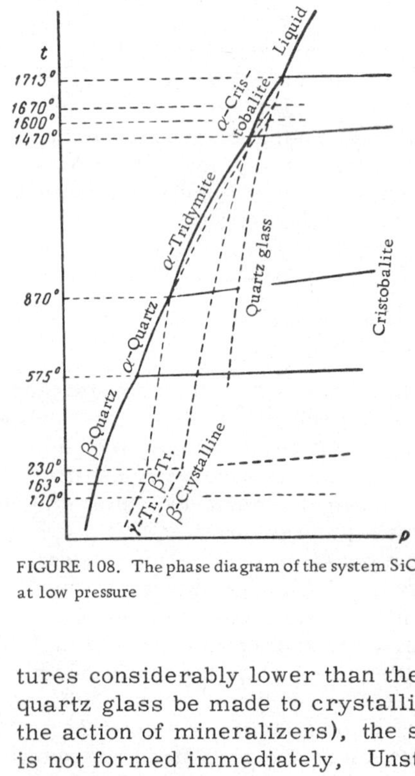

FIGURE 108. The phase diagram of the system SiO_2 at low pressure

As is evident from the diagram, new polymorphous inversions occur in the field of unstable equilibrium; α-cristobalite-β-cristobatite and α-tridymite-β-tridymite (probably followed by γ-tridymite). As these inversions are accompanied by the destruction of the crystal lattice, they proceed without delays and the equilibria are almost enantiotropic despite the metastable conditions, as in the case of quartz. Thus, at low temperatures we usually do not observe cubic cristobalite or hexagonal tridymite but the corresponding orthorhombic modifications.

The mentioned relations cause the inversions of the system SiO_2 to occur in accordance with Ostwald's "step" rule. In reality, the forms of low temperature modifications of cristobalite and tridymite, which do not possess their own stability fields, already comply with this rule. In addition, the formation of high temperature crystalline phases may be observed here during the crystallization of glass at temperatures considerably lower than their stability temperatures. Thus, should quartz glass be made to crystallize at a temperature of 300°C (e. g., under the action of mineralizers), the stable modification in this field — β-quartz — is not formed immediately, Unstable modifications — β-cristobalite and β-tridymite — are the first to form, having lower vapor pressures and correspondingly lower thermodynamic potentials than quartz glass. The former, after forming at so low a temperature, may be preserved in nature in the case of interrupted equilibrium and will not invert to quartz. This case (and especially so for cristobalite) has often been mentioned in literature. There is a basis for considering that the sub-crystalline individua in colloidal opal are also cristobalite and not quartz.

In the section on magma temperatures we have already considered the significance of crystallization of the various silica modifications as geo-
170 logical thermometers. According to the above mentioned, we should be very wary of accepting such conclusions. The data on the upper temperature limit are more reliable, as the formation of low temperature phases

above the temperature of inversion has never been observed. However, a new complication arises here — the influence of pressure. The specific gravity of tridymite differs from that of quartz by 15.7 % (Barth 1952, 1956). This difference is quite considerable and the equilibrium curve (which is almost a straight line) has a steep slope in the field of high pressures (see Figure 108). The difference in specific gravity between cristobalite and tridymite is considerably less (0.5 %). This difference is also rather small in the case of the melt and cristobalite. If one constructs the corresponding curves on the basis of calculations, as shown on Figure 109, I, the tridymite fields wedges out at comparatively low pressures (of the order of 1,500 atms). Further on, due to a similar slope of the quartz — cristobalite curve, the cristobalite field also wedges out.

Experimental investigations of the corresponding inversions at high pressures in the presence of water vapor (which cannot influence the inversion point as water is not included in the solid phases) were carried out by I. A. Ostrovskii et al. in 1959 (Figure 109, II). The invariant point cristobalite-tridymite-quartz, is characterized by a temperature of 1,190°C and a pressure of 1,430 atms. The cristobalite-quartz inversion curve can be traced up to a temperature of 1,360°C. Thus, on raising the pressure, α-quartz may form at temperatures considerably higher than 870°C (inversion to tridymite at low pressures). This should be taken into account when speculating on the temperature of crystallization of quartz in igneous rocks.

The difference in the specific gravities of α-quartz and β-quartz is only 4.9 %. Figure 109 shows the corresponding equilibrium line as determined experimentally. Although the slope of this line is comparatively gentle, the inversion temperature is shifted markedly. Thus, for instance, the inversion temperature at 5,000 atms is 704°C, corresponding to a shift of 131°C, whereas at 10,000 atms it is 815°C, corresponding to a shift of 242°C (Yoder, 1950). However, at 1,000 atms (599° C) and even at 2,000
171 atms (626°C) the shift of the inversion point is comparatively small and does not essentially influence mineralogical thermometry.

Investigations in the field of very high pressures, which were recently carried out (Coes, 1958, MacDonald, 1956, Khitarov, 1957) showed that the silica system has another specific modification called coesite. It has a specific gravity of 3.01 and is similar in optical properties to apatite (Ng = 1.597; Np = 1.594).

Figure 110 shows the inversion curve quartz-coesite.

It should, however, be mentioned that during experimentation under temperature and pressure conditions approximating those of the curve, quartz was found to coexist with coesite in minute quantities almost everywhere; pure coesite can only be obtained at much higher pressures (e. g., 30,000 atms at a temperature of 580°C).

The mentioned results are of great interest when speculating on the state of silica at great depths in various geophysical correlations. The above-mentioned authors indicate that at higher pressures, according to Thompson, an even denser modification of SiO_2 should be anticipated, possessing a structure of the rutile type†. The crystalline structure of coesite has not as yet been investigated, but its specific gravity is too low for that of the

† [This forecast was subsequently affirmed by the synthesis of stishovite possessing a rutile structure and having a specific gravity of 4.35 (Stishov, 1961)].

rutile structure. The authors have put forward interesting ideas which suggest that even forsterite should decompose into separate oxides at high pressures, thus causing a further increase in density. However, this reaction remains uncertain as the magnitudes of density of phases under these conditions are not known with any degree of certainty. In the absence of the above-mentioned decomposition, the appearance of surfaces separating rocks of differing densities can hardly be explained by the phase inversion quartz-coesite alone, as it is difficult to envisage rocks containing free silica at great depths.

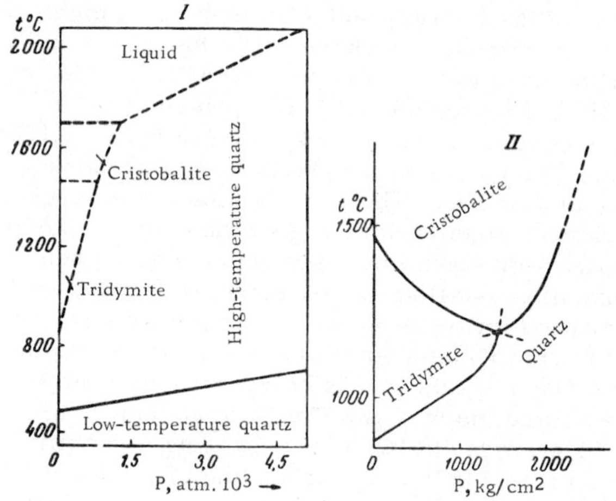

FIGURE 109. Schematic diagram of the system SiO_2 at very high pressures. I - general scheme; II — detail of diagram I (according to Ostrovskii, et al.)

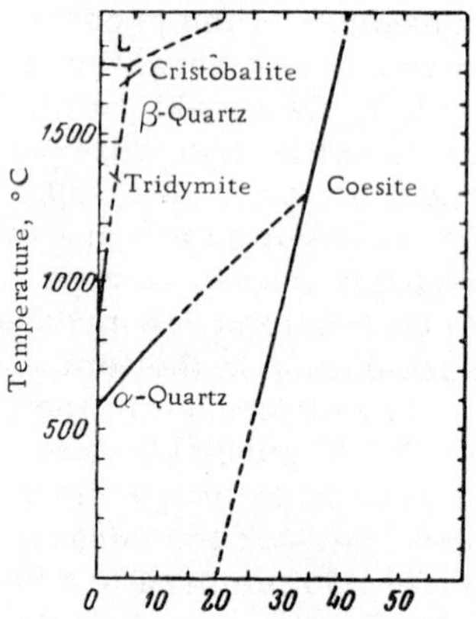

FIGURE 110. The system SiO_2 at high pressures. Equilibrium curve quartz-coesite

The possibility of finding coesite among minerals within the field of our studies is most interesting. The search which was firstly concentrated in jadeitic rocks and eclogites has as yet proved unsuccessful, indicating that the corresponding pressure of 20,000—30,000 atms is usually not attained in the processes of mineral formation. It is true that higher pressures were undoubtedly operative during the formation of diamonds, but free silica is absent in kimberlite, and one may assume that xenoliths of quartz-bearing rocks, including some eclogites, were enclosed within the rock after a sharp drop in pressure, and hence their formation is subsequent to that of the diamonds†.

† [Subsequent to the appearance of this book in Russia both coesite and stishovite were discovered in nature at Meteor Crater, Arizona in 1960].

In contrast to SiO_2, native carbon is a rather rare mineral in igneous rocks. However, taking into account the specific importance of the phase inversions of this system for considering the upper limit of pressure in mineral formations, as well as the magmatic origin of the most precious mineral — diamond — we include the system C in our review.

172 The diamond-graphite relationship has long been of interest to physicists and mineralogists, both from the theoretical and the purely practical points of view. The comparison of the specific gravities of graphite (2.23) and diamond (3.51) has naturally led to the conclusion that the diamond is the stable phase at high pressures.

The first attempt at constructing the graphite-diamond equilibrium diagram on the basis of thermodynamic data was made by Leipunski. Later, calculations of this kind were made by several other authors. As the temperature of formation of diamonds in nature undoubtedly exceeds 1,000°C, the corresponding pressures should be of the order of 40,000 atms. Despite the mentioned consideration many authors hold the view that the diamonds crystallized in nature at much lower pressure, forming a metastable phase under conditions which favor its formation. These specific conditions were sought after in the attempts at synthesizing diamonds in the laboratory. However, all these experiments ended in failure and it was only recently, when the necessary high pressures became technically attainable, that artificial diamonds were created under conditions which agree with the data of the thermodynamic calculations. Although the equilibrium curve graphite-diamond is as yet not sufficiently accurately detailed in all its parts, the general character of the diagram may already be clearly discerned (Figure 111).

Today there is hardly anyone who seriously doubts that diamonds were crystallized as a stable phase at corresponding high pressures in kimberlites, or more precisely the magma which yielded kimberlites. The particular rarity of this mineral is undoubtedly connected with the fact that only in rare cases are we able to observe mineral parageneses which formed under such conditions. As shown further on (see p. 192), the pressures during the formation of metamorphic rocks in parts of the Earth's crust which are usually subject to our investigation have often reached magnitudes of the order of 15,000 atms and more, but have always remained below the values necessary for the formation of coesite, i.e., below 20,000—25,000 atms. Pressures of the order of 40,000 atms and more, which are probably normal for very deepseated hearths of ultrabasic magma, are only in rare cases preserved as the magma rises, thus resulting in the formation of the kimberlite pipes.

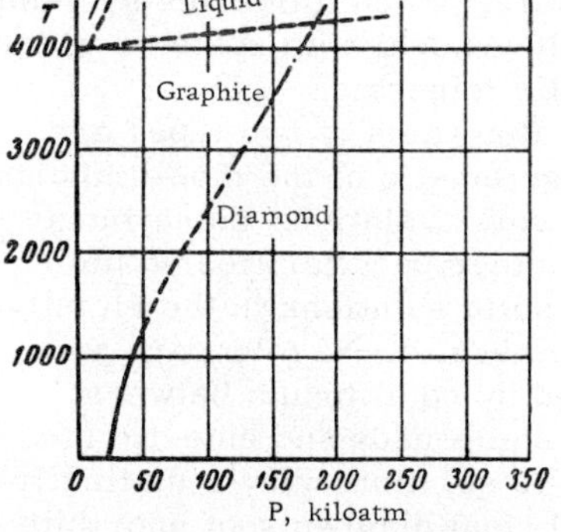

FIGURE 111. The equilibrium curve graphite-diamond

173 CHAPTER IX

BINARY SYSTEMS

GENERAL REVIEW OF BINARY SYSTEMS**

This chapter deals with abundant experimental data on melting diagrams of binary systems which are of great interest for the study of crystallization processes of magmas. Both strictly binary systems and pseudobinary systems (in certain areas) are considered here. Together with well investigated systems we also present such systems for which only preliminary data are available.

According to the order selected by the authors, all problems connected with volatile components will be considered in detail in the third part of this book. However, it has proved necessary to mention the most important data obtained during the study of systems containing water, especially in cases where water is not included in the composition of the solid phases and serves merely as an environment in which one is able to observe phase equilibria below the melting curve.

Taking into account the trend of this book, the following order of representation of material has been selected. Firstly we consider systems which illustrate relations between salic minerals, starting with the systems most important to petrography — plagioclases and Na and K feldspars. The system SiO_2—Al_2O_3 is conventionally also included here. Then follow the systems which correspond to femic minerals, and lastly we deal with systems which illustrate the relations between femic and salic minerals.

In the case of salic minerals, equilibrium diagrams of two types are principally observed. The first type is characteristic of the most important rock-forming minerals showing continuous solid solutions sometimes complicated by polymorphous inversions and other peculiarities on the melting curves, and by the decomposition of solid solutions in the vicinity of the solidus. Here are included the plagioclases, K-Na feldspars and nephelines. The second type is characterized by equilibrium between aluminosilicates and SiO_2 in the presence of compounds and eutectic (and sometimes even peritectic) points, where solid solutions play a distinctly minor role. The second type includes equilibrium diagrams of anorthite with several other aluminosilicates (orthoclase, leucite and nepheline).

In the case of the femic minerals, solid solutions of magnesium and iron components are of specially great interest. These solid solutions characterize the absence of decomposition, and are best illustrated by the system Fe_2SiO_4—Mg_2SiO_4. In the case of other minerals the systems are more complex, being binary. However, the general regularity — the

174 enrichment of the solid phase in the magnesium component as compared to the melt, with the increase of the iron content of the femic minerals as crystallization proceeds — is absolutely certain.

In contrast to the above-mentioned series, solid solutions, in which isomorphous replacement of calcium by magnesium and bivalent irons occurs, are only stable at very high temperatures. The temperature of decomposition in these cases is so high that it usually exceeds that of crystallization of natural magmas. This explains the essential difference between natural minerals and products of crystallization of dry melts.

Solid solutions of other types, which are connected with heterovalent isomorphism, play a considerable role in femic minerals. These solid solutions have only been scantily investigated experimentally, but even the available data are of essential interest to petrography.

The diagrams of systems involving relations between femic and salic minerals are usually related to the eutectic type, but are often complicated by the appearance of phases foreign to the given binary system. In these cases, when the phases crystallize over a small range of concentrations, the system may be considered as being binary with the corresponding reservations. In more complex cases, pseudobinary and multicomponent systems are constructed. Only the most important of these are given in this chapter.

Among the peculiarities of the cited diagrams we draw attention to the marked shift of the eutectic points in the systems femic mineral-albite, as compared to the systems femic mineral-anorthite. A considerable shift, but of a lower order, is also observed during the replacement of the magnesium femic mineral by its iron counterpart in the same type of diagram.

At the end of this chapter in the section devoted to experimental data on phenomena of immiscibility in silicate melts, the general order is broken, and we consider this problem in detail, including data on ternary and multicomponent systems.

The presentation of the material is accompanied by the principal data concerning the physicochemical interpretation of crystallization of the corresponding minerals in magmatic rocks and by a brief review of some general problems of petrogenesis.

SYSTEMS ILLUSTRATING THE RELATIONS BETWEEN SALIC MINERALS

Plagioclases*

The plagioclase series is the best investigated of all the isomorphous mineral series. It fully merits this attention as it is of utmost importance in the study of petrogenesis. Plagioclases are studied both analytically (by considering the relations in rocks) and synthetically — by Day, Iddings and Allen (1905) using the thermal method and by Bowen (1913) using the method of quenching.

The melting curves are of Roozeboom's Type I of mixed crystals (Figure 112). Table 4 presents the physical properties of various members of the plagioclase series obtained artificially.

Knowing the compositions of the liquid and crystalline substances existing together in equilibrium, one may calculate the latent heats of fusion of the components on the basis of thermodynamic equations (Raoult and Clausius).

175 TABLE 4

Name	Composition in molecular ratios †	Interval of melting		Specific gravities		Refractive indices	
		initial	final	crystal	glass	crystal (Ng + Nm + Np)/3	glass
Anorthite	An	1550	1550	2.765	2.700	1.587	1.5755
Bytownite	Ab_1An_5	1465	1521	2.733	2.648	1.577	1.5600
Labradorite	Ab_1An_2	1372	1490	2.710	2.491	1.568	1.5452
Andesine-Labradorite	Ab_1An_1	1287	1450	2.679	2.533	1.558	1.5907
Andesine	Ab_2An_1	1205	1394	2.660	2.483	1.549	1.5166
Oligoclase-Andesine	Ab_3An_1	1175	1362	2.649	2.458	—	—
Oligoclase	Ab_4An_1	—	1334	—	—	—	—
	Ab_8An_1	—	1265	—	—	—	—
Albite	Ab	1100	—	2.605	2.382	1.534	1.4890

† [T.N. In addition to this convention of representing plagioclase compositions, Soviet geologists also make use of plagioclase numbers, which indicate the precentage of anorthite. The translator has preferred to use the Western convention i.e. No. 90 = $Ab_{10}An_{90}$].

Without dwelling on this calculation (the derivation of which may be found in Bowen, 1913), we shall merely present the final results. According to Bowen, the latent heat of fusion of anorthite is 104.2 cal/gm and for albite is 48.5 cal/gm.

The frequent appearance of zoned plagioclase crystals is readily understood from the diagram. The cores of the crystal are always richer in albite than the margins. An inverse relationship, if sharply pronounced, is indicative of anomalous changes in the magma composition during crystallization, e.g., assimilation of limestone, etc. This is an indication of "anomalous rocks" as termed by Fedorov. A less sharply pronounced inverse order may be caused by supercooling. Let us consider the case of a plagioclase of composition $Ab_{50}An_{50}$ on cooling to a temperature of 1,280°C. At this temperature a plagioclase of the same composition as the melt begins to crystallize out. After some time, when a part of the melt has formed crystals, the temperature may rise (due for instance to the liberation of the latent heat of fusion) up to 1,300°. At this temperature the separating crystals are already of composition $Ab_{45}An_{55}$, i.e., a zone of more basic composition will form.

The consideration of the melting curve of plagioclases also reveals the limits within which the composition of the separating plagioclase may differ from that of the residual liquid during the formation of zoned crystals. Let us consider the labradorite $Ab_{50}Ab_{50}$ at an initial temperature of 1,500°C. When the temperature drops to 1,386°C, 50% of the mass will separate as crystals of composition $Ab_{69}An_{31}$ and 50% of the liquid will remain, having a composition $Ab_{69}An_{31}$. If the separating crystals do not react with the liquid, then on the cooling of
176 the latter to 1,282°C half of the liquid, i.e., 25% of the original, will separate as crystals of composition $Ab_{51}An_{49}$ and the other half will remain liquid, having a composition $Ab_{13}An_{87}$. If the remaining liquid follows the same process, 12.5% of the initial liquid will remain, having the composition of almost pure albite — $Ab_{99-98}An_{1-2}$ at a temperature of ± 1,100°C.

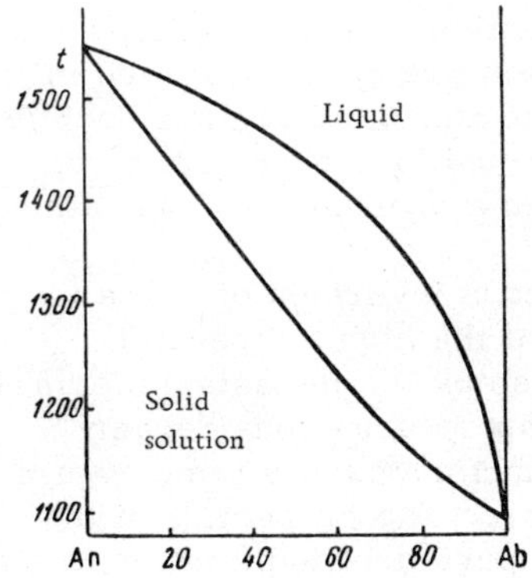

FIGURE 112. Phase diagram of the binary system albite (An)-anorthite (An) (plagioclases) after Bowen.

Thus, we note that under the appropriate conditions a melt of composition $Ab_{50}An_{50}$ may yield a liquid fraction composed of purely alkali feldspar, e.g., albite, the range of crystallization covering a wide interval from 1,400–1,100°C.

[Recently the differences in optical properties of the so-called high temperature and low temperature plagioclases have been determined and thoroughly investigated. The first-mentioned group not only includes such plagioclases as form at rather high temperatures but also (and mainly) those that form under conditions of a sufficiently high vapor pressure, e.g., in volcanic rocks. Plagioclases of plutonic rock, which crystallize sufficiently slowly, are considered as being of the low temperature type. The differences between the above-mentioned plagioclase types are revealed both in the optical orientation† of the acid plagioclases (especially albite) and in a sharp change in the value of 2V. Thus it has been established that as temperature decreases certain changes occur in plagioclases. The physicochemical character of these changes has called for their further consideration.

Tuttle and Bowen (1950) investigated in detail the properties of albite formed from glass, as well as the changes occurring in normal albite after prolonged baking at a temperature of 720°C and more. Both the optical properties and the X-ray pattern changed during this treatment. This led the authors to assume that a true polymorphous inversion takes place here at a temperature of 700°C, and not a gradual change accompanied by drops in temperature during the so-called ordering of solid solutions (when Si and Al atoms occupy definite lattice sites). Such an inversion is not observed in the case of basic plagioclases, thus leading one to assume the

† The optical diagnostics of high temperature plagioclases called for the construction of a special diagram. This diagram was published in 1958 in the form usual for the Fedorov method (Zavaritskii et al., 1958).

presence of a discontinuity in the diagram approximately between $Ab_{65}An_{35}$ and $Ab_{35}An_{65}$. This is apparently also confirmed by X-ray data, which were recently interpreted in such a way as to show a difference in the structures of normal (low temperature) plagioclases in the basic and acid parts of the series. However, a true decomposition of solid solutions and the coexistence of two plagioclases in equilibrium are never observed in natural minerals. This makes the data of Tuttle and Bowen rather doubtful. The presence of a discontinuity is also not affirmed (see page 349) by investigations in the presence of water vapor up to a pressure of 5,000 atms (Yoder, 1957).

MacKenzie (1957) showed that the polymorphous inversion of albite is not only complicated by the sluggish inversion of the high temperature modification to the low temperature one, but also by the metastable formation of the high temperature modification at temperatures considerably below the inversion point. Experiments were performed at a temperature of 700°C and a water vapor pressure of 1 kilobar. The properties of the synthesized albite changed gradually from the temperature modification to the low temperature one as the duration of the experiment was increased. This is apparently associated with a gradual ordering of atoms in the structure. MacKenzie arrived at the conclusion that, as in the case of K-feldspars, the formation of the metastable form takes place at comparative-
177 ly low temperatures, and is followed by a transition into a stable form. According to the views of the author, this may only be explained by a gradational course of inversion during the crystallization of the least stable amorphous phase, e.g., in volcanic glasses. It may scarcely be applied to such processes as pegmatite formation].

K-Na feldspars*

[In contrast to the plagioclases, the system of the alkali feldspars (orthoclase-albite, i.e., $KAlSi_3O_8$-$NaAlSi_3O_8$) was till recently only comparatively poorly investigated. The reason for this lies both in the unusual difficulties encountered in the experimental melting of these feldspars, which always yield glass on cooling, and in the insufficient knowledge of K-Na feldspars in nature. The phase diagrams for K-Na feldspars have only been established from the analysis of relations in natural minerals. This method, however, has nevertheless led to a correct approach to solving the problem, and is of use in considering the history of this problem before presenting the diagrams obtained in the laboratory.

On reviewing the situation in 1926 (when the only available experimental data were on the incongruent melting of orthoclase), A. N. Zavaritskii wrote in the first edition of this book] "Vogt has thus related the system Ab-Or to Roozeboom's Type V; the curve constructed by Miakinen is related to the same type, but with broader limits of miscibility of the K and Na components. Dittler and others determined the possible existence of a continuous series of solid solutions for K-Na feldspars in volcanic rocks. This obliged Herzenberg to accept a melting curve of Type III for K-Na feldspars. Winchell (1925) also presents such a curve.

The decomposition of mixed crystals of already solid K-Na feldspars undoubtedly occurs during the formation of certain microperthites. Another type of perthite intergrowth, developed in the alkali rocks of Norway, was considered by Vogt as a eutectic of K-Na feldspars.

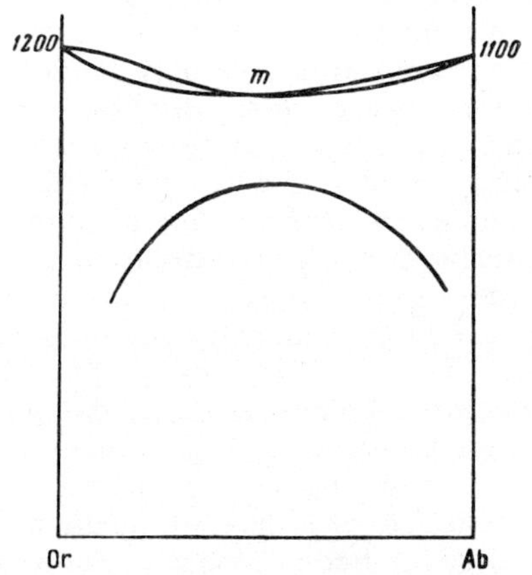

FIGURE 113. Schematic diagram of the system orthoclase (Or) — albite (Ab) after A.N. Zavaritskii (1926) — not according to scale.

All these features of K-Na feldspars are explained by the fact that under various conditions their crystallization takes place at different temperatures. As is known, two substances yielding solid solutions when crystallizing at high temperatures will separate immiscibly when crystallizing at low temperatures (e. g., from aqueous solutions). As an example we may cite K-Na chlorides, which were experimentally investigated by Kurnakov and Zhemchuzhnyi. The ability of increasing the miscibility at higher temperatures is a general property and is affirmed for K-Na feldspars by the study of the compositions of crystals that form at the highest temperatures.
178 The conditions corresponding to the highest temperatures are attained during the crystallization of volcanic rocks lacking mineralizers, the presence of which lowers the temperature of solidification of magmas. It is in such rocks that the continuous transitions in the composition of K-Na feldspars are characteristic. Plutonic rocks crystallize at lower temperatures. At even lower temperatures pegmatites form. Finally, during the enrichment of the melt in mineralizers (mainly water) accompanied by decreasing temperature, a transition to hydrothermal formations is observed. At the same time the limits of miscibility of K-Na feldspars become narrower. These considerations, which may be anticipated, are fully affirmed by studying the analysis of alkali feldspars, as shown by Miakinen. He accepted Roozeboom's melting curve of Type V for the system K-Na feldspar on the basis of the isodimorphism of these minerals. This, however, cannot be considered as proven until the polymorphism of orthoclase and microcline has been determined.

On the basis of investigation of other salts of K and Na (for which the ability to yield isomorphous mixtures during crystallization from a melt has been established as a general property of these metals), and especially on the basis of the existence of the continuous mixture $KAlSi_3O_8$-$NaAlSi_3O_8$ — among the anorthorclases of volcanic rocks, one may assume that in the case of dry melts of albite and orthoclase we are dealing with a melting curve of Type III. As shown on Figure 113, an area of discontinuity occurs at lower temperatures. For the sake of simplicity the area of decomposition of orthoclase into leucite + liquid at high temperatures is omitted on the diagram.

The existence of such a discontinuity is also determined in other compounds of K and Na (e. g., in chlorides). The melting temperatures of orthoclase and albite were accepted as being close to 1,200 and 1,100°C respectively (1,190 and 1,180 respectively, according to Miakinen). In the case of albite the melting point is determined almost exactly. In the case of orthoclase the temperature of 1,200°C corresponds to the inversion point (see the system orthoclase-leucite). A minimum point *m* is located on the continuous melting curve. The melting temperature is only slightly lowered at this point. The lowering of the temperature of solidifcation due to the presence of mineralizers in a magma shifts the melting curve downward, so that the distance between it and the area of decomposition decreases, and finally these areas contact and even intersect one another. Here we enter the field of incomplete isomorphism and crystallization of the eutectic type (Type V). The system $KAlSi_3O_8$-$NaAlSi_3O_8$ should therefore be considered in the presence of water (or more generally any mineralizer which lowers the melting temperature).

Further on we shall consider in greater detail a system of three components, one of which is volatile. Here it is merely necessary to provide a summary phase diagram for elucidating the relations between K and Na feldspars as observed in rocks. We shall make use of a three-dimensional diagram constructed according to the method described hereafter. A third coordinate axis is added to the usual temperature-composition diagram, this axis being perpendicular to the plane of the diagram (Figure 114). The content of the mineralizer in the initial mixture is plotted to scale on this axis. The addition of the mineralizer lowers the melting point. The crystallizating temperature T_o of pure orthoclase is lowered as shown by the curve T_o T'_o. In the case of albite, a similar curve is represented by T_a T'_a. Several sections of the diagram are depicted by the parallel planes, *T*, *Or* and *Ab*. The concentrations of Ab and Or in the liquid and solid phases, as well as the temperature, are projected onto these
179 planes. On each of these sections we obtain individual temperature-concentration diagrams, which express the mutual relations between orthoclase and albite under various conditions of crystallization. Such a combined three dimensional diagram naturally does not constitute a ternary diagram; this series of projections merely presents the most visual concept of differences in crystallization under different conditions.

The decomposition curve of K-Na feldspars in a dry melt P_1 contacts at a certain point the melting curve, which is lowered due to the presence of a mineralizer. This point of contact is a critical point of discontinuity. Below the point V, curves with eutectics occur. Initially, the discontinuity in solid solutions is rather limited (curve *2*). It then increases (curves *3* and *4* on Figure 114). Thus one obtains the contours of a surface limiting the area of discontinuity. The eutectic line *E* is situated on this surface. A definite feldspar composition corresponds to each point on the curve limiting the field of discontinuity. Curves are not produced to the melting point of the K-feldspar, as this point is incongruent and the curves in this area have not been studied.

[Figure 115, I, shows a phase diagram of the system $NaAlSi_3O_8$-$KAlSi_3O_8$, or more precisely the corresponding section of a ternary system according to Schairer (1950). It may be accepted as fully established that here we are dealing with a system with continuous solid solutions and a minimum. However, the minimum curve is very smooth, and the exact determination of the minimum point has proved rather difficult. Within the limits of experimental error, feldspars with 60,65 and 70 % Na component are found to melt at a temperature of 1,063°C. Therefore the value at the minimum point was accepted as 65% (for a system with water the minimum point is shifted to 70%).

Anticipating the mention of the system containing water, which is discussed in detail further on in this book, we shall outline the following changes in the system Ab-Or-H_2O (Bowen and Tuttle, 1950) on the addition of water (Figure 115, II and III): the leucite field is narrowed; at a pressure of 2,000 atms it becomes extremely narrow and at a pressure of 2,500 atms (data of Goranson) it completely disappears. The decomposition curve of solid solutions has a maximum at approximately 660°C (700°C according to Yoder). As the increase in water vapor pressure has almost no influence on the location of the decomposition curve, Yoder shows (as could have been foreseen) that further on the melting curve intersects this curve, i.e., the diagram changes to Type V (at a water vapor pressure of 5000 atms).

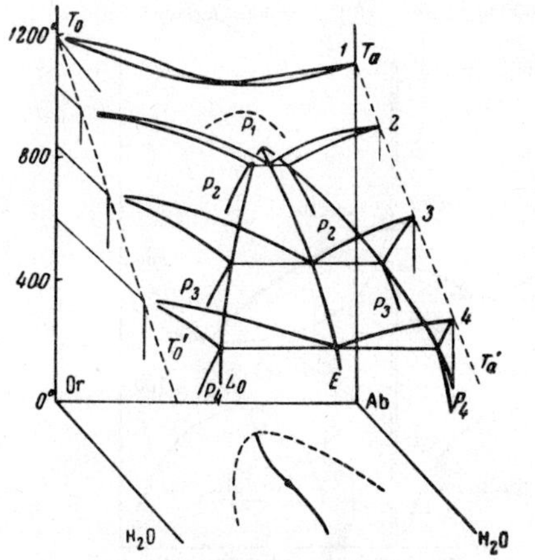

FIGURE 114. Schematic diagram of the system orthoclase (Or) — albite (Ab) — H_2O (after A.N. Zavaritskii, 1926)

On comparing experimental results and interpolated data with Figures 113 and 114, we are convinced of the overall correctness of these prognoses, which were made on the basis of natural observations. Obviously, certain essential corrections must be introduced in quantitative considerations. Even when overlooking the wider field of leucite in the case of dry melts (probably not of great significance in pro-
180 cesses of mineral formation, page 231), we note that the decomposition of feldspars occurs at a much lower temperature than was formerly assumed approximately 1,000°C on Figure 114). Thus, in most intrusive rocks the formation of a continuous series of solid solutions is probably the more stable process during crystallization of the magma, and their decomposition occurs in the post-magmatic stage, which fact has been underestimated during the study of minerals. The formation of primary eutectic perthites as well as primary albite together with K and Na feldspars should be questioned even in alkali rocks.

Up to the present, the nature of the physicochemical inversions of K feldspars has not been fully elucidated. Many investigators are inclined to accept the view of a gradual change in its properties with decrease in temperature; they also attribute the monoclinic character of some feldspars (sanidine and orthoclase) to the pseudosymmetry resulting from microtwinning.

MacKenzie (1957) shows that the inversion temperature from the monoclinic to the triclinic modification is dependent on the conditions under which these feldspars form. In the case of artificial feldspars, this temperature decreases from 1,180°C (for pure albite) to 175°C (for a feldspar of composition $Or_{30}Ab_{70}$), whereas for igneous feldspars the inversion temperature already equals 100°C for a composition $Or_{31.7}Ab_{61.8}An_{10.1}$. If these data, which again indicate the complexity of the process, are in accordance with the occurrence of some triclinic adularia, the formation of cross-hatched microcline in granites undoubtedly occurs at a higher temperature. At present this problem is undergoing further investigation.

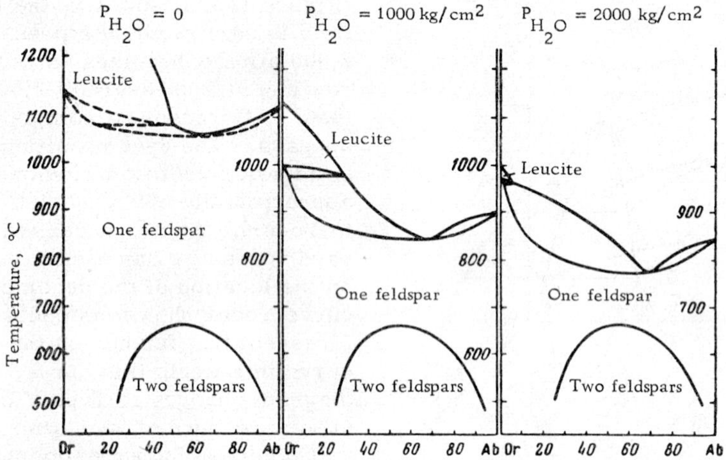

FIGURE 115. Phase diagram of the pseudobinary system $KAlSi_3O_8$ (Or) — $NaAlSi_3O_8$ (Ab) "dry" and in the presence of water at various pressures (after Bowen and Tuttle).

In the meanwhile it remains certain that special properties of high temperature feldspars (in particular the small value of 2V) are preserved during rapid cooling, being thus indicative of a definite condition of cooling of the rock. This is not only observed in relatively young effusive and hypabyssal rocks, but also sometimes in true equigranula granites, usually
181 in the marginal parts of the massif. The increase in the content of the Na component not only assists decomposition, but also facilitates the transition of a feldspar into its low temperature form. The higher temperature K-Na feldspars with a very low content of the Na component will, however, tenaciously preserve their properties even during the slow cooling of metasomatic rocks. It is expedient also to term the high temperature varieties as K-Na feldspars, irrespective of their Na content, thus stressing these feldspars' membership of a continuous series of solid solutions and not their quantitative relations within it.]

The system $NaAlSiO_4$-$KAlSiO_4$ (carnegieite-kalsilite) and special features of nepheline**

The basic features are determined from the character of the system $NaAlSiO_4$-$KAlSiO_4$ which was only recently studied in sufficient detail by Tuttle and Smith (1958). According to its type, this system shows a

continuous series of solid solutions with a minimum (Type III of Roozeboom), but is complicated by several phase transitions (Figure 116). At a high temperature the pure Na component does not crystallize out as nepheline, but as cubic carnegieite, which passes into the triclinic (?) modification showing characteristic twinning in the metastable field at 690°C. In the case of slow cooling, carnegieite inverts to nepheline at a temperature of 1,270°C. A second inversion is assumed at approximately 850°C — into low temperature nepheline, which differs only by virtue of insignificant changes in the lattice parameters. A polymorphous inversion has also been
182 determined for the K component. At 840°C the high temperature rhombic modification inverts to hexagonal kalsilite, which approaches nepheline in its properties. Kaliophylite, as encountered in nature, was successfully synthesized only in the presence of a certain quantity of the Na component and water at a temperature of 1,000—1,100°C, but no corresponding field has been found in this system.

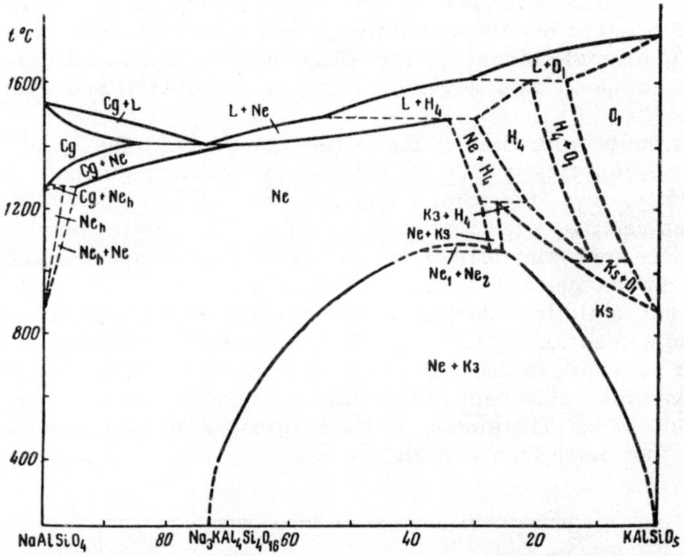

FIGURE 116. Phase diagram of the system $NaAlSiO_4$-$KAlSiO_4$ (after Tuttle and Smith) Dotted lines indicate the assumed phase boundaries; L— liquid; Cg— carnegieite; Ne_h— high-temperature nepheline; Ne— low-temperature nepheline; O_1— orthorhombic modification of $KAlSiO_4$; Ks— kalsilite; H_4— tetrakalsilite, $NaK_3Al_4Si_4O_{16}$.

Finally, the appearance of two additional modifications within the system is assumed: $NaK_3Al_4Si_4O_{16}$ and $KNa_3Al_4Si_4O_{16}$. The former (H_4 on the diagram) is shown in equilibrium with the melt at inversion; the latter — in equilibrium in the field of low temperatures. In all probability one should speak in both cases of phases with superstructures, i. e., with the ordering of solid solutions. As minerals especially enriched in potassium are rare in this system, the composition curve of solid solutions is of extreme interest in this case. It shows a maximum at a comparatively high temperature of the order of 1,100°C, but this maximum as well as the left branch of the curve are shifted markedly toward $KAlSiO_4$, and the nepheline formed is relatively rich in potassium, and approaches the above-mentioned compound at low temperature.

The appearance of a new field of decomposition at low temperatures has been suggested between the mentioned compound and the sodium component. This possibility has not been affirmed by either experimental data or by sufficiently convincing mineralogical findings.

Among the other features, we shall make mention of a certain excess of silica (see page 186) and $CaAl_2Si_2O_8$ (see page 184) dissolved in it.

The systems orthoclase-anorthite and leucite-anorthite**

The system orthoclase-anorthite is not a binary one but a pseudobinary section of the more complex system leucite-anorthite-silica (see Figure 151). The system leucite-anorthite of the above-mentioned ternary system is in itself a simple binary system (Figure 117) and has a eutectic point at a eutectic composition of leucite-55 anorthite-45 (wt %) and a temperature of 1,413 ± 2°C.

Among the peculiarities of the system anorthite-orthoclase (Figure 118) we shall mention firstly that the point of common crystallization of leucite and anorthite, i. e., the point where the cotectic curve contacts the section plane, lies relatively far from the point Or. The albite content in the "eutectic" is approximately 22 %. As the temperature of the melt is lowered, the contents change rapidly and the crystallized mixture is impoverished
183 with respect to albite. During the crystallization of the K-feldspars at the expense of a reaction between leucite and the melt, enriched in silica, the content of anorthite in the melt drops to below 50 %†. The amount of dissolved anorthite component in the solid K feldspar is quite small and does not exceed 2—3 %. The amount of the K feldspar dissolved in anorthite is even less (not more than 1—2 %).

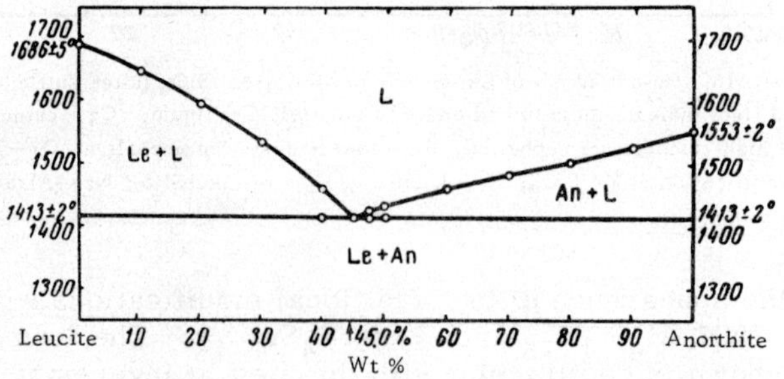

FIGURE 117. Equilibrium diagram of the system leucite (Le)—anorthite (An) after Schairer and Bowen

---

† In a system containing water at an H_2O pressure of 5,000 atms the eutectic Or—An contains approximately 12 % anorthite.

The above-mentioned relations are only of partial significance in interpreting the process of magma crystallization. Usually pseudoleucites are never found in rocks saturated in silica which fact is associated with the presence of water in the magma (see page 179). The quantitative relations in the system explain the well expressed idiomorphism of basic plagioclases as compared to K feldspars in monzonites, which feature shows up even better on the schematic diagram of the ternary system Ab-An-Or (see page 167). Relations in the anorthite-leucite eutectic explain the early crystallization of leucite in basic lavas.

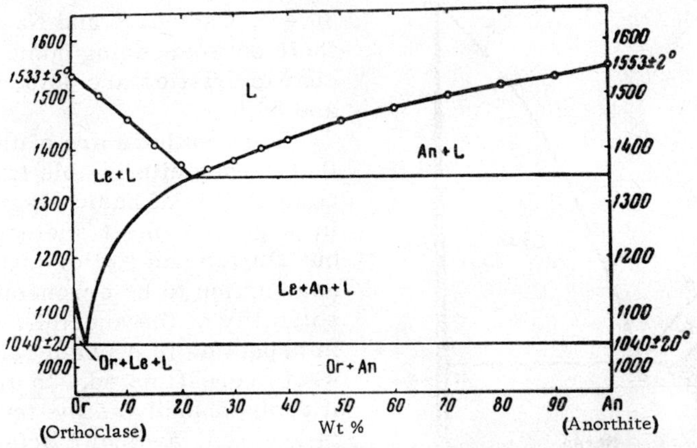

FIGURE 118. The pseudobinary system Or-An (after Schairer and Bowen)

The system nepheline-anorthite

The mutual thermal relations between the sodium and calcium orthoaluminosilicates, represented in nature by the minerals nepheline and anorthite, were investigated experimentally both by means of cooling diagrams and by the method of quenching. The results of the various experiments show good agreement with one another.

Firstly it should be mentioned that $NaAlSiO_4$ is dimorphous. Above the temperature of 1,250°C a triclinic (?) mineral of the same composition separates from the melt in place of hexagonal nepheline and is given the name carnegieite. A melt with the composition of nepheline usually crystallizes with difficulty. During prolonged heating to a temperature of 900—1,000°C, the rather viscous glass crystallizes to yield nepheline. If the melt is heated to a temperature of 1,500—1,600° C, it becomes more fluid and crystallizes more readily on cooling to yield carnegieite.

Using the quenching method, Bowen (1912) obtained such a phase diagram for the system nepheline-anorthite (Figure 119). This curve is of Type V (according to Roozeboom) and is complicated by the inversion of one of the components from one modification to another (viz. carnegieite-nepheline).
184 Characteristic temperatures are shown on the diagram. It is evident from the diagram that nepheline may contain up to 35 % (by wt.) of the isomorphous

admixture of the Ca aluminosilicate, whereas anorthite may only contain 2% of the isomorphous admixture of carnegieite. The weight composition of the eutectic is 54.5 % $Na_2Al_2Si_2O_8$ and 45.5 % $CaAl_2Si_2O_8$: The ratio anorthite : nepheline (of solid solutions) is FG : GM = 10.5 : 52.5 = 1 : 5 (by wt.). The volumetrical relations are very similar. The compositions of other characteristic points on the diagram is as follows: D corresponds to 5 % $CaAl_2Si_2O_8$, and C and B to 18 % and 28.5 % $CaAl_2Si_2O_8$ respectively.

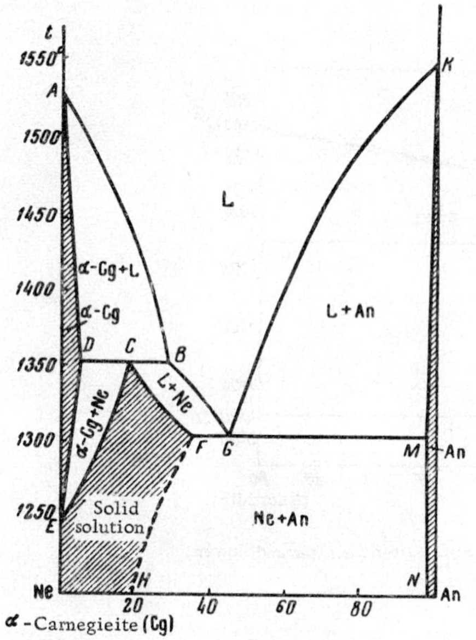

FIGURE 119. Phase diagram of the system nepheline (carnegieite) — anorthite (after Bowen) Cg— carnegieite; Ne— nepheline; An— anorthite.

Bowen gives the following figures (in wt. %) for chemical composition of the eutectic: SiO_2 42.6 %, Al_2O_3 36.4 %, CaO 9.1 % and Na_2O 11.9 % — their corresponding numerical characteristics are 30.6, 12.9, 0 and 56.6

[Some authors are of the opinion that carnegieite is able to enter the composition of basic plagioclases in large amounts ("anemousite"), but Dittler and Keller proved this assumption to be erroneous. The solubility of the anorthite component in nepheline is probably also somewhat overestimated. In all events, it drops sharply as the temperature is lowered, as shown schematically on Figure 119.]

Ginsberg was even in favor of accepting the melting curve of Roozeboom's Type III for this diagram, which however shows no agreement with the isodimorphism of its constituent components as accepted by this author. Bowen and Greig (1925) showed that the isotropic modification of $NaAlSiO_4$ crystallizes above 1,248°C. It is inverted to the triclinic (?) modification, which forms twins below the temperature of 700° C.

Anorthite-silica

The system anorthite-silica is not encountered in nature, but is of great interest as it has been studied experimentally by Rankin and Wright, and later by Andersen. The melts obtained are quite viscous and the melting diagram is only approximate. The melting curve has a eutectic point (Figure 120). The eutectic composition (in wt. %) is 52% anorthite and 48% tridymite. The eutectic temperature (after Andersen) is 1,353 ± 2°C. The chemical composition of the eutectic is: SiO_2 70.4 %, Al_2O_3 19.2 %, CaO 10.4 %, and the numerical characteristics of this composition are 0, 14,0 and 86. [According to recent data of Schairer and Bowen, the eutectic temperature is 1,368 ± 0.5° C and the eutectic composition is 50.5% anorthite

49.5% tridymite (in wt.%). The calculated composition of the eutectic in
185 weight percentage is: SiO_2 71.32, Al_2O_3 18.53 and CaO 10.15.]

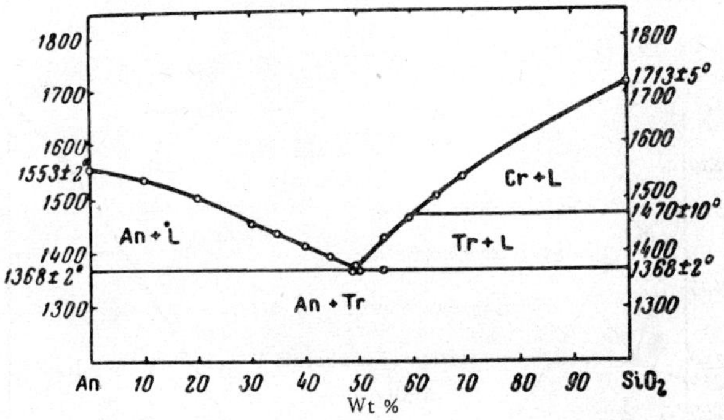

FIGURE 120. Equilibrium diagram of the system anorthite—silica (after Schairer and Bowen)
An— anorthite; Tr— tridymite; Cr—cristobalite; L— liquid.

Feldspathoids and alkali feldspars

The relations between the felspathoids and the feldspars, as well as between alkali feldspars and silica are correspondingly determined by the systems SiO_2 -$NaAlSiO_4$ and SiO_2-$KAlSi_2O_6$. The system SiO_2-$NAlSiO_4$ may be considered in parts due to the incongruent melting of $NaAlSi_3O_8$. The system $NaAlS_3O_8$-$NaAlSiO_4$ was investigated by Greig and Barth (1938) and the system $NAlSi_3O_8$-SiO_2 (as shown on Figure 121) by Schairer and Bowen (1947). These investigations affirmed the conclusions arrived at on the basis of direct mineralogical observations — as to the existence of a eutectic between nepheline and albite (Zavaritskii, 1926). Whereas carnegieite crystallizes during the melting of $NaAlSiO_4$, it passes into nepheline when the SiO_2 content is increased. Carnegieite has not yet been
186 found in nature,which phenomenon is partly explicable by the lowering of the crystallization temperature in the presence of water and partly by the fact that the composition of natural magmas is usually beyond the limits of carnegieite. Both in carnegieite and (in a somewhat larger amount) in nepheline a limited area of solid solutions of the Ab component is observed, the boundaries of which are as yet insufficiently studied. Recently, N. V. Belov even doubted the formula of nepheline and proposed a new structural formula — $Na_4Al_4Si_5O_{18}$ with a higher SiO_2 content. It is well known from mineralogy that the composition of nepheline actually does differ from the formula $NaAlSiO_4$ by virtue of a higher silica content†, which may of course also be explained by a corresponding solid solution. Within the crystal

† According to new data of Tilley (1954, 1958) the solubility of silica in naturally occurring nepheline, rather poor in the potassium component, proved to be comparatively low— of the order of 3% (by weight). The solubility of SiO_2 in $NaAlSiO_4$ is probably also considerably lowered with decrease in temperature, as is considered on Figure 121.

lattice atoms of aluminum may partly be replaced by atoms of silicon, with a corresponding decrease in the number of sodium atoms, thus resulting in a quantitative effect equivalent to a solid solution or $Na_{1-n}Al_{1-n}Si_{1+n}O_4$.

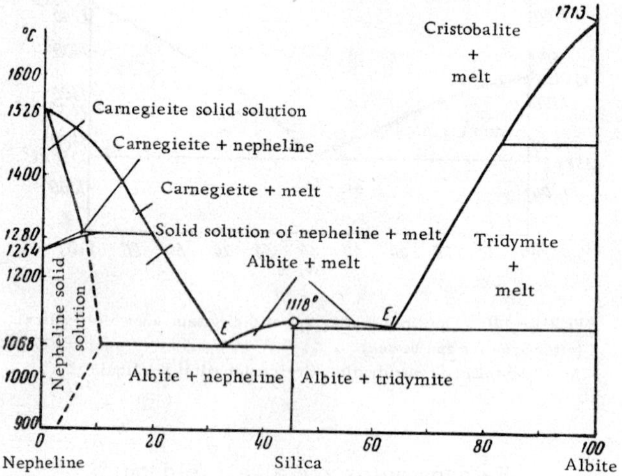

FIGURE 121. The system $NaAlSiO_4-SiO_2$

Between albite and silica (tridymite in a dry melt) there is undoubtedly a eutectic point, the position of which was only approximately known for some time due to the difficulty of crystallization of such melts. Figure 121 presents the most recent reliable diagram of Schairer with a eutectic point corresponding to 68.5 % albite, 31.5 % silica at a temperature of 1,062 ± 3°C.

The considered system undergoes considerable changes as pressure is increased. The mineral jadeite has been known for a long time, and is the sodium analogue of leucite, being related to the pyroxene group. The high specific gravity of jadeite permits one to assume that its formation during the reaction nepheline + albite (with the transition of Al from a fourfold to a sixfold coordination, which is the cause of the increased specific gravity) occurs as pressure is increased (Eskola, 1920).

In 1953 Adams, making use of thermodynamic equations and constants, made the first attempt at constructing an equilibrium curve for the reaction

$$NaAlSi_3O_8 + NaAlSiO_4 = 2NaAlSi_2O_6.$$

The heat of reaction was determined by measuring the heats of solution of the corresponding minerals in hydrofluoric acid. ΔQ = 6,120 cals at 25° C. ΔS was accepted as 14.7 cals/deg at 25° C or 298° K. As usual the entropy was determined by means of measuring the heat capacities down to the lowest temperatures. Hence, at 298° K, $T\Delta S$ = -4,380 calories and $\Delta\zeta$ =-1,740 calories, ie., jadeite should be stable here.

Assuming that $T\Delta S$ remains constant at high temperatures (because ΔC_p approaches zero in this case), one may readily calculate the points on the equilibrium curve (in actuality they are found to lie on a straight line).

As is known, the conditions of equilibrium are determined by the equality of the thermodynamic potentials of the reacting and resultant phases, i. e.,

$$\Delta E - T\Delta S + P\Delta V = 0,$$

where $\Delta E + P\Delta V = \Delta Q$ (the heat of reaction at constant pressure).

The heat of reaction in cases where a gaseous phase participates is
187 usually determined at atmospheric pressure. As the value of ΔV is negligible in comparison with ΔQ (usually far below 1 % at P = 1 atm), the second member may be disregarded and ΔE is regarded as being equal to ΔQ_1 (the heat of reaction at atmospheric pressure). Inserting ΔQ_1 in the equation of pressure which is necessary for attaining equilibrium at a temperature T, we obtain:

$$P = \frac{T\Delta S - \Delta Q_1}{\Delta V}.$$

If ΔV is in cm^3 and P is in bars, ΔF will be in decijoules. In order to convert calories to decijoules we multiply by 41.84, i. e.,

$$P = 41.84\,\frac{T\Delta S - \Delta Q_1}{\Delta V}.$$

The corresponding values of the considered reaction when calculated for two gram molecules of jadeite are: ΔQ_1 = -6,120 cals, ΔS = -14.7 cals and ΔV = - 33.4 cm^3.

For a temperature of 600° C, i. e., 873° K we obtain by inserting corresponding figures the following equation:

$$P = 41.84\,\frac{-873\cdot 14.7 + 6120}{-33.4} = 8300 \text{ bars.}$$

Indirect considerations of the conditions of mineral equilibrium permit one to doubt the figures obtained, which appear to be too low (Sobolev, 1955). On the other hand, a contrary view was expressed stating that jadeite formation in nature does not call for such high pressures (Korzhinsky, 1953). In several cases a very low temperature of jadeite formations was proposed (Yoder, 1952).

Experimental studies of the equilibrium curve (Robertson et al., 1957) showed that this reaction can only be brought about at much higher pressures than assumed by Adams, and that a pressure of more than 12,000 atms is necesarry at a temperature of 600° C (see Figure 228). The inexact determination of ΔQ seems to be the source of all the trouble in this case. According to new data, ΔQ= -3,600 cals. The authors of the mentioned investigations attempted to calculate the temperature at which the decomposition of albite should occur according to the reaction

$$NaAlSi_3O_8 = NaAlSi_2O_6 + SiO_2.$$

The value of ΔQ is accepted as -1,500 cals and the obtained pressure P at 600° C is 21,000 atms, with a possible margin of error of up to 5,000 atms.

Jadeite is not directly encountered in igneous rocks, but is characteristic of a specific facies of metamorphic rocks which are closely related to igneous rocks. The determination of the corresponding pressure is of great importance to petrology. According to the cited diagram (one of the experimental curves is produced) the formation of jadeite at normal pressures is impossible even down to the lowest temperatures. However, this is of no principal importance as jadeite is unstable in the presence of water below a temperature of 600° C (analcite is formed). Thus, it may be claimed that the pressure necessary for the formation of jadeite-bearing rocks is in all events greater than 12,000 atms.

A paragenesis of jadeite with quartz is rarer than that without quartz, but is known in nature, e. g., in Japan. As it is quartz and not coesite which is present in these cases, one may assume that the cited value of
188 minimum pressure for the mentioned reaction is somewhat too high, and the reaction curve should pass at considerably lower pressures than the equilibrium curve quartz-coesite†.

There are also incomplete data on a new polymorphous modification of $NaAlSiO_4$, which is stable at high pressure and comparatively low temperatures. This modification probably has an olivine structure, i. e., a sixfold coordination of Al. However, in the above-mentioned experimental synthesis of jadeite this phase was not obtained; it is probably stable at higher pressures and lower temperatures.

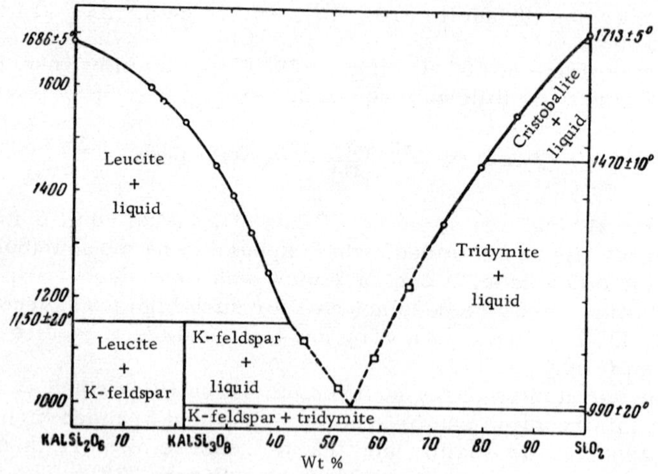

Figure 122. The system $KAlSi_2O_6 - SiO_2$

† After this book had gone to print, an experimental curve was published affirming this assumption (V. S.).

As far back as 1922 Morey and Bowen had determined the incongruent melting of K-feldspar in the system $KAlSi_2O_6$-SiO_2. Certain amendments were later made by Schairer and Bowen (1938, 1955). They also continued to study the system in the area of compositions poorer in silica and showed that a eutectic point existed between leucite and kalsilite (this part of the diagram is not shown on Figure 122).

The leucite field extends considerably beyond the point of the K-feldspars, i.e., into the area of compositions oversaturated with silica. It should be mentioned that in nature leucite is never encountered in rocks oversaturated in silica. Such cases would have been indicated by pseudomorphs after leucite, which are so abundant in rocks undersaturated in silica. This feature may only partly be explained by the action of water, as phenomena associated with the incongruent melting of orthoclase have not been noted in volcanic rocks either. There is a eutectic point between the points $KAlSi_3O_8$ and SiO_2, the location of which is only approximately determined. One should stress that the relationship between K-feldspars and silica at
189 the eutectic point is essentially different from the eutectic proposed for graphic granites — pegmatites (approximately 25% quartz). It is closer to the composition of aplites (40% quartz).

In 1926 A. N. Zavaritskii wrote about the relation between leucite and albite and leucite and anorthite: "the system leucite-albite is apparently unstable under magmatic conditions and is converted into the system nepheline-orthoclase according to the equation

$$2KAlSi_2O_6 + NaAlSi_3O_8 = 2KAlSi_3O_8 + NaAlSiO_4.$$

This follows from the fact that leucite and albite are never encountered together as was already mentioned previously. The system anorthite-leucite probably possesses a eutectic; in the presence of anorthite the more stable system leucite-plagioclase, which is normal in leucite tephrites and basanites, is obtained from the unstable system leucite-albite".

Experimental studies have affirmed these assumptions, but also reveal the greater complexity of the relations between leucite and albite. The incompatability of leucite and albite (and K-Na feldspars rich in sodium) follows directly from the division of fields in the considered ternary diagram (see Figure 149).

The system SiO_2-Al_2O_3**

The considered system is not directly related to the crystallization of magma, because Al_2O_3 is in this case always encountered together with alkalis and CaO, and the minerals which are included in this system are characteristic of metamorphic rocks. However, the appearance of these minerals in some rare types of igneous rocks, as well as their wide distribution in contact metamorphic aureoles and in xenoliths, obliges us to present the principal data of this system.

The diagram of the system SiO_2-Al_2O_3 was adequately studied by Bowen and Greig in 1924 as seen from Figure 123 (presented with later amendments according to the data of the same authors, 1956). It is characterized by the following: the presence of a compound showing incongruent melting; a eutectic point between this component and SiO_2. The very gentle character of the melting curve is striking, thus making it similar to the diagram characterizing decomposition into two liquid phases (see page 205).

Bowen and Greig surprisingly stated that the compound which crystallized from the melt is not one of the usual minerals found in nature — polymorphous modifications of the composition $Al_2O \cdot SiO_4$ — but a compound richer in Al_2O_3 with the composition $3Al_2O_3 \cdot 2SiO_2$.

The data of Bowen and Greig were examined by Toropov and later by Budnikov, Trekhsviatski and Pushakovski as well as by F. Ya. Galakhov (1958). They obtained a congruent melting for mullite (Figure 123, II), although showing a small maximum (Botvinkin, 1955). This seems to affirm the previous data of Shepherd and Rankin (1910).

The mentioned compound was shortly afterwards found as a mineral in argillaceous xenoliths enclosed in basalts of the Isle of Mull (Scotland) and was termed mullite. According to its form, properties and crystal structure, mullite is quite similar to sillimanite, thus rendering its isolation difficult for some time (although it is widely distributed in many artificial products). An essential distinctive feature of mullite is its refractive index, which is lower than that of sillimanite.

190 The refractive indices of sillimanite are: $Ng = 1.677$, $Np = 1.657$, and of mullite: $Ng = 1.654$ and $Np = 1.642$.

The conditions of formation of sillimanite, kyanite and andalusite remained vague for a long time. The first attempt at determining their stability fields by means of thermochemical data was made by Neumann (1925). He produced values for the latent heats of inversion which he claimed that he had determined experimentally by the usual method for silicates — measurements of heats of solution in hydrofluoric acid; according to his data the kyanite-sillimanite inversion is accompanied by the liberation of more than 6,000 cals/g mol and a greater amount of heat (6,700 cals/g mol) should be liberated during the andalusite-sillimanite transition.

To the present day these data are quoted in literature despite their questionable character. Firstly, Neumann, by incorrect use of his data, arrived at the conclusion that sillimanite is the low temperature modification and andalusite the high temperature one, which statement is in conflict with mineralogical data. As regards kyanite, Neumann correctly concluded
191 that at normal pressure this mineral is a metastable modification at all temperatures. Attention was not paid to the fact that all three minerals, and especially kyanite, are completely insoluble in hydrofluoric acid. Hence the entire experiment was impossible. As far back as 1927, Greig showed that on heating both kyanite and andalusite they pass into mullite with the absorption of heat, whereas no such phenomenon is observed during the transition from sillimanite to mullite.

Owing to the considerable confusion arising from the discrepancy between Neumann's data and the heating curves in general and the different methods of presenting these curves, they were reexamined by A. I. Tsvetkov in 1956 (Sobolev and Tsvetkov, 1956), who affirmed the correctness of Greig's curves. In 1940 D. S. Korzhinskii attempted to construct an equilibrium diagram for the unary system $Al_2O \cdot SiO_4$, on the basis of mineralogical data (Figure 124, I). On this diagram the kyanite field wedges out in the area of low pressures, but the minimum pressure necessary for the formation of kyanite is not very high and falls within the range of the first few thousand atmospheres.

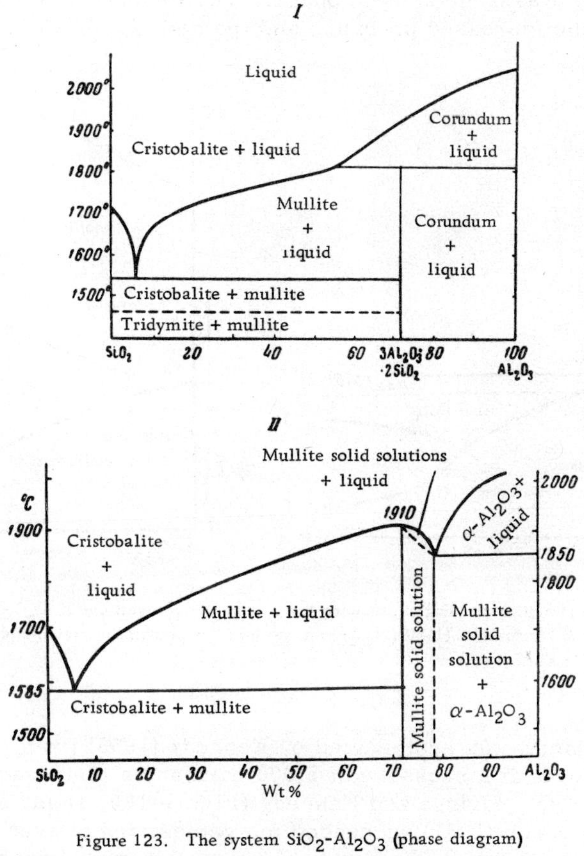

Figure 123. The system SiO_2-Al_2O_3 (phase diagram)

I — after Bowen and Greig; II — after Galakhov

In 1949, Miashira (Thompson, 1955) presented a new schematic equilibrium diagram (Figure 124, II) upon which the andalusite field wedges out in the area of high pressures, and kyanite is stable at normal pressures and low temperatures.

An analogous diagram was presented by Thompson (1955), who made use of data on the entropy of the mentioned silicates and calculated the slope of the equilibrium curve, but was not able to determine the numerical parameters of the points owing to the absence of reliable data on inversion temperatures.

For a long time mullite was the only one of these modifications which could be successfully synthesized experimentally. The investigation of systems with water by Yoder (1952) and Roy and Osborn (1952, 1954) showed that mullite may form at comparatively very low temperatures, starting from

400° C. Neither sillimanite, andalusite nor kyanite were obtained at water vapor pressures of up to approximately 600 atms.

However, D. Roy later (1954) synthesized andalusite within the temperature range 450—700° C and at a water vapor pressure of 600—2,000 atms, and considered it as a metastable phase. The formation of andalusite was facilitated by the increased pressure and the addition of MgO.

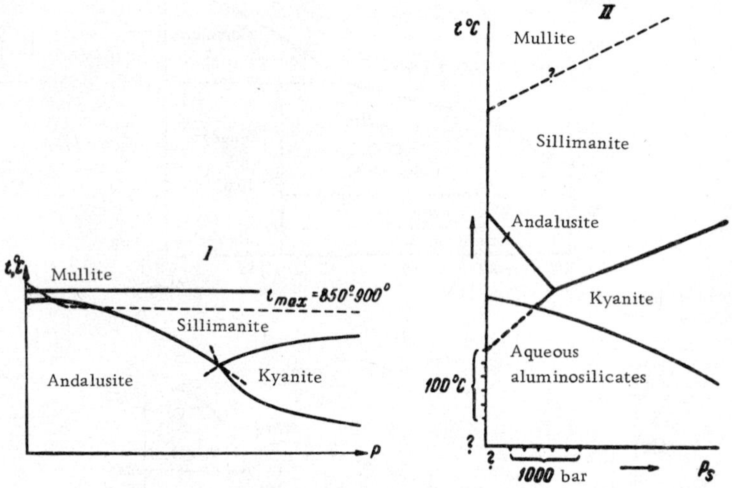

FIGURE 124. Schematic diagram showing the polymorphous inversions of $Al_2O \cdot SiO_4$; I — after D.S. Korzhinskii (1940); t_{max} — maximum temperature of metamorphism; II — after Miashira (1949)

192 True sillimanite was synthesized by Kennedy (1955) from the reaction of mullite and quartz at a pressure of 5,000 bars and a temperature of 600° C.

Finally, in 1956, Griggs and Kennedy (Figure 125, I) and later Clark, Robertson and Birch (1957) succeeded in synthesizing at even greater pressures not only sillimanite but also kyanite, and proposed position of the sillimanite-kyanite equilibrium curve (Figure 125, II). The experiments of Clark et al. were carried out in an atmosphere of nitrogen lacking water. The peculiarities of their result are probably due to this fact. They met with success in synthesizing sillimanite and kyanite at temperatures above 1,000° C, but in this case too inclusions of quartz and corundum were found, whereas at lower temperatures these minerals are formed during the decomposition of kaolinite, within a range where they are unstable.

The obtained data are quite characteristic and permit the use of kyanite as a geological manometer. Assuming that the formation of this mineral in nature usually occurs within the range 600—800° C, we obtain a corresponding pressure ranging from 12—13 thousand atms to 15—17 thousand atms. The latter values apply to the eclogite facies.

Prior to the appearance of experimental data on the lower limit of kyanite stability, this limit had been the source of great controversy. Some authors, however, estimated the necessary pressure as being approximately 10,000 atms (Sobolev, 1949).

Having the equilibrium curve available and producing it to intersect the axis, one may attempt to determine the approximate latent heat of inversion from kyanite to sillimanite. Accepting $t = 0°C$† (i.e., $T = 300°$ K), $P = 5{,}000$ bars, $\Delta S = 3$ cal/mol degree and $\Delta V = 6$ cm^3, we obtain:

$$5000 = 41.84\,\frac{300 \cdot 3 - Q_1}{6}.$$

Thus Q_1 approximately equals 100 cal/mol.

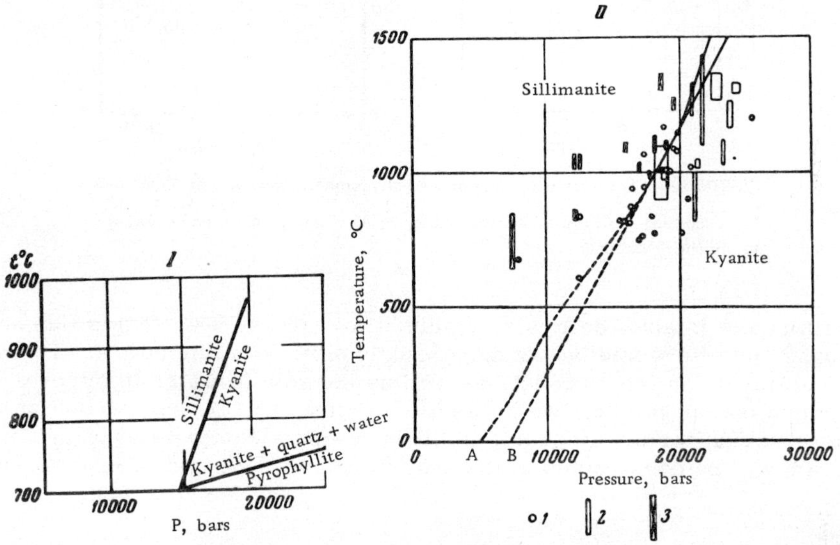

FIGURE 125. Curves of kyanite-sillimanite inversions

I — according to Griggs and Kennedy; II — according to Clark, Robertson and Birch, experimental results: 1 — quartz and corundum; 2 — kyanite; 3 — sillimanite. A — "optimum" experimental curve; B — straight line extrapolation

193 Thus, this inversion is accompanied by the absorption, and not evolution, of a small amount of heat, which increases somewhat during the reaction forming mullite.

If the value of pressure at normal temperature is accepted as being somewhat lower (approximately 3,500 bars according to the data of Kennedy and Griggs), the above-derived value increases to 3,000 cals, but remains comparatively small (as compared with jadeite, page 186).

Figure 126, I, presents a diagram on which Clark et al. attempted to summarize all available data (1957). Accepting the general configuration of the diagram and (obviously) the kyanite-sillimanite curve, which is based on experimental data, we cannot agree with the limits proposed for the fields of andalusite and sillimanite. Andalusite undoubtedly forms in

† [T.N. Apparently an error in the Russin text as 300° K = 27° C and not 0° C as appears in the original].

hornfels at sufficiently high temperatures (700—800° C) and comparatively low pressures (probably of the order of several atms).

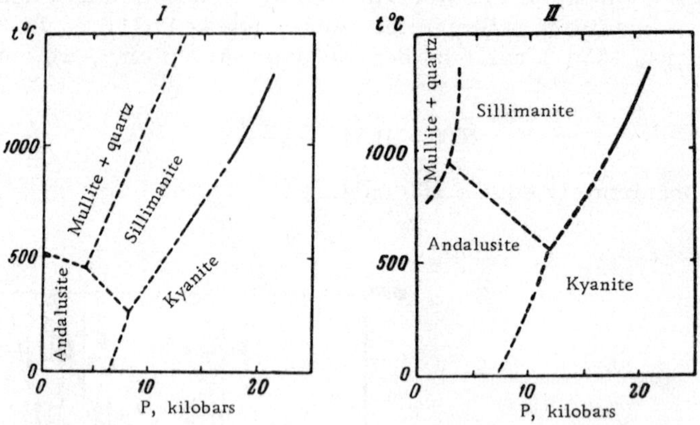

FIGURE 126. Schematic diagram of the polymorphous inversions of $Al_2O \cdot SiO_4$; I — according to Clark, Robertson and Birch; II — ditto, corrected according to mineralogical data

Sillimanite is apparently also stable at low pressures. If this was not the case, mullite would be far more widely distributed in nature. Taking this consideration into account, as well as the role of water in mineral formation (see page 328), we present on Figure 126, II, our variant of the schematic equilibrium diagram, in the hope that a complete equilibrium diagram will be constructed in the near future on the basis of experimental data.

SYSTEMS ILLUSTRATING RELATIONS BETWEEN FEMIC MINERALS

The system CaO-SiO_2

The significance of this system in igneous rocks is limited, as no simple silicates of calcium are known in them. As is known, CaO enters into the composition of aluminosilicates (anorthite, plagioclases) and into a binary compound (CaO and MgO) in the metasilicate diopside. Therefore, we shall produce the diagram of the system CaO-SiO_2, confining ourselves to the following comments (Figure 127).

The calcium metasilicate $CaSiO_3$ is obtained in two modifications:
194 monoclinic wollastonite (α) and hexagonal pseudowollastonite (β). The calcium orthosilicate Ca_2SiO_4 appears in three modifications: α-monoclinic, β-orthorhombic and γ-monoclinic. CaO crystallizes in the cubic system in a manner similar to periclase (MgO). [For general comments on the area of liquation in this and other similar systems see page 205].

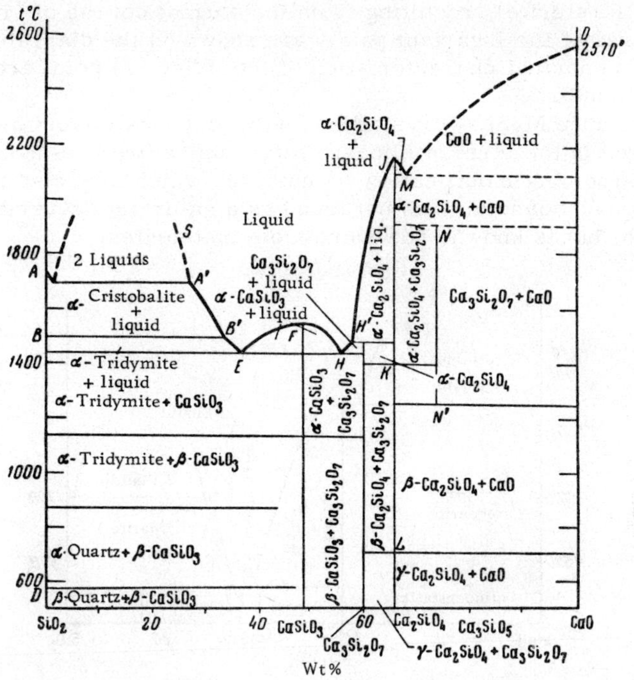

FIGURE 127. The system SiO_2-CaO (after Rankin and Greig)†

The system MgO-SiO_2*

This system, studied by Andersen and Bowen (1914), is of great significance to petrogenesis. Magnesium silicates — enstatite and olivine — are essential constituents of certain igneous rocks, mainly of the basic type.

The diagram obtained by these investigators using the quenching method is shown on Figure 128. As is apparent from the diagram, the two compounds of MgO and SiO_2 are Mg_2SiO_4 and $MgSiO_3$. The former — forsterite — occurs as a maximum on the melting curve and the latter, which at higher temperatures appears as monoclinic clinoenstatite, has a "concealed" maximum. Two eutectics are present in this system: 1) periclase + forsterite and 2) clinoenstatite + cristobalite. The composition of the first-mentioned eutectic is only approximately determined — it contains less than 14 % periclase. The clinoenstatite-cristobalite eutectic contains 12.5 % cristobalite and 87.5 % enstatite. Its numerical characteristics are as follows: 0, 0, 44.7 and 55.3. Due to the presence of the "concealed" maximum and the inflection point on the melting curve, during crystallization of melts of the corresponding composition (in our case containing 60—61 % MgO), resorption of the separated crystals occurs (in

† [T.N. This is apparently an error and should be Rankin and Wright (reference: Am. Journ. Sci., 4th series; 39, 5, 1915].

the case of forsterite), resulting from the normal course of crystallization. Temperatures of the invariant points are shown on the diagram† (see page
195 128). The numerical characteristics of the inflection point are approximately 0, 0, 49 and 51.

The substance $MgSiO_3$ crystallizes polymorphously from melts at high temperatures in the form of clinoenstatite. In nature it is usually found in the orthorhombic modification — enstatite, which may also be attained artificially. Clinoenstatite is not found as a naturally occurring mineral in the Earth, but is known, however, from meteorites.

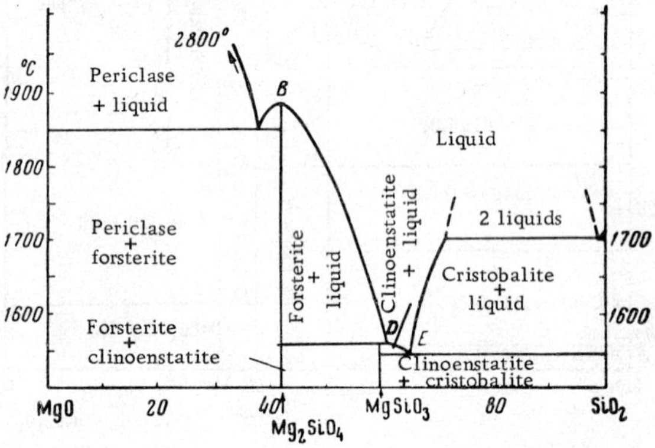

FIGURE 128. Phase diagram of the system $MgO-SiO_2$ (after Andersen and Bowen)

[For a long time the character of inversion remained unclear; suggestions favoring pseudosymmetry were even proposed, as in the case of orthoclase (due to twinning). Although even today this inversion can still not be readily performed reversibly, its enantiotropic character has been ascertained beyond all doubts. However, the inversion has proved to be very sluggish so that the transition from the monoclinic form to the orthorhombic form cannot be observed directly as temperature decreases, and the inversion point has been determined as shown on the diagram††.

As regards the Mg amphiboles (in particular antophyllite), they necessarily contain hydroxyl radicals, i. e., they are formed from a more complex system and the problem of their genesis is considered further on (see page 353)].

† [According to new data of V. D. Nikitin (1941), the temperatures of the most important points, D and E, are increased by approximately 10° C as compared to Bowen's values, but the melting temperature of Mg_2SiO_4 should be lowered to 1,860° C and that of the eutectic point, C, to 1,760° C].

†† [Recently the existence of an additional modification was noted, which is probably unknown in nature and stable only at high temperatures — protoenstatite. Its relation to clinoenstatite is not yet sufficiently clear (see Eitel, 1954)].

The system Fe_2SiO_4-SiO_2**

Investigations of systems containing ferrous iron at high temperatures present great technical difficulties. The above-mentioned system was investigated by Bowen and Schairer as late as 1932 — much later than the systems described previously. All the work was carried out in crucibles of pure electrolytic iron in an atmosphere of nitrogen in order to prevent
196 oxidation. Despite these precautions a small amount of the ferrous iron was oxidized according to the reaction

$$3FeO \rightleftarrows Fe + Fe_2O_3,$$

i.e., the system is actually ternary. However, the admixture of ferric iron is quite small, especially in that part of the system which is rich ir silica (less than 5 %), and may be ignored when constructing the corresponding diagram shown on Figure 129.

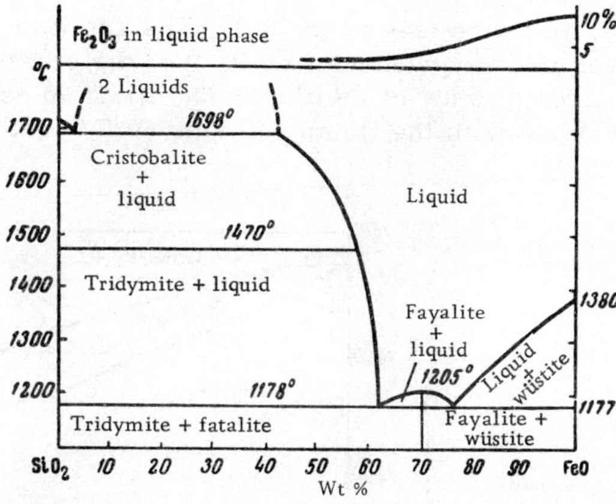

FIGURE 129. The system SiO_2-FeO (after Bowen and Schairer)

As can be anticipated from mineralogical data, the system contains only one compound (corresponding to fayalite), showing congruent melting and the corresponding eutectic points: 1) between fayalite and tridymite at a temperature of 1,178° C and a eutectic composition of 62 % FeO and 38 % SiO_2 (i.e., 88 % fayalite, 12 % tridymite) and numerical characteristics b = 58 and s = 42; and 2) between fayalite and wüstite at a temperature of 1,177° C and a eutectic composition of 76 % FeO and 24 % SiO_2 (i.e., 81 % fayalite, 19 % wüstite).

Bowen described a natural metasilicate of iron which he termed ferrosilite. This allowed for the assumption as to the possible formation of metasilicates at low temperatures (below 660° C). However, such a mineral was never subsequently found, and due to the absence of analytical data (only optical data were presented), its existence should be regarded as rather doubtful.

Olivines and orthorhombic pyroxenes**

The system Mg_2SiO_4-Fe_2SiO_4, despite its great significance, was only rather recently investigated by Bowen and Schairer (1935) as a section of a corresponding ternary system. As had already been presumed by Vogt (1921, 1922) et al. on the basis of purely mineralogical data, we are here confronted with a continuous series of solid solutions with a melting diagram of Type I (Figure 130). The difference in the compositions of the liquid and solid phases is considerable, but somewhat lower than in plagioclases. Zoned crystals are readily obtained. The lesser distribution of zoned crystals in rocks (as compared to plagioclases) may be explained on the one hand by the greater mobility of the ions and on the other by the
197 fact that the zoning is not so evident here due to the rhombic symmetry†.

The section $MgSiO_3$-$FeSiO_3$ proved to be pseudobinary within the same ternary system (Figure 131) due to the incongruent melting of $MgSiO_3$ and the absence of the compound $FeSiO_3$. As already mentioned for the system MgO-SiO_2, at high temperatures a variety of pyroxenes crystallizes which is encountered in rocks as solid solutions together with pyroxenes of the diopside-hedenbergite range (see page 199). The addition of the iron component initially causes the congruency of melting of the corresponding pyroxene and later the incongruency of melting. The limiting composition of a pyroxene in equilibrium with the liquid contains 60.9 % $FeSiO_3$.

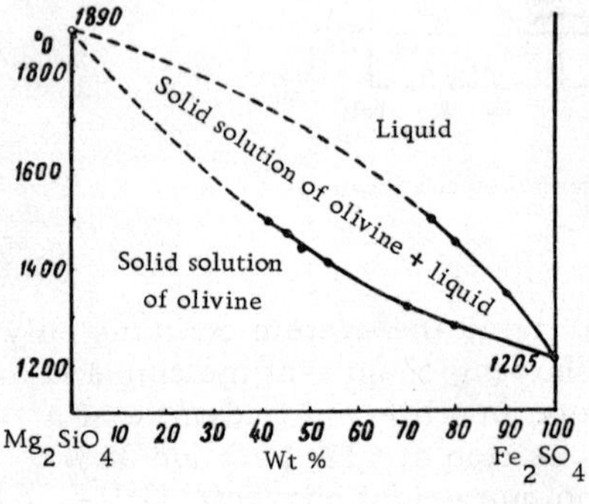

FIGURE 130. The system Mg_2SiO_4-Fe_2SiO_4 (after Bowen and Schairer)

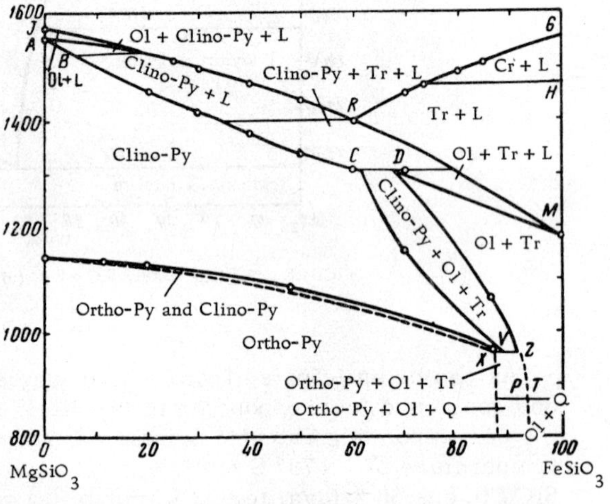

FIGURE 131. The pseudobinary system $MgSiO_3$-$FeSiO_3$

Ol — olivine; Clino-Py — monoclinic pyroxene; Cr — cristobalite; Tr — tridymite; Ortho-Py — orthorhombic pyroxene; Q — quartz

† Some authors have assumed the existence of another heavier modification of Mg_2SiO_4 possessing a spinel structure, formed at high pressures in the depths of the Earth's crust (Abelson, 1959). Such a modification of Fe_2SiO_4 has already been determined.

During investigation it was proved that orthorhombic pyroxenes constitute an enantiotropic polymorphous modification and the corresponding inversion curve has been constructed. Although the inversion from the monoclinic to the rhombic modification was not achieved with the lowering of temperature, the reverse inversion on increasing the temperature showed that each composition has a corresponding inversion point (or more precisely — a very narrow range of inversion). Below this point the inversion does not
198 occur, whereas above this point orthorhombic pyroxene inverted comparatively rapidly to the monoclinic modification. The only knowledge which is available on the space between the curves is that it is very narrow. This is shown on the diagram.

The figure shows that as temperature decreases the series of pyroxenes is considerably widened and the limiting composition of orthorhombic pyroxenes contains up to 85 % ferrous component†. Such rhombic pyroxenes were first described by Sundius from an eulysite in Sweden and were later encountered in certain other rocks, in particular in coarse-grained diabases (diabase-pegmatite) from the River Vilyuya (Sobolev, 1936).

As in the case of intrusive rocks, pyroxenes of the composition (MgFe) SiO_3 are always present as the orthorhombic modification. This may serve to indicate that the temperature of crystallization was in this case lower than the inversion temperature, i. e., 1,150—1,000° C.

The relationship between olivines and pyroxenes is considered in detail below.

The system $CaSiO_3$-$MgSiO_3$**

This system is a part of the ternary system CaO-MgO-SiO_2. The incongruent melting ("concealed" maximum) of clinoenstatite gives rise to the peculiarities of the system which will be discussed during the consideration of the ternary system diopside-forsterite-silica. We shall firstly mention the peculiarities of the part of the curve corresponding to the system diopside-pseudowollastonite (Schairer and Bowen, 1942). On the melting diagram an obvious maximum corresponds to diopside (Figure 132) which is not sharply expressed as this mineral should be considered as a chemical compound of the composition $CaMgSi_2O_6$, the melting temperature of which is 1,391° C. At this temperature the melting of diopside is not accompanied by any decomposition. (Later investigations failed to affirm the presence of solid solutions of the $CaSiO_3$ component in diopside). Diopside and calcium metasilicate form a eutectic of composition 28 % $CaSiO_3$ and 72 % $CaMgSi_2O_6$, which melts at 1,357° C.

The influence of diopside on the polymorphism of $CaSiO_3$ is of great
199 interest. As we have already noted, pseudowollastonite inverts to wollastonite at a temperature of 1,200° C. The addition of diopside, which forms solid solutions with $CaSiO_3$, considerably raises the inversion temperature. With the maximum content of diopside in the solid solution (17 %) it increases to 1,345 ± 10° C. This increase of the melting temperature is associated

† Recently an even richer hypersthene with an $FeSiO_3$ content of 92 % (molecular) was determined by E. N. Ushakova in the crystalline schist of Pobuzh'e (Ukraine).

with the greater solubility of the magnesium component in the low temperature modification of $CaSiO_3$ as compared to the high temperature modification.

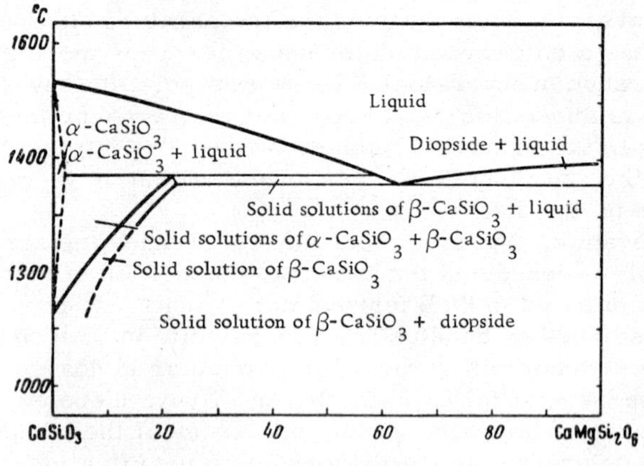

FIGURE 132. The system $CaSiO_3$-$CaMgSi_2O_6$ (after Bowen and Schairer)

The relations between diopside and $MgSiO_3$ are more complex. As we have seen, the compound $MgSiO_3$ has an incongruent melting point. During melting it decomposes into Mg_2SiO_4 and liquid. Therefore one cannot speak of the binary system $CaMgSi_2O_6$-$MgSiO_3$, but is obliged to consider the ternary system Mg_2SiO_4 (forsterite)—$CaMgSi_2O_6$ (diopside)—SiO_2 (silica). This system has been investigated and will be described in detail further on. Here we shall only mention that diopside may contain $MgSiO_3$ (pigeonite) in considerable amounts in the form of a solid solution. The monoclinic variety of $MgSiO_3$ is capable of dissolving considerable amounts of the diopside component.

Figure 133 presents two variants of the corresponding pseudobinary sections of the ternary system Mg_2SiO_4-$CaMgSi_2O_6$-SiO_2 as shown hereafter (see Figure 144). Figure 133, I, is in accordance with formal data, according to which a continuous series of solid solutions is present, and Figure 133, II, is based on recent investigations of Boyd and Schairer, who established the presence of a small miscibility gap. However, this diagram is of a preliminary nature. Figure 133, II, also shows the composition curves of solid solutions, which permit one to assess the temperature of formation of certain pyroxenes. Using this diagram as a basis, one may assume that pigeonites of gabbro diabases, which contain 25% $MgSiO_3$, correspond to temperatures of the order of 1,200° C. However, these figures apply to pure magnesium pyroxenes and are probably considerably lower in the presence of the iron component. Judging from the fact that the mentioned pigeonites often crystallize together with orthorhombic pyroxenes, their crystallization temperature should probably be less than 1,100° C.

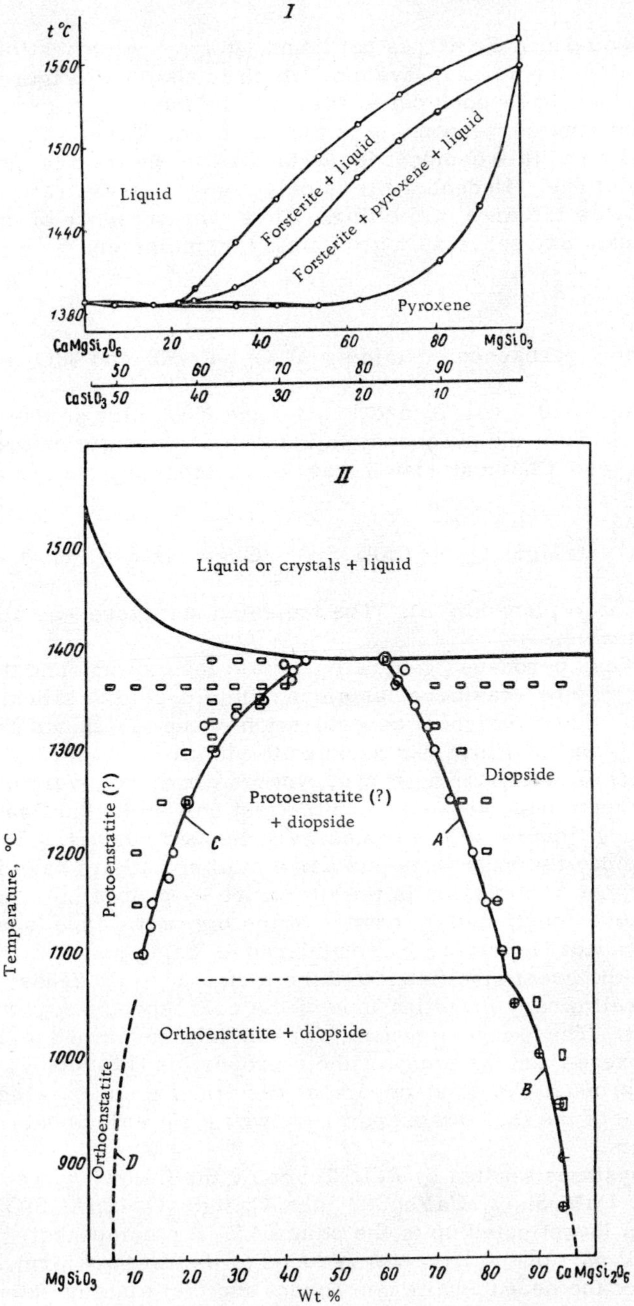

FIGURE 133. The pseudobinary system $CaMgSi_2O_6$-$MgSiO_3$

I — after Bowen; II — after Schairer and Boyd. The squares and circles denote experimentally obtained points in the various fields

The system $CaSiO_3$-$FeSiO_3$**

As the compound $FeSiO_3$ is not found, this system constitutes a pseudo-binary section of a ternary system, which is shown on Figure 134. In contrast to the above-considered system, the melting diagram in this case has no maximum corresponding to the compound $CaFeSi_2O_6$, and the point corresponding to this composition is located in the field of crystallization of solid solutions. Hedenbergite is only stable at temperatures below 940° C — below the area of melting. Thus, the presence of this mineral in certain rocks may serve as a geological thermometer.

Monoclinic pyroxenes containing Al_2O_3, Fe_2O_3 and TiO_2 — augites**

The sesquioxides (Al_2O_3 and Fe_2O_3) and TiO_2 play an essential role in the composition of the pyroxenes in igneous and certain metasomatic rocks. Some time ago Chermak stated that the replacement occurs according to the scheme

$$CaMgSi_2O_6 \rightarrow CaAl_2SiO_6 \text{ (Chermak's silicate),}$$

i. e., MgSi is replaced by Al. The indicated structure was also affirmed by X-ray analysis.

201 Thus, these pyroxenes generally contain less silica, and the aluminum component may be considered as an anorthite devoid of silica. Therefore it is obvious that the highest concentration of the mentioned oxides in pyroxenes is present in rocks unsaturated in silica — leucitic, nephlinitic and (in part) olivine-bearing rocks, whereas in rocks oversaturated in silica the mentioned pyroxenes are present only in insignificant amounts.

For a long time these pyroxenes were termed augites. This term is still used today although Hess and certain other authors have introduced a certain amount of confusion in the literature by proposing to term as augites also the above-mentioned pyroxenes of the pigeonitic type (see page 214). This step cannot, however, be considered as being justified.

Despite the great significance of the mentioned pyroxenes, to the present day only preliminary investigations of the corresponding system have been carried out. These investigations were mainly concerned with the synthesis of pyroxenes and the study of their properties (Tsvetkov, 1951). We shall only produce the most important results of the above-mentioned author, making note of the fact that he had reviewed all previous data pertaining to this problem.

Of the systems studied by A. I. Tsvetkov the following are the two most important: $CaMgSi_2O_6$-$CaFe_2SiO_6$ and $CaMgSi_2O_6$-$CaAl_2SiO_6$. The former system was investigated up to the point 50 % iron component. He noted that a small part of the iron was reduced to the ferrous form, with the formation of the hedenbergite component and that alumina derived from the crucible was introduced into the pyroxenes. The limiting investigated pyroxene consisted of 41 % (molecular) of the component $CaFe_2SiO_6$, which on recalculation to a binary system gives 47 %. The limit of miscibility was not ascertained.

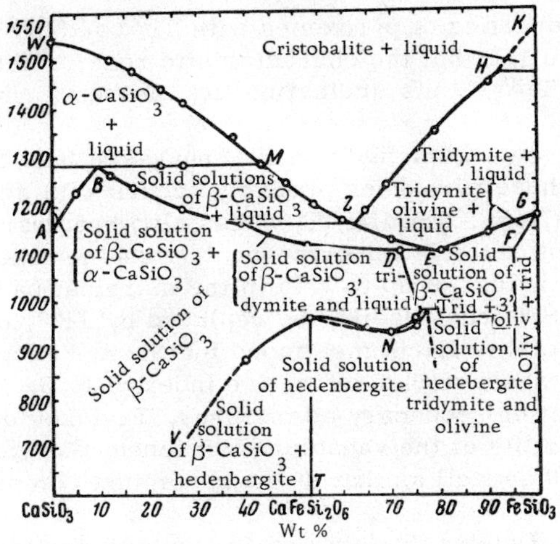

FIGURE 134. The pseudobinary system $CaSiO_3$-$FeSiO_3$

The system $CaMgSi_2O_6$-$CaAl_2SiO_6$ proved to be even more complex, as during rapid crystallization the melt contained more than 50 % melilite, which was only successfully removed after remelting three times.

202 In melts rich in aluminum the final product of crystallization consisted of anorthite, melilite and spinel in addition to the pyroxenes. During the crystallization of the initial mixtures with 40 % and 50 % $CaAl_2SiO_6$ component, pyroxenes of an approximately equivalent composition were formed (with a $CaAl_2SiO_6$ content of approximately 40 %) with a small admixture of the $MgSiO_3$ component (up to 5 %).

A. I. Tsvetkov assumed that in the given system the limiting solubility of Chermak's component in diopside had been attained.

Another essential feature of the obtained pyroxenes which merits attention is their optical properties. Whereas ***cNg*** and ***2V*** increase sharply with the increase of Fe_2O_3 content in iron augites, the value of ***2V*** in the alumina augites remained constant and that of ***cNg*** even decreased somewhat. Thus the increase of ***cNg***, which is considered as the characteristic criterion for augites, does not provide us with any definite idea of the content of the most important alumina component, which may easily be missed during optical investigations.

As regards augites containing trivalent titanium, Tsvetkov mainly produces data on the system $CaMgSi_2O_6$-$CaTiAlSiO_6$. He obtained pyroxenes with a content of up to 40 % titanium component. For systems lacking aluminum, he obtained pyroxenes with a 10 % content of the component $CaTi_2SiO_6$ and arrived at the conclusion that in the absence of aluminum trivalent titanium enters the pyroxenes with greater difficulty.

Ti^{4+} has been investigated in the system $CaMgSi_2O_6$-$CaTiAl_2O_6$ and the limiting content of the latter component was found to be approximately 30 %.

In melts with 40 % $CaTiAl_2O_6$ component perovskite is present together with pyroxenes, which have optical properties similar to those mentioned above, i. e., the limit of solubility has probably been attained in this case.

The optical properties of pyroxenes with Ti^{3+} and Ti^{4+} proved to be comparatively similar, but the characteristic rose-violet coloration is associated with Ti^{3+}, thus indicating its presence in natural augites.

Data on this system $CaMgSi_2O_6$-$CaTiFe_2O_6$ are less convincing, and the formation of these pyroxenes probably occurs with great difficulty.

The system $CaMgSi_2O_6$-$CaMgTi_2O_6$ has also been partially examined. With a 10 % content of the latter component, zoned pyroxenes were obtained with *cNg* reaching values of up to 1.75 in the marginal part, thus affirming Barth's view that Si^{4+} can possibly be replaced by Ti^{4+} in a fourfold coordination. In actuality Barth mentioned that he had succeeded in obtaining homogeneous pyroxenes with a refractive index of 1.73.

While investigating secondary admixtures, Tsvetkov determined the considerable solubility of the vanadium component CaV_2SiO_6 in pyroxenes (greater than 10 %) as well as that of the chromium component $CaCr_2SiO_6$ (up to 15 %).

Thus, although Tsvetkov's data provide us with certain essential new ideas on the formation of augites and their properties, they do not allow for the construction of the corresponding phase diagram of binary systems (melting points and the relations between liquid and solid phases were not determined). The investigation of these binary systems should be undertaken in the near future.

Melilites**

Some idea on the crystallization of these comparatively rare but nevertheless important magmatic minerals is obtained from the system $Ca_2MgSi_2O_7$-$Ca_2Al_2SiO_7$ as investigated by Fergusson and Buddington
203 (1920) and corrected by Osborn. In this case we are dealing with a comparatively simple diagram of Roozeboom's Type III, i.e., a continuous series of solid solutions with a minimum point on the melting curve (Figure 135). This readily explains the presence of zoned crystals with various sequences of zoning as determined by the optical properties of the minerals. One should, however, bear in mind that natural melilites always contain an admixture of another component, probably $NaCaAlSi_2O_7$, i. e., they have a somewhat more complex composition than shown in the investigated system.

THE RELATIONS BETWEEN FEMIC AND SALIC MINERALS**

Previously (see page 194) we considered the relations between femic minerals and silica. In some cases orthosilicates form compounds of the metasilicate type with silica, which compounds form a eutectic with SiO_2. In other cases (for ferrous iron) the metasilicate is not formed, and eutectic relations occur directly between the orthosilicates and silica. Further on we shall consider systems in which femic and salic minerals of the aluminosilicate type are included.

The investigated systems show two types of relations. Equilibrium diagrams with eutectic points are related to the first type: anorthite-diopside (Figure 136; Bowen, 1915), albite-diopside, albite-fayalite (Figure 137; Bowen and Schairer, 1937) and leucite-diopside (Figure 138; Bowen, 1937). Attention should be drawn to the change in quantitative relations at the eutectic point. In systems with anorthite and leucite, the eutectic point
204 lies approximately in the middle and is possibly nearer to the femic constituent. The salic mineral in the system is often the first to crystallize, and melts which approach the eutectic, contain 50 % or more of the femic mineral. In the system albite-diopside, a sharp shift of the eutectic point toward albite is observed (see Figure 148), and the diopside content at this point is less than 3 % so that little can be seen from the diagram, even if drawn to scale. In a system with ferrous femic minerals (fayalite), the eutectic with albite contains a somewhat larger percent of the femic mineral (18 % instead of 3 %) which is nevertheless yet small. When referring to the incongruent melting of orthoclase one cannot speak of a simple eutectic. However, the crystallization of leucite already starts at a content of 40 % diopside, and eutectic relations are only similar to those in the system albite-diopside if a large excess of silica is present.

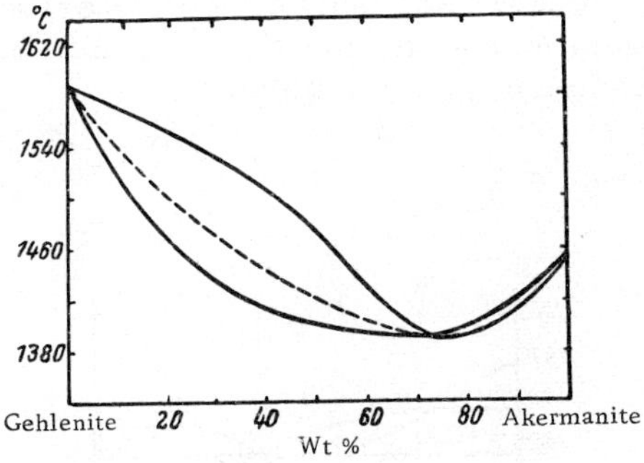

FIGURE 135. The system gehlenite-akermanite (after Osborn and Schairer)

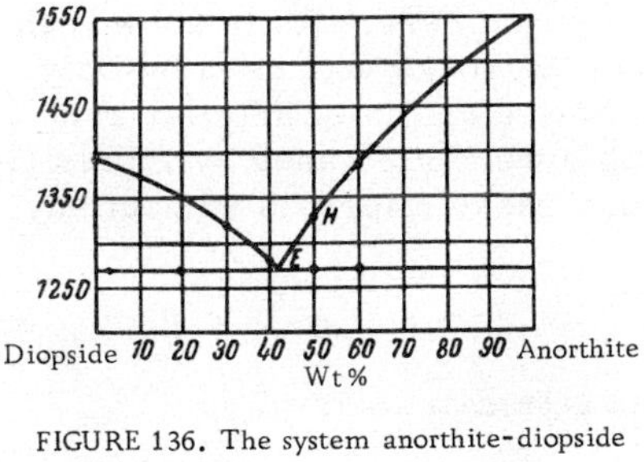

FIGURE 136. The system anorthite-diopside (after Bowen)

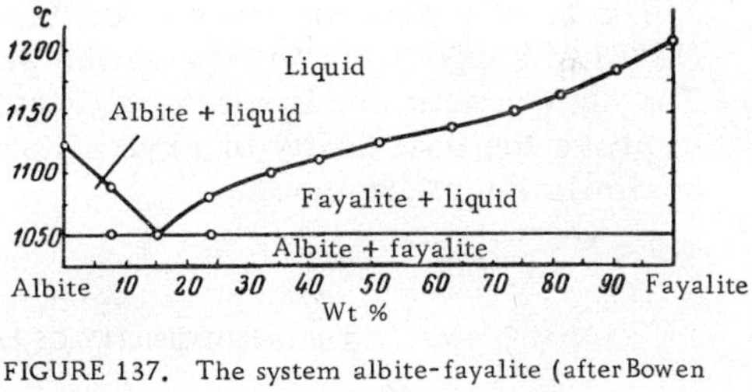

FIGURE 137. The system albite-fayalite (after Bowen and Schairer)

The regular change in the content of femic minerals in rocks, accompanied by a change in the basicity of the plagioclase, permits one to assume that the corresponding magmas were close to the eutectic composition (for greater detail see page 279).

For the sake of accuracy it should be noted that the above-mentioned system deviates somewhat from a true binary one owing to a small content of Al in the pyroxenes which is usually overlooked when constructing this diagram. The well-expressed pseudobinary type of relationship is characteristic of the systems anorthite-forsterite, and especially diopside-nepheline (also nepheline-fayalite, see Figure 153). In this case phases crystallize, the compositions of which may be represented by more complex systems (ternary and quaternary), and a two-dimensional diagram is only capable of showing a corresponding section (in the plane of which the compositions of certain phases are not depicted). As shown on Figure 146 (see later) spinel constitutes such a phase in the system forsterite-anorthite. Spinel later reacts with the liquid and disappears completely at equilibrium.

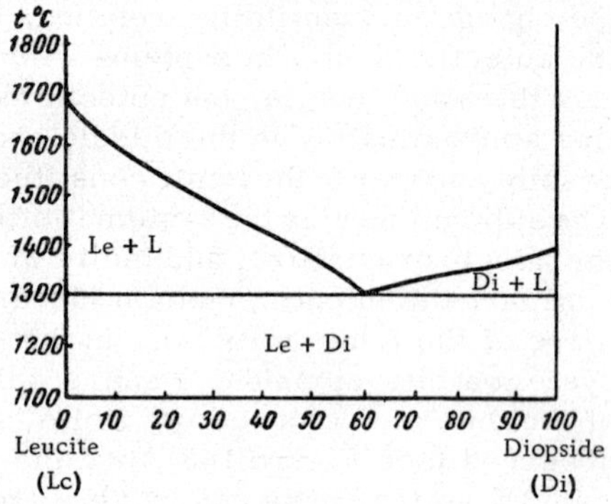

FIGURE 138. The system leucite-diopside (after Bowen and Schairer)

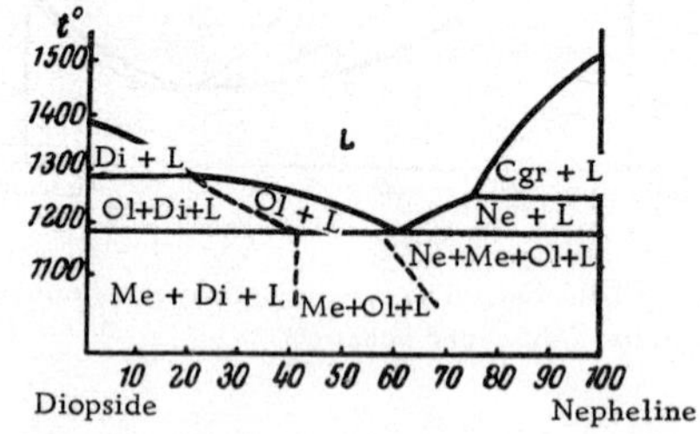

FIGURE 139. Schematic diagram of pseudo-binary system diopside-nepheline (after Bowen). Temperature of complete consolidation unknown

Di - diopside; Ol - olivine; Cgr - carnegieite; Me - melilite; Ne - nepheline; L — liquid

In the system diopside-nepheline (see Figure 139) even more complex relations are encountered. Calcium-rich minerals — monticellite and melilite — crystallize as the intermediate phase here. As is known, the presence of these minerals in certain alkaline rocks was used by Daly (1936) to support his theory on the assimilative origin of alkaline rocks.
205 The above-mentioned relations (Figure 139) were also used by Bowen (1934) to prove the possibility of crystallization of these minerals without any assimilation of limestone.

THE IMMISCIBILITY OF LIQUIDS IN SILICATE MELTS**

Immiscibility in binary systems

The immiscibility of silicate melts with iron sulfide and sulfides of other heavy metals has been known for a long time from metallurgical data; an hypothesis was proposed as to the liquation origin of certain sulfide ores.

Although considerations on the possible immiscibility of natural silicate melts have appeared in the works of certain petrographers (especially F. Yu. Levinson-Lessing), the first experimental data were only obtained by Greig in the year 1927. He stated that in a series of binary systems, especially of the type SiO_2-RO, the formation of two immiscible liquids was observed, one of which was relatively rich in silica (the content of the other oxide did not exceed 3 %), while the other was richer in the second component (RO), but nevertheless contains a considerable amount of silica (of the order of 60—70 %), i. e., considerably more than in the corresponding metasilicate. The first liquid, which is sometimes termed the paramelt, is very viscous, whereas the second (the orthomelt) is relatively fluid. It was noted that such a split into two liquids occurs in the following systems of significance to petrography: SiO_2-MgO; SiO_2-FeO; SiO_2-CaO.

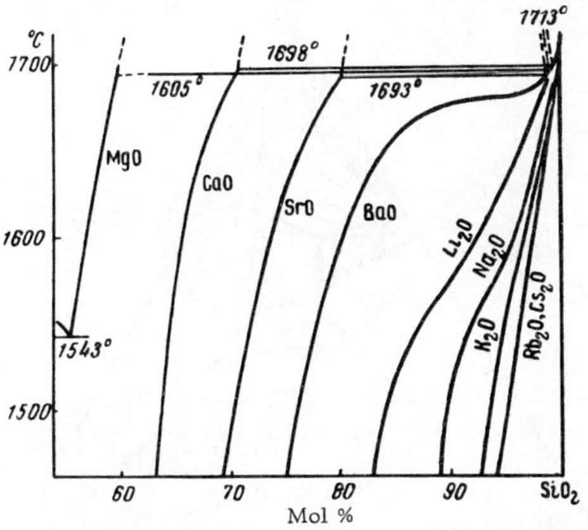

FIGURE 140. Liquid immiscibility in the systems SiO_2-RO and SiO_2-R_2O (after Kracek)

Splitting also occurs in the system SiO_2-Fe_2O_3, but in this case the second liquid is of a different type; it consists mainly of Fe_2O_3 with a small (up to 5 %) content of silica.

On the other hand, it has been determined in the systems SiO_2-Al_2O_3, SiO_2-Na_2O and SiO_2-K_2O that complete miscibility is present and that the splitting of liquids is completely absent. If several diagrams are compared (Figure 140) it becomes evident that a gradual change in the cations is accompanied by a gradual change on the diagram. Diagrams of the intermediate type — the system SiO_2-BaO (see Figure 140) — are characterized by the appearance of a smooth inflection on the melting curve, which appears to replace the range of immiscibility. It is interesting to mention that a similar inflection is also observed in the system SiO_2-Al_2O_3.

The obtained data were greeted with enthusiasm by the supporters of the theory of magma liquation, who explained the formation of various rocks by the differentiation of magma in the liquid phase. However, the relations in more complex ternary systems showed that the conclusions reached were somewhat premature, and the experimental data serve rather as evidence negating the possible liquation of magma.

206 Immiscibility in ternary systems

Ternary systems of the following type are most important in petrography: the first component is silica; the second component is an oxide which forms an immiscible liquid with silica; the third component is an oxide, which is completely miscible with silica in the liquid phase. An example of such a system is shown on Figure 141. A characteristic feature of all such systems is the extraordinary narrowness of the areas of immiscible melts. The addition of only a small amount of the third component results in the formation of only one liquid phase.

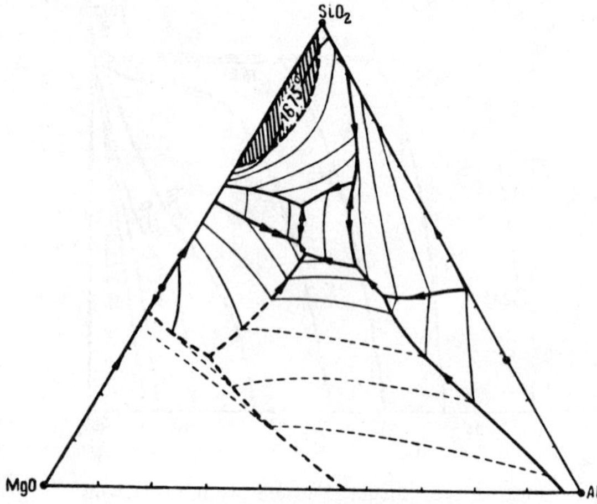

FIGURE 141. An example of liquid immiscibility (liquation) in a ternary system. The immiscibility field is shaded

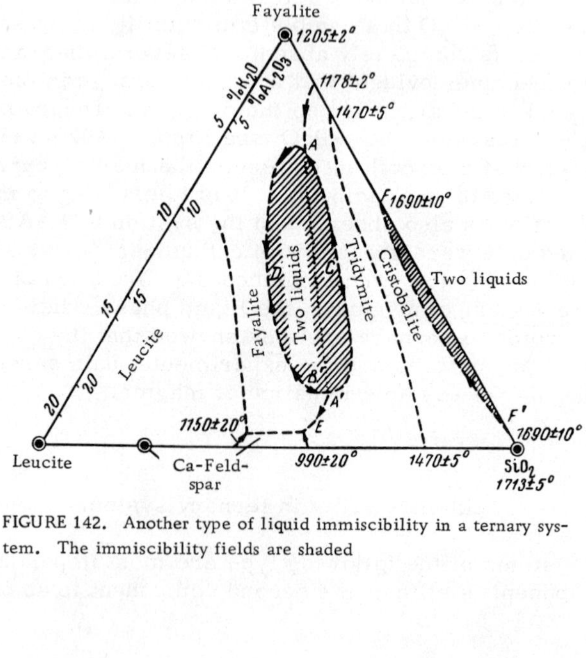

FIGURE 142. Another type of liquid immiscibility in a ternary system. The immiscibility fields are shaded

207 Recently certain amendments were introduced after the investigations of Roedder (1951), who discovered another area of liquid immiscibility in the system K_2O-FeO-Al_2O_3-SiO_2. The most important part of this system, indicating the area of immiscibility of two liquids, is shown on Figure 142. In ternary systems, in which the third component is also immiscible with silica in the liquid phase, a continuous field of immiscible liquids appears, which intersects the concentration triangle (see page 211, Figure 144).

Certain new data on the increased area of immiscibility in ternary systems due to the participation of SiO_2 and phosphates are presented by Fischer.

Immiscibility in multicomponent systems and the problem of liquid immiscibility in natural magmas

As already mentioned above, data on the formation of two liquids in certain silicate systems were accepted as a definite proof of such phenomena in the magma. Greig (1928) presented data on the ternary system which raised objections to the above-mentioned assumptions, and produced a diagram (Figure 143), on which several systems of the above-mentioned type were compiled and lines were drawn, limiting the areas of immiscibility. He assumed there are no changes in multi-component systems and that the area of liquid immiscibility is not widened. On the same diagram were plotted the analyses of certain igneous rocks recalculated to the corresponding system. These analyses, including those of certain rare types of rocks, show up on the diagram as points far beyond the limits of liquid immiscibility. In actuality, rocks relatively rich in silica always contain a considerable amount of alkalis and alumina, i. e., they are located near the side SiO_2-(K_2O, Na_2O, Al_2O_3) of the triangle, whereas rocks poor in alkalis and alumina (ultrabasic and similar rocks) contain considerably less silica than the other liquid (orthomelt).

Roedder's investigations oblige us to make an essential amendment to the mentioned diagram, as the area of immiscibility is located on that part of the diagram upon which points of analysis of rocks are found. It is true that all the mentioned rocks will differ markedly from the system shown on Figure 143 due to the presence of other components which are not represented on this figure. The influence of these components is as yet unknown.

However, long before Roedder carried out his experiments, Greig's hypothesis had met with other objections: it was shown that his conclusions could not be accepted as being correct until such time as investigations of silicate systems containing water and other volatiles had been carried out, as these investigations may essentially change the behavior of melts and cause the formation of two liquid phases. These considerations were strongly supported by experiments of D. P. Grigor'ev, who noted a typical splitting into two liquids in the case of complex silicate melts, rich in fluorine (fluorine content of the order of 3—5 %). Projections of the points representing melts in which D. P. Grigor'ev (1935, 1937) obtained immiscible liquids, were plotted on Figure 143. It is evident that

these points lie within the area of compositions of igneous rocks. Despite the objections raised in literature, D. P. Grigor'ev's data may be considered as completely trustworthy as regards the obtained facts. However, it is obvious that such a high content of fluorine is never found in natural magmas. Despite the general similarity in the behaviors of fluorine and
208 water, they are known to differ in various cases. It is true that certain authors have considered the absence of splitting in certain silicate melts as being merely apparent — due to the high viscosity of the liquid; fluorine causes splitting as it lowers the viscosity, and in this respect could also be replaced by other mineralizers.

Here we shall not dwell in detail on data dealing with investigation of silicate systems containing H_2O. We shall confine ourselves to a brief summary of the conclusions pertaining to the problem. At present it has been proved that the solubility of water in silicate melts similar to natural magmas is quite low (less than 10%) even up to the highest pressures, and that areas of immiscibility of silicate liquids exist in the presence of water, whereas they are absent in dry melts. Immiscibility in the system Na_2O-SiO_2-H_2O is often spoken of. However, this system differs essentially from natural magmas, and the second liquid is in actuality already an aqueous solution and not a melt.

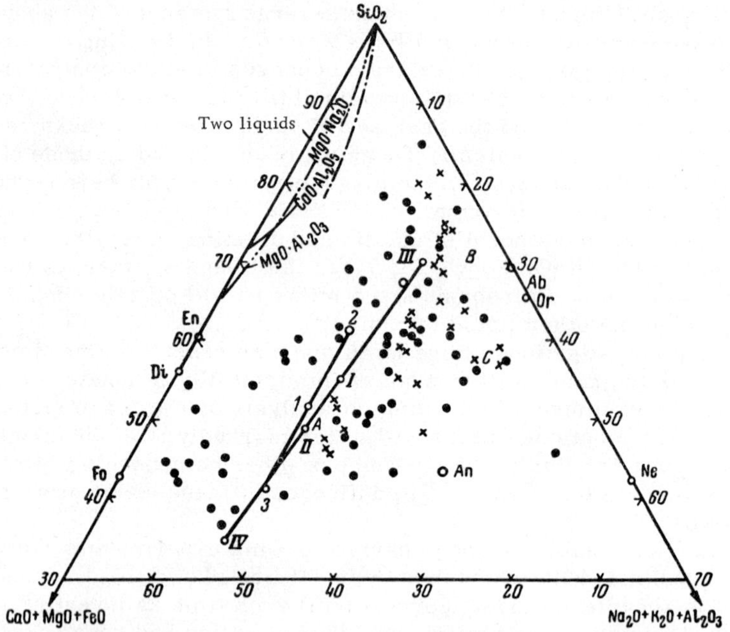

FIGURE 143. Summary diagram of liquid immiscibility in silicate melts (after Greig with amendments after A. S. Ginsberg)

A — basalt, gabbro; B — rhyolite, granite; C — phonolite, nepheline syenite. Filled-in circles represent igneous rocks; blank circles represent minerals; blank circles on dark lines represent the composition of fluorine fluxes in two experiments of D. P. Grigor'ev. 1 — initial flux; 2-3 — glasses obtained on liquation. I — initial composition; II — composition of the light upper layer; III — summary composition of the lower emulsion layer; IV — composition of the heavy fraction of the emulsion Di — diopside; Ab — albite; Ne — nepheline; An — anorthite; En — enstatite; Fo — forsterite; Or — orthoclase

Thus, on summarizing available material one may arrive at the conclusion that although experimental data have not proved beyond all doubts that natural magmas cannot or do not split up into two immiscible liquids,
209 these data appear to favor this conclusion, but can in no way be regarded as a proof of the phenomena of magma liquation.

In addition to the utilization of experimental data we may arrive at indirect conclusions on the physicochemical processes in the Earth's crust in general, and in magma in particular, by studying the facts of petrography and mineralogy. This material was used by petrographers in claiming the presence of magma liquation, and such textures as appeared to form as the result of crystallization of an emulsion of two liquids in the initial stage of its formation were regarded as an indirect proof of liquation — when one liquid forms spheroidal bodies within the other. Variolitic and spherulitic textures were regarded as examples of these (F. Yu. Levinson-Lessing, 1934, 1935).

In all mentioned examples, however, crystalline formations are present. It is well known that such spheroidal aggregates may appear merely as the result of the specific course of the crystallization process in a homogeneous liquid medium. Naturally, the existence of emulsions could be proved directly by their presence in glasses. Many of the volcanic rocks possess a glassy groundmass. If phenomena of liquation were sufficiently widespread, it would be natural to assume that in some cases a flow of lava should occur which would correspond to the initial stage of splitting. However, despite a long and thorough search, no spherulites of one type of glass were ever encountered in glass of another type. Such glasses were claimed to have been found, but on closer inspection the spheres always proved to be spherulites, i.e., products of devitrification of glass.

Various authors have provided other data which negate the possibility of liquation of magmas. Confining ourselves to the above-outlined facts, we may already arrive at the following important conclusions: if two immiscible liquids do form in natural magmas, this process is in all events a rare one and cannot be of primary significance in the differentiation of magma and the formation of igneous rocks.

Despite the fact that the formation of two-silicate liquids has long been a source of controversy, most authors accepted without any doubts the examples of splitting into two liquids, as shown in the peculiar sulfide ores of the Kola Peninsula, Noril'sk, Sudbury (Canada) and others. The sulfide portion is represented by almost pure pyrrhotite with copper and nickel sulfides, and the silicate portion — by ultrabasic or basic rock. The drops of the sulfide liquid appear to have concentrated at the bottom part of the ore body due to their greater density. It should, however, be mentioned that both in the Soviet Union and abroad, this explanation of the genesis of the above-mentioned deposits has met with serious objections, and several if not all of these sulfide deposits have been ascribed to a metasomatic origin. Considerations on the liquation genesis of certain iron ore and phosphate deposits are even more controversial (Fischer, 1950).

CHAPTER X

TERNARY SYSTEMS

GENERAL COMMENTS ON TERNARY SYSTEMS**

As in the section on binary systems, not all available experimental data on silicates are presented here, but only those systems which are of greatest importance in the interpretation of the crystallization processes in natural magmas. Accordingly, an entire series of investigated systems was omitted (see Eitel's handbook for comparison). Such systems as $MgO-Al_2O_3-SiO_2$ and $CaO-MgO-SiO_2$ are described in brief. At the same time various areas in other systems are given a most detailed treatment with regard to their crystallization processes, and some comparative data are cited from petrology and mineralogy.

As a rational classification of ternary systems has not yet been evolved, we adhere to the sequence adopted in the final chapter of A. N. Zavaritskii's book "Introduction to the Petrochemistry of Igneous Rocks", and have confined ourselves to the introduction of only a few of the recently determined systems.

As in the previous chapter, we shall present together with strictly ternary systems also such systems as are pseudoternary in individual areas (mentioning the corresponding restrictions).

DIOPSIDE-FORSTERITE-SILICA*

The ternary system diopside-forterite-silica was investigated in detail by Bowen (1914, 1915). As the basic material for his experiments, he used especially purified calcium carbonates and MgO, which he obtained by the calcination of the corresponding carbonate. Equilibria were investigated by the quenching method, followed by a microscopic examination for the determination of the phases. The melting temperature of forsterite was measured optically in an iridium oven.

The binary system $Mg_2SiO_4-SiO_2$, which had been investigated earlier (see Figure 128), is characterized by the presence of the incongruently melting compound $MgSiO_3$. The peritectic point was found at a temperature of 1,567° C.

The system $CaMgSi_2O_6-SiO_2$ has a simple eutectic. The melting temperature of diopside is 1,391°C; the eutectic temperature at 16 % SiO_2 content is 1,580° C; on quenching, a liquidus inflection was determined at 1,470° C, corresponding to the tridymite-cristobalite inversion.

The system $CaMgSi_2O_6$-Mg_2SiO_4, with a eutectic composition of 12 % forsterite and 88 % diopside, has a eutectic temperature which is only 4 to 5 degrees lower than the melting temperature of diopside.

211 More than 80 fusions were performed in order to determine the limits of the fields. The system was maintained at equilibrium for 1/2—1 or 2 hours, depending on the acidity of the melt. The obtained diagram is shown on Figure 144, where Bowen's original data are supplemented by later data of Greig on liquation. Four fields are present here, in which forsterite, tridymite, cristobalite and pyroxene are crystallized. The temperature surfaces which form the fields of forsterite and silica slope steeply from the apices of the diagram toward the eutectic curve. The surface of the pyroxene field, which is bounded by the curves, slopes very gently downward from clinoenstatite (1,557° C) toward diopside up to the eutectic point at 1,362° C, after which it rises somewhat toward diopside†. In this field a continuous series of pyroxenes crystallizes between diopside and clinoenstatite.

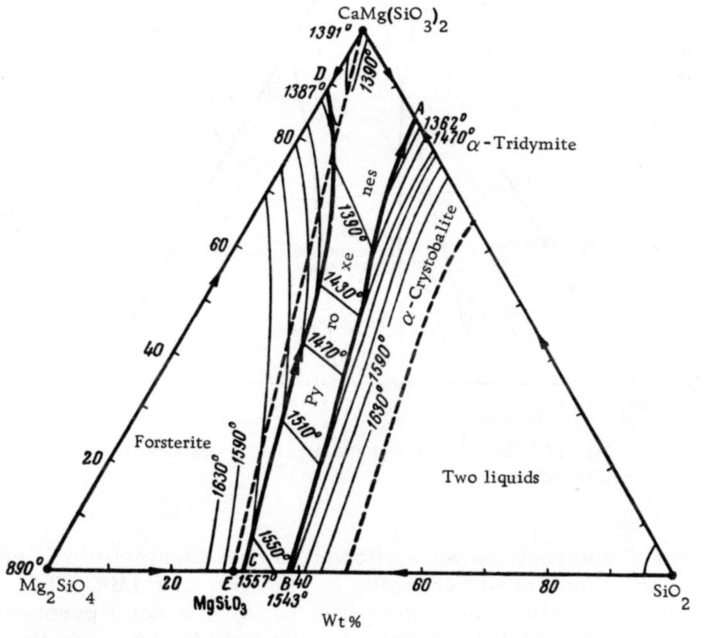

FIGURE 144. The system diopside-forsterite-silica (after Bowen)

On Figure 133 (see previously) is presented a section of the pseudobinary system $Mg_2Si_2O_6$-$CaMgSi_2O_6$, which corresponds to the series of solid solutions of pyroxenes††. Within this series, as the content of

† [On the boundary line the temperature decrease is shown up to the forsterite-diopside eutectic. However, the difference in temperatures is rather small, and the presence of a minimum point should be considered as more likely.]

†† [As already mentioned above, according to the recent data of Boyd and Schairer (Ann. Rep., 1957) this section has a somewhat different form (see Figure 133, II). If these data, which are of yet of a preliminary nature, are later affirmed, the considerations on crystallization within the ternary system presented in this section will hold true for only the lower part of the diagram. For compositions falling within the area of the immiscibility gap, two types of pyroxenes will be in equilibrium with the liquid.]

$MgSiO_3$ is increased the extinction angle ***cNg*** decreases from 38.5° to 22°; the refractive index decreases from 1.676 to 1.654; the birefringence decreases from 0.030 to 0.009. The angle of the optic axis decreases from
212 59° in the plane {010} in diopside to 0° at approximately 40 % diopside, and with a higher content of $MgSiO_3$ it increases up to 20° at 80 % $MgSiO_3$ and up to 53° for pure $MgSiO_3$, but in a plane perpendicular to {010}.

In view of the fact that the area between the diopside-clinoenstatite line and the eutectic curve is extremely narrow, one is obliged to widen this area, for the sake of clarity of geometrical construction in describing the crystallization path of the melt, by distorting the figure, but naturally preserving the general character of distribution of the curves (Figure 145).

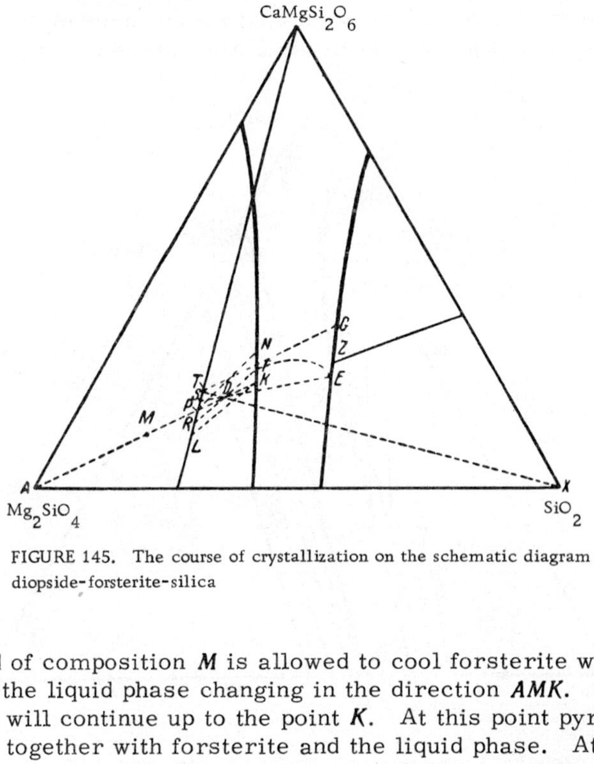

FIGURE 145. The course of crystallization on the schematic diagram diopside-forsterite-silica

If a liquid of composition *M* is allowed to cool forsterite will begin to crystallize, the liquid phase changing in the direction ***AMK***. The separation of forsterite will continue up to the point *K*. At this point pyroxene is already stable together with forsterite and the liquid phase. At this moment composition *M* lies on the *AK* side of the phase triangle ***AKL***. On further crystallization, composition *M* will for a certain time cross this field, being located within the triangle, and an equilibrium of three phases is obtained — forsterite, a pyroxene of variable composition and the liquid phase. The composition of the pyroxene, which continuously separates during this stage, may readily be determined if one knows the corresponding location of the phase triangles. When the composition of the liquid corresponds to *K* a pyroxene of composition *L* begins to crystallize out: *KL* is the side of the phase triangle. On further cooling, the composition of the crystallized and crystallizing pyroxenes changes from *L* toward diopside; the composition of the liquid phase moves along the curve to *N*. Simultaneously,

forsterite begins to redissolve. The quantitative relations of phases at any particular temperature may be readily determined geometrically from the location of the point **M** within the phase triangle. At the moment when the composition of the liquid phase reaches **N**, the composition of the separated pyroxenes is **P**. At the point **N** the liquid phase disappears and the whole consists of pyroxene of composition **P** and forsterite; **NP** is the side of the phase triangle at this temperature. At this moment the composition **M** also falls on the side of the phase triangle and then moves beyond
213 it. The proportion of pyroxene to forsterite is **MA** : **MP**. The composition of the pyroxene is determined according to the location of the point **P** on the pyroxene line.

If the composition of the original melt had been that of the pyroxene **P**, forsterite would be the first to crystallize as before: it would have continued to crystallize up to point **K**. On further crystallization, a pyroxene of variable composition starts to crystallize from **L** to **P** and forsterite will starts to redissolve.The liquid phase will change from **K** to **N**, and at point **N** it will be used up completely, point **P** reaching the apex of the phase triangle **APN**. Simultaneously the forsterite will be completely redissolved, and the whole will solidify as the pyroxene **P**.

If the composition of the original had been that of the point **D**, a somewhat different picture is obtained. Forsterite would crystallize out first as before, the composition of the liquid phase changing up to the point **K**. At point **K** a pyroxene of composition **L** starts to crystallize. During the further course of crystallization, the composition of the liquid phase will change up to point **F** and the composition of the pyroxene — up to point **R**, being accompanied by the simultaneous solution of forsterite. When the three-phase border **FR** passes through the point **D** only the pyroxene **R** and the liquid phase will remain in equilibrium in the system. At this moment, composition **D** will lie beyond the phase triangle and will cross from the three-phase to the two-phase field. Further cooling causes the crystallization of the pyroxene, which changes in composition from **R** toward **S**. The liquid phase naturally changes in such a way that its point of composition always remains on the straight line between the composition of the separated pyroxene and the initial composition **D**. At the same time the composition point of the liquid leaves the curve **KF** and, on entering the pyroxene field, it intersects the curve further on. The moment it reaches point **E** on the eutectic curve bordering the fields of pyroxene and cristobalite, the crystallizing pyroxene will be of composition **S**. The eutectic curve corresponds to the simultaneous crystallization of pyroxene and cristobalite: i. e., composition **D** again enters a three-phase field, but this time of different minerals (phases). Pyroxene, cristobalite and liquid are the phases in equilibrium in this field. The straight line **SE** is a side of the phase triangle **SEX**. On further cooling, pyroxene and cristobalite crystallize out together from the melt. The composition of the pyroxene changes toward **T**; the composition of the liquid phase changes along the curve from **E** toward **G** At a temperature corresponding to the point **Z**, cristobalite inverts to tridymite. When the pyroxene reaches the point **T**, the liquid phase, which has reached a composition **G**, disappears, and the melt is completely crystallized as pyroxene **T** and tridymite **X**. **GT** is a side of the phase triangle **GTX** at this temperature; at this moment composition **D** lies on the side of the phase triangle. It now consists only of pyroxene and tridymite. The ratio of pyroxene to dridymite is ***DX*:*DT***. The above-mentioned phenomena of crystallization in melts of the system diopside-forsterite-silica occur only in the case of perfect equilibrium between all phases at all temperatures. In the case of imperfect equilibrium, or if one of the crystallizing minerals is removed from the melt, the crystallization process proceeds differently. In such a case only the crystals which are separating at any particular instant are in equilibrium with the liquid, but those that have already crystallized may not be so.

At each moment we deal with a new initial melt. When crystallization takes place in this manner, a liquid of composition ***M***, ***P*** or ***D*** will behave as follows: forsterite will crystallize out first, the composition of the liquid changing to ***K***. At this point a pyroxene of composition ***L*** will begin to separate. As the separated forsterite is removed from equilibrium
214 during the continuous separation of a pyroxene with a variable composition, the composition of the liquid crosses the pyroxene field and meets the pyroxene-silica eutectic curve. The crystals of the separated pyroxene have a composition varying between ***L*** and ***T***. When the composition of the liquid phase reaches the boundary of the silica field at any point ***G***, crystalline silica (in this case tridymite) begins to crystallize together with the pyroxene. On further crystallization, the composition of the separating pyroxene will change continuously, the liquid phase will change along the curve and crystallization will terminate at the diopside-tridymite eutectic point, provided that an even infinitesimal amount of liquid will remain at this stage.

The processes which occur in the considered system somewhat elucidate crystallization phenomena in natural silicate melts — magmas. This is of particular importance as we are now in possession of an explanation of the formation of zoned pyroxenes. The existence of the diopside-enstatite and diopside-clinoenstatite series was also observed in igneous rocks. Wahl described a pyroxene of such a series by the term enstatite-augite. The change in optical properties of these natural minerals is completely analogous to their change in the diopside-clinoenstatite series. The resorption of olivine crystals, which is often observed in igneous rocks, is explained by the normal course of crystallization of the magma with a retarded equilibrium.

[However, in the majority of rocks the relations of pyroxenes are complicated by several factors not taken into account in the considered system; decomposition of solid solution, the inversion of the pyroxene $MgSiO_3$ into the orthorhombic modifications and the partial replacement of magnesium by bivalent iron.

The ionic radii of Mg^{2+} and Ca^{2+} are so different that the exchange of these ions with the formation of a continuous series of solid solutions is possible only at very high temperatures. On cooling, the solid solutions are decomposed, thus causing a discontinuity in the series of mixed crystals. The decomposition is further complicated by an inversion at a temperature of approximately 1,150° C (see further on the system MgO-FeO-SiO_2) of $MgSiO_3$ to the orthorhombic modification. Under natural conditions the melting temperature is considerably lowered due to the influence of a number of factors, the foremost being the water content. Therefore, irrespective of the character of crystallization from melts of a ternary system (compare Figures 133, I and II), a continuous series of solid solutions is not possible here. Two pyroxenes differing considerably in their compositions will crystallize simultaneously. One of them is a monoclinic pyroxene with a $MgSiO_3$ content of the order of 25—30 % pigeonite and the other is a magnesium pyroxene. The latter may be either orthorhombic with a low Ca content or (at higher temperatures during the crystallization of gabbro diabases) monoclinic clinoenstatite containing 30—40 % diopside content. If the content of the iron component is either absent or very low, reactional relations will obviously exist between these two pyroxenes, as was already assumed by Bowen; the Mg-rich pyroxene is the first to crystallize, followed by the Ca-rich pyroxene. However, these relations change with the increase of the iron content.

The orthorhombic pyroxenes with fine lammelar intergrowths of monoclinic pyroxenes observed in nature and sometimes mistaken for polysynthetic twinning, as well as intergrowths of orthorhombic pyroxene within monoclinic pyroxene, are apparently the result of subsequent exsolution.]

215 Bowen claims that the process of settling of separated crystals (olivine, pyroxene) into the lower layers of the cooling magma is of great importance to petrogenesis. This results in the enrichment of the lower layers in the more refractory minerals, whereas the upper layers are predominated by the liquid. Over a sufficient time range the upper and lower layers attain widely differing compositions, and the crystallization of each proceeds along its own individual course. The lower layer will naturally crystallize considerably earlier and at higher temperatures of the magma chamber, at a time when the upper layers are still mobile. During disturbances in the Earth's crust, intrusions of the upper, more acid magma into the lower, already solid, basic rocks may be expected in such regions.

Finally, we shall mention the numerical characteristics of certain specific points: 1) the diopside-silica eutectic — 0, 0, 42.8 and 57.2 and 2) the diopside-forsterite eutectic — 0, 0, 52.2 and 47.8.

ANORTHITE-FORSTERITE-SILICA

This system was investigated by O. Andersen (1915) in the geophysical laboratory of the Carnegie Institution (Washington). The results of the investigations are shown on the diagrams (Figures 146 and 147). The sides of the ternary diagram are binary systems. The systems forsterite-silica and anorthite-silica have already been considered previously. The system anorthite-forsterite is not really a binary system. In its central part there is a gap where spinel separates, the formation of which will be described in the quaternary system CaO-Al_2O_3-MgO-SiO_2. This gap extends over a range from $Ab_{45}An_{55}$ to $Ab_{13}An_{87}$. The lateral parts are binary systems with eutectics. The area of the diagram of the ternary systems is divided into several fields limited by eutectic curves and curves of inflection. The location of fields is apparent from the diagrams and schemes on Figures 146 and 147. The field ***AKD*** contains points of composition during the crystallization of which spinel separates. In the field ***AKB*** forsterite is the first to crystallize and is followed by spinel. In the field ***DKC*** anorthite is the first solid phase to separate; within the triangle ***CKB*** spinel is the first to separate. The lines limiting these fields are not eutectic curves. They only provide one with an idea of which phase is the first to separate. As our diagram cannot represent the composition of the spinel, we shall leave this field on the diagram, ***AKD***, without further consideration.

On considering the diagram, we note that a part of it, namely the area limited by the straight lines ***MA***, ***AQ***, ***QF*** and the curve ***FM***, presents the ternary system anorthite-clinoenstatite-silica; each pair of components forms a simple eutectic, and the system as a whole has one common ternary eutectic point ***N***. The course of crystallization within the limits of this part is unusual for eutectic mixtures. The final stage of crystallization of any composition within the limits of this field is the attaining of the ternary eutectic point ***N***.

Crystallization in another part of the diagram — ***DKAMF*** — is somewhat more complicated. Continuous interaction between the separated crystals and the liquid phase takes place here. In all parts of this area forsterite will crystallize sooner or later during crystallization, and will subsequently react with the liquid phase, being wholly or partially redissolved.

In the area ***AKM*** anorthite is the first phase to separate, the liquid phase
216 changing up to the eutectic curve ***KM***. On reaching this curve, it starts to separate anorthite and forsterite simultaneously and moves along the curve to point ***M***. At ***M*** clinoenstatite becomes stable, and a reaction occurs between forsterite and the liquid phase: forsterite is dissolved. Clinoenstatite separates in its stead. At the same time, the crystallization of anorthite continues. The termination of crystallization depends on the position of the point of initial composition. If the composition lies within the field ***AKE*** (Figure 147), the liquid phase will be exhausted before forsterite is com-
217 pletely dissolved, and crystallization will terminate at ***M***, yielding anorthite, forsterite and clinoenstatite. If the initial composition lies on the straight line ***AE***, the liquid phase and forsterite are exhausted simultaneously, and the final products of crystallization will be anorthite and clinoenstatite. If the initial composition lies within the field ***AEM***, forsterite will be completely dissolved, separating clinoenstatite, and the residual liquid phase will solidify along curve ***MN***, separating anorthite and clinoenstatite. Crystallization terminates at the eutectic point ***N***; the eutectic will consist of anorthite, clinoenstatite and silica (tridymite).

Within the limits of the field ***DKMF***, forsterite is the first phase to separate. The crystallization of the next phase is dependent on the position of the point of initial composition within this field. The change of the liquid phase during the crystallization of forsterite proceeds toward ***KM***, and when this curve is reached anorthite starts to crystallize simultaneously with forsterite. Further on the composition of the liquid phase will change along ***KM***. At point ***M*** a reaction occurs between forsterite and the liquid phase, and clinoenstatite is separated as the result. The further path of crystallization depends on the location of the composition point relative to the clinoenstatite-anorthite curve (***EA***). If the point lies within the field ***DKLg***, the liquid phase is first exhausted, and the final products of crystallization are forsterite, anorthite and clinoenstatite. If the composition point lies on the straight line ***EA***, the disappearance of the liquid phase and forsterite occurs simultaneously, and the final result will be clinoenstatite and anorthite. If the composition point lies within the area ***LgM***, forsterite will be the first to disappear and on termination of the reaction the liquid phase, on cooling, will change along ***MN***, separating anorthite and clinoenstatite simultaneously. Crystallization terminates at point ***N*** with the formation of a eutectic mixture — anorthite + clinoenstatite + tridymite.

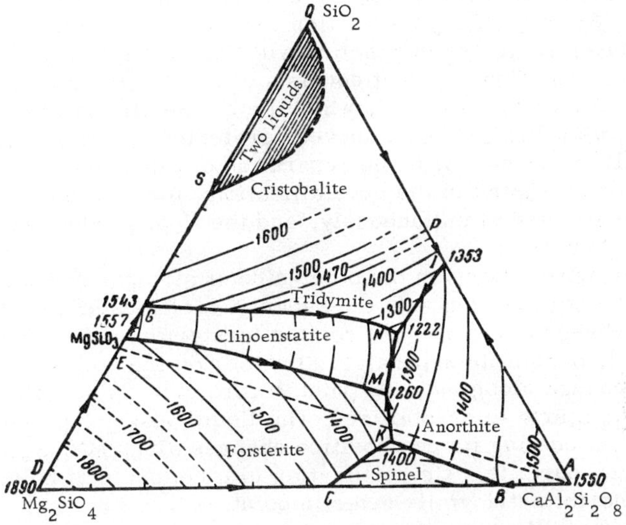

FIGURE 146. The system anorthite-forsterite-silica (after Andersen with amendments by Greig)

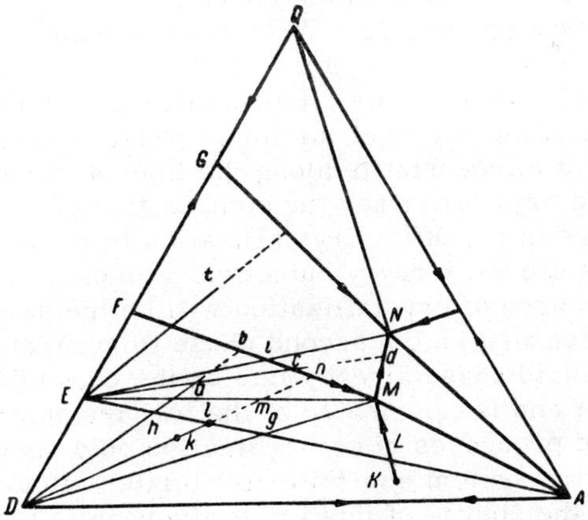

FIGURE 147. The scheme of crystallization in the system anorthite-forsterite-silica

The crystallization of composition points within the field ***DME*** is also dependent on their position relative to the straight line clinoenstatite-anorthite (***EA***).

A composition ***k***, which lies nearer to forsterite, will crystallize as follows (see Figure 147): forsterite is the first phase to crystallize, the liquid phase changing along the straight line ***kn***. During the change of its composition from ***n*** to ***M***, forsterite is dissolved and clinoenstatite is separated. At point ***M*** crystallization will be terminated when all the liquid phase has been exhausted in reaction with forsterite and in the separation of anorthite. The final product consists of forsterite, clinoenstatite and anorthite. For compositions ***l***, which lie on the straight line ***EA***, the course of crystallization is as above: forsterite separates up to ***n***, and is then partially resorbed with the separation of clinoenstatite. After ***M*** is reached anorthite joins in the crystallization; forsterite and the liquid phase are exhausted simultaneously, and the final product will be clinoenstatite and anorthite.

Any composition lying on the other side of straight line ***EA*** will also commence crystallization with the separation of forsterite; the liquid phase will change along ***mn***. At point ***n*** forsterite starts to react with the liquid and clinoenstatite separates. Further on the composition of the liquid phase will change along ***nM***. At point ***M*** forsterite is completely redissolved and anorthite starts to crystallize. The liquid phase is present in excess and on further cooling its composition changes along ***MN***, separating clinoenstatite and anorthite. Crystallization is terminated at point *N* — the clinoenstatite-anorthite-tridymite eutectic point.

In the area ***EFM*** crystallization proceeds somewhat differently. On cooling an initial composition ***a*** forsterite will again be the first phase to separate, the liquid phase changing along ***ab***.

From point ***b*** clinoenstatite starts to crystallize, forsterite being simultaneously dissolved. Further on the composition of the liquid phase will
218 change along the straight line ***bc***. When the composition of the liquid phase reaches *c* (which lies on the extension of line ***Ea***) all the forsterite is dissolved. On further cooling, only clinoenstatite and the liquid phase are in equilibrium. The composition of the liquid phase leaves the curve ***FM*** and crosses the field of clinoenstatite along the line ***cd*** to point ***d***. At this point anorthite starts to crystallize and the composition of the liquid will further on change along the curve ***MN***. Crystallization is terminated at point *N* with the formation of a simple ternary eutectic. For compositions lying within the field ***EFi*** the course of crystallization will be the same, with the only exception that silica will be the second phase to separate, as the composition point of the liquid phase already lies on the curve ***GN***. Crystallization proceeds along ***GN*** and is terminated at the ternary eutectic point ***N***.

The considered processes of crystallization only occur at perfect equilibrium of the entire system anorthite-forsterite-silica. If perfect equilibrium is absent the course of this crystallization is somewhat different when the separated crystals do not react with the liquid to separate clinoenstatite. Naturally, this difference only applies to the compositions of those areas where a reaction takes place between the separated crystals and the liquid phase. It only concerns the field where interaction occurs — the field ***DKAMF***, and will have no influence whatsoever on the field ***AMFN***, where crystallization proceeds along eutectic curves toward the ternary eutectic point, and the separated crystals are of no significance to equilibrium. Let us consider a composition lying in the field ***DEg***. Crystallization starts with the separation of forsterite, the liquid phase changing up to the curve ***FM***. If forsterite is removed from the melt during crystallization, its further course does not proceed along ***FM***, but crosses the clinoenstatite field, with the separation of the latter only. Crystallization is terminated at point ***N*** with the eutectic crystallization of anorthite, clinoenstatite and tridymite. This will be the course of crystallization, for instance, when forsterite crystals settle to the lower layers of the melt; the same will occur when forsterite crystals are isolated from the melt due to overgrowths of clinoenstatite. The resorption and settling of forsterite were observed during laboratory investigations. Rims of pyroxene enveloping cores of resorbed olivine grains are a usual phenomenon in rocks. The final product of such a crystallization has a somewhat anomalous character: crystallization terminates at the eutectic point ***N*** where, together with other minerals, SiO_2 is separated. Thus we find forsterite (olivine) and tridymite (quartz) existing together. Natural rocks of such a type are only formed under exceptionally favorable conditions; they are rare, but are known and have been described from various places.

The numerical characteristics of special ternary points on the diagram are as follows (see Figure 146): ***K*** — 0, 15.2, 29.0 and 55.8; ***M*** — 0, 13.6, 24.7 and 61.7; ***N*** — 0, 12.5, 16.2 and 71.3

ANORTHITE-ALBITE-DIOPSIDE (THE HAPLOBASALTIC AND HAPLODIORITIC SYSTEM)

This system was studied by Bowen (1915). The investigation was carried out by the quenching method. An equilibrium diagram was constructed from the results (see Figure 148). The sides of the concentration triangle are binary systems: 1) albite-anorthite (Ab-An) — a system of solid solutions of Roozeboom's Type II; 2) anorthite-diopside (An-Di) — a simple eutectic system with a eutectic point at 1,270° C and 42 % anorthite; 3) albite-diopside (Ab-Di) — also a eutectic system with a sharp shift of the
219 eutectic point toward the more fusible component — albite. The numerical values and location of the last-mentioned eutectic are presented by Bowen on a purely hypothetical basis, as the eutectic was not reached due to the viscosity of the melt. The temperature of the eutectic is 1,085° C (the initial melting of the mixture diopside + albite); the eutectic composition is 97 % albite, 3 % diopside.

In the ternary system the entire area of the concentration triangle is divided into two fields, one of which contains liquids in equilibrium with plagioclase and the other — liquids in equilibrium with diopside. These fields are separated by a boundary eutectic curve E_1E_2, the points of which represent the composition of the liquid phase capable of coexisting in equilibrium with plagioclase and diopside simultaneously. The shape of the temperature surface of equilibrium is shown by isotherms (of incipient crystallization). We are already familiar with such diagrams where two out of the three components form a continuous series of solid solutions of Roozeboom's Type I, while the third component is incapable of forming mixed crystals with them. This case was described in sufficient detail in the general section. The introduction into the system of components which form solid solutions complicates the picture of melt equilibria (as we have already seen).

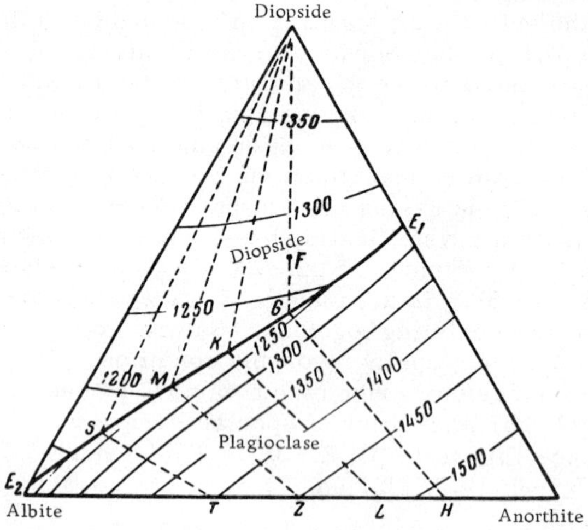

FIGURE 148. The system anorthite-albite-diopside (after Bowen)

Knowledge of the isotherms on the diagram is insufficient for determining the state of any composition at any particular temperature. It is also necessary to be acquainted with the composition of the plagioclases separating at this temperature in various parts of the triangle. This calls for a knowledge of the position of the phase triangle determining the composition of the liquid, diopside and plagioclases in equilibrium at the given temperature. Knowing the apices of this triangle, one may determine all the elements necessary for characterizing the state of any composition at a given temperature. The phase triangle is constant at each temperature. After determining the apices, one may quantitatively characterize crystallization processes in perfect equilibrium in the diopside field. Thus, the melt ***F*** (50 % Ab_1An_1 and 50 % diopside) starts to crystallize at 1,275° C with the separation of diopside (see Figure 148), the liquid phase changing along the straight line ***FG*** toward ***G***. At 1,235° C, when the liquid has the compo-
220 sition ***G***, a plagioclase of composition ***H*** (Ab_1An_4) starts to crystallize; point ***H*** is determined by the ***GH*** side of the phase triangle, the apex of which is ***G***. On further lowering of the temperature, the composition of the liquid changes toward ***M***, following the curve E_1E_2. The composition of the separated plagioclase changes continuously. At 1,218° C, when the liquid has a composition ***K***, the composition of the plagioclase changes from H to ***L*** (Ab_1An_2). Finally, at 1,200° C the liquid is exhausted; its last drop is of the composition ***M***; the plagioclase changes its composition up to ***Z*** (Ab_1An_1). ***MZ*** is a side of the phase triangle which passes through the composition point Ab_1An_1, and determines, at point ***M***, the temperature of complete solidification of melt ***F*** or any other mixture containing a plagioclase of

composition Ab_1An_1 and diopside. Similarly, for a melt of composition 60 % Ab_2An_1 and 40 % diopside, the temperature of initial crystallization of diopside is 1,252° C; at 1,218° C (point ***K***) a plagioclase of composition ***L*** (Ab_1An_2) begins to crystallize; at 1,176° C the liquid is exhausted and the final product of crystallization is diopside and the plagioclase Ab_2An_1. For mixtures in the plagioclase field, the temperature of the liquid phase moves toward the boundary curve E_1E_2 during the separation of plagioclase crystals, but does so along the curved and not the straight path of crystallization. Each composition has its own curve, which must be determined separately (see Figure 46, III and Figure 77, I).

The composition of the liquid phase was determined by Bowen from the refractive index of the glass after quenching. In order to find the composition point of this liquid phase at a given temperature, he determined the refractive indices of glass of various compositions and plotted lines of equal refractive index (isofracts) of glass on a diagram. The composition of the liquid phase is obtained from the intersection of the corresponding isotherms and isofracts. The composition of the separated plagioclase in equilibrium with the liquid phase lies on the intersection of the line passing through the total composition of the liquid phase and the AbAn side of the diagram. Knowing the refractive indices of glasses of the considered mixtures at various temperatures, one may trace the path of crystallization of these mixtures within the plagioclase field. For example, a mixture of 70 % Ab_1An_1 and 30 % diopside starts to crystallize at 1,302° C with the separation of a plagioclase of composition Ab_1An_1. At a temperature of 1,230° C, the composition of the liquid phase is described by a point lying between ***G*** and ***K***, and the composition of the separated plagioclase — by a point lying between ***L*** and ***H***; diopside starts to separate. At a temperature of 1,216° C (corresponding to point ***M***), the liquid phase is of composition ***M*** and the plagioclase composition is represented by point ***Z*** (somewhat more basic than Ab_1An_1); ***ZM*** is a side of the phase triangle at 1,216° C. On further cooling, the composition of the liquid phase proceeds along the eutectic curve, diopside and plagioclase crystallize simultaneously and at a temperature of approximately 1,200° C the liquid phase disappears. The composition of the plagioclase is now Ab_1An_1.

If the separated crystals are removed from the melt by one means or another, conditions of equilibrium are disturbed: interaction between the crystals and the liquid ceases, and plagioclases of a different composition begin to crystallize from the liquid phase. The entire process of change in composition of the liquid phase proceeds along a different curve. A further removal of crystals will again cause the residual liquid to become an initial point of a new curve. The composition of the plagioclase in each concrete case is determined by the composition of the initial liquid phase.

An analogous process is observed under conditions of incomplete equilibrium in the crystallizing system, when the rate of cooling is relatively fast. The separating plagioclase does not have time to react with the liquid phase if the drop in temperature is sufficiently rapid, and plagioclases of a different composition crystallize from the liquid and grow around cores
221 of the older, more basic plagioclases, thus resulting in the formation of zoned crystals.

From the point of view of petrogenesis, the system albite-anorthite-diopside is of great interest. The following interesting features come to light after considering the above-described system.

Meriting first mention is the absence of a eutectic point of determined composition between diopside and any intermediate plagioclase Ab_mAn_n, as well as between a plagioclase and a complex pyroxene melt, e.g., augite. Instead of a eutectic point, in this case we have a eutectic line. Here, as distinct from the case of a eutectic point, there is no obligatory and unavoidable final phase of a determined composition for each initial composition of the system.

Another, no less important feature is the elucidation of the relation between the content of the alkali feldspar and the melanocratic minerals in the magma. The more acid the plagioclase, the less is the amount of the melanocratic mineral needed for eutectic relations. The connection between the plagioclase composition and the relative amount of the melanocratic mineral in a rock, which feature had been noted for some time already, is thus given an explanation.

The following are the numerical characteristics of the eutectic points: anorthite-diopside — 0, 10, 35.1 and 54.9; albite-diopside — 13.7, 0, 2.1 and 84.2.

CRYSTALLIZATION DIFFERENTIATION OF BASALTIC MAGMA AND ASSIMILATION†

[The investigations of the above-mentioned ternary systems have played a major role in the history of petrology. On the basis of their chemical properties, Bowen arrived at his theory of crystallization differentiation of magma and also considered the phenomenon of assimilation. Despite all the opposing views denying the universality of such processes, Bowen's concepts are undoubtedly a true reflection of important general regularities, and are thus still regarded as significant today. Owing to the function of this book, we shall not dwell upon the detailed considerations of complex problems of recent petrology, and shall only briefly touch on those phenomena, the explanation of which follows directly from the systems considered above.]

One of the first minerals to crystallize from a basaltic magma is olivine. On considering the binary and ternary systems in which olivine (forsterite) and silica participate, a general feature is noted, arising from the incongruent melting of enstatite (pyroxene in general). The separation of olivine continues until the moment that the melt becomes saturated in silica — up to the composition of a metasilicate. It is only when the liquid composition attains a certain degree of oversaturation that the redissolving of the separated olivine commences, accompanied by the formation of pyroxene in its stead. As we have seen, when the equilibrium between the separating crystals and the liquid is disturbed due to the removal of olivine crystals from the melt, this may cause the formation of a melt oversaturated in silica from an initial melt undersaturated in silica, and free SiO_2 may separate. The removal of olivine crystals may occur due to their gravity

† [Material from the corresponding chapters of Part III of the 1926 edition is given in this section, together with minor addenda.]

settling within the liquid. The settling of separated olivine crystals was also observed during the study of artificial melts of composition $mMg_2SiO_4 + nSiO_4$ in crucibles.

Thus, according to Bowen, the incongruent melting of $MgSiO_3$ offers the possibility of obtaining rocks containing both olivine and free silica (in the form of quartz) from the same melt of basaltic magma.

222 During the consideration of crystallization in the system anorthite-forsterite-silica, we noted that the settling of olivine crystals may lead to differentiation, yielding a mixture of forsterite, pyroxene (clinoenstatite) and anorthite at the bottom — a basal layer of peridotitic composition may even be obtained — whereas a mixture of pyroxene (clinoenstatite), anorthite and free silica (tridymite) is obtained in the upper parts. This simple system elucidates the formation of acid rocks containing free quartz from an initial basaltic magma. The conclusions following from the study of the system plagioclase-diopside (the haplobasaltic magma of Bowen) are of some importance. We have seen that during the crystallization of a mixture of plagioclase $Ab_{50}An_{50}$ and diopside, a plagioclase close to $Ab_{20}An_{80}$ is separated, and the melt is correspondingly enriched in the albite component. If equilibrium is continuously maintained between the crystals and the liquid during crystallization, the last drops of liquid will contain a plagioclase with a composition of approximately $Ab_{75}An_{25}$. If the initially crystallized plagioclase grains are removed from the melt, either by settling or by the formation of zoned crystals, a liquid with the composition of a rather acid plagioclase may be obtained. As follows from the course of the eutectic curve (see Figure 148), the content of diopside in the residual liquid will rapidly decrease simultaneously. On further cooling, the subsequent stages of the changing residual melt will yield ever increasingly leucocratic rocks.

Thus, if the physicochemical equilibrium between the melt and the separating crystals is disturbed (due to settling or other phenomena) during the crystallization of basaltic magma, the residual liquid may attain a composition in which an excess of free silica will be present together with a predominantly alkali feldspar. This residual liquid is none other than a granitic magma. Thus, according to Bowen, the crystallization differentiation of basaltic magma leads to a granitic magma. This process is chiefly conditioned by the settling of the early separated crystals and the subsequent prevention of their reaction with the liquid. Certain favorable conditions are necessary for the removal of crystals from the liquid. In comparatively small intrusive bodies the rate of cooling is of essential importance. As Bowen has stressed, a basaltic magma on cooling rapidly will yield a diabase. If cooling is slower and accompanied by settling of the earlier formed pyroxene and basic plagioclase, i.e., the enrichment of the liquid in the more alkali feldspar, diorite will be the differentiation product formed in the upper part of the magma chamber. Under conditions more favorable to differentiation — with the sufficient enrichment of the liquid in molecules of alkali feldspar and water — quartz-diorite, granodiorite and finally granite may be obtained. The role of water in the formation of biotite has already been mentioned previously†.

† [T.N. In the original 1926 edition the description of the effect of water on biotite formation apparently preceded the section on crystallization differentiation. As the order was changed in the present edition this reference is entirely out of context].

Under appropriate crystallization conditions of basaltic magma, the development of zoned crystals may lead to the formation of interstitial micropegmatites.

In some cases the separation and settling of olivine during the first stages of crystallization of basaltic magma lead to an increase in the amount of the acid products of differentiation. Slower cooling may cause the separation of olivine from an initial magma which yields pyroxene and plagioclase on more rapid cooling. In other cases (see anorthite-forsterite-silica diagram) spinel may crystallize. The accumulation of the early separated crystals at the bottom of the magma chamber sometimes leads to the formation of a spinel-bearing peridotite.

223 The course of crystallization of a normal basaltic magma always trends toward the formation of biotite granite, but this final stage is only reached if cooling is sufficiently slow. With more rapid cooling, diorite and quartz diorite may result. In view of this, the dioritic marginal facies of granites may be explained by the fact that cooling was more rapid in the marginal parts than in the central parts, which reached the next stage of differentiation. According to the above explanation, being the last residue and cooling at the lowest temperature, granitic magma is differentiated in the upper part of layered intrusive masses, especially in batholiths. The wide distribution of granites on the Earth's surface is explained by the fact that erosion has exposed these upper parts of magmatic masses, which either cooled in situ or were tapped off to form separate granite massifs. This wide surface distribution does not correspond to the actual volume of granitic material in the Earth's crust. On calculating the average composition of igneous rocks found on the Earth's surface, we do not arrive at the composition of a primary magma. The composition arrived at corresponds to a magma more acid than a primary one, due to the predominance of acid differentiation products in the upper, exposed parts of batholiths and other granitic intrusions.

Obviously, all the above said should at present be regarded as a general scheme, which is actually much more complicated in nature, but it is clear from the above that the differentiation of a basaltic magma may lead to the formation of a granitic magma, and that this is a natural process which does not call for any extraordinary conditions in order to be fulfilled.

[At present, one is scarcely entitled to speak of the universality of this theory, i. e., that all granites originate as the differentiation products of basaltic magma. However, the attempts at completely refuting the phenomena of crystallization differentiation in nature have not been substantiated. Geological literature cites numerous unquestionable examples both of settling of olivine and of its accumulation in the lower parts of intrusion (Tomkeieff, 1934), the upper parts of which often contain free quartz, as well as the formation of squeezes of granitic composition during the crystallization differentiation of basaltic magma (Sobolev, 1936). Such squeezes, which cool as veins, have the same composition and structure as interstitial micropegmatites in diabase.

Thus, the general regularity of phenomena and the possibility of forming a granitic magma from a basaltic magma by crystallization differentiation are established facts which should be borne in mind. On the other hand, it has been proved that during the crystallization differentiation of a basalt markedly undersaturated in silica, not only is excess SiO_2 not observed in the residual melt, but in some cases even felspathoids accumulate together with the feldspars. Depending on the initial composition of the basaltic magma (and sometimes on the course of its differentiation, or the assimilation of other rocks), the residual melt, though always trending toward enrichment in alkali feldspars and becoming increasingly leucratic, will at different times contain varying amounts of silica.

The above-considered systems and mentioned general regularities permit one to trace phenomena occurring as the result of assimilation of country rock by the magma]. There is no sufficient reason for assuming that the superheating of magma plays a major role in assimilation. This obliges one to consider cases of solution and assimilation of a foreign material in magma when it is saturated — shortly after incipient crystallization has started. Obviously, if the added material interacts
224 with the liquid changes will occur, both in its composition and in that of the liquid. Let us consider a simple case: a melt of plagioclase composition has reached saturation, the separation of plagioclase crystals having just started. The separated plagioclase is somewhat richer in anorthite than the liquid. Let us add a certain amount of a more basic plagioclase (preliminarily heated to the temperature of the liquid) to this liquid. The system consisting of the liquid and the added plagioclase cannot be in equilibrium at the given temperature, as the added plagioclase will start to dissolve, and the melt will crystallize a plagioclase which is in equilibrium with the liquid at this temperature. These crystals are more acid than the added material, but more basic than the liquid; the newly formed plagioclase is derived partly from the substance of the initial liquid and partly from that of the more basic plagioclase. The relative amount of the liquid will change. Thus, although the liquid was initially in a saturated state, the addition of the new substance led to its being dissolved and to the separation of crystals of a different composition from that of the liquid. However, we cannot consider this as a solution in the same sense as the solution of salt in water for instance. It involves interaction between the liquid and the added substance. The amount of crystalline matter is increased, and a certain amount of heat is evolved. Thus, we are dealing with an exothermic process.

Naturally, we cannot add a more acid plagioclase to a saturated plagioclase melt of the same temperature, as the former, being the more fusible, will be liquid at this temperature. If, however, the more acid plagioclase is added at a lower temperature (when it is already solid — at its melting temperature), the following occurs: the added crystals and the liquid will attain an intermediate temperature, accompanied by the heating and melting of the added crystal and the cooling of the initial liquid. As the latter is in a state of saturation, cooling will lead to crystallization; the heating and melting of the added crystals is due to the heat liberated during crystallization. The heat capacity of silicates is negligible in comparison with their latent heat of fusion, and the melting of the added plagioclase calls for approximately the same amount of heat as is evolved during the crystallization of the same quantity of newly crystallized plagioclases. When equilibrium is attained, the liquid is richer in albite than both the added plagioclase and the crystals which are in equilibrium with this liquid. The amount of liquid increases somewhat relative to the case of crystallization of the liquid up to the same equilibrium temperature without the addition of the more acid plagioclase. The process involving the solution of the added solid material is an endothermic one.

All that has been said of plagioclases should also hold true for other series of solid solutions (as is the case in most rock-forming minerals). On addition of the more refractory member of the series, a decrease in the amount of liquid occurs, this process being exothermic. On the addition of the more fusible member of the series, the amount of liquid increases somewhat, the process being endothermic. In the latter case the liquid changes its composition, approaching the more fusible end of the series.

Obviously, the process is not so simple in the case of very complex multicomponent solutions (e. g., basaltic magma), but the phenomena remains essentially the same. In similarity to the manner in which the
225 phenomena of crystallization of members of the plagioclase series reflect upon more complex systems in which these plagioclases participate, the features of other isomorphous series also maintain their significance. The essence of this phenomenon is, as we have seen, the interaction between the liquid and the crystals.

In addition to the considered systems of solid solutions, the phenomenon of interaction between solution and crystals† also occurs in igneous systems, and the most simple type of this reaction is encountered in the system Mg_2SiO_4-SiO_2.

Let us consider this simplest case (see Figure 128). A saturated melt (clinoenstatite +silica), intermediate between the composition corresponding to the inflection point on the melting curve and the eutectic point, will be undersaturated with respect to added forsterite. The latter will start to dissolve, and the composition of the liquid becomes more impoverished in SiO_2 than the melt saturated in clinoenstatite at the same temperature, i. e., the melt becomes oversaturated in clinoenstatite, which immediately separates. Thus, the conversion of the added forsterite to clinoenstatite occurs. This is not simply the impoverishment of SiO_2 necessary for the reaction $Mg_2SiO_4 + SiO_2 = 2MgSiO_3$. The separation of clinoenstatite is accompanied by a change in the composition of the liquid along the melting curve, and the amount of the separating clinoenstatite is considerably larger than would have been obtained from forsterite according to the above-recalled equation.

If clinoenstatite is added to a melt saturated in forsterite (at a temperature of 1,557° C, which is the inversion temperature of clinoenstatite), this addition will cause a certain cooling of the melt. Clinoenstatite will subsequently be dissolved due to the heat liberated during the crystallization of forsterite, which crystallization starts as the saturated solution cools. The final outcome is that the added clinoenstatite passes into solution, thus increasing the amount of liquid, and this process is accompanied by a simultaneous separation of forsterite.

Thus, when forsterite is added to a melt saturated in clinoenstatite, a conversion of forsterite to clinoenstatite occurs, and inversely on addition of clinoenstatite to a liquid saturated in forsterite, the former is dissolved

† [T.N. The original text erroneously states "interaction between solution and liquid"].

and the latter separates. During the normal course of crystallization, as is known, the earlier separated forsterite afterwards inverts to clino-enstatite. Thus, on addition of a mineral of the first generation (forsterite) to a liquid saturated in a mineral of the second generation, the inversion of the former takes place; in the case of addition of minerals of the second generation to a liquid saturated in minerals of the first generation, an increase in the amount of liquid occurs, which attains a composition solidifying at lower temperatures than the initial liquid.

Relations similar to those existing between forsterite and clinoenstatite are also known for many other minerals: pyroxene inverts to hornblende, and hornblende inverts to mica. Thus, if fragments of pyroxenic rock are introduced into a magma from which hornblende is crystallizing normally, they will be dissolved in the magma and converted into amphibole. Such phenomena are usual during the intrusion of pyroxenites by plagiaplites and pegmatites.

After the above-mentioned considerations, one is able to understand the main features of interaction between magma and older igneous rocks. We have seen that the most widely distributed igneous rocks of the normal calc-alkali type are located within the following series: peridotitic rocks, gabbros and basalts, diorites and andesites, granodiorites and dacites and granites and rhyolites. These rocks are mentioned in the order of transition from earlier to later stages of differentiation. On comparing the
226 mineralogical composition of these rocks, we note two features. The rocks of the later stages of differentiation are characterized firstly by more acid plagioclases and secondly by melanocratic minerals, which are formed due to the reaction of the magma with melanocratic minerals of earlier stages of differentiation: augite in gabbro is replaced by amphibole in diorite and by mica in granodiorite and granite. These main features determine the course of the process on interaction between the magma and the earlier formed igneous rocks.

Two cases should be distinguished: 1) a magma of a late stage of differentiation acts upon a rock of an earlier stage, e. g., a granitic magma acting on inclusions of gabbro, and 2) a magma of an early stage of differentiation acts upon a rock of a later stage, e. g., basaltic magma acting on granite.

From previous considerations it becomes apparent that granitic magma is, strictly speaking, unable to dissolve intrusions of gabbro (or of other rocks of earlier stages of differentiation); it reacts with minerals of these intrusions, converting them into other minerals. Basic plagioclases are replaced by more acid ones, pyroxene is converted to amphibole and subsequently to mica, etc. These conversions may be accompanied by the disintegration of the inclusions themselves and the dissemination of the reaction products throughout the granitic mass so that they are scarcely discernible from the usual composition of the rock. The inclusions are completely "assimilated", but are not dissolved.

If, on the other hand, a basaltic magma acts on granite inclusions, they become a part of the liquid, which process is accompanied by the separation of normal mineral constituents of basalt. The liquid (especially if the separating minerals are removed from it either due to settling or some other reason) changes in composition, approaching that of the inclusions, and its further differentiation proceeds along an individual course, finally yielding granite. The natural process of differentiation of basaltic magma, which finally results in the formation of granitic magma, is intensified by the action of granitic inclusions. As a result, the amount of granitic differentiation products from an initial basaltic magma is considerably increased.

According to Daly, many granites have resulted from the solution of older granites by basaltic magma, followed by their subsequent differentiation.

[Despite all objections to the overwide application of Daly's theory, a great amount of data illustrating the above-described phenomena have accumulated in the literature of petrology. Numerous examples are known (Sobolev, 1936), starting with the case described by A. A. Polkanov (1913) of the rare melting of granites by the intrusive basaltic magma on the coast of the White Sea. During this process the granitic melt was in part mixed with the basaltic melt and in part remained liquid after the solification of the basic rocks, forming veins within them. In the cases of residual, incompletely assimilated xenoliths of basic rocks within granites, their conversion is considerably complicated by metasomatism, which occurs due to the influence of the introduced volatiles. A special chapter of A. N. Zavaritskii's monograph on the Berdyaushi pluton (1937) is devoted to these problems as well as to the question of hybridization of igneous rocks in all its manifestations].

THE SYSTEM SILICA-NEPHELINE-KALSILITE**

This system, which is of very great significance to petrography, is un-
227 fortunately far from being completely studied. Only the preliminary data of Schairer and Bowen (1935) are available (as well as Ann. Rep. 1956), which mainly deal with the liquidus surface in the area of compositions relatively poor in silica. Melts with a high silica content are extremely viscous and cool to form glass. The sides of the concentration triangle (Figure 149) represent binary systems: 1) $NaAlSiO_4$-$KAlSiO_4$ (carnegieite-kalsilite) — a system of solid solutions with a small discontinuity and a eutectic point (Roozeboom's Type V), complicated by polymorphous inversion (see page 181); 2) $NaAlSiO_4$-SiO_2 (carnegieite-silica) — a system with a congruently melting compound $NaAlSi_3O_8$ (albite) and two corresponding eutectics ***C*** and ***B***, complicated by the polymorphous inversions of carnegieite to nepheline and cristobalite to tridymite (see page 185); the albite-tridymite eutectic was determined only approximately; 3) $KAlSiO_4$-SiO_2 (kalsilite-silica) — a system with two compounds (see page 188), one of which ($KAlSi_2O_6$ — leucite) melts congruently, and the other ($KAlSi_3O_8$ — orthoclase) melts incongruently. Correspondingly, there are two eutectic points, ***F*** and ***G***, and one peritectic point ***H***.

Within the triangle on Figure 149, the section along the albite-orthoclase line is a pseudobinary system from ***G′*** to Or and a binary system from Ab to ***G′***; the feldspar almost undoubtedly forms a continuous series of solid solutions with a minimum approximately corresponding to the composition 65 % albite and 35 % orthoclase. Below the carnegieite-nepheline inversion temperature, the kaliophyllite-nepheline series also probably yields continuous solid solutions with a minimum, complicated by the polymorphous inversion of the rhombic modification $KAlSiO_4$ (kalsilite) to the hexagonal modification, and possibly also by a second inversion from one hexagonal modification to another.

Without considering these polymorphous inversions, we may distinguish on the diagram (Figure 149) four fields of crystallization of solid phases: 1) the field of solid solution of nepheline; 2) the leucite field; 3) the field of feldspar solid solutions; 4) the field of silica crystallization.

Point ***K***, where the fields of three solid phases meet, is an invariant peritectic point. Here, leucite reacts with the liquid to form nepheline + feldspar. If all three phases were purely potassic we would have a case where the liquid does not participate in the decomposition reaction (see Figure 62, VI). However, in actuality, leucite, which contains almost no sodium, is here in equilibrium with nepheline which is quite rich in sodium, and with K-Na feldspars. In order to determine the course of crystallization in the system, a triangle of solid phases in equilibrium with the liquid should be constructed on the bases of experimental data. Such a triangle has, however, not been constructed as yet, and no data are even available for the position of any nepheline-K feldspar conodes.

The first precise information on the natural association of K-Na feldspar and nepheline appears in Tilley's work (1954). Whereas the nepheline richest in potassium contains 45 % $KAlSiO_4$, the limiting content found in parageneses with feldspars is of the order of 25 %.

Later, V. S. Sobolev (1956), using analyses cited in Rosenbusch's work on the leucite-nepheline dolerite of Vogels Berg (Hesse, Germany), made the first attempt at constructing the corresponding paragenetic triangle. Whereas a point corresponding to approximately 10 % Na content was taken for the composition of leucite (according to the limiting composition, as
228 the cited analysis is not satisfactory), a composition of 50 % Na component was taken for the feldspar and 70 % Na component — for nepheline.

Thus, the considerable slope of the conode was proved, but at the same time it was noted that the analyses made by Knopp in 1865 were not sufficiently reliable, probably due to faulty sampling of the material, and that in practice the conode may have an even steeper slope. Tilley (1958) resampled the same rocks from Vogels Berg, examined the nepheline and feldspar which had been previously analyzed by Skun, and determined with greater precision the location of the considered phase triangle. The composition of nepheline was found to be $Ne_{76.2}Ks_{19.5}An_{2.8}Q_{1.5}$, and that of the feldspar — $Ne_{78.4}Ks_{20.0}Q_{1.6}$, i. e., ignoring the secondary components and making approximations, we obtain a nepheline with 80 % of the sodium component and a feldspar with 25 % of the same. Thus, the slope of the conode proves to be much steeper than was apparent from the preliminary data and the difference in the compositions of the nepheline and the feldspar is as much as 55 %.

Taking into account Tilley's above-mentioned data, as well as the influence of the calcium component, the presence of which should increase the difference between the feldspar and the nepheline in the rock, we accept a nepheline composition of 75 % sodium component for the triangle.

It is interesting to mention that in 1956 MacKenzie (1956) for the first time presented experimental data on one pair of equilibrial compositions

of feldspar and nepheline, but these experiments were carried out under somewhat different conditions, namely, in the presence of water and within a temperature range of 625—675° C: feldspar — 54 % sodium component; nepheline — 96 % sodium component. The determination of the composition
229 was carried out according to X-ray analyses. Although the mentioned values are not strictly accurate and cannot be used without certain reservations, they do affirm the steep slope of the corresponding conode.

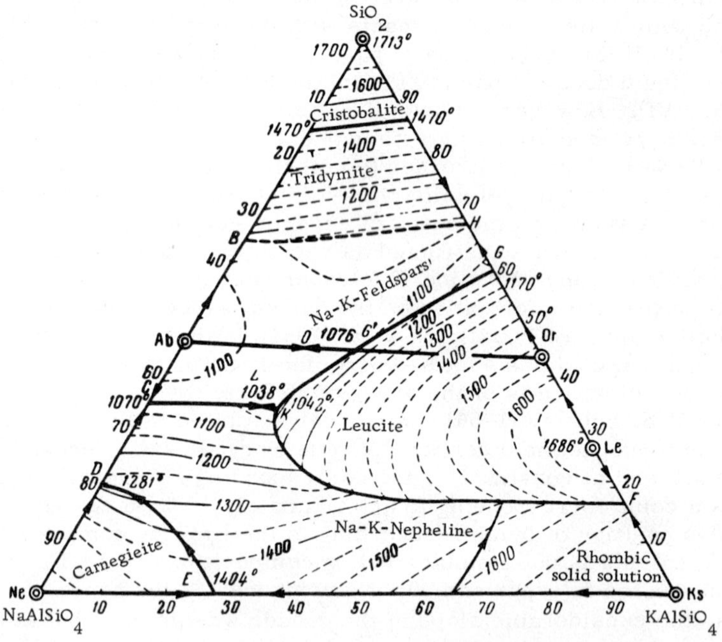

FIGURE 149. System silica-nepheline-kalsilite (according to Bowen and Schairer)
Tr — tridymite; Ab — albite; Ne — nepheline; Ks — kalsilite; Le — leucite

The question of the change of the minimum point of feldspar with the change in silica content of the system has not been satisfactorily solved. However, it is necessary to take two facts into account: 1) neither silica nor nepheline form solid solutions in K-Na feldspars. Hence, the addition of these minerals should not influence the behavior of feldspars; 2) the addition of water, which causes a much more marked decrease in the temperature of crystallization of feldspars (see page 180) than do the previously mentioned two minerals, has almost no influence on the shifting of the minimum point. Hence, the assumption may be made that the composition of K-Na feldspars at the minimum points along the line Ab-Or also corresponds to 65 % Ab and 35 % Or.

Accepting this assumption, the minimum point on the line *BH* is fully determined, as the composition of SiO_2 remains constant and the given point should lie on a straight line joining the SiO_2 apex and the minimum point of the feldspars (see Figure 149).

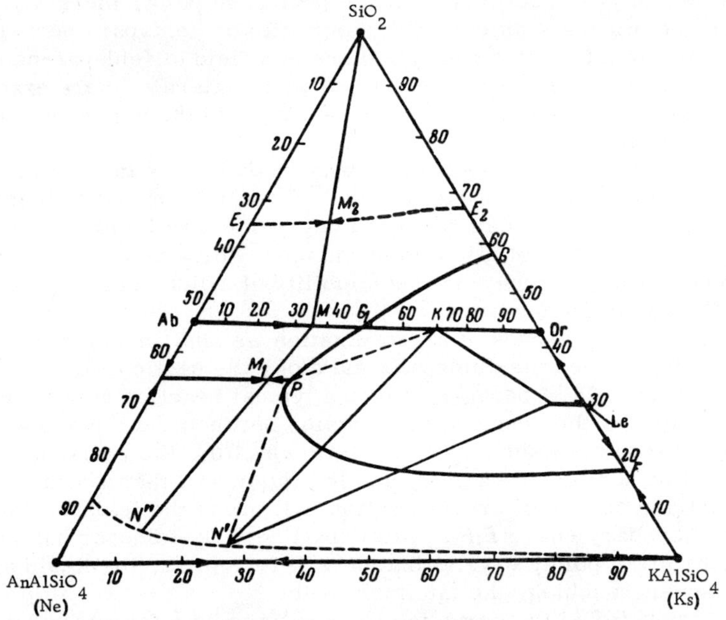

FIGURE 150. A scheme of the fields of crystallization and paragenesis in the system silica-nepheline-kalsilite

Ab — albite; Or — orthocase; Le — leucite

It should be stressed that this point of minimum is usually shown schematically on diagrams (as no accurate determinations are as yet available for this area) for compositions richer in the sodium component (approximately 50 %).

The minimum point on the curve, which borders the fields of nepheline and feldspar, is also only approximately determined, as its temperature within this area is almost the same as the temperature of the reaction point.
230 Several authors, including Tilley (1958), do not indicate this point on the diagram. However, the corresponding nepheline-feldspar conode should necessarily pass through this point, as the phase triangle is here converted into a straight line.

Thus, the composition of the feldspar should change at least by 30 % during the transition from the reaction point to the minimum point, which should undoubtedly correspond (even in the case of a very gently sloping curve) to an essential change in the composition of the liquid. Taking all available data into account, we may approximately trace the position of the mentioned point, accepting the following compositions for the solid phases: nepheline — 90 % sodium component, feldspar — 65 % sodium component, i. e., again assuming that the minimum for feldspars is not shifted in the presence of nepheline.

Taking the mentioned amendments into account, it is easy to divide our triangle (see Figure 150) into fields of paragenesis, which are stable during the final stage of the crystallization of the liquid. Above the line Ab-Or we have an area of silica (tridymite) and solid solutions of feldspars. Below the line the relations are more complex. In the above-constructed phase triangle, which is in equilibrium at the peritectic point, there will be a field of three minerals (of constant composition): feldspar, nepheline and leucite. To the left of this triangle there is a field of feldspar-nepheline paragenesis, and to the right, in the two quadrilaterals (more exactly the trapeziums), leucite and feldspar will be stable in the upper one, and leucite and nepheline — in the lower one.

Finally, in the lower part of the triangle, there is a fairly limited field of ternary solid solutions of nepheline, i. e., a field of monomineralic paragenesis. Tilley's latest data are not in agreement with earlier data on the considerable solubility of silica in nepheline. For a nepheline containing 20—25 % potassium component, the solubility of silica does not exceed 3 %, as is reflected on the diagram.

Among the peculiarities of crystallization we shall mention that the curve limiting the leucite-feldspar field along the section GG_1 (i. e., the silica-feldspar field of paragenesis) is a typical reaction curve. Feldspar is formed here as the result of the reaction between the liquid and leucite. In all initial points, which are in the area (Or) GG_1, the redissolving of leucite will be completed earlier, and the liquid will then reach point G, after which the course of crystallization will cross the feldspar field and reach the boundary curve E_1E_2. After the line SiO_2-feldspar has passed through the given point, crystallization will terminate, the liquid approaching, but never reaching, the minimum point.

The section G_1P of the curve is characterized by a fan-shaped distribution of the feldspar-liquid conodes (in equilibrium with leucite), i. e., here we have a case when the character of crystallization is dependent on the initial composition. Initially, the amount of leucite decreases and then subsequently increases. In the area where the conodes intersect, leucite is dissolved completely, followed by the separation of the feldspar, and the course of crystallization will return to the boundary curve, after describing an arc. Following this a simultaneous crystallization of feldspar and leucite will occur, either until the liquid completely disappears, or up to the peritectic point (refer to page 239).

According to the location of the point within the phase quadrilateral, leucite is preserved up to the peritectic point P, where it becomes unstable and is replaced by K-Na feldspar and nepheline due to reaction with the liquid. Depending on the location of the given system within the triangle of solid phases, the liquid is either exhausted or the decomposition of leucite is complete and the simultaneous crystallization of feldspar and
231 nepheline will take place. As point M_1 is not a eutectic point, but a minimum point, crystallization will terminate on the line, and the final liquid will not reach point M_1.

Considering the significance of the mentioned system to petrology, Bowen (1937) termed it a system of residual melts and drew attention to an interesting feature of the liquid surface, on which there is situated a rather wide, long trough between the points M_1, M and M_2, in the field of temperature of approximately 1,050°. Within the boundaries of this trough-shaped depression are located the points of analyses of most leucocratic effusive rocks, from liparites to phonolites, as was shown graphically by Bowen for the lavas of East Africa. According to Bowen, this area is the final limit toward which magma melts trend during crystallization differentiation.

However, the mentioned correspondence in the upper part of the diagram (for rocks oversaturated in silica) is true only in the case when the minimum point on the cotectic curve is taken, as is generally the case, as 50 % content of the sodium component. It is possible that this feature in itself played a certain role in the determination of the minimum point, as precise experimental data were not available in this case.

After introducing our amendment (see previously the location of the minimum point) the typical liparites lie to the right of the line passing through the SiO_2 apex and the minimum point. It appears to us that such a location is in complete agreement with the concept of crystallization differentiation, after taking into account the role of the plagioclases as considered on page 243.

According to the opinion of many petrographers, the point ***P***, which is termed the pseudoleucitic reaction point, elucidates the magmatic formation of pseudoleucite — a pseudomorph after leucite consisting of K-feldspar and nepheline. However, a considerable increase in volume should accompany such a reaction, and the disintegration of pseudomorphs is possible, thus accounting for the proposed postmagmatic character of this reaction in igneous rocks (Bilibin, 1947). As already mentioned, the incongruent melting of orthoclase and the considerable penetration of the leucite field into the area of melts oversaturated in silica are apparently of no essential significance in interpreting the relations of these minerals in igneous rocks. Leucite has not been found together with free silica or with glass oversaturated in silica, which phenomenon is probably explained by the character of the diagram on lowering the temperature. On the other hand, it should be mentioned that Bowen and certain other authors ascribed great importance to the formation of leucite from a magma oversaturated in silica, with a subsequent crystallization and differentiation, in their explanation of the origin of certain alkali rocks.

THE SYSTEM LEUCITE-ANORTHITE-SILICA**

This system was investigated by Schairer and Bowen in 1947, and represents the second face of the tetrahedron of salic minerals: silica-nepheline-kaliophyllite-anorthite. Considered on its own, however, it is only of limited significance to petrology.

The sides of the concentration triangle on Figure 151 are binary systems: 1) Lc—An (leucite-anorthite) — a system with a eutectic; 2) Lc—Q (leucite-silica) — a system with a compound (Or — orthoclase) which melts incongruently, i.e., with corresponding peritectic and eutectic points; 3) Q—An (silica-anorthite) — a system with a eutectic.

232 Four fields of crystallization are located within the triangle: leucite, orthoclase, anorthite and silica (tridymite and cristobalite). In addition there are two invariant points: a peritectic point corresponding to the reaction $L + Lc \rightarrow An + Or$, and the ternary eutectic point, where orthoclase, anorthite and silica are in equilibrium with the liquid.

The tracing of the course of crystallization of any given point of the above-considered system presents no difficulties, and shall not be dealt with here.

Only those melts which are undersaturated in silica have relations similar to those in igneous rocks, i. e., those melts which lie to the left of the line Or—An. They provide us with a concept of the change in eutectic relations between leucite and anorthite with different silica contents. Anorthite and basic plagioclases are generally not found in equilibrium with free silica in igneous rocks.

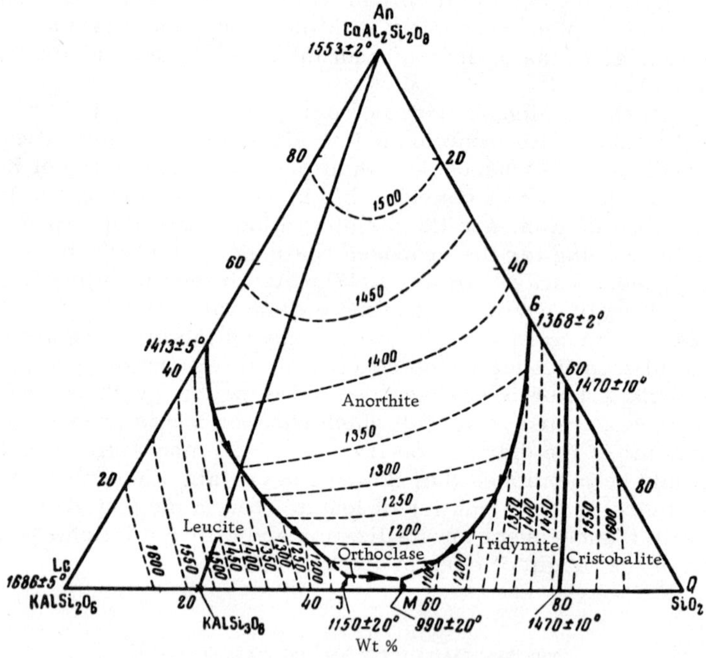

FIGURE 151. The system leucite (Le)-anorthite (An)-silica (Q) (after Schairer and Bowen)

THE SYSTEM DIOPSIDE-LEUCITE-SILICA**

This system was investigated by Schairer and Bowen (1938), after preliminary work had been done by Bowen in 1937.

The sides of the concentration triangle on Figure 152 are represented by the following systems: 1) Di—Lc (diopside-leucite) — a system with a eutectic; 2) Di—Q (diopside-silica) — a system with a eutectic complicated by the polymorphous inversion of silica (cristobalite-tridymite) and decomposition into two immiscible liquids; 3) Lc—Q (leucite-silica) — a system with an incongruently melting compound, which has been considered on numerous previous occasions.

233 As in the case of the previous system, four fields may be distinguished within the triangle (not taking into account the division of the silica field), and two invariant points are present: a peritectic point (liquid+leucite → diopside +orthoclase) and a eutectic point (liquid → orthoclase+diopside+tridymite).

On considering the relations in igneous rocks, the changes in the eutectic relations between leucite and diopside with the increase in silica content, as well as the location of the eutectic point, are of interest. The latter, in similarity to the location of the albite-diopside eutectic, permits one to consider the compositions of acid igneous rocks, and in particular granites, as approaching that of the eutectic despite the small content of femic mineral.

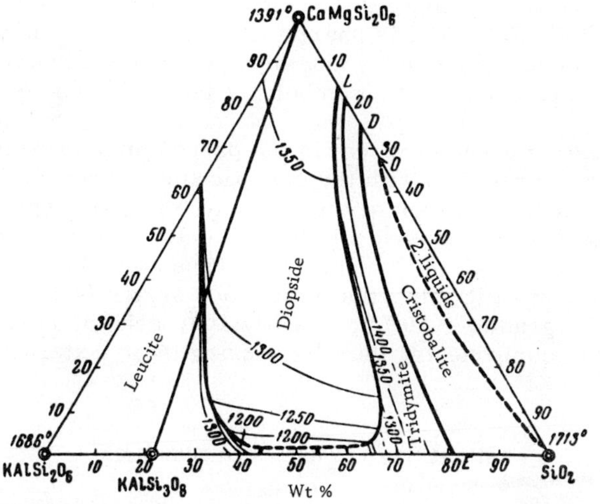

FIGURE 152. The system diopside-leucite-silica (after Schairer and Bowen)

On considering the diopside-leucite eutectic, we note a marked change in relations between nepheline and diopside. In contrast to the latter system, the system leucite-nepheline proves to be a simple binary system, the leucite content at the eutectic point being less than 40 % (by weight). Owing to this it will start to crystallize at a very early stage from melts rich in femic minerals, and it is quite often present as phenocrysts in basaltic rocks, whereas nepheline usually appears only in the matrix of such rocks. As is seen from Figure 139 (see previously), the relations between nepheline and diopside prove to be much more complex, and the crystallization of nepheline may only start when its content reaches a concentration of 60 %, and even then will be accompanied by the separation of melilite and olivine.

This feature of the diagrams fully explains the difference between the sodium and potassium series of volcanic rocks according to the theory of crystallization differentiation, as appears most strikingly on A. N. Zavaritskii's diagram (page 276). The volcanic series of rocks rich in sodium (lines 7, 8, 9 and 10), falling in the area of high values of ***b***, i. e., rocks which are rich in femic minerals, differ only slightly with respect to their ***a*** value from normal basalts, and further on they abruptly leave
234 the line representing non-alkali rocks as ***b*** is lowered. Lines 5 and 6,

which represent potassic rocks containing leucite, follow a completely different course, as here, already at high values of *b*, we have a sharp increase of *a*, which is associated with the early crystallization of leucite together with femic minerals.

The early crystallization of leucite also explains another feature of leucite-bearing rocks on the basis of analysis of petrochemical diagrams. This applies to the MgO:FeO ratio, which is very high in alkali rocks rich in potassium, thus further stressing the gently sloping vectors in the left part of the diagram. In contrast to this, when rocks become enriched in sodium, the MgO:FeO ratio is sharply decreased. The early crystallization of leucite occurs, obviously accompanied by the crystallization of earlier, i.e., more Mg-rich, members of isomorphous series of femic minerals.

It is interesting to mention that in the pseudobinary system diopside-orthoclase the intersection with the cotectic line corresponds to a composition with a fairly high content of the potassic feldspar (approximately 60 %), whereas in the system diopside-albite, the eutectic point corresponds to a composition of 97 % albitic component. Thus, the quantitative relations between femic and salic minerals in the rock series (shonkinite — 60 %; syenite — 25 %; granite — 10 %) generally approach cotectics with a certain excess of femic components, thereby causing their textural peculiarities.

THE SYSTEM FAYALITE-NEPHELINE-SILICA**

This system was investigated by Bowen and Schairer (1937). Here we encounter for the first time an example of a ternary system in which a ferrous oxide participates. All the previously mentioned general peculiarities noted in the study of such systems (see page 195) also apply here.

The sides of the concentration triangle (Figure 153) are constituted by the following systems†: 1) Ne—Q (nepheline-silica) — a system with a congruently melting compound (albite), already encountered above; 2) Fa—Q (fayalite-silica) — a system with a eutectic point, complicated by the inversion of cristobalite to tridymite in the field of liquid immiscibility; 3) Ne—Fa (nepheline-fayalite) — a system complicated by the formation of hercynite ($FeO \cdot Al_2O_3$) — a phase which is not in agreement with the boundaries of the three considered components, i.e., the system is pseudoternary within this area, and the corresponding lines only represent the borders of the field.

The relations within the system are quite simple. With the exception of the special field of hercynite and complications connected with the decomposition of the liquid phase and the inversion of cristobalite to tridymite and carnegieite to nepheline, we have here four fields of crystallization: 1) nepheline, 2) fayalite, 3) albite and 4) silica. The boundary lines meet at two ternary eutectic points: *P* (nepheline-albite-fayalite) and *O* (albite-tridymite-fayalite).

A section along the line Fa—Ab (fayalite-albite) may be considered as a binary system with a simple eutectic, the quantity of fayalite in this eutectic

† On the figure our system is represented as a part of the more general system $FeO\text{-}NaAlSiO_4\text{-}SiO_2$.

equalling 16 %. The ternary eutectic points are only approximately determined, but the content of the fayalite component in the liquid remains approximately the same.

After the appearance of data on systems with ferrous oxide, many assumptions were made as to their essential difference from analogous
235 systems with magnesium oxide. As we shall see further on, such assumptions have been largely affirmed. However, with regard to the problem of eutectic relations between alkali aluminosilicates and femic minerals, the content of the fayalite component in the albite-fayalite eutectic is considerably higher than in the eutectic with magnesium minerals (e.g., albite-diopside), but the magnitude of the values and the general conclusions remain the same: at the eutectic of albite with femic minerals (and also at the ternary eutectic of albite and silica), the point is in this case also sharply shifted toward the aluminosilicate, being thus in agreement with the low content of femic minerals in granites and syenites. On the other hand, the essential increase of the content of femic mineral in the eutectic, accompanied by the replacement of magnesium by iron, is of significance to petrology. The figures thus obtained more closely approximate the relations in igneous rocks.

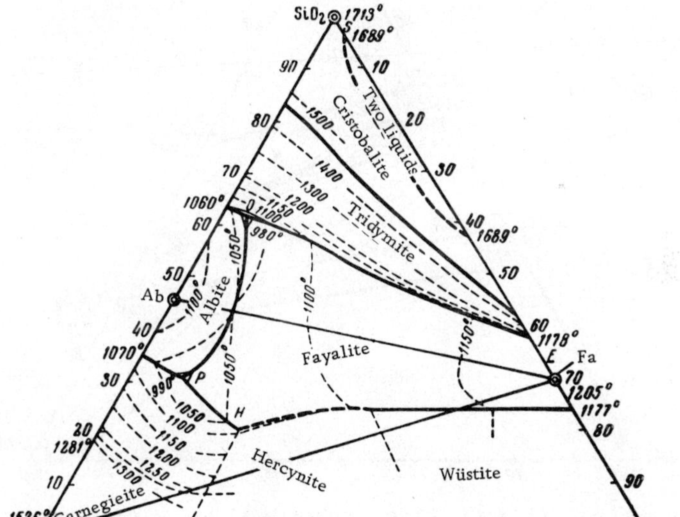

FIGURE 153. The system nepheline-silica-FeO, and its part — the system nepheline-silica-fayalite (after Bowen and Schairer)

Ab — albite; Fa — fayalite

THE SYSTEM FORSTERITE-FAYALITE-SILICA**

Some time later than the investigation of other important silicate systems, the study of systems containing ferrous oxide was started, one of which has already been considered above. The given system

which is of great significance to petrology, was investigated by Bowen and Schairer (1935). The experiments were carried out in a pure iron crucible kept in an atmosphere of nitrogen. Certain specific features of such systems have already been considered in the description of binary equilibria (see page 195). The described system constitutes a part of the more general system SiO_2-MgO-FeO (Figure 154). However, the minerals of the series MgO-FeO (magnesowüstite) in igneous rocks constitute an extraordinary rarity and are only found under specific conditions, thus permitting us to restrict ourselves to the consideration of the upper silicate part only.
236 The sides of the concentration triangle are formed by binary systems: 1) Fo—Fa (forsterite-fayalite) — a continuous series of solid solutions of Roozeboom's Type I; 2) Fo—Q (forsterite-silica) — a system with an incongruently melting compound — $MgSiO_3$ (clinoenstatite), and a eutectic complicated by the inversion of cristobalite to tridymite and a field of two liquids; 3) Fa—Q (fayalite-silica) — a system with a eutectic point, but lacking a compound analogous to clinoenstatite.

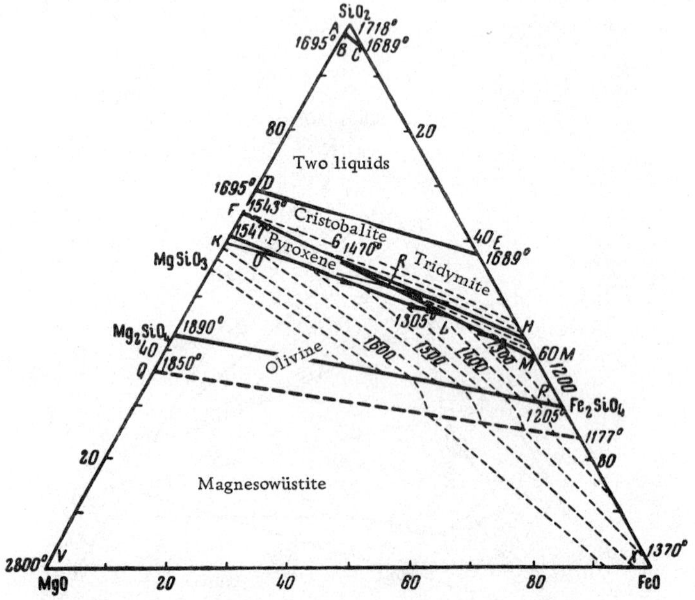

Figure 154. The system MgO-FeO-SiO_2 (after Bowen and Schairer)

In the ternary system (Figure 155), in addition to the field of two liquids and the line separating the fields of tridymite and cristobalite, three fields of crystallization of various solid phases and liquid are apparent: silica (cristobalite and tridymite), pyroxene and olivine. In similarity to olivines, pyroxenes constitute phases of variable composition. However, in contrast to olivines, the pyroxenes form only a limited series of solid solutions, ranging from pure $MgSiO_3$ to a pyroxene with approximately 55 % $FeSiO_3$ component. A pyroxene of this composition is in equilibrium

with the liquid and two other solid phases (olivine and tridymite) at an invariant peritectic point at a temperature of 1,305°. The boundary curve ***KL***, separating the field of olivine from that of pyroxene, approaches this point ***L***, as does the boundary curve ***FL***, which separates the field of pyroxene from those of tridymite and cristobalite, i. e., here the narrow field of pyroxene crystallization wedges out, and the curve ***LM***, separating the fields of crystallization of olivine and tridymite, enters the area of lower temperatures. On the concentration triangle, one may also readily distinguish the areas of final paragenesis of solid phase, which form during
237 crystallization from a melt (not taking into account the subsequent inversions in the solid phases below the melting point — for further details see below) under conditions of complete equilibrium: 1 — silica (cristobalite or tridymite), pyroxene; 2 — tridymite, olivine; 3 — olivine, pyroxene; 4 — tridymite, pyroxene, olivine. During the paragenesis of two solid phases (the liquid constituting the third phase), one degree of freedom remains, i. e., changes in composition of one of the solid phases are possible. In those areas where only one solid phase has a variable composition (1 and 2), the compositions of the pyroxene or olivine may easily be found by drawing a straight line through the given point of the initial composition and the apex SiO_2. In the fields of olivine and pyroxene, for each composition of the former there should be a corresponding composition of the latter (as there is only one degree of freedom). The conjugate compositions are connected by conodes (sides of the phase triangle), which, in similarity to the case of equilibrium between the liquid and a solid phase of variable composition, are determined experimentally. In the considered system pyroxene differs from olivine by its lower iron content (approximately 5 % iron component).

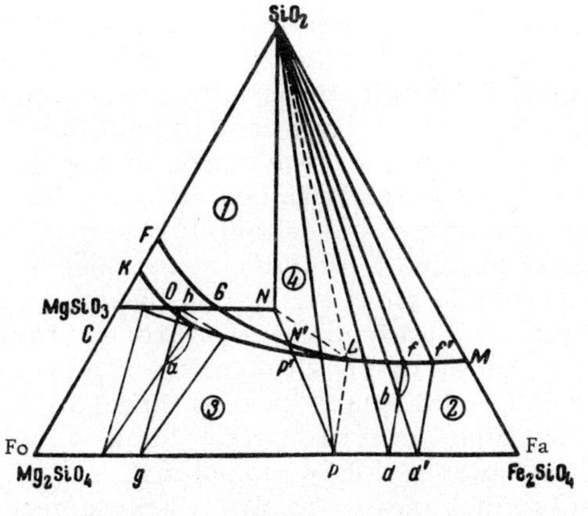

FIGURE 155. Schematic equilibrium diagram of the system fosterite-fayalite-silica (scale is not accurate)

In the area of equilibral coexistence of three minerals (tridymite, olivine and pyroxene) there is no degree of freedom, i. e., both olivine and pyroxene have fixed compositions, which are defined by points ***N*** and ***P***. The solid phases of this triangle are in equilibrium with a liquid of composition ***L***. The mentioned paragenetic areas are limited by straight lines which correspond to compositions of either two-mineral or one-mineral paragenesis, solid solutions, pyroxene or olivine.

Thus, one may easily determine from the diagram the character of the first crystallizing solid phase for each composition, and the final paragenesis of minerals in equilibrium with the last drop of liquid. However, despite the simple outline of the diagram, the determination of the course of crystallization is not easily made. This is associated, on the one hand, with the possibility of crystallization of two series of solid solutions, and, on the other hand, with the change in the character of behavior of the pyroxene. Initially (following from the MgO-SiO_2 side) the composition line of the pyroxene ***CN*** lies in the field of crystallization of olivine, i. e., the pyroxene melts incongruently, then it crosses the field of pyroxene (con-
238 gruent melting), and, finally, it again leaves the field of pyroxene, entering the field of tridymite. Here the melting of the pyroxene is again incongruent. However, the first mineral to separate from the corresponding melt is not olivine, but tridymite, i. e., silica will crystallize from melts undersaturated in it.

The field of crystallization of olivine is of greatest interest to petrology and we shall therefore start by considering it. As the olivine has a variable composition, its crystallization will here proceed in the same way as in the previously considered system of plagioclase crystallization. In comparison with the normative composition of the liquid, the settled crystals will be enriched in the more refractory magnesium component. The liquid is correspondingly enriched in the iron component, and the solid phase should continuously change its composition. If we are to consider a certain given initial point, we should draw an olivine-liquid conode through it (in equilibrium with pyroxene or tridymite), using a series of already constructed conodes on the diagram. The course of change of the liquid is illustrated by a curve, which has its origin at the initial point and returns to the given conode on the boundary curve (in the case of point ***a*** — the curve olivine-pyroxene; in the case of point ***b*** — the curve olivine-tridymite). The precise construction of these paths is rather difficult, but the curves are always convex toward the side representing lowering of temperature.

A second solid phase joins the olivine at the boundary curve. If this phase is tridymite (area 2, point ***b***), a simultaneous crystallization of two solid phases — olivine and tridymite — will proceed further on, accompanied by continuous change in the composition of the olivine up to the point where the side of the triangle — (SiO_2)***d*** — passes through the initial point, i. e., crystallization of the liquid is terminated. In our example, the composition of the liquid phase is determined by the point ***f'***.

Crystallization is much more complex when the second phase is a pyroxene, but the consideration of the course of crystallization is made somewhat easier by taking into account the final paragenesis. As can be predicted, in area 3 the melt will cool as a mixture of olivine and pyroxene, the compositions of which are determined by ***g*** and ***h*** in our example. Along the segment ***OCK*** the final products of crystallization should be pyroxene and cristobalite, i. e., all the olivine should disappear. After the dissolving of the olivine the composition point of the liquid will leave the boundary curve and cross the pyroxene field, while the course of crystallization will again have the form of a curve, convex toward the more fusible ferruginous component. Its location is determined according to the pyroxene-liquid conode in equilibrium with cristobalite, which passes through the point of initial composition (in similarity to the system diopside-forsterite-silica). Crystallization terminates when the given point falls upon the silica-pyroxene line of the phase triangle.

In area 4 the final products of crystallization are olivine, pyroxene and tridymite, and, finally, in the triangle **PL**(SiO_2) the final product in area 2 should consist of olivine and tridymite, i. e., pyroxene will have disappeared completely. The latter occurs at the peritectic point, where the reaction

$$L + (Mg, Fe)SiO_3 \rightleftarrows SiO_2 + (Mg, Fe)_2 SiO_4$$

occurs, the outcome of which depends on the location of the given point in phase quadrilateral.

However, the mentioned sequence and the final result do not determine all the peculiarities of the crystallization course in the given field. The olivine-pyroxene boundary line has an initial configuration, such that the tangent to it (here the direction of the tangent actually coincides with the direction of the boundary line) passes beyond the phase triangle, i. e., the crystallization of pyroxene proceeds due to the reaction between the liquid and the olivine. Further on, after point ***O***, it enters the phase tri-
239 angle, i. e., here a simultaneous crystallization of olivine and pyroxene occurs from the liquid. Thus, on a certain area, as pyroxene crystallizes the amount of olivine initially decreases and subsequently increases, accompanied by continuous changes in composition of both solid phases. As a result, a peculiar case might arise here, when the earlier formed olivine will entirely disappear, and the point of the liquid, after leaving the triangle, will trace a certain course within the field of two-phase equilibrium (pyroxene-liquid) and again return to the boundary curve, now crystallizing pyroxene and olivine simultaneously. Such crystallization is characteristic of systems close to point ***O***, in the area where liquid-pyroxene conodes (in equilibrium with olivine) intersect each other from the conode ***OP′***, coinciding with the line of pyroxene composition, to the conode ***OB***, coinciding with the olivine-pyroxene boundary line (Figure 156). Here each point (e. g., point ***x***) twice touches the olivine-pyroxene side of the phase triangle. The first time it reaches this side at the moment it departs from the phase triangle, i. e., on the disappearance of first generation olivine. Afterwards, the course of change in the liquid composition, after describing an arc ***b′b″*** in the pyroxene field, returns to the boundary curve at point ***b″***, i. e., at the moment when the side of the phase triangle is reached for the second time and the system again enters the triangle.

For systems falling within the crystallization field of pyroxene (see Figure 155), the course of change of the liquid composition also describes a curve, convex toward lowering of temperatures, the position of which is determined by the corresponding conode: the conodes pyroxene-liquid in equilibrium with silica — for the area above the straight line $(MgSiO_3)$***N***, and the pyroxene-liquid conodes in equilibrium with olivine — for the area below this straight line (with the exception of a small specific area in the corner at point ***G***). A second solid phase appears on the boundary curve, i. e., silica (cristobalite or tridymite) or olivine. The further course of crystallization may be readily determined from the area of the final paragenesis. Within the area ***KOGF***, the final products of crystallization are two solid phases (pyroxene and silica), after the given point has reached the corresponding side of the phase triangle. Within the area ***OGP'N'*** the melt cools to form olivine and pyroxene and within the area ***N'P'L***, the composition of the liquid reaches the invariant point, where tridymite appears, and the final product is either three minerals (area 4) or all the pyroxene disappears and crystallization of olivine and tridymite continues along curve ***LM***. In the small specific area around point ***G***, tridymite initially joins the pyroxene and then later disappears, and the liquid, after crossing the pyroxene field, reaches the pyroxene-olivine boundary curve.

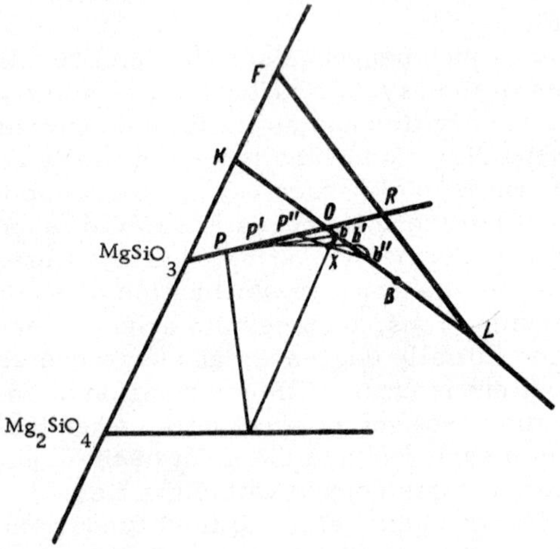

FIGURE 156. A detail from the system fosterite-fayalite-silica in the region of intersection of conodes

240 Taking into account the above considerations, there is no need to consider the field of silica crystallization in the same detail (as is recommended for checking whether the method is understood). Attention should only be drawn to the fact that, as from a certain moment, the simultaneous crystallization of pyroxene and silica on the boundary curve is replaced by reactional crystallization (crystallization of pyroxene due to the reaction between tridymite and the liquid). This may result in the complete disappearance of tridymite and the transition of the system into the area of two-phase equilibrium, involving the crossing of the pyroxene field and the crystallization of olivine. The final products of crystallization may be pyroxene and olivine (area ***OGP'N'***) or crystallization may reach the invariant point, where tridymite reappears and pyroxene may even disappear.

The section along the line $(MgSiO_3)N$, up to the Fe_2SiO_4-SiO_2 side of the triangle (see Figure 131, on page 197), is of special significance in this system. In order to be in equilibrium with the melt, this section is pseudobinary over most of its range, with the exception of a small area ***OR***, where a pyroxene separates directly from the melt, and the system behaves as a binary one, i. e., the composition of the liquid constantly remains in the plane of the considered section. The inversions at temperatures below the area of stability of the liquid phase are of especial interest to petrology, and will thus herewith be considered.

As already mentioned, a monoclinic modification of pyroxene $MgSiO_3$ crystallizes from the melt, whereas in igneous rocks an orthorhombic pyroxene of the corresponding composition is almost invariably present. As no reversible inversions were obtained, the concept of the monotropic character of inversions prevailed, and a view as to the pseudo-orthorhombic crypto-twinned character of the natural rhombic pyroxenes was even put forward. At present there are no longer any doubts to the fact that orthorhombic pyroxenes constitute a truly orthorhombic modification. The monoclinic-orthorhombic inversion on decrease in temperature was not experimentally obtained. However, on heating orthorhombic pyroxenes, it was possible to determine a strictly defined temperature for each composition and to construct a corresponding curve. Below this curve, an even rather lengthy halt was not able to bring about an inversion, whereas at higher temperatures inversion was achieved over a shorter period of time. This permitted Bowen to claim that the inversion is reversible, but has a high degree of inertia, thus causing supercooling of the high temperature modification. Such a conclusion is in agreement with the formation of orthorhombic pyroxenes from artificial melts on the addition of mineralizers (e. g., as obtained by D. P. Grigor'ev in systems with fluorine), which lower the temperature of melting.

Another essential peculiarity of the section, following from observations of natural orthorhombic pyroxenes and their inversions, is the considerable shift of the triangle (SiO_2) ***NP*** on lowering the temperature in the areas of ferruginous compositions, i. e., the widening of the field of olivine-pyroxene rocks at the expense of quartz-olivine rocks. The composition of the orthorhombic pyroxenes reaches 90 % iron component. However, the ratios shown on the diagram (for low pressure) may be doubted to a certain degree, as olivine with 60 % iron component was found in rocks together with tridymite and orthorhombic pyroxene, and not the monoclinic modification, as follows from the diagram. The orthorhombic pyroxene enriched in iron is only rarely found, probably being formed at increased pressures.

Later on Bowen described a pure ferrosilite, which he claimed to have
241 obtained under special conditions below 660°. The analysis of the mineral is not presented and it proved impossible to decompose it into fayalite and silica, thus making the correctness of the determination doubtful.

On describing the diagram, Bowen and Schairer mentioned its importance to petrology, indicating that the obtained data explained the presence of iron-bearing olivine together with quartz, as well as the possibility of change in the order of crystallization of olivine and pyroxene with a variation of iron content. However, they stressed that similar deviations from the usual order occur only in extraordinary cases and are of no significance to petrogenesis. It should be noted that Bowen, having ignored the role of iron in femic minerals in his first publications, continues to underestimate the role of femic minerals even after the study of several systems containing ferrous iron had been made and despite petrological data. This role will be considered in greater detail further on (see page 248).

In addition to the relations between pyroxene, olivine and quartz, the investigated diagram also provides one further criterion for the evaluation of the upper limit of crystallization temperature of igneous rocks from magma — the curve of the orthorhombic pyroxenes (1,070—1,170°).

THE SYSTEM SiO_2-CaO-MgO*

Parts of this system — diopside-forsterite-silica and diopside-calcium metasilicate — have already been considered above. In addition to these compounds, which are found in nature as minerals, we encounter in the
242 considered system (Figure 157) several silicates of calcium and partly of calcium and magnesium, the fields of which occupy the left part of the diagram (Ferguson and Merwin, 1919, with amendments — see Eitel, 1954).

FIGURE 157. The system CaO-MgO-SiO_2 (after Ferguson and Merwin)

MW — merwinite; Mtc — monticellite; Di — diopside; Ak — akermanite; CS — $CaSiO_3$; C_3S_2 — $Ca_2Si_2O_7$; C_2S — Ca_2SiO_4; C_3S — Ca_3SiO_5; MS — $MgSiO_3$; M_2S — Mg_2SiO_4

Among the minerals we shall also mention monticellite, which melts incongruently [(the formation of solid solutions is possible up to the content of 10 % Mg_2SiO_4) and merwinite — $Ca_3Mg(SiO_4)_2$. Akermanite — $Ca_2MgSi_2O_7$, as a constituent of the mineral melilite, participates, showing congruent melting, and yields with diopside a simple binary system of the eutectic type.

In order to evaluate the temperature of formation of metamorphic rocks, the inversion point of a simple triclinic (pseudo-monoclinic) wollastonite to the hexagonal modification (pseudo-wollastonite) at 1,200° is of interest. However, the solubility of the magnesium component in wollastonite is higher than in the high temperature modification (up to 17 % diopside) and the temperature of inversion is raised to 1,345 ± 10°.

The given system is of almost no significance above the considered section (the system forsterite-diopside-silica) for the understanding of crystallization processes of igneous rocks, and will not be dwelt upon in detail].

THE SYSTEM $CaO-FeO-SiO_2$*

[The system was investigated by Bowen, Schairer and Posnyak (1933). As according to the conditions of equilibrium, there is no possibility of separation of the partial system hedenbergite-fayalite-SiO_2, which is analogous to the system forsterite-diopside-SiO_2, we are obliged to consider the entire system in greater detail].

The binary system $CaSiO_3$-$CaFeSiO_4$. This system may be considered as a binary one, the diagram of which is shown on Figure 158. The eutectic temperature of 1,203° is only 5° lower than the melting temperature of $CaFeSiO_4$, and the composition of the eutectic contains 80 % of this component.

The existence of a compound $Ca_2FeSi_2O_7$, which is only stable at temperatures below 775 ± 25°, is interesting. Above this temperature it decomposes into $CaSiO_3$ and $CaFeSiO_4$.

The pseudobinary system $CaSiO_3$-$FeSiO_3$. The mentioned system is not binary as $FeSiO_3$ is not formed from the melt as a crystalline phase. The diagram of this system is shown on Figure 134 (see previously). We shall not consider this system in detail. The most interesting feature is the formation of hedenbergite, which will herewith be considered.

[The most important feature of this section, as compared to the analogous section $CaSiO_3$-$MgSiO_3$, is the greater extent of solid solutions, which is directly associated with the greater similarity of the ionic radii of Ca^{2+}
243 and Fe^{2+} as compared to that of Mg^{2+}. If, due to the participation of magnesium, solid solutions are only formed between the double salt $CaMgSi_2O_6$ and $CaSiO_3$, no formation of double salts will occur at the melting temperature, and the series of solid solutions of pseudo-wollastonite and wollastonite extends up to the limiting point of stability of metasilicates (we shall bear in mind that the silicate $FeSiO_3$ cannot possibly exist at the melting temperature). Hedenbergite proves to be stable only at temperatures below 965°. At this temperature solid solutions are yet possible, both in the direction of enrichment in $FeSiO_3$, as is known beyond all doubt,

and in the more questionable direction of $CaSiO_3$. On the inversion curve of solid solutions of hedenbergite to solid solutions of wollastonite, a type of maximum is noted at point *H* which corresponds to the congruent melting point of the compound. To the right there is a characteristic transition curve, analogous to the melting curve of solid solutions with a rather gentle minimum. However, above the corresponding transition curve, corresponding solid solutions of wollastonite or olivine with tridymite are present instead of a liquid phase].

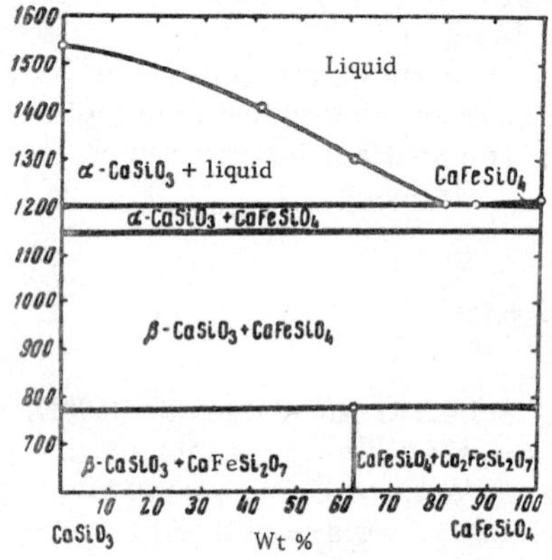

FIGURE 158. The binary system $CaSiO_3$-$CaFeSiO_4$ (after Bowen, Schairer and Posnyak)

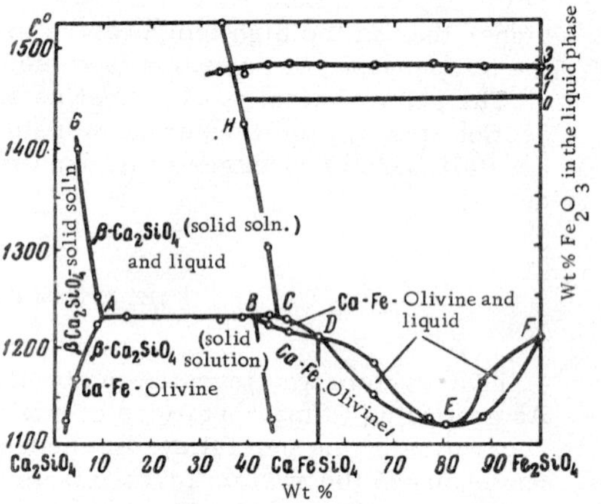

FIGURE 159. The binary system Ca_2SiO_4-Fe_2SiO_4 (after Bowen, Schairer and Posnyak)

The binary system Ca_2SiO_4-Fe_2SiO_4. In comparison with the system Ca_2SiO_4-Mg_2SiO_4, it is distinguished by a wider development of solid solutions (Figure 159), and in particular forms a complete series from point *F* to point *D* (a melting curve with a minimum). However, in contrast to the previous section, a double salt of the monticellite type is already stable at the melting temperature, below which only a usual decomposition curve of solid solutions should exist.

Isotherms and phase triangles. In phase triangles, defined by a diagram of the type shown on Figure 160, in addition to the liquid phase, members of two types of solid solutions participate (orthosilicates and metasilicates) as well as compounds of constant composition — tridymite, pseudo-wollastonite, Ca_3SiO_7 and others (below the line of orthosilicates).

Using the isotherms of phases and the basis of general principles, as presented in the first part of this book, one may represent the ratio of phases at each given temperature.

In order to determine the position of the phase triangles in the case of
244 solid solutions, one should know the compositions of the solid solutions in equilibrium with the liquid. The tables represent data on these compositions (in wt %), as obtained on the basis of optical measurements in three cases of equilibrium.

a) Solid phases: tridymite and solid solution of wollastonite [(100—*n*) $CaSiO_3$*n*$FeSiO_3$].

TABLE 5

Liquid			Wollastonite, *n*	Temperature, °C
CaO	FeO	SiO_2		
25.5	20.0	54.5	12	1264
24.0	23.0	53.0	15	1240
20.0	30.5	40.5	32	1177

b) Solid phases: pseudo-wollastonite $CaSiO_3$ and solid solution of wollastonite [(100—*n*) $CaSiO_3$*n*$FeSiO_3$].

TABLE 6

Liquid			Wollastonite, *n*	Temperature, °C
CaO	FeO	SiO_2		
28.0	27.5	44.5	5	1240
29.5	33.5	37.0	3	1200

c) solid phases: solid solution of wollastonite [(100—*n*) $CaSiO_3$*n*$FeSiO_3$ and Ca—Fe olivine [(100—*y*) Ca_2SiO_4*y*Fe_2SiO_4].

TABLE 7

Liquid			Wollastonite, *n*	Olivine, *y*	Temperature, °C
CaO	FeO	SiO_2			
25.5	38.5	36.0	7	60	1161
21.5	42.5	36.0	15	63	1130
18.0	45.5	36.5	40	72	1097
17.0	46.0	37.0	50	80	1093

The course of crystallization. Using general principles it is easy to follow the course of crystallization in various cases on Figure 160. The authors of the investigation consider typical examples in detail. For our purposes it is sufficient to make several general remarks.

The path of crystallization depends on the location of the initial point in relation to the boundaries of the various fields of initial crystallization and lines of solid solutions.

Let us consider a composition located within the metasilicate α-$CaSiO_3$ field (*GHOPN*) and below the line *WZ* (but above the imaginary line *WO*).

Crystallization starts with a separation of α-$CaSiO_3$, the composition of the liquid phase leaving *W* along a straight line until reaching line *HO*. There a solid solution of wollastonite appears, the composition of which is represented by a point on the straight line *WZ*. The apices of the phase triangle are: this point, point *W* and the composition point of the liquid (i. e., a point on the curve *HMO*). Further on, during crystallization, the
245 composition point of the liquid will move along this curve to *O*, the other apex of the triangle will move along the straight line *WMZ* (toward *M*), and the third, *W*, will not move. α-$CaSiO_3$ (*W*) will dissolve and the amount of liquid decreases.

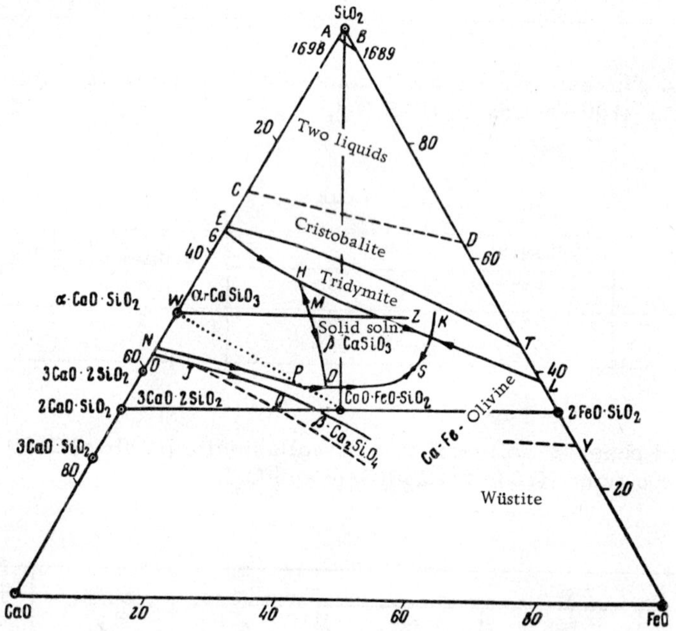

FIGURE 160. The system CaO-FeO-SiO_2 (after Bowen, Schairer, and Posnyak)

As the final outcome, α-$CaSiO_3$ should dissolve completely, the entire system should cool as two solid solutions — β-$CaSiO_3$ and Ca—Fe olivines, and the liquid will disappear along the curve *OK*. The dissolving of α-$CaSiO_3$ may occur either at point *O* or before the liquid reaches point *O*, depending on the location of the given point. In the latter case the path of the liquid will cross the field *MOK*.

If the point of the overall composition lies above the line *WMZ*, the final product will be a solid solution of β-$CaSiO_3$ and tridymite. Depending on the initial overall composition, a second phase may join α-$CaSiO_3$, being either β-$CaSiO_3$ or tridymite. α-$CaSiO_3$ is dissolved at point *H* with the exception of a small area in the vicinity of line *MH*, where its solution is already completed, and the path of the liquid crosses the β-$CaSiO_3$ field.

For compositions within the field ***HKO***, crystallization starts with the formation of a solid solution of wollastonite and is terminated by the formation of a mixture of this solid solution, either with tridymite or with olivine (solid solution), depending on the location of composition in relation to the straight line ***WZ***.

The compositions in the vicinity of point ***K*** are of interest, as their crystallization finally causes the formation of a liquid. The composition of the latter, moving along ***HK*** or ***LK***, reaches the inversion point ***K***. Here the separating tridymite is dissolved, either partially or completely, depending on the location within the phase quadrilateral (which is not shown on the diagram).

246 When crystallization is insufficiently slow for the continuous maintaining of equilibrium, zoned crystals may appear and the final product will consist of phases not corresponding to equilibrium.

We shall again draw attention to the absence of eutectic crystallization in the system. The lowest temperature corresponds to the minimum on the curve ***KSO***, in the fields of solid solutions of olivine and wollastonite.

Within the lower right-hand part of the field ***LKOP***, olivine is the first phase to separate and crystallization is terminated by the formation of a mixture of solid solutions of olivine and wollastonite.

[The significance of the above-considered system to petrogenesis is firstly due to the location of the decomposition point of hedenbergite. This once again enables us to determine the upper temperature limit of formation of certain metamorphic rocks. In igneous rocks pyroxenes contain considerable amount of magnesium, and we should be very wary of drawing comparisons, as we do not know the limiting ratios of iron and magnesium at which the decomposition of the corresponding double salt ceases below the melting temperature.

The increase of the iron content in wollastonite, as in the case of magnesium, raises the inversion point, thus rendering this point impractical as a geological thermometer. However, one should bear in mind that the temperature of formation of metamorphic rocks is probably much lower than the inversion temperature, even in the case of pure wollastonite.

The course of crystallization of the melts may be of interest when considering assimilation phenomena of limestone by magmas rich in iron, a phenomenon which sometimes occurs during the formation of alkali rocks. However, under natural conditions such crystallization always occurs at low temperatures (due to the presence of volatiles) below the decomposition point of hedenbergite, and one may observe considerable changes here in the course of crystallization].

THE SYSTEM SiO_2-Fe_2O_3-$Na_2O \cdot SiO_2$**

This system was investigated by Bowen, Schairer and Williams (1930). Only the behavior of compositions in certain parts of the concentration triangle was studied in detail. The diagram on Figure 161 allows for the understanding of the relations of the phases present, this being facilitated by the absence of solid solutions. We shall only dwell upon the behavior of the compounds $Na_2O \cdot Fe_2O_3 \cdot 4SiO_2$—$NaFeSi_2O_6$ (acmite-aegirine) which are of direct interest to petrology.

As is evident from the new diagram, aegirine is a triple compound with an incongruent melting point. The point of the corresponding composition is located within the hematite field of crystallization. Thus, if we consider a melt of aegirine composition, the phase Fe_2O_3 is the first to crystallize, the composition changing along the line (Fe_2O_3) *T* up to the boundary curve. Here we meet with a specific point — a maximum point, and the reaction liquid + hematite = aegirine will occur while the composition of the liquid phase remains constant. In our particular case the liquid and hematite are exhausted simultaneously, and the final product is aegirine. If the initial composition is taken on the line (Fe_2O_3) *T* and not at point *T*, then, depending on its location, the liquid will either disappear giving the final product of aegirine and hematite, or hematite is exhausted and the point crosses the field of aegirine up to the next boundary curve where the system ceases to be binary and its crystallization terminates at the ternary eutectic point E_1.

247 If we consider an initial point within the crystallization field of hematite above the straight line $(Fe_2O_3)T$, the crystallization of aegirine will take place at the boundary curve, accompanied by a change in the liquid toward point R_1. Depending on the location of the initial composition, several cases may arise: 1) hematite may be exhausted before the ternary eutectic point is reached, the composition of the liquid passing along the straight line within the aegirine field; 2) the liquid reaches point R_1 where crystallization takes place (hematite + aegirine + tridymite); 3) at point R_1 hematite disappears before the liquid, and crystallization continues along the boundary curve to the eutectic point E_1.

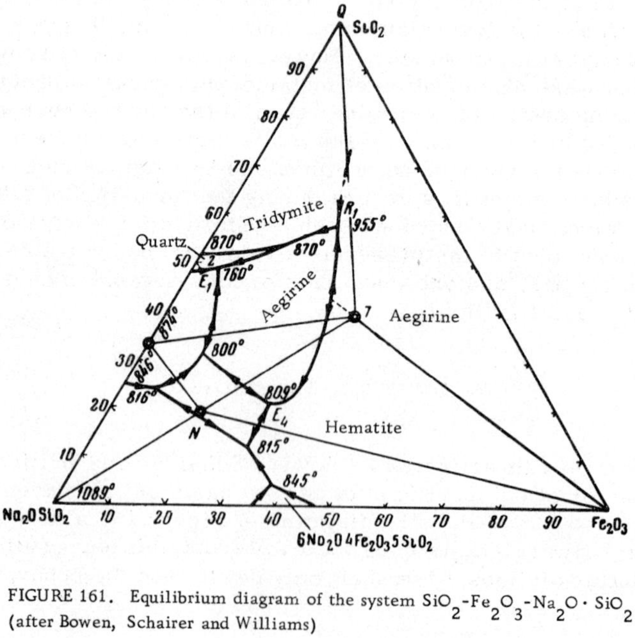

FIGURE 161. Equilibrium diagram of the system SiO_2-Fe_2O_3-$Na_2O \cdot SiO_2$ (after Bowen, Schairer and Williams)

If the initial composition is located on the boundary curve below the straight line (Fe_2O_3) T, the composition of the liquid will change toward the ternary eutectic point E_4. Thus, only two cases are possible here: either hematite is exhausted before this point is reached and the composition of the liquid crosses the aegirine field, or crystallization terminates at the eutectic point. The first case, however, has several variants, depending on the boundary line reached during the crystallization of aegirine.

Simple silicates of sodium or ternary compounds of the considered system which are richer in sodium than aegirine, are never encountered in nature. However, this does not signify that all compositions of this system will lie within the triangle (Fe_2O_3) TQ or on the boundary line (Fe_2O_3) T. There is no doubt as to the fact that the residual melt is often rather enriched in sodium, but the formation of the corresponding silicates in the presence of water is impossible, and the sodium compounds are either removed by solution or a reaction takes place between the sodium silicate and several other magma components with the formation of various rare silicates. A mineral (arfredsonite), more complex than those in the investigated system, is also formed containing ferrous iron and water, the sodium content exceeding that of iron. The possible formation of residual
248 melts enriched in sodium silicate as a result of crystallization differentiation follows from the incongruent melting of aegirine.

THE SYSTEM $CaMgSi_2O_6$-Mg_2SiO_4-$CaAl_2Si_2O_8$**

The given system (Figure 162), which is of essential importance for the understanding of crystallization processes in basic rocks, was only investigated in 1952 by Osborn and Tait. As already mentioned previously (page 215), the forsterite-anorthite side of the triangle is not a binary system because the crystallization of spinel occurs here, the decomposition of which is not shown in this plane. The corresponding field of crystallization of spinel is also present in the triangle, and this area should not be considered. However, the entire system beyond the limits of crystallization of spinel is not strictly ternary, merely presenting a section of a quaternary system CaO-MgO-Al_2O_3-SiO_2, as under these conditions of crystallization the pyroxene formed is not diopside, but contains a small amount of alumina (i. e., it approaches augite), and its composition does not lie within the plane of the triangle.

If we overlook the mention deviation, the given system (excluding the field of spinel) has an exceptionally simple character, containing neither compounds nor solid solutions, and crystallization terminates at the ternary eutectic point. Whereas pyroxene and anorthite are present here in approximately equal amounts, the content of olivine is rather small (7.5 %), which is in good agreement with the concept of the early crystallization of this mineral in basic rocks.

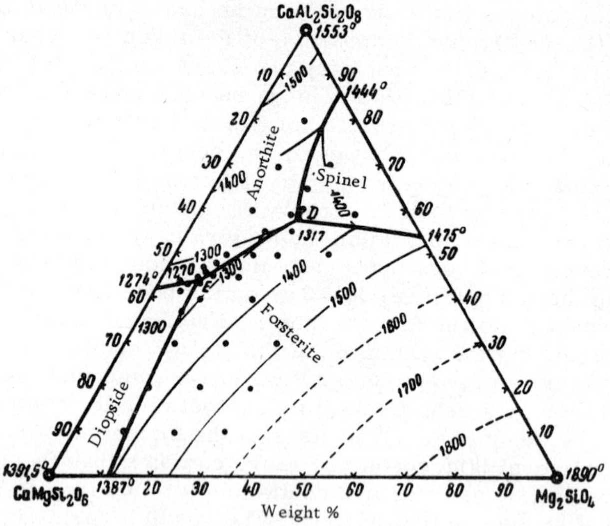

FIGURE 162. The system diopside-forsterite-anorthite (after Osborn and Tait)

THE SYSTEM DIOPSIDE-ANORTHITE-AKERMANITE**

This system, investigated by de Wys and Forster (1958), is of certain
249 significance in the consideration of rare alkaline rocks and is at the same time an example of a most simple ternary diagram with a eutectic point (Figure 163).

The previous data of Osborn and Schairer (1941), stating that akermanite is an unstable phase below 1,325°, were not affirmed. Akermanite should only decompose into wollastonite and monticellite at much lower temperatures (approximately 700°).

This system is thus a characteristic section of a quarternary diagram Al_2O_3-CaO-MgO-SiO_2 at temperatures approaching the area of melting.

The authors of the investigation assume that the content of alumina in diopside as well as of the gehlenite component in akermanite are insignificant, as otherwise they would have caused complications in the given system. However, the work does not cite direct investigations of solid phases and the distributions of boundaries of solid solutions.

THE SYSTEM MgO-Al_2O_3-SiO_2**

This system was investigated by Rankin and Merwin (1918), amended by Bowen and Greig (1924) and more recently revised by Forster (1950, 1951). As it is only of secondary importance for the consideration of the genesis of igneous rocks, we shall only consider it in its general aspect (Figure 164).

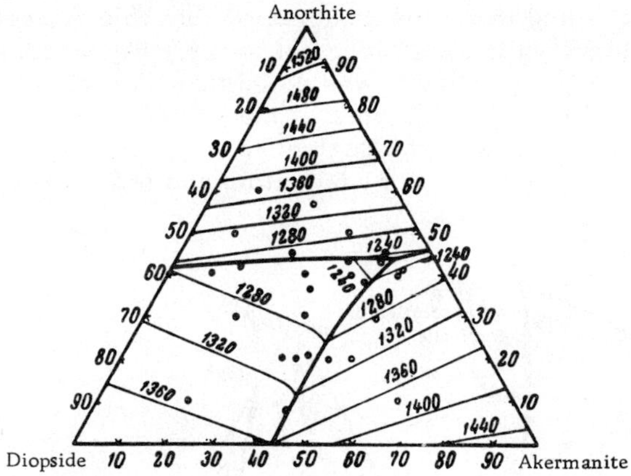

FIGURE 163. The system diopside-anorthite-akermanite (after de Wys and Forster)

As already mentioned, there is only one incongruently melting stable compound — mullite — in the binary system SiO_2-Al_2O_3. The minerals encountered in nature — kyanite, sillimanite and andalusite — do not obviously appear in the ternary field, as their appearance is in no way connected with the change in composition (see previously, page 191). The system is characterized by the crystallization of the important rock-forming mineral —cordierite, which is only rarely encountered in igneous rocks oversaturated in alumina. As is seen from the diagram, this compound melts incongruently and its composition point is located within the mullite field. The mullite-cordierite boundary line is undoubtedly characterized by a reaction crystallization — the solution of mullite occurs here, accompanied with the crystallization of cordierite. The lines separating the fields spinel-cordierite, forsterite-cordierite, clinoestastite-cordierite and tridymite-cordierite are characterized by simultaneous crystallization.
250 For cordierite the only ternary eutectic is attained during crystallization of cordierite-tridymite-clinoenstastite. The other ternary points are peritectic.

According to recent investigations the rare mineral — sapphirine, which is sometimes found in ultrabasic rocks as a reaction rim surrounding spinel, may also constitute a stable phase during the crystallization of melts of the given system.

Equilibria, approaching those of the given system, may be attained in natural magmas oversaturated in alumina, the origin of which is usually attributed to contamination with alumina-rich sedimentary rocks. In these cases a cordierite of undoubtedly magmatic origin is sometimes actually formed, which fact can scarcely be considerd as proved with regard to minerals even more enriched in alumina (kyanite-sillimanite-andalusite). As is seen from Figure 164, the field of garnet (pyrope) is absent. Experimental investigations (see page 354) showed that pyrope is also not formed at lower temperatures in the presence of water up to a pressure of 2,100 atms. The author of the experiments (Yoder Jr.) attempted to deny the role of high pressures in the formation of pyrope and other similar minerals in rocks and made the assumption that this mineral is formed at lower temperatures with a deficiency of water. These considerations were criticized (Sobolev, 1955) and it was noted that the field of stability of pyrope should be located in the area of much higher pressures. It was recently determined (Bowen and England) that pyrope crystallizes from a melt at pressures greater than 21,600 atms and temperatures greater than 1,510°.

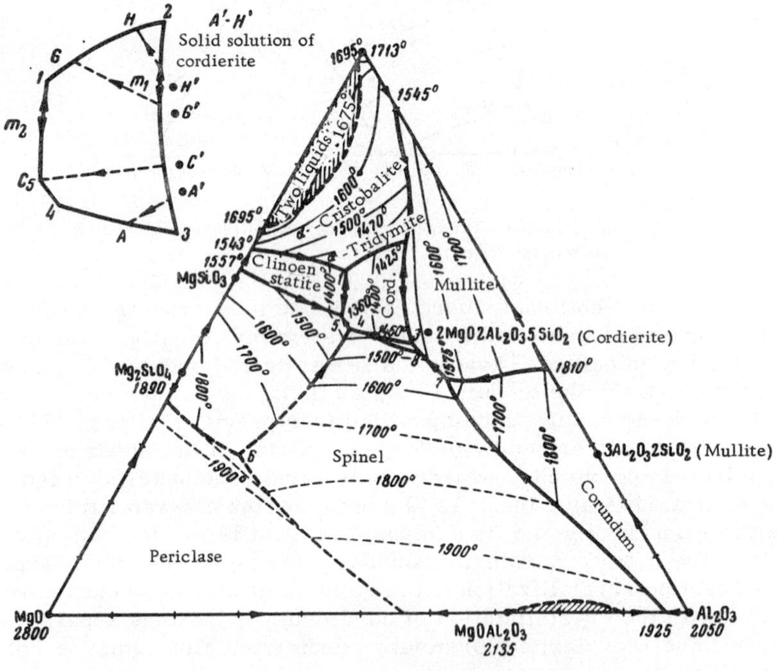

FIGURE 164. The system $MgO-Al_2O_3-SiO_2$ (after Rankin and Merwin, with amendments by Forster)

251 THE SYSTEM $FeO-Al_2O_3-SiO_2$**

In similarity to the previous system this system was only investigated by Schairer and Yagi in 1952 (Figure 165). In addition to the usual difficulties encountered in systems with ferrous iron, others arise in the area of crystallization of cordierite, which was successfully obtained only after appropriate seeding of the melt.

Differing from the previous system due to the absence of the pyroxene field, the given system is again characterized by the incongruent melting of ferrous cordierite and the absence of a corresponding garnet (almandine). The crystallization field of cordierite becomes very narrow. Both the cordierite-mullite and cordierite-spinel lines correspond to reactional crystallization.

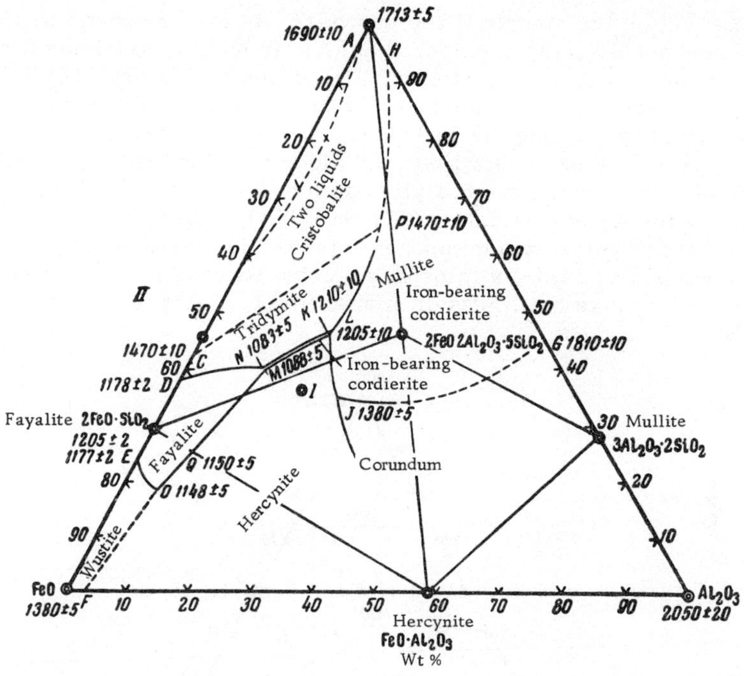

FIGURE 165. The system SiO_2-FeO-Al_2O_3 (after Schairer and Yagi)

I — composition corresponding to almandine; II — composition corresponding to Fe_2SiO_3

The mentioned features are in agreement with observations of magmatic cordierite comparatively rich in iron in certain igneous rocks. As regards the primary magmatic almandine, the existence of which has now been proved (e. g., phenocrysts in glassy dacite of Transcarpathia), its crystallization probably occurs at considerable pressures.

Experiments carried out at temperatures lower than the melting temperature showed that the formation of almandine calls for much lower pressures than does pyrope, and apparently it may even be obtained at normal pres-
252 sures. The latter statement is rather doubtful and experimental data cannot be used for speculating on the pressure of formation of almandine in our rocks for the following two reasons: the exact temperature of crystallization is unknown; the small admixture of the magnesium component should have a great effect on the conditions of crystallization of garnet.

THE SYSTEM $CaAl_2Si_2O_8$-$NaAlSi_3O_8$-$KAlSi_3O_8$**

The system anorthite-albite-orthoclase is of extreme importance for the understanding of crystallization of igneous rocks, in the majority of which feldspars are present as a major constituent. Nevertheless, this system is as yet insufficiently investigated due to the difficulties of crystallization of various melts. The investigations carried out to date have not yet provided a complete idea of the crystallization of feldspars in nature, because the presence of the other components (especially water) has a marked effect on the course of the process. As the compound $KAlSi_3O_8$ melts incongruently, the considered system is merely a section through a more complex quaternary system. Figure 166 shows crystallization isotherms after Franco and Schairer (1952), and fields of feldspar and leucite are distinguished. Along the $NaAlSi_3O_8$-$CaAl_2Si_2O_8$ side a continuous series of plagioclases crystallizes and along the side $NaAlSi_3O_8$-$KAlSi_3O_8$ a series of K-Na feldspars crystallizes up to the field of leucite, having a minimum point at approximately 65 % $NaAlSi_3O_8$. Besides the fact that a small content of potassium component is noted in plagioclases, and of the calcium component in K-Na feldspars, these features are not shown on the diagram because the limits of miscibility were not investigated experimentally.

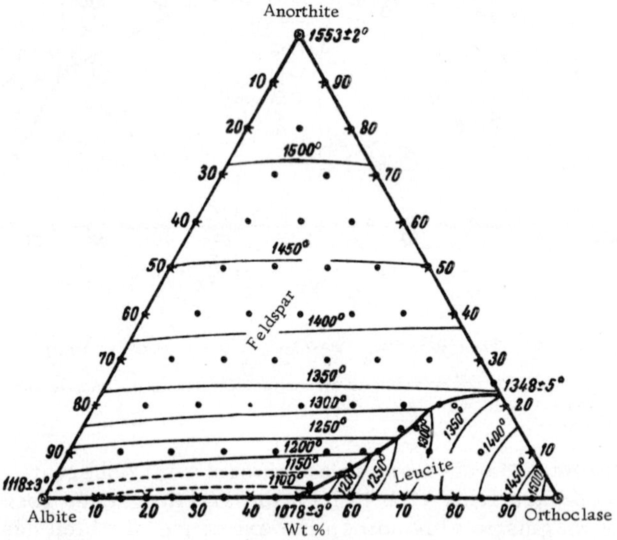

FIGURE 166. Crystallization isotherms of the pseudoternary system anorthite-albite-orthoclase (after Franco and Schairer)

253 As already mentioned, the incongruent melting of the potassium feldspars is never manifest under natural conditions, which fact is probably associated with the marked narrowing of the leucite field when water is present in the systems. This field disappears and $KAlSi_3O_8$ melts congruently at a water vapor pressure of approximately 2,500 atms or more.

Figure 167, I, presents a schematic diagram of the mentioned system at considerable water vapor pressures, under conditions where the potassium feldspar already melts congruently, but the intersection of melting curves with composition curves does not yet take place, i. e. , a continuous series of solid solutions crystallizes (at a water vapor pressure below 5,000 atms). During this process the temperature of crystallization is sharply lowered by more than 300° in comparison with the isotherms shown on Figure 166. The eutectic point of the binary system anorthite-orthoclase is calculated on the basis of experimental data for a system with a water

vapor pressure of 5,000 atms; the anorthite content is 12 %. As already
254 shown, the solubility of the anorthite component in potassium feldspar is low at melting temperatures — in all events less than 5 %; the limits of solid solutions may only decrease as temperature is lowered, which fact is taken into account on our diagram. The miscibility of potassium feldspar in anorthite is even lower and almost approaches zero. The low content of anorthite in the eutectic readily explains the early crystallization of plagioclase in the presence of potassium feldspar, which finds expression in the characteristic monzonitic texture.

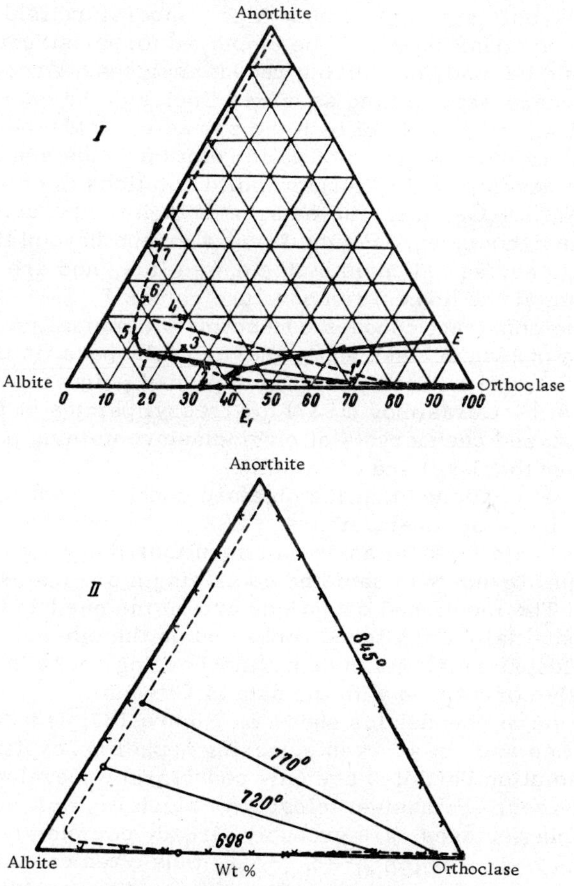

FIGURE 167. Schematic equilibrium diagram of the system orthoclase-albite-anorthite

I — at a water vapor pressure below 5,000 atms, after V. S. Sobolev (1956), with addenda and amendments; II — at a water vapor pressure of 5,000 atms, after Yoder.

In handbooks (e. g., Winchell, 1953, page 253) more or less wide areas of stability of feldspars are presented as constructed by various authors on the basis of analysis of natural minerals. However, not only the analysis of pure feldspars are taken into account, but also those of the so-called perthites and antiperthites, the origin of which is accepted as being the result of decomposition of solid solutions. The latter is, however, not always essential, and the decomposition of solid solutions is often accompanied by metasomatic phenomena which are partly connected with the replacement of potassium feldspar by plagioclase. On the other hand, the substance may not always be pure and may contain various inclusions. It should be taken into consideration that relations may change in more complex systems, in particular in the presence of nepheline (see page 240).

Therefore, only analyses of pure high temperature feldspars not having undergone decomposition should be employed for constructing diagrams, and use should be made of reliable recent analyses. According to these data, the borders were plotted as dotted lines and the corresponding points marked. Along either side of both the potassium feldspar and the plagioclase lines very narrow strips of solid solution fields are obtained, and a considerable development of ternary solid solutions is only possible in the corner $NaAlSi_3O_8$, i. e., in the area of high temperature plagioclases and anorthoclases. Points ***3*** and ***4*** extend beyond the usual fields of solid solutions, as indicated by the dotted line, and are probably related to equilibrium at the highest temperature values T.

In order to construct conodes connecting the mutually crystallizing compositions of plagioclases and potassium feldspars (in the absence of nepheline), only one sufficiently reliable pair of points is available, which was obtained by A. P. Gerasimov (1937) for trachyliparites of Mount Beshtau, where he analyzed phenocrysts of plagioclase containing potassium (which he termed anorthoclase) and of sanidine.

It should be mentioned that the obtained conode is not in agreement with the lines produced by other authors. Thus, for instance, Oftedahl (1952), using indirect data (without analysis), mentions that plagioclase $Ab_{65}An_{35}$ will be in equilibrium with sanidine containing more than 45 % sodium component. The mentioned deviations are on the one hand explained by the use of material from altered rocks and on the other hand — by a marked change in feldspar relations in nepheline-bearing rocks in which the conodes approximately correspond with the data of Oftedahl.

The location of conodes (as shown on Figure 167, I) in systems with an excess of silica and, in all events, lacking nepheline, explains the well known fact that exsolution perthites are only considerably developed in rocks with acid plagioclases. Potassium feldspars, which crystallize together with more basic plagioclases, are probably already completely devoid of sodium component as from $Ab_{70}An_{30}$-$Ab_{60}An_{40}$, thus often causing a halt in the inversion from the high temperature to the low temperature modification.

255 Data on the average composition of plagioclases and potassium feldspars in granites (Chirvinskii, 1953) are in agreement with the obtained location of conodes. The composition of the plagioclase (excluding albite) corresponds to $Ab_{78}An_{22}$, and that of potassium feldspar — 26 % albite component, the latter value being somewhat raised due to albitization (the conode shown by a dotted line on Figure 167)†.

† Yoder (1957) presents the first results of experimental investigation of a corresponding quaternary system with water at a water vapor pressure of 5,000 bars. It is interesting to note that under these conditions he already indicates an immiscibility gap of the K-Na feldspars. Two conodes were drawn (see Figure 167, II), the location of which is in good agreement with the above-mentioned values.

If a diagram of the given system is constructed at a higher water content, its shape will change as decomposition of the K-Na feldspar solid solutions occurs at temperatures below 700° (Figure 167, II). However, the petrographic material apparently proves that the decomposition of feldspar is always postmagmatic. Doubts may only arise in the case of alkali rocks, but even in this case all of the albite is most probably secondary after potassium feldspar, and in some cases may be after plagioclases.

SOME PECULIARITIES IN THE CRYSTALLIZATION OF THE SYSTEM $MgSiO_3$-$FeSiO_3$-$CaSiO_3$**

As we have already noted, this system is also not strictly ternary in the area of melts rich in iron (crystallization of fayalite from melt $FeSiO_3$). However, the main difficulty encountered in explaining experimental data is that the phenomena of crystallization in the rock, occurring at low temperature, differ essentially from those occurring in dry melts: instead of unlimited series of solid solutions of monoclinic pyroxenes ($MgSiO_3$-$CaMgSi_2O_6$), in igneous rocks we have limited solid solution of pigeonites crystallizing together with orthorhombic pyroxenes. Solid solutions of $FeSiO_3$-$CaSiO_3$ are totally absent in the rock and hedenbergite is the stable phase which replaces them. Owing to these peculiarities, no complete experimental investigation has been carried out to date, but numerous attempts were made at analyzing the course of crystallization of pyroxenes in igneous rocks on the basis of petrographic data.

These last-mentioned data are quite complex and are not always sufficient due to the absence of chemical analyses of minerals, thus allowing for various interpretations. In most basic and intermediate effusive rocks with porphyritic textures, the following relations are observed: both orthorhombic pyroxene (rather poor in the calcium component) and monoclinic pyroxene of the pigeonitic type (containing up to 25—30 % (Mg, Fe)SiO_3 component) are present as phenocrysts; orthorhombic pyroxene is absent in the matrix and the monoclinic pyroxene has different optical properties, showing that a solid solution is present here with a much higher content, greater than 50 %, of (Mg, Fe)SiO_3 (see Tsuboi, 1932). In some cases the composition of the monoclinic pyroxene of the matrix becomes even poorer in the calcium component and contains more than 70—80 % (Mg, Fe)SiO_3.

Such relations are apparently most easily explained by the change in crystallization conditions. The pyroxene of the matrix obviously crystallizes at a higher temperature, under conditions approaching the crystalliza-
256 tion of dry melts. The possibility still exists that the inversion from the orthorhombic to the monoclinic modification is directly dependent on pressure (higher pressures probably widen the area of stability of orthorhombic pyroxenes). In addition, no direct experimental data are available. On the other hand, as in the case of all femic minerals, an enrichment of the melt in the iron component undoubtedly occurs during the crystallization of both orthorhombic and monoclinic pyroxenes, which is sometimes proved by the presence of zoning. However, as in the case of olivine (see page 196), this zoning is more rarely encountered than that in plagioclases. As shown on Figure 131 (see previously), the inversion temperature from the

orthorhombic to the monoclinic modification is lowered with the increase of the iron content; the critical point of the solid solution curve should be affected in the same way.

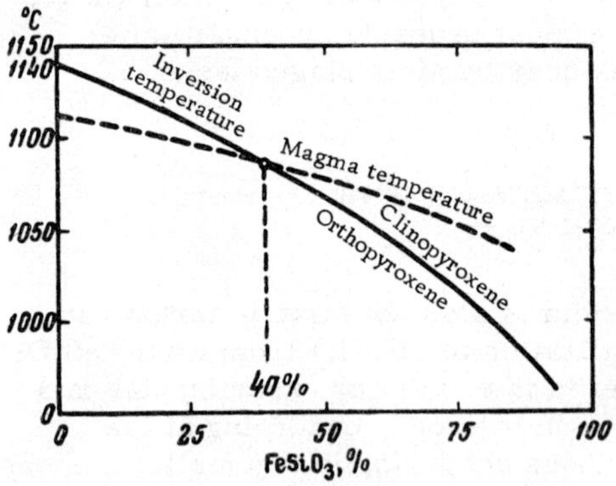

FIGURE 168. Crystallization scheme of orthorhombic and monoclinic pyroxenes (after Hess)

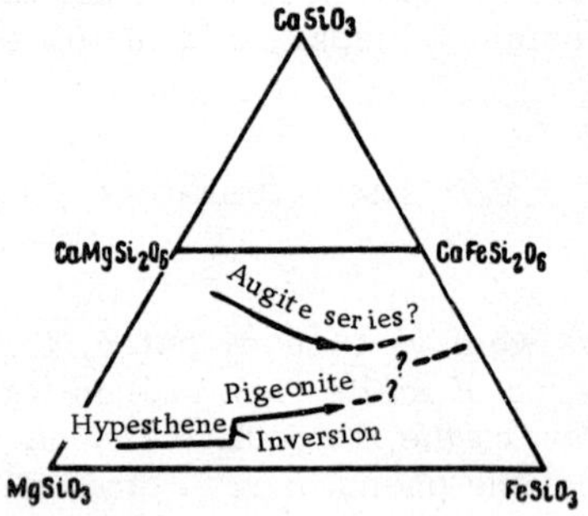

FIGURE 169. Crystallization scheme of pyroxenes in a ternary system (after Hess)

Basing his facts on the mentioned relations, Hess (1949) attempted to explain in a different manner the course of crystallization of pyroxenes in magma, assuming that what occurs is simply an intersection of the magma temperature curve with the pyroxene inversion curve (Figure 168). He also outlined the courses of crystallization in a ternary system (Figure 169). T. Barth (1956) correctly noted that we are in practice not dealing with the inversion temperature of orthorhombic to monoclinic pyroxene, but with temperatures of decomposition of solid solution, which should be somewhat higher and dependent on the iron content. H. Kuno (1955) made certain amendments to the effect that the inversion point should obviously depend on the character of the magma, because the temperature of crystallization of the magma should decrease with increasing acidity.

The mentioned considerations on porphyritic rocks do not appear convincing for the following reasons: 1) the replacement of two pyroxenes by one always corresponds to the transition from phenocrysts to the matrix;
257 2) the iron content of the pyroxenes in the phenocrysts not only varies in different rocks (generally increasing from basic to acid rocks), but also in rocks of the same acidity. Thus, in the andesites of Transcarpathia the iron content of hypersthene changes from 25 % to 50 %, thus not falling in the limits of the presented scheme.

The relations between the pyroxenes in hypabyssal rocks of the gabbro-diabase type are much more complex. One may already note in phenocrysts of andesite and basalt that a type of reaction relation, manifest by the overgrowth of monoclinic pyroxene on orthorhombic pyroxene, is present here, in addition to the simultaneous crystallization of hypersthene and monoclinic pyroxene. In hypabyssal rocks crystallization of zoned monoclinic pyroxenes often occurs, showing a sharp change in calcium content.

Two different cases were noted: the increase in calcium content from the core to the marginal parts of the crystal (Wahl, 1908) and inversely an overgrowth of pyroxene with a (Mg, Fe)SiO_3 content exceeding 70—80 % (upon usual pigeonite with 25—30 % (Mg, Fe)SiO_3 (Sobolev, 1936)). The latter case was noted in the Siberian trapps where the orthorhombic pyroxene also crystallized later than the monoclinic one, and reaction relations undoubtedly occurred. Relations opposite to those of Bowen's reaction series were determined in several other cases, this change being undoubtedly associated with the change in content of the iron component (Sobolev, 1936). The usual sequence (orthorhombic-monoclinic-pyroxene) is characteristic for rocks with lower FeO:MgO ratios (less than 40 % iron component), and the reverse sequence — with higher ratios.

The mentioned complication shows that in hypabyssal rocks crystallization occurs under conditions approaching those of dry melts, i. e., the formation of continuous series of solid solutions or, in all events, the greater miscibility of pyroxenes than in effusive rocks. There is no doubt that the order of crystallization of monoclinic and orthorhombic pyroxenes depends on the possible formation of olivine and the time of its reactional inversion.

Finally, we shall touch upon one additional phenomenon of intrusive rocks, which serves as evidence of decomposition of solid solutions. Signs of exsolution of monoclinic pyroxenes are observed as regularly oriented lamellae of hypersthene parallel to (100). In orthorhombic pyroxenes they are expressed by very fine lamellae of monoclinic pyroxene parallel to (100) and correspondingly of orthorhombic pyroxene parallel to (010). In several cases the intergrowths of lamellae are so fine that they were described as polysynthetic twinning, and in some cases they give the impression of "oblique extinction" in hypersthene in sections parallel to the *c* axis.

GENERAL CONCEPTS OF CERTAIN QUATERNARY SYSTEMS**

As has already been mentioned, a diagrammatic representation of quaternary systems is rather difficult as one is obliged to use a three-dimensional figure, even for the consideration of paragenetic relations. Up to the present no complete experimental investigation of the most simple quaternary silicate systems has been carried out. Therefore, this chapter will be restricted to the schematic consideration of the most interesting of these systems.

THE SYSTEM SiO_2-$NaAlSiO_4$-$KAlSiO_4$-$CaAl_2Si_2O_8$

One of the most important quaternary systems is the so-called salic tetrahedron, which was used on several occasions by E. S. Fedorov for considering physicochemical regularities in the formation of igneous rocks. He selected the following components for this system: silica, nepheline, anorthite and leucite. We shall extend this system somewhat by inserting kaliophyllite instead of leucite. One of the first schemes of crystallization of this system was provided by A. N. Zavaritskii during his description of the leucite rocks of Ishim (1936). As experimental data on partial systems included within this quaternary system have subsequently become available, the general scheme should be changed (Sobolev, 1959).

In view of the complexity of the system, we shall separately consider the diagram of paragenesis of minerals at the moment the liquid disappears (Figure 170), i. e., at the temperature of melting, and the melting diagram. Certain minor simplifications were made in the construction of both diagrams, namely: the polymorphous inversions of silica, kaliophyllite and nepheline were not taken into account; the solubility of silica in nepheline was overlooked as well as the solubility (although very small) of the calcium component in feldspars and the sodium component in leucites. The solubility of the anorthite component in nepheline and the nepheline component in anorthite influence the character of paragenesis and should be taken into account.

Within this tetrahedron (see Figure 170) a tetrahedral part is first distinguished in which the feldspars are stable together with free silica-tridymite, and in rocks usually with quartz.

Conodes on the boundary section $KAlSi_3O_8$-$NaAlSi_3O_8$-$CaAl_2Si_2O_8$, the
259 construction of which was considered previously (see page 253), present an idea on the composition of K-Na feldspars which are in equilibrium with the corresponding plagioclase and SiO_2.

In the area lying below this boundary section either undersaturated aluminosilicates — felspathoids — appear, or leucite, or solid solutions of nephelines (we neglect the early crystallization of carnegieite and kaliophyllite showing other structures as they usually undergo inversion).

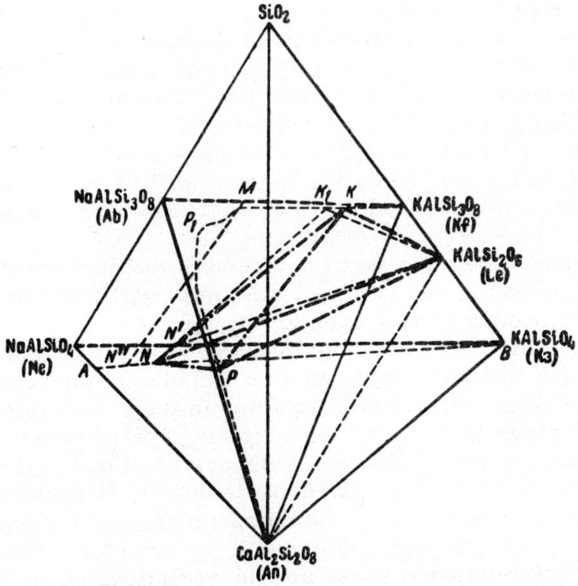

FIGURE 170. Paragenetic diagram of the system SiO_2-$NaAlSiO_4$-$KAlSiO_4$-$CaAl_2Si_2O_8$

As already mentioned, nepheline may usually dissolve considerable amounts of the anorthite component. The laws of crystal chemistry permit us to claim that in kaliophyllite such solubility is, to all intents and purposes, absent and addition of potassium should readily decrease the limiting content of the anorthite component in nepheline. As the lowering of temperature will also cause a corresponding lowering of the solubility of the second component in limited solid solutions, the limit of solubility of anorthite in nepheline, which crystallizes together with K-Na feldspars, should also be considerably lower than in the system nepheline-anorthite. Exact figures are not available in this case and we conventionally accept the maximum solubility of $CaAl_2Si_2O_8$ in $NaAlSiO_4$ as being equal to 15 % (for parageneses with alkali feldspars), and for the series of nephelines this limit is shown schematically by the dotted line *AB* on the nepheline-kaliophyllite-anorthite face of the tetrahedron. After connecting this curve to the feldspar line by means of conodes, we shall obtain a small area of two-mineral parageneses (calcium nepheline-K-Na feldspars) above the base of the tetrahedron nepheline-kaliophyllite-SiO_2, and by drawing the corresponding conodes for leucite we shall obtain an area of parageneses (calcium nepheline-leucite) and an area of three-mineral parageneses (calcium nepheline-leucite-K-Na feldspars).

260 These areas are rather small and can be overlooked if the addition of calcium does not sharply change the position of the conodes of the above-mentioned face of the tetrahedron. One may foresee the influence of $CaAl_2Si_2O_8$ on the location of the conodes, as the calcium component is only dissolved in nepheline and only in its sodium component at that. Le Chatelier's rule provides us with a direct concept of the nature of this influence: the increase in concentration of the calcium component in the system should lead to that reaction which causes the precipitation of this component in minerals, i. e., it should favor the increase of the sodium component in nepheline. In the paragenesis of nepheline with feldspar this may only occur as a result of an ion exchange between Na^+ and K^+ in the feldspar and in nepheline, i. e., the conodes will have steeper slopes and the differences between the relations of Na_2O and K_2O in nepheline and feldspar will increase.

We had already taken this fact into account when interpreting Tilley's data on nepheline containing calcium. In the parageneses with plagioclases the shift of the conodes will be even greater.

On the other hand, the equilibrium between the potassium feldspar and plagioclase in rocks containing nepheline is also somewhat changed, apparently due to the selective solution of nepheline in the anorthite component of plagioclase. Tröger (after A. N. Zavaritskii, 1941) presents eight analyses of nepheline-bearing rocks, for which the compositions of the plagioclase and the potassium feldspar are given. As is known, in such rocks the plagioclase is as a rule zoned and only the composition of the marginal part may be considered as being in equilibrium with the potassium feldspars. Only in two of the eight mentioned rocks are the variations of the plagioclase composition given, e. g., in tautirite† (No. 533 after Tröger or No. 538 after A. N. Zavaritskii) plagioclases ranging from $Ab_{60}An_{40}$—$Ab_{80}An_{20}$ are in equilibrium with sanidine containing 54 % sodium component, and in sodalite-tephrite (No. 569 after Troger and No. 457 after A. N. Zavaritskii) plagioclases ranging from $Ab_{45}An_{55}$—$Ab_{70}An_{30}$ are in equilibrium with sanidine containing 40 % sodium component. The points corresponding to the mentioned parageneses may be used for constructing conodes. In the other rocks, where the zoning of the plagioclase was not indicated, we either obtain results in agreement with the above-mentioned data (two rocks) or a composition is obtained which is even richer in the sodium component, sanidine, that should be expected when taking into account the basicity of the earlier formed plagioclase nucleus.

On comparing the cited figures with the diagram on Figure 167, we note that the conodes for these rocks are considerably shifted in comparison to the quartz-bearing rocks. These data are in much better agreement with the conodes on the diagrams of Oftedahl (1948) and Barth (1956). Both systems of conodes were drawn on the feldspathic section (see Figure 170).

From the study of the parageneses of igneous rocks it is known that leucite is only encountered together with basic plagioclases. The limiting composition of plagioclase found together with leucite may be considered as that of labradorite — $Ab_{50}An_{50}$. If the plagioclases are more acid in composition, leucite should react with the albitic component with the

† This rock is termed in handbooks as a leucite tautirite, which is however in poor agreement with the cited data.

formation of nepheline and potassium feldspar. Bearing in mind the amendments introduced, one may accept a nepheline with an 80 % content of the sodium component and a sanadine with a 20 % content of sodium component for the mentioned paragenesis. After connecting the mentioned points we shall obtain a corresponding tetrahedron (Le) ***KNP*** for the quaternary paragenesis.

261 Further on in the considered part of the system areas of three-mineral paragenesis are encountered: leucite-K-Na feldspar-plagioclase (more basic than $Ab_{50}An_{50}$); leucite-nepheline-plagioclase (more basic than $Ab_{50}An_{50}$); K-Na feldspar-nepheline-plagioclase (up to $Ab_{50}An_{50}$).

The leucite-K-Na feldspar-plagioclase area of paragenesis is the easiest to distinguish. It has the form of a pyramid with four sides, the apex of which is leucite and the base of which is the quadrilateral (An) ***PK(Kf)*** which lies in the plane of the feldspar section. For each composition of the plagioclase (from anorthite to labradorite — $Ab_{50}An_{50}$) there is a definite corresponding composition of sanidine.

Two other three-mineral areas have a more complex form, as the compositions representing true series of solid solutions do not lie in one plane. As only one degree of freedom is present in a three-mineral paragenesis, if the composition of one of the mineral changes, the compositions of the other two will change regularly, and the area of the three-mineral paragenesis will be controlled by all three phase triangles.

In the case of a paragenesis between leucite, nepheline and plagioclase, the initial phase triangle will be (Le) (Ks) (An) on the kalsilite-anorthite-silica face of the tetrahedron. The final phase triangle with the most acid plagioclase in this case corresponds to the face (Le) ***PN*** of the tetrahedron, which determines a four-mineral paragenesis. As the composition of leucite is accepted as remaining constant, two sides of all intermediate phase triangles will pass through a single point. Thus, our paragenetic area will also be limited by the triangles (Le) (An) ***P*** and (Le) ***N*** (Ks). Once again a type of four-sided pyramid with (Le) as its apex is obtained, but the base is not flat because the points (An), ***P***, ***N*** and (Ks) do not lie in the same plane. The third side of the phase triangle, which connects the conjugate compositions of nephelines and plagioclases, describes a complex spiral surface enclosing the area of the three mentioned minerals when the triangle (Le) (Ks) (An) is displaced to the position (Le) ***PN***.

The area corresponding to the parageneses of plagioclase with nepheline and K-Na feldspar starts with the triangle ***KPN***, which is also a phase of the four-mineral tetrahedron. Further on the phase triangle is shifted in such a way that the plagioclase changes its composition from labradorite $Ab_{50}An_{50}$ along the dotted line of feldspathic solid solutions up to the minimum point ***M***, where the composition of both feldspars becomes identical. Thus, the final phase triangle is reduced to a straight line ***MN″***. On tracing the shifting of the sides of the phase triangle, we see that the area of the mentioned parageneses is limited by the trapezium ***KMN″N*** (the K-Na feldspar-nepheline side) and the area $\boldsymbol{KMP_1P}$, the form of which approaches that of a triangle (the sanidine-plagioclase side). This side of the phase triangle is an extension of the above-mentioned spiral surface dividing the area of three-minerals from the area of coexistence of plagioclase-nepheline, which further on passes into the area of anorthoclase-nepheline.

With an unlimited content of the calcium component we shall obtain intermediate locations of the phase triangle from (Le) *KN* to (Le) K_1N', the total of which will limit the area of three-mineral parageneses (leucite-sanidine-nepheline). They will only differ from the parageneses in ternary systems by their content of calcium and nepheline.

In the two-mineral area nepheline-plagioclase the existence of various
262 minerals (up to a limiting composition) is possible. This follows directly from the Phase Rule as two degrees of freedom are present here.

If certain compositions within the given area are considered and can be expressed as points on the diagram, the following constructions are necessary for the determination of the minerals: 1) the drawing of a plane through the given point and one of the straight lines corresponding to the position of the mineral, e. g. , the nepheline line. This plane will intersect the plagioclase line at a certain point, the position of which will determine the composition of the plagioclase. After this we shall draw a straight line through the given point and point x, producing it to intersect the nepheline line. The obtained position of the latter point y will determine the composition of nepheline, and after measuring the distance from the given point to points x and y, one may readily calculate the quantitative relations between these minerals (compare Figures 170 and 175).

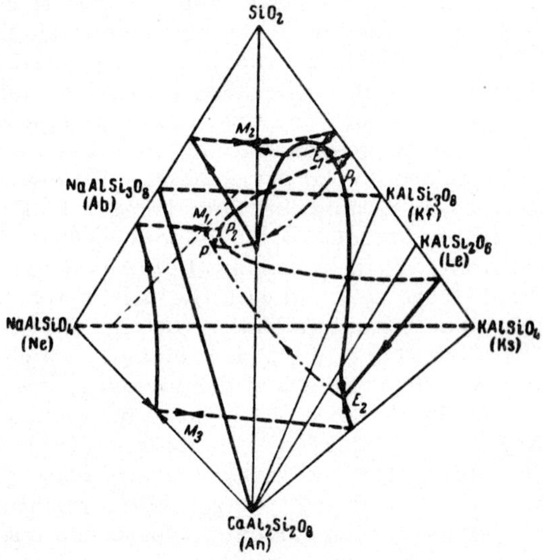

FIGURE 171. Schematic phase diagram of melting and courses of crystallization in the system SiO_2-$NaAlSiO_4$-$KAlSiO_4$-$CaAl_2Si_2O_8$

The dot-dash line represents the equilibrium line within the tetrahedron

Figure 171 represents a schematic melting diagram and courses of crystallization of the considered system, where, in addition to the above-mentioned simplifications, the solubility of anorthite in nepheline is also not shown.

Equilibrium lines are taken from diagrams representing the face $NaAlSiO_4$-SiO_2-$KAlSiO_4$ on Figure 149 and the face SiO_2-$CaAl_2Si_2O_8$-$KAlSiO_4$ on Figure 151 (see previously). On the latter face, the data on the system anorthite-kaliophyllite and on the location of the eutectic point are absent (and there are no doubts that these relations are eutectic). Correspondingly, the kaliophyllite-leucite-anorthite ternary eutectic point E_2 is only shown approximately. After introducing the above-mentioned simplifications, a cotectic line, representing mutual crystallizations of solid solutions of nepheline with anorthite, and having a minimum point M_3, should undoubtedly occur on face $NaAlSiO_4$-$KAlSiO_4$-$CaAl_2Si_2O_8$, and is shown approximately. Finally, two cotectic lines should appear on face
263 SiO_2-$NaAlSiO_4$-$CaAl_2Si_2O_8$: plagioclase-nepheline and plagioclase-silica, with the course of crystallization directed from the anorthitic plagioclase to the albitic one. These lines connect the eutectic points on the faces of the tetrahedron.

As is evident from the above-considered paragenetic diagram, there should be one invariant point P within the tetrahedron where four minerals are in equilibrium: leucite, plagioclase, K-Na feldspar and nepheline. This point may be obtained from the intersection of four curves representing the corresponding three-mineral parageneses. The three-mineral crystallization curve (leucite-K-Na feldspar-plagioclase) should issue from the peritectic point P_1 (leucite-K-Na feldspar-anorthite). The curve leucite-nepheline-plagioclase issues from the eutectic point E_2 (leucite-kaliophyllite-anorthite). The curve leucite-nepheline-K-Na feldspar also issues from a corresponding peritectic point P_2. In the case of the fourth curve (plagioclase-nepheline-K-Na feldspar), the corresponding invariant point is not present on the face of the tetrahedron, and this curve intersects the face of the tetrahedron (see page 227) at the minimum point M_1 (nepheline-K-Na feldspar).

If we take into account the behavior of leucite in ternary systems, as well as the fact that in any system the lowering of the basicity of plagioclase corresponds to a lowering of temperature, we may draw the corresponding arrows and accept the fact that the given invariant point has a peritectic character.

The curved surfaces which connect the above-mentioned lines correspond to melts in equilibrium with two minerals, and the corresponding areas between these surfaces — with one mineral.

The comparison of this diagram with Figure 170 permits us to assess the cases in which crystallization will terminate upon certain surfaces and curves and in which cases it should necessarily reach the invariant point.

In the area of compositions oversaturated in silica, invariant points are absent. Only a three-mineral equilibrium line is present here: plagioclase, K-Na feldspar and SiO_2. This line issues from the eutectic point E_1 (anorthite-orthoclase-SiO_2) and terminates at the minimum point M_2 (K-Na feldspar-SiO_2).

The association of basic plagioclases with potassium feldspars is not characteristic of igneous rocks and only in certain cases is the corresponding line reached during the final stages of crystallization of monzonite — for relatively basic plagioclases ($Ab_{50}An_{50}$-$Ab_{60}An_{40}$). Here the

quantitative relations between plagioclases and K-Na feldspars are still similar to those in the system with anorthite, and the main part of the plagioclases is separated during the pre-eutectic period of crystallization, thus giving rise to the characteristic monzonitic texture. The mentioned line is of great significance to petrography as it is undoubtedly reached during the crystallization of granites with comparatively acid plagioclases $Ab_{80}An_{20}$-$Ab_{75}An_{25}$. As the potassium feldspar, which is in equilibrium with this plagioclase, is poor in the sodium component (of the order of 20 %), and the amount of the plagioclase in the cotectic melt is not very high, the projection of this point on the kaliophyllite-nepheline-SiO_2 face will be located much further to the right than the so-called "trough" (see page 231), namely to the right of the line connecting the minimum points $\boldsymbol{M_2M}$. In actuality, after introducing the change in the position of the minimum point $\boldsymbol{M_2}$ on the feldspar-SiO_2 crystallization curve (see page 229), the majority of the rhyolite points, and correspondingly also of the granites,
264 will occupy this very position, which feature is well explained by the presence of an anorthitic component. This shift of points into the "potassium area" is somewhat restricted by the fact that in most granites there is an essential excess of plagioclase in relation to the cotectic curve, as is noted from textural relations.

THE SYSTEM $CaAl_2Si_2O_8$-$NaAlSi_3O_8$-Mg_2SiO_4-SiO_2 AND BOWEN'S REACTION SERIES

The relation between salic and femic minerals in the most widely distributed igneous rocks, previously considered in the description of ternary systems, lends itself to diagrammatic representation by means of a quaternary system as is shown schematically on Figure 172. As in other cases, certain simplifications were introduced here and the area of spinel was omitted (see Figure 146), as within this the system becomes quinary.

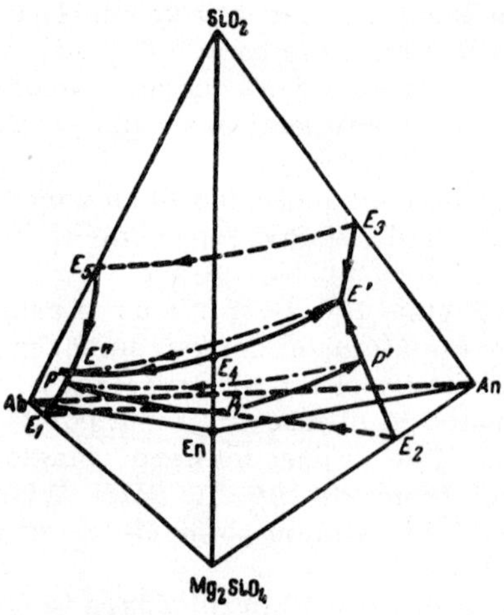

FIGURE 172. Schematic diagram of the system $CaAl_2Si_2O_8$-$NaAlSi_3O_8$-Mg_2SiO_4-SiO_2
Dot-dash line—the line of equilibrium within the tetrahedron

Within the tetrahedron only the $CaAl_2Si_2O_8$-SiO_2-Mg_2SiO_4 face has been well investigated. A eutectic point $\boldsymbol{E_1}$ should exist within the system $NaAlSi_3O_8$-Mg_2SiO_4 at which the relations between albite and forsterite are probably similar to those for the system albite-diopside, but in all events, this point is displaced considerably towards albite as compared to the system albite-fayalite (see Figure 137). After making such assumptions one may construct schematic diagrams for other faces of the tetrahedron. The face $NaAlSi_3O_8$-Mg_2SiO_4-SiO_2 will be qualitatively similar to the above-mentioned diagram, and with the replacement of anorthite by albite only the quantitative relations will change.

The face $CaAl_2Si_2O_8$-$NaAlSi_3O_8$-Mg_2SiO_4 should bear a great similarity to the well-investigated system $CaAl_2Si_2O_8$-$NaAlSi_3O_8$-$CaMgSi_2O_6$ according to the character of the diagram. Finally, the two eutectic points E_3 and E_5 are known to exist on the face $CaAl_2Si_2O_8$-$NaAlSi_3O_8$-SiO_2, and all that remains to be done is to schematically draw the cotectic line as in the case of the previously mentioned quaternary system.

Within the tetrahedron there are two lines representing equilibrium between three solid phases and the melt, corresponding to two possible types of three-mineral parageneses: plagioclase-Mg_2SiO_4-$MgSiO_3$ and plagioclase-$MgSiO_3$-SiO_2. The first of these lines connects the peritectic points upon the faces P' and P'' of the tetrahedron, and is in itself a peritectic line, as it entirely lies in the part of the tetrahedron corresponding to the final paragenesis with SiO_2. The other line connects the eutectic points E' and E'' and is a cotectic line. The character of crystallization along the peritectic line within the tetrahedron can be readily established by comparing the course of crystallization in the systems $CaAl_2Si_2O_8$-Mg_2SiO_4-SiO_2 and $CaAl_2Si_2O_8$-$NaAlSi_3O_8$-$CaMgSi_2O_6$.

A reaction between olivine and the melt occurs here with the formation
265 of pyroxene, which reaction may lead to the disappearance of olivine and the transition of the melt to a surface of crystallization of two minerals, provided that the corresponding quantitative relations are present. At the same time a continuous reaction takes place between the melt and the plagioclase, the latter becoming increasingly acid, and is accompanied by an increase in the plagioclase : femic mineral ratio.

The given system, or more precisely, the above-mentioned ternary systems clarifying the course of crystallization of this quaternary system, was accepted by Bowen as the basis for his reaction principles and for the scheme known as the reaction series.

The possibility of two types of reactions between the melt and the earlier separated crystals has already been mentioned: step reactions corresponding to the decomposition of minerals in systems with incongruent decomposition, and continuous reactions characteristic of series of solid solutions. On comparing the investigated silicate systems with petrographic data, Bowen arrived at the conclusion that such reactional relations between the crystallizing minerals and the melt constitute the basic law governing crystallization of igneous rocks (1922, 1934).

It is interesting to mention that more than thirty years prior to the publication of Bowen's paper, E. S. Fedorov had actually formulated the reaction principle and stressed on numerous occasions the importance of reactional relations (however, only for reactions of the continuous type). He wrote: "Minerals of certain groups, and first of all the groups of feldspars, pyroxenes and amphiboles, are able to adapt themselves to the gradually changing composition of the magma, themselves undergoing a parallel change in composition". As regards the reactional relations in non-alkaline igneous rocks, Bowen proposed the famous crystallization scheme corresponding to two crystallization series, as is shown on Figure 172. Whereas the plagioclases change according to a scheme of continuous reactions, Bowen maintains that femic minerals are involved in a "step" crystallization series. In the latter only the "first step" follows directly from experimental data. Relations between the other minerals are based on direct observations of textures of igneous rocks, and in particular of reaction rims. The change in the character of minerals is accompanied by a simultaneous change in the quantitative relations, which follows from the system anorthite-albite-diopside.

The relations in the final stage of crystallization, when the femic mineral may to all intents and purposes be neglected, are illustrated by the system of "residual crystallization" in the field of quartz-bearing rocks, and have been considered in greater detail in the above descriptions of quaternary systems. Bowen's original scheme also included muscovite, which is omitted on Figure 173 as it is probably always a secondary mineral.

As is known, the above-mentioned scheme served as Bowen's basis for the development of the broad hypothesis of crystallization differentiation, according to which all igneous rocks were formed in a similar way, having originated from the crystallization of a single basaltic magma. This hypothesis, which was met with wide acclaim by most petrographers, was also the subject of lively discussion*. For greater details we refer the reader to courses of petrology and the wide range of specific literature on the con-
266 sidered problems, and we shall only mention here that it may now be claimed with certainty that crystallization phenomena are of a more complex character and that magmas of various compositions may be formed in various parts of the Earth's crust. However, crystallization differentiation is undoubtedly of very great significance in magmatic evolution. The possible formation from a basaltic magma of ultrabasic rocks on the one hand (by means of accumulation of earlier separated minerals), and of granites on the other hand, although on a small scale (during the evolution of the residual melt), is now a proved fact.

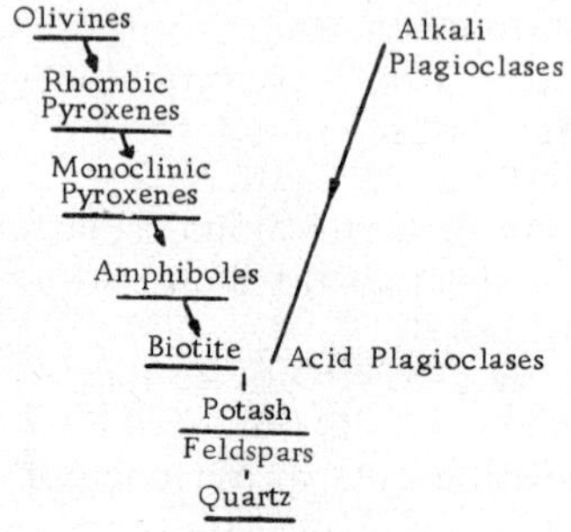

FIGURE 173. Bowen's reaction series

In addition to the geological arguments put forward against the universal application of crystallization differentiation, Bowen's crystallization scheme itself was also subjected to severe criticism. Thus, V. N. Lodochnikov mentioned the unfounded nature of the femic series, as the reaction rims which were used by Bowen could possibly be explained as being postmagmatic. D. P. Grigor'ev (1939) largely refuted this argument after synthesizing fluorine-bearing hornblendes and biotite, thus almost completely reproducing Bowen's reaction series of femic minerals directly by crystallization from a melt. In this case the fluorine actually replaced the hydroxyl group and a much greater similarity to magmatic phenomena was observed than in the case of liquation.

* It is interesting to note that one of the first criticisms of the generality of Bowen's hypothesis was made by Fedorov (1916) in his well-known abstract devoted to the experimental investigations of the most important ternary systems. While stressing the great importance of these investigations, which diagrammatically illustrate the course of crystallization differentiation, to which Fedorov himself always ascribed the greatest significance, he indicates that Bowen underestimated the assimilation of country rocks by magma.

Several facts presented by other authors (Lodochnikov, 1936; Sobolev, 1936 et al.) were of even greater interest; sometimes the sequence of femic minerals changes as well as the relations between femic and salic minerals. The significance of these contradictions to Bowen's scheme and their explanation will be presented further on.

A SYSTEM OF THE TYPE $CaAl_2Si_2O_8 - NaAlSi_3O_8 - Mg_2SiO_4 - Fe_2SiO_4$

During the investigation of igneous rocks great attention was always paid to the accurate determination of plagioclase composition by means of optical data, whereas in the case of femic minerals, sparse data are often only provided, which usually do not allow one to assess the essential peculiarities of their composition. Such disregard of the femic minerals follows from the wide application of the above-mentioned scheme, termed "Bowen's reaction series", where the continuous change in the composition of the plagioclase is contrasted with the step series of femic minerals. However, in the simplest case, each femic mineral presents a series of solid solutions of the magnesium and iron components; as the melt changes the composition of the minerals should change correspondingly, the residual melt always being enriched in the iron component. Various reactions between the mentioned relations (continuous and step) gave rise to different variants in crystallization schemes, thus explaining in part some of the objections to Bowen's scheme (Bowen and Schairer, 1935), e.g., the sequence of crystallization of orthorhombic and monoclinic pyroxenes. Bowen in his later investigations already mentioned this point, but was
267 nevertheless inclined to underestimate the change in femic minerals in igneous rocks.

Figure 174 presents a simplified scheme showing the relations between femic minerals and plagioclase in the case of continuous reactions. Experimental data are only available for certain faces and edges of the tetrahedron. Secondary phenomena, connected with the appearance of spinel, which are not included in the corresponding components, are omitted. The quaternary diagram, constructed by means of corresponding extrapolation, is a schematic one but does not give rise to any doubts as to the correctness of interpretation of the qualitative relations. A diagram of this type was presented for the first time by Wahl (1929) and was later considered in greater detail by V.S. Sobolev (1936).

Since the formation of only two phases of variable composition is possible in this system, two degrees of freedom are present in all parageneses at constant temperature and pressure. Thus, a plagioclase of any composition may be in equilibrium with any olivine (or in another similar system — with another femic mineral). The mineralogical composition of any point within the tetrahedron may be determined by the above-considered method (see page 33). We shall draw a plane (Figure 175) which passes through the given composition point *a* and one of the lines corresponding to solid solutions, e.g., the olivine line. This plane intersects the composition line of plagioclase at a certain point *x*. Through this point in the given plane we draw a line which passes through the composition point and extends up to the composition line of olivine. The points x and y correspondingly determine the composition of the minerals comprising the rock and according to the lever rule we determine the quantitative relations of the minerals:

$$\frac{C_{plagioclase}}{C_{olivine}} = \frac{ay}{ax},$$

where C is the percentage of the corresponding mineral.

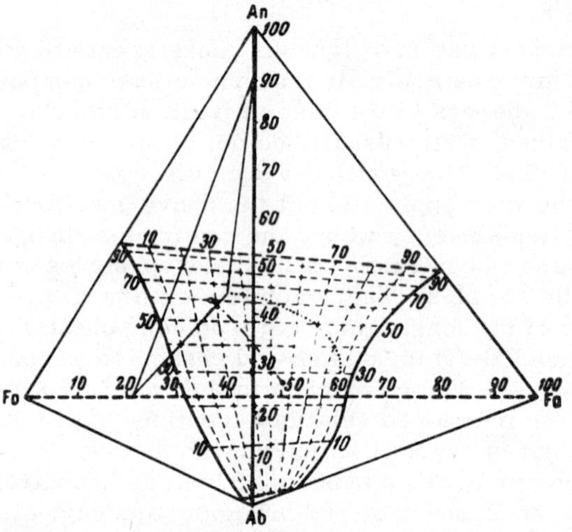

FIGURE 174. Schematic diagram of the system anorthite (An) — forsterite (Fo) — albite (Ab) — fayalite (Fa)

Dotted line — changes in mineralogical composition in the Siberian trapps after V.S. Sobolev; lines with dots and dashes — changes in the mineralogical composition in the series gabbro-diorite-granite after P.N. Chirvinskii

268 Within the tetrahedron (see Figure 174) both invariant points (where four minerals could be in stable equilibrium with each other) and lines of three-mineral equilibria are absent. Lines of two-mineral equilibria on the faces of the tetrahedron are connected by a surface (within the tetrahedron) on which are located all melts simultaneously in equilibrium with two minerals. This surface divides the tetrahedron into two parts: that in which the crystallization of plagioclase occurs, and that in which the crystallization of olivine occurs.

It has already been mentioned previously that when equilibrium is attained in a system, the coexistence of any composition of the minerals is possible depending on the initial composition. However, crystallization is always accompanied by a certain trend: as temperature decreases a simultaneous increase in the acidity of the plagioclase and the iron content of the olivine occurs in the residual melt.

If we consider a point on the mentioned surface, we may trace the corresponding course of change of the melt within the limits of this surface, and here as in other systems two limiting cases may exist: 1) the course of change of the melt is accompanied by a continuous equilibrium reaction with the earlier separated crystals and 2) the course of change of the melt is not accompanied by such reactions, i. e., the crystals are continuously isolated from the melt. In the first case the first mineral to crystallize will depend on the location of the given point. In the second case it is independent of the initial composition, and hence the lines of the second type may be drawn upon the corresponding surface as is done in the case of ternary systems.

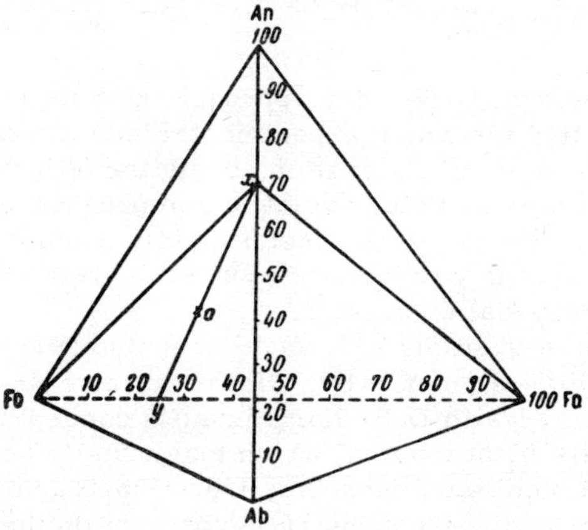

FIGURE 175. Scheme for determining the mineralogical composition of the system anorthite (An) — albite (Ab) — forsterite (Fo) — fayalite (Fa)

At present no experimental data are available which may be used for the construction of the corresponding courses of crystallization. However, some of these lines may be traced on the basis of petrographic investigations. Thus, on Figure 174 "the crystallization path" was drawn for the Siberian trapp. In this case a very rapid change in the composition of the olivine is typical, together with a comparatively small change in the composition of plagioclase, thus causing a sharp increase in the FeO: MgO ratio in the residual melt and rendering this ratio the most sensitive indicator of the position of the rock within the differentiation series. A similar tendency is also characteristic of other rock series, e. g., for the intrusions of Skaergard in Greenland (Wager and Deer, 1939), and is typical of magmatic complexes of platforms (Sobolev, 1937).

In contrast to the mentioned complexes, normal igneous rocks of folded regions show a different tendency: here the change in the composition of the plagioclase occurs much more rapidly, and the iron content of the femic minerals increases much slower; the femic minerals on the whole comply with step reaction sequence. The corresponding line, representing the
269 change in composition of minerals, is also conventionally drawn on the boundary surface, although here the system is obviously more complex, as

the femic minerals replace each other in various rocks, and biotite and not olivine is found to coexist in equilibrium with acid plagioclase. In order to plot the mentioned line, average mineral compositions of igneous rocks were used by P. N. Chirvinskii (1953). The same regular increase in the iron content of femic minerals as the acidity of the rock increases was diagrammatically presented by A. N. Zavaritskii (1950) during the analysis of the petrochemical diagram of average compositions of igneous rocks.

COMMENT ON MULTICOMPONENT SYSTEMS

Although individual rocks may approach the relations in ternary or quaternary systems for which experimental data are available, magma, as a whole, is a more complex system consisting of seven components even after excluding water and other volatile components as well as secondary constituents (including Fe_2O_3). Insufficiently considered generalizations have led to several contradictions between actual parageneses of igneous rocks and the data of physical chemistry.

A description of systems with such large numbers of components cannot be represented diagrammatically. However, new possibilities have arisen, which are connected with D. S. Korzhinskii's concepts (1946) on the specific role of the alkalis in the magma, which may usually be related to the number of so-called mobile components, i. e., those components the behavior of which depends on their concentration and not on their mass per unit volume of melt†. Thus, the magma is presented as a partially open system from which certain components may be removed and others introduced — mainly alkalis (in addition to the volatiles). This is possible in the case when the magma is saturated with water and the diffusion of hydrous so-called "trans-magmatic" solutions occurs within it, the equilibrium with which leads to the "mobile character" of the alkalis.

The mobility of the alkalis permits one to construct a diagram according to Schreinemaker's method, on which logarithms of concentration and potentials of K_2O and Na_2O are correspondingly plotted on the abscissa and the ordinate. In systems of six components (MgO and FeO are considered as a single component, which holds true for certain limits of their ratios), invariant points will appear on the diagram corresponding to the parageneses of six minerals, equilibrium lines of five minerals and areas of four-mineral equilibrium. In order to construct such diagrams, natural six-mineral parageneses of igneous rocks should be considered. Knowing the compositions of the corresponding minerals, one may calculate the position of the equilibrium lines of the five minerals.

D. S. Korzhinskii (1946, 1957) presented methods for carrying out the corresponding calculations and also considered in detail the most important six-mineral node, corresponding to the granites—plagioclase $Ab_{67}An_{33}$, potassium feldspar, quartz, monoclinic pyroxene, hornblende and biotite. Here we shall only mention that the construction of this diagram visually

† It should be mentioned that A. N. Zavaritskii and D. S. Korzhinskii were not in full agreement on this matter, but the former author did not express his views in print.

demonstrates the partial importance of Bowen's theory in petrology. It has already been mentioned previously that Bowen did not consider the variation in iron content in femic minerals, which often causes a sharp
270 change in the sequence of crystallization of minerals, e.g., monoclinic and orthorhombic pyroxenes. D. S. Korzhinskii showed that paragenetic relations also depend on the changes in alkali concentration in the magma. An increase in their content leads to the replacement of CaO in the anorthitic component of plagioclase and its transference to femic minerals. This results from the fact that either hornblende or monoclinic pyroxene crystallize instead of biotite in the case of equilibrium with potassium feldspar and acid plagioclase.

An attempt at studying the parageneses of igneous rocks more completely was made by V. S. Sobolev (1949), basing his facts upon the concepts of D. S. Korzhinskii. Two diagrams were constructed — one for rocks with a higher FeO : MgO ratio, and the other for rocks with a lower ratio and a series of invariant six-mineral associations which include both quartz-bearing rocks and undersaturated rocks with olivine and nepheline.

Although the presented diagrams are of a preliminary nature and many nodes should yet be revised, the construction permits one to elucidate a series of most important peculiarities of igneous rocks. Thus, it was shown in the case of quartz-bearing rocks with potassium feldspar that the increase in concentration of K_2O leads to the formation of hypersthene (or with a high iron content — olivine) instead of biotite. This apparent paradox is explained if one bears in mind that the amount of aluminum in biotite is always essentially higher than that of potassium (in atomic ratios) and the reaction biotite + quartz = potassium feldspar + hypersthene (or olivine) is always accompanied by the absorption of potassium.

In the case of parageneses without potassium feldspar, it was shown that the formation of hornblende at the expense of monoclinic pyroxene, hypersthene and basic plagioclase occurs at an increased concentration of Na_2O, thus elucidating the relations between pyroxenes and amphiboles in andesites (i.e., amphibole-andesites are formed from magma with a higher activity of Na). The same also applies to ultrabasic rocks, but in the absence of free SiO_2 the formation of hornblende occurs at lower concentrations of Na_2O as compared to rocks oversaturated with silica.

The entire series of paragenetic relations was also clarified in the case of nepheline-bearing rocks. A most interesting feature is the distinctly limited field on the diagram of leucite-bearing rocks which is bordered by the line of the reaction: nepheline + potassium feldspar + K_2O = leucite + + Na_2O.

The concepts on the mobility of alkalis in magma may be directly applied to the consideration of magmatic processes in the most simple equilibrium diagrams. Thus, for instance, D. S. Korzhinskii noted that an increase in the concentration of Na_2O (which is the fourth component in this case) in the system diopside-albite-anorthite should favor the formation of plagioclases and increase the area of the corresponding field by shifting the albite-diopside eutectic point and the cotectic line.

There is a pressing need in the near future to consider in detail and comprehensively the parageneses of igneous rocks by means of constructing the corresponding "multi-cluster" diagrams, and to compare them with experimental data on a new basis, using methods similar to those of A. N. Zavaritskii (1950) on petrochemical diagrams. Such investigation will permit us to introduce order into the classification of igneous rocks and will allow for a new approach to the solutions of most important geological problems.

271 CHAPTER XII

EXPERIMENTAL DATA AND DIAGRAMS OF NATURAL ASSOCIATIONS OF IGNEOUS ROCKS

In the book "Introduction to Petrochemistry" † data on the compositions of igneous rocks were considered as well as their natural associations. Our primary aim was to present these compositions in such a way that one may grasp the overall features, indicate and clarify the features of these generalities and elucidate the regularities observed. Such features and generalities were actually noted. The two most important of them are geometrically expressed firstly by the fact that the directions and lengths of the vectors, representing the chemical compositions of igneous rocks, change regularly depending on their location within the diagram, and secondly, by the fact that on subdividing the great number of compositions of igneous rocks into subgroups, which correspond to natural associations, these subgroups are expressed on the diagram as swarms of vectors oriented in a certain direction, where the location of these bands shows a certain regularity.

Both of these features illustrate the phenomenon which is designated by the broad concept of magmatic differentiation yielding igneous rocks. The process of differentiation is a complex natural one. In this chapter we shall compare the results of this process with the results of the most simple physicochemical phenomena which have been investigated experimentally, and the causes and conditions of which are well known. The most simple physicochemical phenomena may be compared with processes of igneous rock formation, despite all the differences due to the complexities of both types of phenomena and the conditions under which they occur.

The process of formation of an igneous rock can be essentially summarized as the inversion of a homogeneous silicate melt to an aggregate of crystalline minerals. The most simple physicochemical phenomena which can be compared with this complex natural process, are those of crystallization of silicate melts consisting of the same components as the igneous rock, but in a much lesser number.

† [The present chapter is a summary of the final part of A. N. Zavaritskii's book "Introduction to Petrochemistry of Igneous Rocks" (second edition, 1950). The diagrams of the considered systems have been provided previously with the description of the corresponding systems. The summary omits some of the auxiliary diagrams of these systems which are not presented diagrammatically in the usual petrochemical diagrams, as these diagrams were not directly used for comparing experimental data with petrochemical diagrams of rocks].

272 We shall limit ourselves to the consideration of those silicate systems which have been most investigated up to the present. These are the following ternary systems:

1. Albite-anorthite-diopside.
2. Anorthite-forsterite-silica.
3. Diopside-forsterite-silica.
4. Silica-kalsilite-nepheline (carnegieite).
5. Silica-anorthite-leucite.
6. Silica-diopside-leucite.
7. Silica-fayalite-nepheline.
8. Silica-lime-magnesia.
9. Silica-magnesia-ferrous oxide.
10. Silica-lime-ferrous oxide.
11. Silica-sodium silicate-ferric oxide.

In order to compare, on the one hand, those regularities which occur during the crystallization of these most simple systems, and of those regularities which are expressed in the variability of chemical composition of igneous rocks formed by the crystallization of magmatic silicate melts on the other hand, we select the following obvious method: by expressing the peculiarities of the diagrams of the most simple silicate systems which are now under consideration within the same coordinate system as that used for representing the various chemical compositions of igneous rocks. In order to do this it is necessary to calculate the numerical characteristics of specific points which are designated on the diagrams by letters and numbers. New diagrams should be constructed from these numerical characteristics, which describe features of the considered simple systems in the same parameters used for considering the peculiarities of the chemism of igneous rocks. The diagrams thus obtained are compared with one another and may also be combined and compared with other diagrams of the chemism of igneous rocks.

The features of the new diagrams, which may be constructed on the basis of the numerical characteristics of specific points (tables with the description of the systems), are briefly summarized. Such a diagram is presented on Figure 176, where the designation of points is in correspondence with thermal diagrams as cited above.

1. Albite-anorthite-diopside. This system, which is of the greatest importance to petrography, possesses the following obvious properties: as it does not contain potassium, the n coefficient remains constant — equal to 100. In the complementary numerical characteristic, which corresponds to the femic constituent of the rock, the ratio of the parameters $f' : m' : c' = 0 : 50 : 50$.

On Figure 176 the solid lines represent projections of the cotectic curve of this sytem on the planes *ASB* and *CSB*. The mentioned curve connects the eutectic points anorthite-diopside and albite-diopside; this line may readily be distinguished on the diagram as it is designated by the numbers *1, 2, ..., 5, 6*. The projection of the first eutectic on the plane *CSB* lies on the line connecting the points An and Di. The projection of this eutectic is located on the plane *ASB* at axis *SB*. The projection of the other eutectic point, i. e., the eutectic albite-diopside, on the plane ***CSB***, coincides with the apex ***S*** and on the plane ***ASB*** it is located on the ***SA*** axis in the vicinity of point Ab. The vectors are not shown on the diagram; their projections on the plane ***ASB*** are at an angle of 45° to the coordinate axes, and on the plane ***CSB*** the projections are vertical.

2. Anorthite-forsterite-silica. In this system, which is also most important for petrography, alkalis are absent in the composition and there-
273 fore we only deal with the projection on plane ***CSB***. The parameter ***a*** of the principal numerical characteristic is equal to zero. In the complementary characteristic ***f′*** : ***m′*** : ***c′*** the parameter ***f′=c′=0*** and ***m=100***.

The endpoints of the corresponding lines of the diagram on plane ***CSB*** are also plotted — as solid lines (see Figure 176).

3. Diopside-forsterite-silica. As this system includes neither alkalis nor anorthitic lime, there are no projections of points or lines of this diagram on the planes ***ASB*** and ***CSB***. All of its points are located on the ***SB*** axis, and only the value and the direction of the vectors on the surface ***ASB*** will express the difference between these points. In the principal numerical characteristics, both parameters ***a*** and ***c*** equal zero. In the complementary numerical characteristic ***f′*** : ***m′*** : ***c′*** the value of ***f′***=0 and the values of ***m′*** and ***c′*** vary.

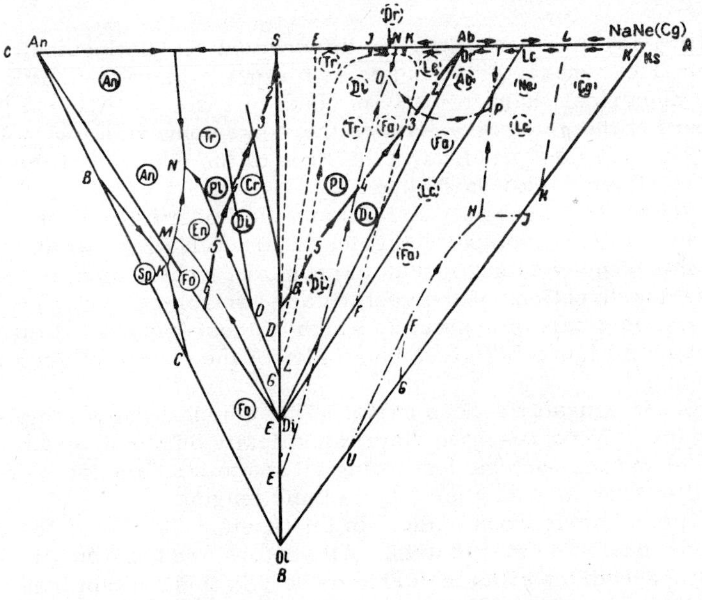

FIGURE 176. A diagram of important silicate systems within the coordinates of the principal numerical characteristics (after A.N. Zavaritskii)

On Figure 177 vectors are plotted according to our coordinate system, which represent the compositions corresponding to the above-mentioned points of diagrams. The endpoints of the vectors are joined by lines in an order corresponding to the sequence of change in composition during crystallization.

4. **Silica-kalsilite-nepheline (carnegieite).** This most important diagram does not sufficiently lend itself to direct diagrammatic representation upon our coordinates. As both the femic constituents and the feldspar line are absent in this system, all the points of the diagram are located on the ***A*** axis. The lengths of the vectors are equal to zero (point vectors), but these vectors will nevertheless have different directions.

5. **Silica-anorthite-leucite.** The femic constituents are absent in this system, as is also sodium. Therefore all parameters in the complementary
274 numerical characteristics are equal to zero. In order to represent the relations between points of this diagram, it is convenient to make use of a construction on plane ***ASC*** (usually not used as it makes the comparison with usual petrochemical diagrams more difficult).

6. **Silica-diopside-leucite.** The numerical characteristics of this system are distinguished by the following features: parameter $n=0$ and the parameter of the complementary numerical characteristic $f':m':c'=$ $=0:50:50$.

As parameter $c=0$, the diagram is located on the plane ***ASB***. All vectors will have the same length and direction, corresponding to the composition of diopside. The lines of the diagram are shown on Figure 176 as broken lines.

7. **Silica-fayalite-nepheline.** In this system the distinguishing features of the numerical characteristics are as follows: the parameter c of the principal numerical characteristic is equal to zero, $n=100$, and $f'=100$. The vectors of the given diagram are therefore converted into points.

On Figure 176 (see previously) the lines of the diagram of the considered system are shown as dot-dash lines.

8. **Silica-lime-magnesia.** Alkalis and the feldspar line are absent in this system. In the principal numerical characteristics $a=c=0$. All initial points of the vectors of this diagram are located on axis ***SB***. The magnitude and directions of the vectors are variable.

[System 3 is a part of system 8, which is most important to petrology, and the diagram on Figure 177 gives us an idea of the change of vectors in this principal area].

9. **Silica-magnesia-ferrous oxide.** This comparatively simple system cannot be directly represented diagrammatically on our diagram (see Figure 176), as the vectors describing all its points coincide with axis ***SB*** and differ from one another only in their lengths.

10. **Silica-lime-ferrous oxide.** In this system the initial points of the vectors are located on the ***SB*** axis. All vectors are horizontal.

11. **Silica-sodium silicate-ferric oxide.** In this system instead of obtaining the numerical characteristics $acbs$ we obtain $a\bar{c}bs$. Obviously, this system is described as a projection on plane ***CSB*** (see Figure 176).

[For the sake of petrochemical comparisons, it is of interest that the corresponding residual melts show a certain trend towards decrease in silica content and increase of alkalis and ferric oxides].

After considering the experimental systems, one may arrive at the following conclusions: of all the diagrams reconstructed according to the mentioned method and expressing the compositions within the coordinates (*acbs*) (*f′m′c′*) and *n*, the ones of greatest interest are those which coincide with the projection planes *ASB* and *CSB*, as well as the cotectic line of the system diopside-plagioclase. On considering the diagram of Figure 176 (see previously) we may distinguish a system of lines which generally resemble the lines on the diagram of Figure 178. On the latter diagram, variation lines are shown which represent the typical direction of change in chemical composition of rocks forming natural associations. On Figure 176 the lines indicate directions in which the composition of the residual liquid changes when crystals are separated from various synthetic silicate melts. Attention should be paid to the similarity in direction of both types of lines, and one may therefore speak quite objectively about the chemical compositions of various rocks in natural associations varying in approximately the same directions as do the compositions of residual melts of crystallizing silicate liquids. Thus, the consideration of the chemism of rocks may lead to the formulation of a law, which was stated some time ago, of p a r a l l e l i s m o f c r y s t a l l i z a t i o n a n d d i f f e r e n t i a t i o n, as derived on the basis of the study of mineral compositions of igneous rocks and the sequence of crystallization of their minerals.

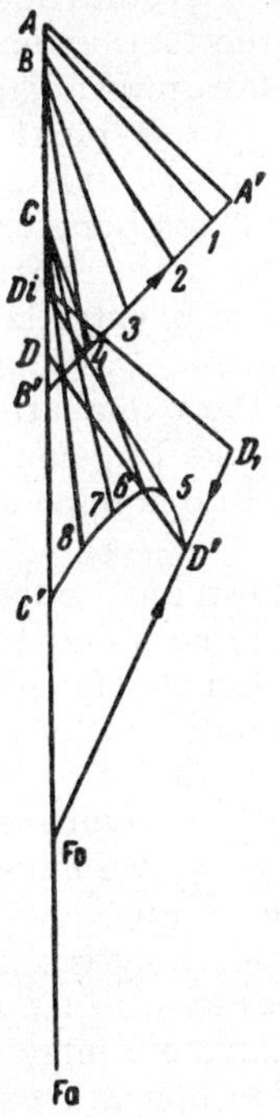

FIGURE 177. Vectors of the system diopside-forsterite-silica (after A. N. Zavaritskii)

At present we may only speak of the approximate similarity of variation lines on the diagram (see Figure 178) with the three-phase lines of the ternary diagram as shown on Figure 176. This firstly depends on the fact that the investigated experimental systems are nevertheless rather dissimilar to rocks encountered in nature. The variation lines of natural associations of rocks on our diagram are represented by curves which are drawn in three dimensions and do not lie in the same plane. The lines on
275 Figure 176 lie either in the planes of the coordinates *ASB* and *CSB* or (the pyroxene-plagioclase curve) on the plane Di-Ab-An. Nevertheless, one may note some very characteristic features of similarity between both diagrams.

If one superimposes on Figure 176 a swarm of vectors representing the association of volcanic rocks typical of the Hawaiian Islands (Zavaritskii, 1950), one may note the coincidence between the swarm of vectors of the Hawaiian Islands and the cotectic line diopside-plagioclase. This line intersects other lines of the diagram in a manner similar to that in which line 7 on Figure 178, representing the Hawaiian lavas, intersects the swarm of lines *1—5*.

In certain alkali associations we observe the so-called pantellerite trend which is expressed on the diagrams by the curvature of variation lines in their upper parts towards the apex *S*. A similar feature may be noted on Figure 176 where the peculiarity is connected with the location of ternary eutectic points.

The diagrammatic representation of the parallelism of crystallization and differentiation, as shown on Figure 176, allows for the further understanding of the significance of curves. Comparing the latter with Figure 176, it is evident that one may ascribe a certain direction to typical variation curves (see Figure 178) as shown by the direction of arrows on Figure 176. This direction corresponds to the drop in the temperature of the liquid, which changes its composition during the process of crystallization, and together with it a decrease in the temperature of crystallization of rocks with the corresponding change in their composition. Having arrived at this conclusion, one may proceed to consider the explanation of the principal feature of chemism diagrams of igneous rocks in general, which consists in the change of direction and length of vectors depending on their locations on the diagram.

276 We have noted that on progressing along the lines shown on Figure 178 toward the bottom right-hand corner, the vectors become both shorter and rotate in a counter-clockwise direction. The direction and length of a vector expresses the nature and quantity of the femic constituent of the rock, and in artificial silicate melts one may express peculiarities of systems by means of vectors (e. g. , systems 8 to 10). When representing a characteristic point of these systems by means of vectors, we have noted that they change their lengths or directions, and when examining the dependence of this change on the change in temperature, one may readily note that the anti-clockwise rotation of the vector and the decrease in its length generally correspond to a decrease in temperature. Thus, one may state that on progressing along variation lines towards lower temperatures we shall pass to vectors which change their lengths and directions in compliance with the decrease in temperature. Naturally the correspondence in this case is also only approximate due to the considerable difference in the composition of natural magmatic melts and the experimentally investigated systems.

However, in the variation lines to the extreme left of the diagram, which approach the Mount Pelé type, a reverse rotation is noted in the vectors of the upper part of the series. On decreasing in length the vectors first become vertical, and are then rotated so as to slope to the left, thus indicating the oversaturation of the composition with respect to alumina. It has already been clarified that this is connected with the change in mineralogical composition: augite is replaced by hypersthene and pigeonite, and later on by hornblende and mica. As we have noted, these features may also be explained by the crystallization phenomena of igneous rocks and is also associated with the lowering of temperature caused by the accumulation of volatile components in the final stages.

On considering the similarity between Figures 178 and 176, the convex form of the variation curves on the projection of ***CSB*** is also explained. This convex form corresponds to the combination of those lines which are noted on Figure 176.

Thus, after considering the peculiarities of diagrams presenting the course of crystallization of silicate systems, the principal regularities in the distribution of directions of vectors and their lengths on diagrams are elucidated. The empirically stated law of parallelism of crystallization and differentiation is reflected in these regularities.

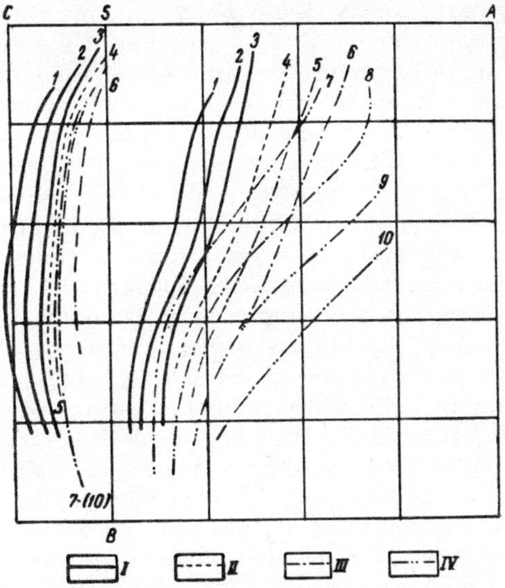

FIGURE 178. Summary petrochemical diagram of the types of natural associations of volcanic rocks (after A. N. Zavaritskii)

I — associations of the alkaline-earth type; II — transitional type; III — alkali associations, enriched in potassium; IV — alkali associations enriched in sodium

277 As is known, the theoretical basis for this law is found in the hypothesis of crystallization differentiation. Notwithstanding the significance ascribed to this phenomena and the degree to which this hypothesis can be borne out by the theory, the parallelism between the crystallization of igneous rocks and the differentiation of magmas (which phenomenon has actually been observed) remains a prime principle for the understanding of the diversity of igneous rocks.

We have previously considered examples showing variability of igneous rocks forming natural associations, the course of differentiation leading to such variability and the sequence in which rocks of a given association are formed.

On passing from one vector to another on the diagram, following the order in which rocks appear (the chemical composition of which are described by these vectors), we shall obviously move along a strip which is usually formed by the vectors describing compositions of rocks belonging to a single natural association. However, this path may have various directions: on passing from one vector to another, we may either ascend or descend on the diagram.

By characterizing different types of natural associations by means of variation lines (see Figure 178) and comparing these lines to courses of crystallization in the most simple silicate systems, we arrived at the conclusion that variation lines may show a trend, i. e., they may be directed upward corresponding to the decrease in crystallization temperature of magmatic melts, from which igneous rocks are formed, which changes their composition in the direction of the variation line. Therefore, on arranging various rocks of a given association according to their age, their sequence will either coincide with the mentioned direction of the variation line, or will proceed in the opposite direction. Thus, a concept is obtained of one additional geometrical criterion of the diagrams representing natural associations of igneous rocks, namely, complying and opposing directions of sequences of rocks according to their age in relation to the variation line. Magmatic differentiation expresses itself as a geological process in time by the consecutive appearance of various rocks. On considering the change in the magmatic melt under the influence of changing physicochemical conditions, one may distinguish two opposing trends, the diagrammatic representation of which was mentioned above. These two trends may be differently termed. Using the Greek words already applied in crystal optics†, one may speak of homodromic and antidromic sequences of rocks according to their age. The first sequence will correspond in general to replacement of more basic rocks by more acid ones, and the second on the contrary — to the replacement of acid rocks by more basic rocks. In the first case temperatures of crystallization of magmatic melts, corresponding to consecutively appearing rocks, pass from higher to lower ones and in the second case, on the contrary, magmas which crystallize at lower temperatures precede those which crystallize at higher temperatures.

278 As is known, in the case of formation of igneous rocks under plutonic conditions in complex intrusive massifs, we are usually dealing with a homodromous sequence of differentiation: basic rocks precede the more acid ones. In natural complexes of volcanic rocks the consecutive replacement of lavas of various compositions is more complex. As the previous chapters have dealt with examples of volcanic formations ††, we should now mention several possible assumptions on the sequence of eruption of lavas of different compositions.

The mere fact that volcanoes during their prolonged existence erupt lavas of varying composition at different times, is an obvious indication of the prolonged existence of a magma chamber which feeds these volcanoes and of the fact that the differentiation process occurring in this magma chamber also occurs over a long period of time. According to the existing concepts of magmatic differentiation, the composition of the magma changes in such a way during this process that its residue becomes increasingly fusible, thus being able to exist in the liquid (amorphous) state at lower and lower temperatures. This determines the variability of intrusive rocks with time, as has already been mentioned. The influence of all factors of differentiation is apparently directed towards attaining a state of nonhomogeneity in the magmatic basin, the acid, less dense part being located at the top and the more basic, denser part at the bottom. We have

† In crystal optics the terms "homodrome" and "antidrome" are used to designate the direction of movement of the end of the isogyre in relation to the direction of rotation of the stage. When they move in one direction we speak of "homodromous movement" and when in the opposite direction — "antidromous movement".

†† [TN — Apparently most of the references in this chapter to so-called previously mentioned data apply to Zavaritskii's book "Introduction to Petrochemistry of Igneous Rocks", and not to the present work].

already considered the examples of Vesuvius and Klyuchevskaya where gravitation differentiation is self evident. The study of volcanoes shows that differentiation occurs during periods of quiescence, and one may assume that the complexity in the sequence of eruptions of various lavas, as compared to the sequence of formation of intrusive rocks, should be associated with the fact that eruptions interrupt the normal course of differentiation in the magma chamber.

On comparing all these concepts one may regard two possible types of manifestation of the differentiation process in the variability of lava composition.

1. Eruptions occur from a differentiated magma chamber, which is more acid at the top and more basic at the bottom. The subsequent exhaustion of such a chamber from the top downwards during a series of eruptions leads to a corresponding increase in the basicity of the lava and hence to the antidromic shape of the vectors on our diagram.

2. Eruptions occur at various intervals, during which the magma in the chamber increasingly changes its composition in accordance with normal magmatic evolution. Each subsequent eruption yields products corresponding to ever-later stages of magmatic evolution. The composition changes from basic to acid, and a homodromic shift of vectors is observed on the diagram. One may assume that the diverse sequence of lava eruptions of varying compositions depends on the relation between the rate of the differentiation process on the one hand and the frequency and intensity of the disruptive eruptions on the other hand. After long periods of quiescence the volcanoes erupt products of a later stage of differentiation.

In addition to such changes in the composition of lavas, expressed in the homodromic or antidromic shift of vectors along a certain variation line, changes occur in the composition of the products of eruption, which are expressed on the diagrams by the shifting of the variation lines themselves, as was most clearly noted in the example of Vesuvius. Many hypotheses may be proposed to explain the cause of such shifting which
279 determines the very type of natural associations of igneous rocks, but this subject can as yet not be considered as sufficiently clarified. All these assumptions are related to the development of a single magma chamber or in other words—of a separate volcanic center. In a certain volcanic area, where a series of such volcanic centers are found, the magma chambers which feed them will be at various stages of evolution. One may arrive at this conclusion from the fact that various volcanoes of the same area sometimes simultaneously erupt lavas of completely different compositions.

This may also be noted in Kamchatka where the Klyuchevskaya volcano yields basaltic lavas, the Avacha — andesite, and the Karymskaya volcano — dacite. In studying the chemism of entire volcanic regions, one often has to be satisfied with a general picture of the diversity of volcanic products. The task of the diagrams in this case is to provide a more complete picture of this diversity.

Part III

280 # VOLATILE CONSTITUENTS OF MAGMA

GENERAL COMMENTS*

The difference between the above-considered artificial silicate systems and natural magmas is not only due to the greater complexity of composition of the latter. The fact that the magma solution contains volatile-gaseous substances, which are liberated on cooling, is most important. Therefore, as already indicated, the chemical composition of the igneous rock is not identical to that of the magma from which it originated. Until recently this circumstance restricted experimental petrography as it was only possible to investigate haplomagmatic systems, which have comparatively little similarity to natural magmas. However, during recent years technical difficulties involved in the study of such systems have been surmounted, and our knowledge of these systems increases from year to year, thus providing for a deeper and more complete understanding of the formation of igneous rocks as well as other processes connected with magmatism.

The presence of volatiles in the magma is proved both by direct observation during volcanic eruptions and by indirect data following from the study of intrusive rocks.

Taking into account the functions of this book, we will not dwell in detail on facts concerning the contents and composition of volcanic gases and refer the interested reader to the corresponding references—in particular to the work of I. A. Ostrovskii (1956). Here we shall confine ourselves to a very brief consideration of this problem insofar as it is relevant to our subject.

The most important component of the "volatile" part of the magma is water. Its content in some volcanic glasses (completely fresh and homogeneous) reaches 8-10% (in pitchstone). In the volcanic gases of Vesuvius (after Gautier) the water content reaches 68%, and in the gases of Katmai (after Allen and Ziess) — 98.1%.

The presence of water in the magma (although only in small quantities) is directly proved by the primary crystallization of amphiboles and biotite.

At present the assumption prevails that the water content of a granitic magma is quite high (up to 10 %) and is somewhat lower ($\sim$ 5—6 %) in basic magma. We shall return to this problem further on. However, we shall mention that another point of view exists, according to which water, in similarity to other volcanic gases, is of secondary origin, being derived from country rocks. This second point of view doubts the existence of true magmas with a considerable content of volatiles at great depth.

Carbon dioxide is the second most important volatile constituent. Despite the fact that oxygen is always present in considerable amounts in the gases of fumaroles, due to admixture of air, free oxygen is absent in the volatile part of the magma and, on the contrary, the reduction

281 products of water and CO_2 are present — H and CO. Finally, a direct analysis of both volcanic gases and products of sublimation affirm the presence of volatile compounds of chlorine, sulfur (mainly in the form of H_2S) and fluorine in the magma. Magmatic gases also contain hydrocarbons (mainly methane), nitrogen and inert gases.

The role of the volatile components, especially water, in the crystallization of magma consists first of all in the lowering of the melting point — by approximately 40° per 1 % dissolved H_2O. This phenomenon is of greatest significance when considering problems of both crystallization of magmatic rocks and of the formation of the magma itself. The available experimental material gives these considerations a well-founded scientific basis.

However, the volatiles do not only play a passive role as mentioned above. They enter into the composition of certain primary magmatic minerals. Hornblende and biotite of igneous rocks constitute examples of undoubtedly primary minerals containing water. Magmatic nosean (containing SO_4), sulfides and possibly other minerals also belong to this category; a theory even exists as to the formation of magmatic calcite.

Lastly, during the crystallization of minerals from magma, the equilibrium in the dissolved gaseous mixture is controlled by reactions of oxidation and reduction of iron (at the expense of oxygen formed from the dissociation of water and carbon dioxide, see further on page 370), thus directly influencing the formation of all femic and ore minerals.

Water and other volatile components play an even greater role when separated from the magma during metamorphic processes, e. g., during autometamorphism (the subsequent change within metamorphic rocks) and metamorphism in general. Finally, the formation of most mineral ore and other deposits is closely connected with the post-magmatic solutions.

In presenting the material of the third part of this book, we shall adhere to the scheme used in the first two parts. We shall firstly consider the general problems of physicochemical investigations of equilibrium diagrams with the participation of volatile components, with the possible formation of a gaseous phase. We shall in part revert to the method used in Part I by considering equilibria in condensed systems. On the basis of the considered general regularities we shall present the most important factual and experimental data which allow for the deeper understanding of both the processes of formation of igneous rocks and, in part, the phenomena of metamorphism, which cannot be completely disregarded when considering the petrology of igneous rocks.

CHAPTER XIII

TYPES OF BINARY AND TERNARY SYSTEMS CONTAINING VOLATILE COMPONENTS

Firstly we shall consider the most general features of cooling of melts containing volatile components. It is known that magma represents a melt — a solution of many nonvolatile components which contain various volatile constituents. On the one hand the presence of volatile substances influences the sequence of separation and the course of the crystallization of the nonvolatile components. This influence is of primary importance. On the other hand we note that the composition of the emanations changes during the process, and it is therefore at least necessary to consider in general the peculiarities of these changes.

As in the case of dry melts, one may obtain a rather clear visual concept of the state of the system from the geometrical representation as diagrams, but the restrictions imposed by three-dimensional representation limit the use of such methods. Only binary and ternary systems can be readily interpreted geometrically. Using this method when dealing with systems containing volatile components, we are obliged to confine ourselves to the consideration of the following combinations:

1. Binary systems: one volatile and one nonvolatile.
2. Ternary systems: a) one volatile and two nonvolatiles; b) two volatiles and one nonvolatile.

In these most simple systems various cases of mutual relations between the components may arise: 1) components do not form compounds or solid solutions (mixed crystals); 2) components form chemical compounds, and 3) components form solid solutions. The second case corresponds to the formation of hydrates and other compounds with water of constitution (e. g., biotite in igneous rocks). The third case is typified by the formation of zeolites.

The method of considering all these cases by means of diagrams is essentially the same as the one used during the consideration of "dry systems", but with the presence of a volatile component one can no longer disregard the influence of pressure, which is of no importance in dry systems. Pressure has a great influence on inversion processes, if one of the phases of the system is a gas. In this case one should use a phase diagram which shows the relations between the concentrations $x_1, x_2, \ldots, x_n$, temperature T and pressure P. These relations can only be expressed in three dimensions for a binary system, where the concentration variable is x_1 only. For such a diagram we are able to present the projection on the coordinate plane on which we may readily discern the univariant lines

283 of equilibrium and invariant points, as well as sections of the diagram parallel to one of the coordinate planes. Sections parallel to the plane *Px* are most often provided, i. e., sections showing constant temperature (isotherms) and parallel to the plane *TP* i. e., sections showing constant pressure (isobars). A complete three-dimensional representation of the diagram is impossible in the case of ternary systems, and one is obliged to limit oneself to certain sections or projections of the diagram [in most cases a three-dimensional diagram at constant pressure (isobaric) is constructed, on which temperature is plotted at the apex of a trihedral prism, as in the case of all other melting diagrams. Similar, constant temperature (isothermal) diagrams are also used, on which pressure is plotted at the apex of the prism. On transition to a two-dimensional representation on triangles, resort is again made to contours in order to represent the relief of the three-dimensional surfaces — in this case they represent lines of equal pressure — isobars].

The state of a system containing only one component, depending on the different temperatures and pressures, may be represented on a planar diagram as shown on Figure 2 (see page 20).

The state of the system is conditioned by a point with coordinates ***P*** (pressure) and *T* (temperature). Depending on the phase field in which this point lies (*S, L* or *G*), the system will either represent a solid body, a liquid or a gas. If the point lies on the boundary of two fields a two-phase system will result, e. g., for points on the line separating fields *G* and *L* we have a system of liquid and gas; on the boundary between *L* and *S* — of liquid and a solid body. At the ternary point ***D*** where the fields of *S, L* and *G* meet, gas, liquid and solid coexist simultaneously. The boundary curve between *S* and *L* represents the change in melting temperature with change in pressure. The melting temperature is only slightly dependent on ***P*** and is obviously almost parallel to the coordinate axis. After briefly recalling the features of *PT* diagrams, we shall consider a diagram representing a system of two components, one of which, ***A***, is considerably more fusible and volatile than the other — ***B***.

THE DERIVATION OF ISOBARIC PHASE DIAGRAMS FOR BINARY SYSTEMS †

These diagrams may be derived from the consideration of the relations of thermodynamic potentials in the systems at various temperatures, as in the case of the derivation of diagrams for the crystallization of solutions. This method, used by van-Rijn-van-Alkemade, is widely known and is presented in many textbooks of physical chemistry — particularly when dealing with the phase theory.

If one of the components is volatile and may exist in a gaseous state, one is obliged to include a gaseous phase into the consideration of the

† [This section is taken from A.N. Zavaritskii's paper "The main physicochemical problem in the process of pegmatite-formation" (1944) In compiling the following sections of this book, material from this paper as well as A.N. Zavaritskii's book (1926) were used to various degrees and certain amendments were introduced].

relations of the thermodynamic potentials. This feature is not taken into account during the derivation of crystallization diagrams. The course of derivation remains the same. In each considered case it is also necessary to know the conditions of mutual solubility of the components in various phases. As is known, the crystallization diagram of a system consisting of immiscible components differs from that of a system with components forming solid solutions.

284 During the derivation of the isobaric diagram of a binary system in which one of the components is volatile, we shall limit ourselves to the case where both components yield neither compounds nor solid solutions and their liquid solutions do not show phenomena of splitting (liquation)†.

The derivation of the phase diagram from the thermodynamic potential diagram is significant because it has a strict mathematical derivation, following the rule that the thermodynamic potential should have a minimal value in an equilibrial system — being only one of the manifestations of the second law of thermodynamics.

In the case of our diagram, we start from a diagram on which thermodynamic potentials of various conditions of a system of two components are represented at constant pressure.

As the components in the liquid phase are miscible in all ratios according to the above-stated condition, which feature is also always true for gases, the potential curves for the liquid and gaseous phases will be convex downwards throughout their lengths. As the components are completely immiscible in the solid phase, their potentials are represented by the ordinates of the diagram at the composition points *A* and *B* representing the pure components. The potentials of the mechanical mixtures of *A* and *B* are the ordinates of the straight line joining the ordinates of the pure components. At comparatively high temperatures the gaseous phase is the most stable one. The thermodynamic potential of the gaseous phase will lie below the curve of the liquid phase and the potential points of the solid phases. At lower temperatures the system, which contains a sufficient amount of the nonvolatile component *B*, will be more stable in the liquid phase. The thermodynamic potential curve of the liquid phase for these compositions of the system will lie below the thermodynamic potential curve of the solid phase, and the area of such compositions widens as the temperature is lowered. At even lower temperatures, when crystallization starts, the thermodynamic potential of the crystallizing component will be lower than that of its liquid phase. The relations between thermodynamic potentials vary during cooling of the system. On considering the properties of the diagram, revealed during the mentioned relative shifting of the curves, a phase diagram of the system may be derived — namely the composition-temperature diagram.

In order to accomplish this it is very important to know the following: we consider the process in which the liquid and gaseous phases differ from

† The system mentioned is the one considered by Vogt. Niggli dwells mainly upon it in his considerations of the physical chemistry of systems containing volatiles. Obviously, the diagrams, considered by the mentioned authors and others engaged in the application of physical chemistry to the problems of petrology, are not those considered initially by Vogt and Niggli, but were derived long before them by physical chemists (such as Roozeboom, Schreinemaker, Smith, and others), and should scarcely be termed "Vogt-Niggli diagrams".

one another. In this case each of the phases has its own thermodynamic potential curve. This is only possible if the pressure does not exceed a certain constant value, and it is only under these circumstances that it is possible to distinguish between the liquid and gaseous phases. At pressures exceeding this value it is impossible to distinguish between the liquid and gaseous phases and only one, fluid phase exists. Hence, only one thermodynamic potential curve, representing this fluid phase, will
285 appear on the diagram. The consideration of the process of crystallization is in this case not distinct from that of a homogeneous liquid.

We shall start by considering systems at a pressure allowing for the independent existence of both liquid and gaseous phases. This is the case to which the "Vogt-Niggli diagram" applies.

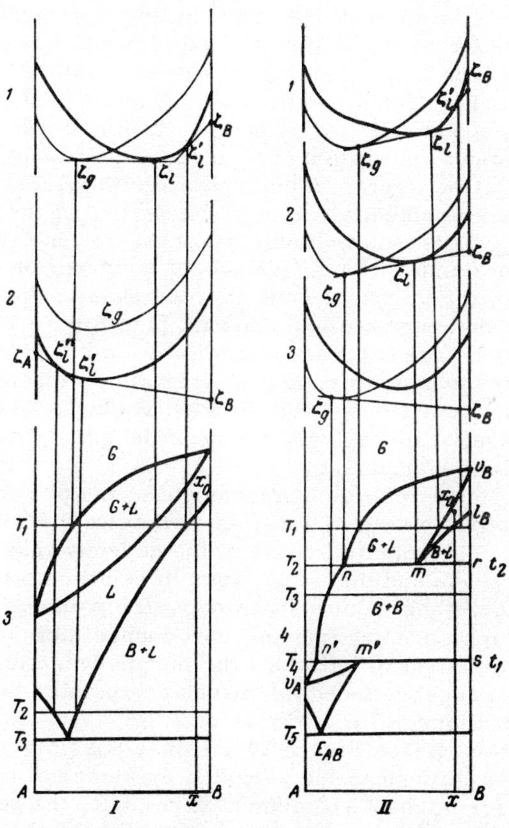

FIGURE 179. The derivation of isobaric sections of a *PTX*– diagram of a binary system using thermodynamic potentials

We shall consider the relations of thermodynamic potentials in our system at a certain temperature T_1, which is somewhat lower than the crystallization temperature of the pure nonvolatile component B. On the

top of the diagram (Figure 179) the relations between the thermodynamic potential curves of the phases of the system are shown at different temperatures. The relations corresponding to temperature T_1 are shown by the numeral 1. The thermodynamic potential of the pure component B lies below the end of the potential curve of the liquid phase. At this temperature the pure volatile component is stable in the gaseous phase. The thermodynamic potential curves of the liquid phase (darker lines) and of the gaseous phase (lighter lines) intersect. When such relations of thermodynamic potentials exist, it is possible to draw tangents to the potential curves as shown on
286 Figure 179. One of these tangents, drawn through the thermodynamic potential point ζ_B of the solid phase of the nonvolatile component B, contacts the thermodynamic potential curve of the liquid phase at the point ζ_l'; another tangent contacts the curve of the liquid phase at point ζ_l and with the curve of the gaseous phase at point ζ_g. The sections of the tangents ζ_B ζ_l' and ζ_l' ζ_g' represent the thermodynamic potentials of a mixture of two phases: the first — a mixture of a solid phase and a liquid corresponding to the composition ζ_l' and the second — a mixture of the liquid phase of a composition corresponding to the point ζ_l and a gaseous phase of a composition corresponding to the point ζ_g. Obviously, depending on the overall composition of the system at a given temperature, the most stable state (proceeding on the diagram from right to left) will be the one with the lowest thermodynamic potential value of the system at the given composition: from point ζ_B to ζ_l' — solid + liquid, from ζ_l' to ζ_l — one liquid; from ζ_l to ζ_g — liquid of composition ζ_l and gas of composition ζ_g; from ζ_g to the extreme left of the diagram — one gaseous phase. We now consider the composition-temperature diagram (see Figure 179, I-3), upon which these relations are expressed as follows: on drawing a horizontal straight line corresponding to a given temperature we note that this straight line passes through four fields (right to left): $B+L$, L, $L+G$, G, where B is the solid phase, L is the liquid and G is the gas. These relations are the same on both diagrams of Figure 179.

On further cooling a relative displacement of the potential curves of the liquid and gaseous phases occurs, as well as of the potential points of the solid phases. The potential curve of the gaseous phase rises above that of the liquid phase and the latter rises in relation to the potential points of the solid phases. Correspondingly, the points of contact, designated by ζ with corresponding indices, will change their position; they will be shifted from right to left along the thermodynamic potential curves. During this relative displacement of curves and points, two cases may arise as shown in Figure 179, I and II.

1. Points ζ_l' and ζ_l (see Figure 179, I) may possibly maintain their separate identities throughout the period of existence of the gaseous phase. In other words, during the displacement to the left, the point ζ_l' on moving along the thermodynamic potential curve of the liquid phase does not reach the point ζ_l. At sufficiently low temperatures the liquefaction of the gaseous phase of the pure component A occurs, followed by the crystallization of component A at temperature T_2 (see Figure 179, I-2) and finally the eutectic crystallization of A and B.

The separate location of points ζ_l' and ζ_l or, in other words, the presence of the segment $\zeta_l' \zeta_l$ during the existence of a gaseous phase gives rise to the appearance of a pair of curves on the diagram which represent the compositions of the liquid and the gas and limit their fields of coexistence. This feature is completely analogous to that of the presence of a similar segment of tangents and thermodynamic potential curves of liquids and solid solutions, as their crystallization gives rise to the liquidus-solidus pair of curves upon the composition-temperature diagram, which limits the field of coexistence of liquid and solid solutions.

The fact that the tangent from ζ_B to the liquid curve contacts this curve only leads to the appearance of a separate crystallization curve of the eutectic type on the diagram. Between the fields $B+L$ and $G+L$ there is a continuous field L, which is conditioned by the section of the thermodynamic curve of the liquid phase between points ζ_l' and ζ_l. The phase diagram of a volatile-nonvolatile binary system has such a form at high pressures, when the stability fields of the liquid phase is enlarged.

287 2. During its movement along the thermodynamic potential curve of the liquid phase, to the left, point ζ_l' may possibly reach point ζ_l, which is shifted in the same direction. The tangents $\zeta_B \zeta_l'$ and $\zeta_l' \zeta_g$ will become fused into one straight line passing through point ζ_B and which is tangential to the thermodynamic potential curves of both the liquid and gaseous phases. Three phases will then be in equilibrium. According to the Phase Rule the system becomes invariant, and an inversion occurs at constant temperature. The liquid phase will disappear, being transformed into a mixture of the solid phase B and gas the composition of which corresponds to the potential point ζ_g (see Figure 179, II-2). The location of the points of contact of the straight line $\zeta_B \zeta_l \zeta_g$ determines the location of the corresponding points on the composition-temperature diagram (see Figure 179, II, for temperature T_2). The field of liquid L, which narrows as temperature T_2 is approached, wedges completely out at this temperature due to the fact that the section of the thermodynamic potential curve $\zeta_l' \zeta_l$, which became shortened as point ζ_l' approached point ζ_l, finally becomes zero when point ζ_l' reaches point ζ_l.

At lower temperatures (e. g., at T_3) the tangent from point ζ_B contacts the thermodynamic potential curve of the gaseous phase at point ζ_g, and the thermodynamic potential curve of the liquid phase is located entirely above the straight line $\zeta_B \zeta_g$ (see Figure 179, II-3). On the composition-temperature diagram we enter at this moment into the field of coexistence of the gas and solid phase B. The field $G+B$ is limited from above by a horizontal straight line corresponding to the temperature of invariant inversion which was mentioned above.

With further cooling the point of intersection of the liquid and gaseous thermodynamic potential curves is shifted further to the left, but the liquefaction of the pure component A should finally occur, i. e., the liquid potential curve should drop below that of the gas on the left side of these curves. Therefore the shifting to the left of the intersection point of these curves must lead to a state when the liquid potential curve, which is convex downwards throughout its length, will again be tangential to the straight line $\zeta_B \zeta_g$ at a certain temperature T_4 (Figure 179, II-4). At this moment the system will again consist of three phases, and hence a new invariant inversion will take place. On the thermodynamic potential diagram this and subsequent moments are not considered in this case for the sake of simplification of the entire scheme. The reappearing liquid phase differs from the one existing at high temperatures, which was rich in the nonvolatile component *B*, and will mainly consist of the volatile component *A*. This process involves the liquefaction of the gaseous solution of *B* in *A*, and further cooling to temperature T_5 will lead to eutectic crystallization (freezing) of this solution (see Figure 179, II-4).

The relations in the second case considered, which are shown on Figure 179, II, occur when the pressure is not so high as to preserve solutions containing any amounts of the volatile component *A* in the liquid state during the cooling of the system.

[The obtained sections directly illustrate the influence of pressure on the relations of boiling and melting phenomena in such a system. The first section, which represents high pressure, is shown in such a way that the boiling points are considerably higher than the melting points (see Figure 179, I). The melting diagram here has the well-known shape of a eutectic. Boiling point diagrams are principally analogues of diagrams of solid solutions, the only difference being that the lower curve gives the composition of the liquid, whereas the upper one gives the composition of the gas.
288 As is known, Roozeboom constructed his melting diagrams on the basis of these diagrams and the regularities which follow from them (Konovalov's laws).

In the case where complete miscibility of the components in the liquid state is assumed as well as a marked difference in their boiling points, the diagram will be related to Type I (Figure 179, I — curves *G* and *L*). The boiling point curve is here separated from melting curves by a one-phase liquid field, i. e., for any composition boiling starts in a pure liquid without any solid phase.

On lowering pressure the melting points of both components will change only slightly. On considering the section at P_2 we note that the melting curves are only slightly shifted. As is seen on Figure 180, the boiling points behave completely differently, being lowered radically as pressure is lowered. Obviously, the pressure P_2 may be selected by us in such a manner that the boiling points of the pure components will lie even higher than their melting points, but in this case the boiling curve will intersect the melting curve. In the physical sense this means that phenomena of boiling will occur in a crystallizing melt, and at temperatures T_2 and T_4 (see Figure 179, II) three phases will be in equilibrium — solid *B*, liquids *m* and *m'* and gases *n* and *n'*. The curve is interrupted between the intersection points and the gas is in direct equilibrium with the solid phase.]

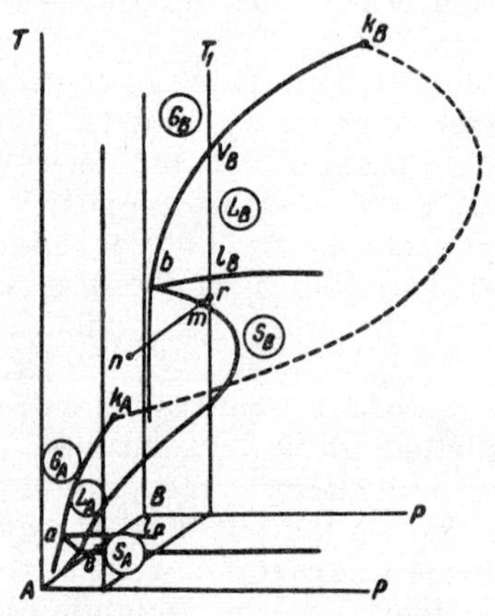

FIGURE 180. General view of a diagram of a binary system lacking critical phenomena in the area of crystallization

PTX - DIAGRAM IN THE ABSENCE OF CRITICAL PHENOMENA*

This diagram may be obtained by joining several series of isobaric sections. Those properties of the thus obtained three-dimensional *PTX*-diagram, which follow from the properties of *PT*-diagrams of unary systems forming our binary system, should be taken into consideration. The *PT*-diagram of a unary system, presenting the state of this system at various pressures and temperatures, is well known from elementary textbooks of physics and chemistry. On Figure 180 the darker lines represent such diagrams for the volatile and nonvolatile components separately. These diagrams are shown at the ends of the section *AB*, the points of which are used to represent the composition of the binary system from *A* and *B*. The planes of these components on the *PT*-diagram are taken as being perpendicular to *AB*.

On the diagrams of the components, the borders of three fields of various states of these unary areas are seen. They are designated by letters and indices: S_B — designates the field of stability of the solid phase of the more refractory component *B*, L_B — is the field of the liquid state of the same component and G_B — is the field of the gaseous phase. In the case of component *A* the fields of these same states are designated by S_A, L_A, and G_A. The boundary lines between these fields represent the coexistence of two phases and they meet at points of simultaneous existence of three phases or so-called ternary points. On the presented diagram
289 (see Figure 180) ***b*** designates the ternary point of component *B* and ***a*** designates the ternary point of component *A*. The boundary lines, as is known, represent the curves of variation of melting temperature (bl_B, al_A) and temperature of vaporization (bk_B, ak_A) as dependent on pressure. The boundary line between the liquid and the gaseous phase terminates either at the points k_A, or k_B which are the critical points of component points *A* and *B* respectively. The critical points are determined by the temperature-pressure coordinates which in this case are the critical temperature and the critical pressure†.

† It is here necessary to recall certain well-known facts, as A. E. Fersman in his book "Pegmatites" claims that the critical point of water is raised by pressure. He writes: "The critical point corresponds to 374° at 219 atm saturated vapor pressure. An increase in pressure raises it to 400-425° (? — A. N. Zavaritskii), and the participation of dissolved substances may, according to W. Foshag, raise it even more". On page 48 we find the following statement: "A great increase in pressure inhibits the liberation of the volatile components at high temperatures to a great extent, maintains them in the residual melt, elevates the critical point of water, lowers the point Q and lessens the area of the fluid field until it almost completely disappears. In this case the pegmatitic process is expressed by rather sharp changes in the magma and the superheated aqueous solutions". All this perplexes the reader even more. Both the critical point and the point in the three-dimensional diagram are determined by invariable temperature-pressure coordinates. They are points only because their coordinates do not change. If the change in one of the coordinates (pressure) also causes a change in the other (temperature), the point now becomes a line: the "fluid" field not does become smaller as pressure is increased, but on the contrary, it widens.

At the transition from unary to binary systems which contain the unary systems as their constituents, according to the Phase Rule the binary system will have lines corresponding to the points of the unary system and surfaces corresponding to the lines of the unary system. These lines or surfaces will issue from corresponding points or lines of the unary diagrams which are the constituents of the binary system. They connect these points of the unary diagrams contained in the binary system.

In the case considered at present, attention should be paid to the curves connecting points k_B and k_A, and the ternary points ***b*** and ***a***. In systems in which critical phenomena are observed, the curve $k_A k_B$ is the line of intersection of the surfaces which limit the field of stability of the "fluid" from the areas of separate existence of liquid and gaseous phases; as we shall see further on, it is the line of curvature of this surface which limits the "fluid" field.

The curve† connecting the ternary points where three phases coexist simultaneously, consists of two branches ***be*** and ***ae***. Each of these branches represents the change in the state of the system when it is formed by three phases: either solid ***B*** + liquid + gas, or solid ***A*** + liquid + gas; in the binary system, where the pressure also varies, in the case of three phases we have a univariant system. The two branches meet at the invariant point ***e*** corresponding to the eutectic point the existence of which depends on the fact that the components ***A*** and ***B*** do not form solid solutions. Four phases coexist simultaneously at this point ***e***: solids ***A+B+*** liquid + gas. It is
290 a unique point — an inflection point on the three-phase line ***bea*** — similar to the eutectic point on the melting curve of a binary system at constant pressure.

It is now especially important to consider the properties of three-phase lines. The coordinate ***P*** (pressure) on moving along this line from ***b*** to ***a***, initially increases up to a certain maximum, then decreases on approaching ***e*** and hence passes to ***a***. This property may be easily understood from the consideration of the expression $\frac{dP}{dT}$ for a binary system. Without dwelling upon it here, we shall merely present Eitel's considerations, by means of which he explains the obvious necessity for such a maximum ***P*** on the curve ***ab***. "In saturated molten-liquid solutions the vapor pressure of the volatile component ***A*** increases as temperature is raised. On the other hand at point ***b*** the vapor pressure is practically zero and the addition of component ***A*** lowers the temperature of melting but also increases the vapor pressure. Both of these conditions together lead to the appearance of a maximum pressure at which the increasing vapor pressure due to

†[In the general case of a three-dimensional diagram we should have three curves which meet at one point and correspond to the conjugate change in composition of three phases with changing temperature and pressure. On Figure 180 for the sake of simplification only one curve is shown for the liquid phase. The curve for the composition of the gaseous phase, which should pass through ***n***, is omitted here (see Figure 181), as the solid phases *B* and *A* have constant compositions under these conditions. The corresponding curves lie in the coordinate planes *B* and *A*, coinciding with the projections of the corresponding curves of the liquid and gas on these surfaces].

increased temperature is combined with the vapor pressure which increases with a drop in temperature (in melts richer in component ***B***)".

Knowing the location of melting curves and curves of vaporization of the components, as well as the course of the three-phase lines, it is easy to determine the location of all parts of isobaric sections of the general ***PTX*** -diagram of the binary system, the form of which was derived by us previously from the consideration of the relations of thermodynamic potentials. By constructing a series of such isobaric sections for various pressures we obtain a complete picture of the stability areas of the system in states which differ by the number of phases and their compositions.

Let us now refer to the diagram on which the location of one of the isobaric sections is shown (see Figure 180). The three-phase line intersects this section at point ***m*** which is identical to the point designated by the same letter in Figure 179, II-4. The same letters ***m, n*** and ***r*** on Figure 180 correspond to points designated by the same letters on Figure 179, II-4. The line ***rmn***, which corresponds to invariant inversion at constant pressure, is a horizontal line. As pressure is changed it will describe a cylindrical surface which passes through the three-phase line and the generatrices of which are parallel to the ***AB*** axis.

Figure 181 shows the same relations. Two sections are presented here corresponding to the pressures P_1 and P_2. The lines of the first isobaric section are shown as solid lines and those of the second as broken lines. The cylindrical surface, which is formed by the displacement of the line of invariant inversion ***mnr*** during the transition from one section to another, is shaded. Pressure P_2 is lower than the vapor pressure of the volatile component at its ternary point ***a***, and ***P*** is higher than this pressure. At high pressures, exceeding the value of maximum pressure for the point of the curve of three phases ***bea*** (see Figure 180) or $bm_2m_1q_1q_2e$ (see Figure 181), the plane of isobaric sections will not intersect the cylindrical surface, the generatrices of which are lines ***rmn***. This occurs when the pressure is so high that it holds the volatile component in solution through the period of crystallization of the nonvolatile component from the solution. In this case the isobaric section of the ***PTX***-diagram has the form shown on Figure 179, I-3. On Figure 181 this section is not shown so as not to overcomplicate the diagram. On comparing the phase conditions of the system on Figure 179, II-4, with the location of isobaric sections on Figure 181, we obtain a clear idea of
291 these areas into which the entire space is divided and which include points of all possible conditions of a system of two components, one of which is volatile.

On the top left side of the diagram (see Figure 181) we find the area of the gaseous phase of the systems. To the left it is limited by the coordinate plane of the diagram of component *A* and to the right by the surface formed by the line v_1n_1 when displaced during the transition from one isobaric section to the other. The area of coexistence of two phases (liquid and gas) is limited to the left only by the mentioned surface formed by the movement of curve $v_1 n_1$ and to the right by the surface obtained due to the displacement of the curve v_1m_1 ; from below it is limited by the cylindrical surface resulting from the displacement of the horizontal straight line m_1n_1. The surfaces limiting the area liquid+gas intersect at edges, which are represented by the curve bv_2v_1 (the vaporization curve of component ***B***), the three-phase curve m_1m_2 and the curve n_1n_2. Further to the right is located the area of the liquid phase of the system which predominantly consists of the more refractory (nonvolatile) component and the volatile one which is dissolved in its melt; this liquid may be considered as a prototype of a magma up to the beginning of crystallization (compare with page 293). Its area of stability is limited to the left by a surface formed by the dislocation of v_1m_1, from the bottom by the surface of incipient crystallization of component ***B*** formed by the displacement of the line of incipient crystallization T_1m_1, and to the right by the coordinate plane of component ***B***. A wedge-shaped area of coexistence of a liquid melt and crystals of ***B*** is located below this area. It is limited by the surface of incipient crystallization Tm, the cylindrical surface mr and the coordinate plane of component ***B***. The edge T_1T_2b represents a line of crystallization temperatures of the pure component ***B*** as dependent on pressure.

The cylindrical surface, formed by the displacement of line $r_1m_1n_1$, represents the boundary of coexistence of two phases: gas + solid ***B***; the system will be in this state on the side of the boundary surface facing the coordinate plane of the *TX*-diagram, i. e., at lower pressures.

At lower temperatures new areas appear which are associated with the inversion of the gaseous phase to a liquid rich in the volatile component. This process is the prototype of the formation of postmagmatic thermal solutions. Here again, an area of coexistence of liquid and gaseous phases
292 appears (the straight line $u_2q_1s_1$ being the section for P_1), below which is observed a one-phase area of liquid and an area of coexistence of a liquid and solid ***B***.

[Several areas also exist which are connected with the appearance of solid crystals ***A***; which are, however, of no interest to us, as, in the case considered by us, the situation where the volatile component crystallizes is never reached (e. g., up to the freezing of water during the crystallization of silicates).

A three-dimensional phase diagram of a system is usually represented as the projections upon three coordinate planes ***TX***, ***PX*** and ***TP***. Such projections of curves of intersection of the surfaces of a three-dimensional diagram are shown on Figure 182. Here are shown some of the peculiarities which were omitted on Figure 181. On considering the projections we may make the following remarks: on the projection ***TX***(II) we find a melting curve ae_1b with the eutectic point e_1. For each composition of the liquid represented by points of this curve there is a definite corresponding composition of the vapor which will be richer in the volatile component than is the liquid. The phase of composition e_2 will correspond to the composition of the eutectic liquid. Thus, in addition to the composition curve of liquid ***L*** we shall also have a composition curve of the vapor ***G***, namely ae_2b, indicated by a dotted line.

On projection ***PX***(III) two curves are also present: one for the liquid ($bM_1e_1'a$) and the other for the vapor ($bM_2e_2'a$). On considering the system at various pressures we shall in each case obtain the composition of a vapor which will be richer in the volatile component than the liquid. The curves of pressure have maximum points (M_1 and M_2) and inflection points corresponding to the eutectic composition.

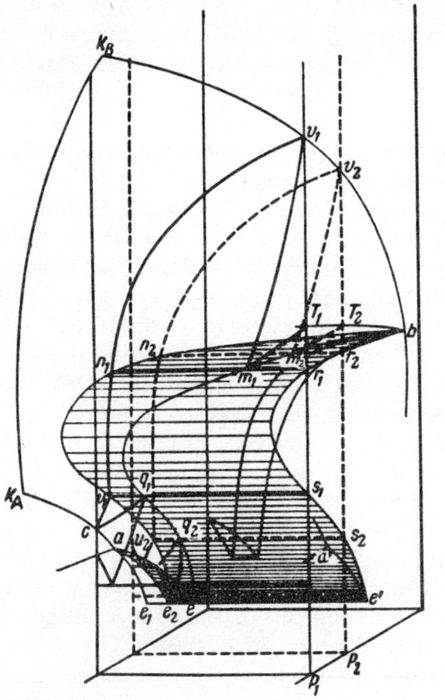

FIGURE 181. Derivation of a ***PTX***-diagram of a binary system by means of isobaric sections in the case of absence of critical phenomena

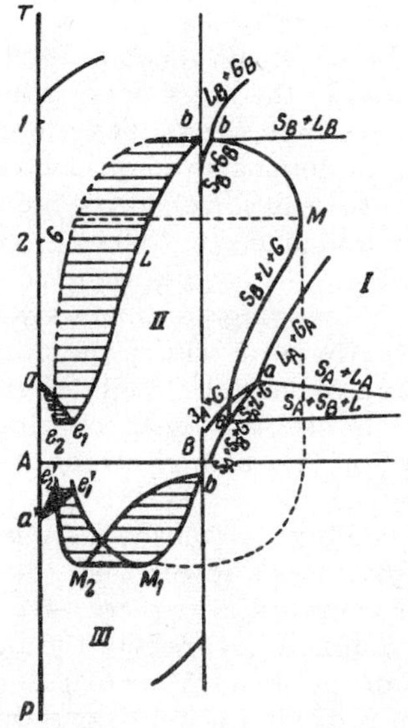

FIGURE 182. Projections of the diagram of a binary system with a volatile component lacking critical phenomena in the field of crystallization

On projection ***PT***(I) are located points ***a*** and ***b*** — ternary points of components ***A*** and ***B***. These points are designated on Figure 2 (see page 20) by ***D***. The three branches of this figure correspond to the following curves at point ***b***: $L_B + G_B$; $S_B + L_B$ and $S_B + G_B$, and at point ***a*** these curves will be: $L_A + G_A$; $S_A + L_A$ and $S_A + G_A$ (see Figure 182).

In addition, at points ***a*** and ***b*** the corresponding lines $S_B + L + G$ and $S_A + L + G$ meet. These lines connect the mentioned points with the point ***e*** which is their analogue for eutectics.

The curves $S_B + L + G$ and $S_A + L + G$ correspondingly represent the projections of the three-dimensional curves bm_1e_1 and bn_1e_2 (see Figure 181). The curve $S_B + L + G$ as well as bm_1e_1 have a maximum ***M*** which corresponds to the maximum M_1M_2 of the projection ***PX***. Each of the curves of projection ***PT*** corresponds to the state of the system at which phases corresponding to the letters designating these curves exist; where S_A is the solid phase of component ***A***; S_B — the solid phase of component ***B***; ***L*** — the liquid phase; ***G*** — vapor; point ***e*** is the invariant (i. e., lacking any degree of freedom) quaternary point, where S_A, S_B, liquid L and vapor G of a definite
293 composition coexist; the curves designated by three letters correspond to univariant systems of three phases, e. g., $S_B + L + G$ corresponds to a composition of a system of solid S_B, liquid L and vapor G, etc.

Now we must consider phenomena during the cooling of a binary system with a volatile component at constant pressure. We shall start from a melt — a solution of the volatile component in a quantitatively predominating nonvolatile component. The composition of such a solution is represented on Figure 179 by the point x. In the beginning of the process the system represents a homogeneous liquid (prototype magma); its corresponding point is located in the area designated on the diagram by x_0.

At very high pressures the change in the states of such a system on cooling is represented on the left of the diagram (see Figure 179). When temperature is lowered the mentioned point of the system descends along the vertical line x_0x. When it reaches the boundary line between the fields L and $B+L$, the crystallization of component B starts, the composition of the melt changing continuously along the mentioned boundary line up to the eutectic point where the component A will join B in crystallization (in this case freezing). Such a combination of processes for a magmatic system is essentially theoretical (imaginary). It is now important for us to mention that at high pressures, for which the diagram of the type shown on Figure 179, I-3 is valid, a continuous change in composition occurs, and hence also a continuous change in the properties of the solution.

At pressures lower than the maximum, corresponding to the points of the three-phase curve *bea* (see Figure 180), i. e., when the isobaric section of the diagram intersects the cylindrical surface of boiling, the change of states of the system occurs as presented on Figure 179, II-4. The considered point x_0, on descending, reaches the boundary line l_Bm between the areas L and $B+L$; crystallization of B starts; the composition of the melt changes along the boundary line between L and $B+L$ up to point m. Here a gaseous phase appears together with crystals and a liquid of composition m; the composition of this gaseous phase at a given temperature is shown by point n. In a binary system consisting of three phases $B(r)m$ and n an invariant inversion occurs at constant temperature. The liquid m disappears giving rise to crystals $B(r)$ and gas n, and finally the entire system becomes a mixture of a solid phase (crystals) and gas present within this crystalline aggregate. The gaseous phase is a solution of the nonvolatile component B in the volatile component A. It is important to mention that this gaseous solution will be saturated in the nonvolatile component B, and hence it will attain equilibrium with the crystals B, as was the case of liquid m, this liquid having been exhausted during crystallization at the invariant point m. On further cooling the crystallization of B will proceed, but in this case from a gaseous solution possessing properties different from those of the liquid from which this component initially crystallized. In this case the crystallization process of B will not be continuous but will occur in two stages differing sharply from one another according to the properties of the solutions from which component B crystallizes. During the crystallization of B from the gaseous solution the composition of the latter will change along the line nn'.

Of special interest is the phenomenon which occurs in the system when the melt, rich in component B, cooling at a constant pressure and separating crystals B, attains the composition m. Up to this moment the separation of crystals B is not essentially different from the usual crystallization
294 during cooling. During the separation of the solid phase B, the liquid

naturally becomes enriched in the volatile component ***A***. After reaching the composition corresponding to point ***m***, no further enrichment of the liquid in the volatile component ***A*** is possible, and it separates as a gaseous phase involving a part of ***B*** together with it in a dissolved form. The liquid maintains its composition ***m*** and at this constant composition and at constant temperature the gaseous phase separates from it simultaneously with the crystallization of ***B***. The liquid starts to vaporize and boils in a manner similar to that in which a solution boils on heating, but in this case, on the contrary, it occurs during cooling.

Such "boiling" has therefore been termed retrograde boiling, or the second boiling point, in contradistinction to the first boiling point where vaporization starts on raising the temperature.

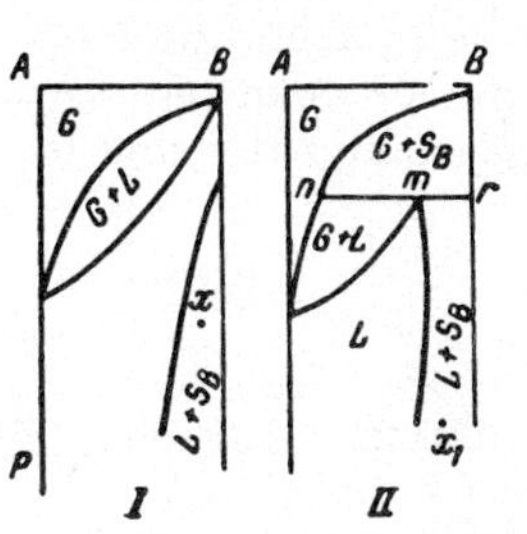

FIGURE 183. Main types of isothermal sections of a binary system with a volatile component

This first boiling point on elevation of temperature or, on the contrary, the point of liquefaction on cooling, corresponds in the considered system to the temperature of the lower boundary ***n'm's*** of the field ***G+B*** (see Figure 179, IV-4). Thus, the area ***G+B*** is limited by two horizontal lines corresponding to the first and second boiling points. As we have seen, these lines are the generatrices of the cylindrical surface which constitutes the frontal limit (from the side of higher pressures) of the field ***G+B*** on the three-dimensional ***PTX***-diagram. This cylindrical surface may therefore be termed a surface of vaporization or boiling. A part of it, lying above the generatrix, corresponding to the maximum pressure for points of this surface (i.e., above the three-phase line ***bea*** on Figure 180) is a surface of retrograde boiling (surface of second boiling points); another part, located below this generatrix, is a first boiling point surface.

As we have seen, retrograde boiling, in similarity to "usual" boiling, occurs at various temperatures depending on the pressure, namely: the increase of temperature lowers its pressure in contrast to the "usual" boiling on heating when increased pressure elevates the boiling point. The compositions of the liquid and of the vapor that separates during retrograde boiling also change as temperature is varied. The curves of the diagram represented on Figure 181, and the projections of the curves be_1 and be_2 (see Figure 182) provide us with an idea of the variation of these compositions as dependent on temperature, which in its turn is determined by the pressure. It is very important to mention that after crystallizing, the liquid is entirely "boiled out" during retrograde boiling, and the obtained gaseous phase, which predominantly consists of the volatile component ***A***, represents a gaseous solution (vapor) saturated in component ***B***, being hence in thermodynamic equilibrium with crystals of ***B***.

The process of cooling and crystallization of magma at greater and lesser depths with the restrictions mentioned further on (see page 335) illustrates the considered section of the diagram at constant pressure ***P***. The process of magmatic eruption is primarily connected with a decrease in pressure; this is the case of a section of a diagram at constant *T*.

[Figure 183 shows two similar isothermal sections. Figure 183, I, shows a section which is designated on Figure 182 by *I*, i. e., at a temperature exceeding that of the invariant point of the component *B*. Here the system which consists of a liquid and a melt (point *x*) on lowering of pressure initially passes through the area of the liquid phase, i. e., crystals
295 should melt, and then on further decrease of temperature the liquid should boil and the separation of the gaseous phase should occur.

Figure 183, II, shows a section at a lower temperature (see Figure 182, II). Here for a point x_1, where the system again consists of two phases — liquid and crystals, on decreasing the pressure the solution of a certain amount of crystals will again occur initially, but in the given system this process will not proceed to its end. On the phase segment *mr* boiling of the system occurs and the separation of the gaseous phase is accompanied here by rapid crystallization. After all the liquid has been exhausted, the solid phase will commence to crystallize from the gas (naturally this is only possible in a closed system).]

A SYSTEM IN WHICH CRITICAL PHENOMENA OCCUR IN SATURATED SOLUTIONS*

Although in all the previously considered systems the difference in the volatility of the two components was regarded as being considerable, this difference is actually not so great as to cause the appearance of critical phenomena in the area of separation of the solid phase. This case is considered here in order to provide a natural link between systems of this type of melts and those where volatile components have a critical temperature considerably lower than the temperature of melting of the nonvolatile component and the solubility of the nonvolatile component is considerably small, thus giving rise to the appearance of critical phenomena in the presence of the solid phase.

As far back as 1879 Hannay and Hogarth raised the question whether only liquids have the ability of dissolving solid substances or whether there is complete continuity in this respect. In the first case after the critical temperature has been attained, the solid substance should crystallize. In the second case a homogeneous system should also exist above the critical temperature. The investigations of the above-mentioned authors showed that a series of solid bodies are also held in solution in various solvents above the critical temperature. These investigations were not given much attention and not much other research was done in this direction due to the technical difficulties encountered.

On heating a saturated solution of a salt in a volatile solvent at elevated temperatures, the solution becomes unsaturated. On further elevation of temperature and with a correspondingly increased pressure within the vessel (tube) containing the solution, at a certain moment we reach a temperature of critical phenomena at which all the liquid is converted into gas. Initially the meniscus disappears and a cloud of vapor is formed which later also disappears, and the entire tube is filled with gas, but the dissolved substance does not separate out during this process. If the solution was colored (e. g. , an anhydrous solution of $CoCl_2$ in ethanol, possessing a blue color) the gaseous solution will preserve this same coloration.

Very interesting experiments were carried out by Greig, Merwin and Shepherd (1933). Samples of various rocks enveloped in platinum foil were heated in vacuo (about 0.002 mm Hg) in a fused silica glass tube at a temperature of 600—1,200°. In the atmosphere of volatile constituents included within the rocks, a transfer of silicic acid occurred in the gaseous state, which was precipitated in the colder parts of the tube as minute crystals of cristobalite (see the diagram of the modifications of SiO_2 on page 169). After the volatile constituents had been removed from the rocks by means of calcination, no further transfer of SiO_2 was observed. Later, several investigators made analogous observations of the transfer of certain refractory substances, and especially of silica, by water vapor.
296 Certain quantitative relations have even been obtained and will be considered further on.

The temperature of critical phenomena in such a solution is higher than the critical temperature of the pure solvent. The critical point is thus raised on the addition of the solute (in analogy to the boiling point). From the investigations of Centnerschwer (1903) it follows that this elevation complies with Raoult's law. If Δt is the elevation of the temperature, S — the amount of solid substance introduced into L grams of the solvent, M — the molecular weight of the solute and K — a constant coefficient, then

$$\Delta t = K \frac{S}{ML};$$

this formula is valid for dilute (unsaturated) solutions.

If one refers to a phase diagram of a binary system, one may draw a projection of the critical curve connecting temperatures and pressures on the projection of plane PT. In the cases already considered this curve lies above the saturation curve $S_B + L + G$ (see Figures 180, 181). If critical phenomena occur in saturated solutions, an intersection of the critical curve and curve $S_B + L + G$ will take place (see Figure 184). The points of intersection of these two curves Q and P determine the gap in the saturation curve $S_B + L + G$ and are projections of the critical points at which the liquid and vapor phases are identical.

The curves Pk_A and Qk_B of the three-dimensional diagram are critical curves which intersect the surface of saturation with respect to component B at points P and Q. The projections of the lines of the three-dimensional diagram on planes TX, PX and TP are shown on Figure 185. On the projection PT we note two points of discontinuity P and Q which simultaneously lie on the three-phase line $S_B + L + G$ and on the critical curve $L + G$. Thus they are critical points of a saturated solution.

At points of the curve $k_B k_A$, i. e. , at critical points of the system, the properties of the liquid and gaseous phases become identical. On Figure 181 the properties of liquid and gaseous phases at constant temperature and pressure are expressed by different points: m (liquid) and n (gas). At the intersection point of the curve $k_B k_A$ with the cylindrical surface of boiling, these points (m and n) should coincide. They will approach each other as the mentioned intersection is neared, and at point Q (see Figure 184), at the upper of the two possible points of intersection, they will meet one another and will coincide with this point Q. The same will also occur at the other, lower intersection point P of the curve k_Bk_A with the cylindrical surface. Thus, the curves m_1m_2 and n_1n_2, which are two independent curves on the given diagram (see Figure 184), will meet at points Q and P and a continuous belt — part of the cylindrical surface of boiling which is included between the curves corresponding to the composition of the liquid and the vapor (see Figure 181) — will be split into two separate closed parts as shown on Figure 184.

The area $L+G$ of separate existence of gas (vapor) and liquid will be closed frontally at the critical point Q and from above by the critical line Qk_B in the case shown on Figure 181, where the critical line passes beyond the cylindrical surface of boiling and we disregard the area of very high pressures through which it passes; on Figure 184 this closure of the area 297 $L+G$ falls within the limits of our diagram. The surfaces limiting this area laterally pass through the lines m_1m_2 and n_1n_2 and meet curve Qk_B, being fused here into one limiting surface which appears as if bent along the curve Qk_B, forming a type of fold. This curve is therefore also termed the curve of inflection points.

Points Q and P on the plane PT are projected as Q' and P', and the intersection of the line k_Bk_A with the cylindrical surface of boiling is projected as the intersection of the projection of the critical curve and the curve of the three-phase line. One may readily discern on the TX projection that at points P and Q the branch of the diagram corresponding to the composition of the vapor passes continuously into the liquid branch and that both branches have a common horizontal tangent at these points.

Let us now consider the sections of the diagram which correspond to definite pressures and temperatures. As previously, the locations of these sections are marked on the PTX-diagram (see Figure 185) by the numbers 1—4.

On considering the sections on the PTX-diagram perpendicular to the P axis, i. e., at constant pressures (Figure 186), one may note many features which are common to a binary system with a volatile component but lacking critical phenomena. We only note novel features in the area of the critical points P and Q. On comparing the sections of the considered diagram with those of cases lacking critical phenomena in saturated solutions (see Figure 179), we may easily compare analogous areas of state of the system and note the differences on the diagrams.

298 [At high pressures (see Figure 185, I) the section is essentially similar to that of a system lacking critical phenomena in the area of crystallization. In this case two critical points are shown on the section, at which the areas of coexistence of the liquid and gaseous phases is closed. However, a similar closure may also be observed at sufficiently high pressure in sections of the system shown on Figure 181. Crystallization of the solid phase occurs continuously without boiling up to the eutectic point.

When the pressure corresponds to the critical point Q we obtain the relations shown on Figure 185, II; areas $G+L$ and $F+S_B$ contact each other at point Q, and further on the area $L+S_B$ appears with the composition of the solution rich in B (see Figure 185, III). It is this type of isobaric section which constitutes the novel feature when compared with the sections of the system lacking critical phenomena in the area of crystallization. On the one hand we have characteristic retrograde boiling and on the other — the lower point of liquefaction is absent. The curve separating the area ***F*** from $F+S_B$ usually shows a characteristic bend which causes a certain temporary increase in the solubility S_B with decrease in temperature. This bend can be readily explained by the position of the ***PX***-diagram (see Figure 185, III). At point ***n*** we have a typical gaseous phase and at lower temperatures the solution already possesses all the properties of a liquid. However, the transition points at which liquefaction might have occurred, accompanied by a considerable increase in solubility (see Figure 183), are absent here. The change in properties occurs gradually and as the properties of the solution gradually approach those of the liquid, the solubility of component ***B*** increases, being expressed by the characteristic bend.

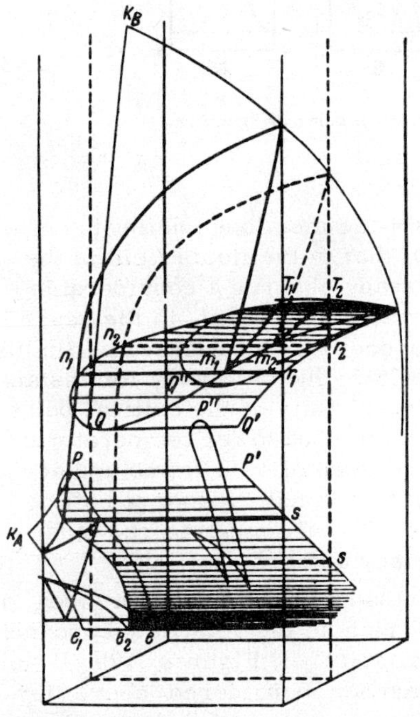

FIGURE 184. ***PTX***-diagram of a binary system in the presence of critical phenomena

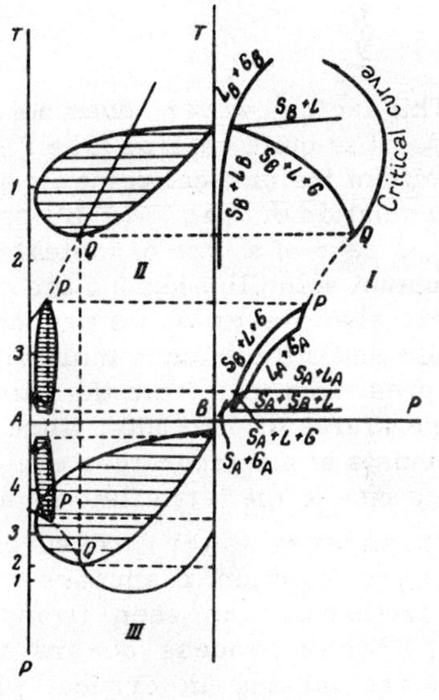

FIGURE 185. Projection of a diagram of a binary system with critical phenomena in the area of crystallization

The area of retrograde solubility (increase of solubility with decrease in temperature), which is located between the two bends of the solubility curve, differs quantitatively both from the gas and the liquid. Therefore, if one singles out the specific intermediate state by terming it a fluid (see page 325), one of the criteria for such a distinction is the property of retrograde solubility. It is true that in this case we are obliged to extend our fluid area somewhat below its lower critical pressure. In the vicinity of the critical point of isobaric sections (see Figure 186, IV; see also further on Figure 218) a corresponding bend will be noted on both the liquid and gaseous curves, i. e., the curve of the fluid state here appears to be dislocated at the boiling point.

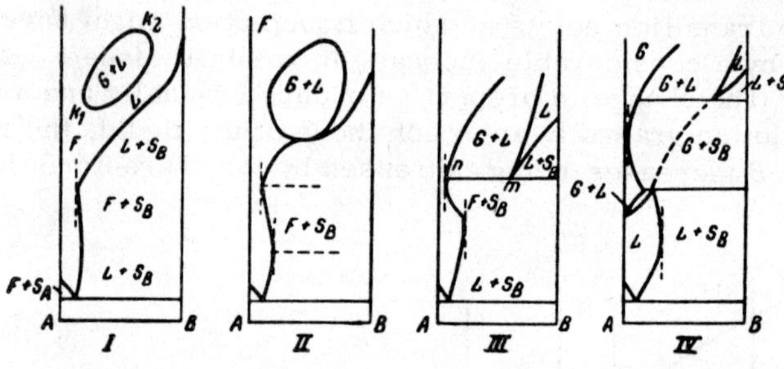

FIGURE 186. Isobaric sections of the diagram shown on Figure 185.

This fact, however, does not contradict the mentioned definition of a fluid. It is quite natural (see Figure 210) that in the liquid field in the vicinity of the critical point we should already observe a considerable accumulation of "gas" in the liquid and a residual "liquid" in the gas,
299 i. e., a case of a type of mutual solubility occurs, which passes into unlimited mutual solubility beyond the critical point. Such a liquid, for instance water, also preserves its properties in the vicinity of the critical point (boils) and differs much more from usual water at lower temperatures and pressures than from the same at a pressure of 1,000 atms and at temperatures of 400—500°, although the latter is not capable of boiling and exists at a temperature exceeding the critical temperature, thus giving rise to the term fluid as used by many authors.

At an even lower pressure (lower than the critical) an area of a second solution *L* appears which is rich in the volatile component, and further on, as seen from the similarity of Figures 179, II and 186, IV, the process occurs in the manner considered above for the case lacking in critical phenomena in saturated solutions].

Sections of the three-dimensional diagram of the considered binary system which are drawn perpendicular to the *T* axis are shown on Figure 187. Case I is analogous to that considered above (see Figure 183, II). For values intermediate between two critical temperatures (Figure 187, II) the diagram splits into only two fields — the unsaturated "fluid" solution and the coexisting solution and solid phase S_B. The *S*-shaped bend of the boundary line is also characteristic in this case.

On comparing the above-considered binary system, showing a discontinuity of the three-phase line due to its intersection with the critical curve, with the previously considered system lacking such a discontinuity, we note a great similarity. The general character of the diagram is similar in both cases. The principal difference between these systems lies in the fact that if critical phenomena occur in saturated solutions, an area of temperature is created, in which solution and vapor may coexist under certain known conditions and known pressure. A fluid phase is formed, having properties approximately similar to those of a liquid at high pressures and to those of a gas at low pressures, with gradual transitions under intermediate conditions. These solutions are much more mobile than a solution rich in ***B*** (which may be compared to a magmatic melt containing water). They possess those properties which many investigators have ascribed to the residual pegmatitic melt. This problem will be dealt with in greater detail further on

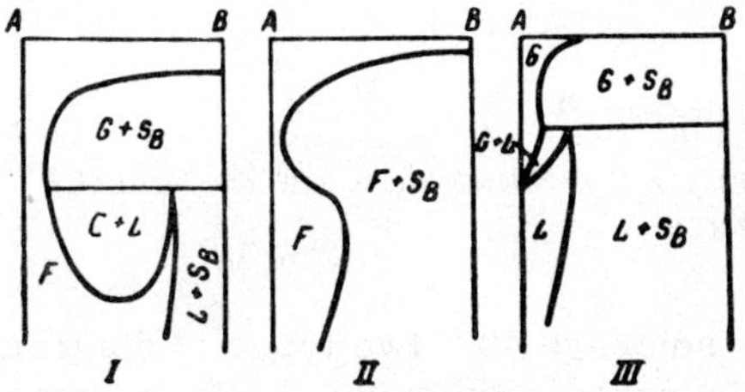

FIGURE 187. Isothermal sections of the diagram shown on Figure 185

[The presented scheme, in similarity to the diagram of the binary system with a volatile component lacking critical phenomena (see Figure 182), illustrates the general character of phenomena without concrete quantitative limitations both for critical points and for point ***M***.

The relation between the critical curve and the three-phase equilibrium curve depends on the solubility of the nonvolatile component in the volatile one. With a considerable solubility and its rapid increase with increase in temperature, critical phenomena cannot be observed in saturated solutions (see Figure 182). With an insignificant solubility of the nonvolatile component, which is only slowly increased on heating, we obtain a system of the type shown on Figure 185.

300 As is shown further on in the case of aqueous solutions of entire series of salts and silicates of potassium, critical phenomena are absent in the area of crystallization and P_{max} is rather small — of the order of a few hundred atmospheres The critical pressures for other systems with other salts, e. g. , the system H_2O-Na_2SO_4, also lie within similar limits. However, the first investigations of aqueous systems containing vapor-forming silicates already showed that here we are dealing with entirely different values. The solubility of silicates in water is insignificant and the lower critical point actually coincides with the critical point of water, whereas the upper critical point is not only not attained, but is not even approached at pressures as high as 5,000 atms. In this connection a qualitatively new type of diagram was proposed (Nikolaev, 1945), lacking an upper critical point. This diagram has given rise to much discussion. At present it has become evident that this type of diagram, which appears to be the limit

to which the system trends (see Figure 185) when the upper critical point shifts toward infinity, should at the same time be considered as a system with immiscible liquids in the areas of high pressures. This is considered in greater detail on page 307].

THE CASE WHEN THE VOLATILE COMPONENT FORMS A COMPOUND WITH THE NONVOLATILE COMPONENT**

As is known from previous mention (see page 49), two types of diagrams are possible in the area of melting of the considered case: 1) for a compound showing congruent melting and 2) for a compound showing incongruent melting. If the temperatures of melting and boiling of the components differ radically from one another, compounds with incongruent melting points will predominate, and we shall limit ourselves to the consideration of systems of this type. As is in the former systems, we accept for the boiling curve a type without a maximum and a minimum.

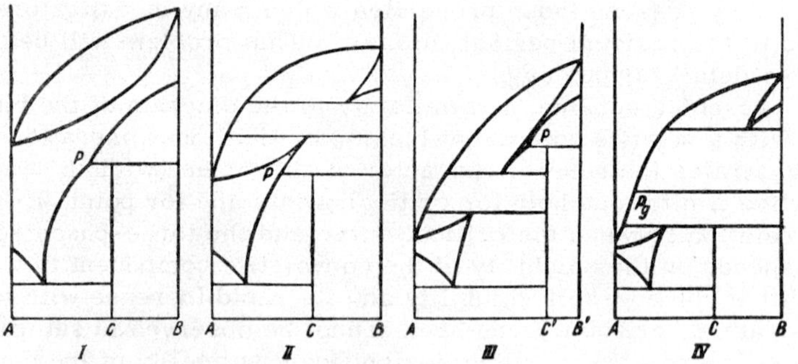

FIGURE 188. Typical isobaric sections of a diagram of a system in which the volatile component forms a compound with the nonvolatile component

Without touching upon the critical phenomena at this point, we shall compare the possible isobaric sections (Figure 188) of the mentioned type of system at high and low pressures. The pressure on Figure 188, I, was chosen in such a manner that the boiling curve lies above the melting curve, thus corresponding to diagrams considered in Parts I and II of this book. On Figure 188, II and III, isobaric sections are shown for lower pressures at which the intersection of the melting and boiling curves occurs. Two variants are possible: intersection either occurs above the peritectic point
301 (Figure 188, II), or below this point (Figure 188, III); on further lowering of pressure we obtain a section in both cases in which the peritectic point falls into the area of gaseous solution (Figure 188, IV).

On passing to the construction of the projection of equilibrium diagrams, we shall mention that in comparison with the *PTX*-diagram of a system lacking critical phenomena and compounds as considered above (see Figure

180), in this case we should encounter new types of three-phase curves: a liquid curve which is in equilibrium with the solid phases ***B*** and ***C***, and a gaseous curve in equilibrium with the solid phases ***B*** and ***C***. In both cases the equilibrium is of a reactional type as the formation and decomposition of compound ***C*** proceed according to the scheme: $B+L \rightleftarrows C$ or $B+G \rightleftarrows C$.

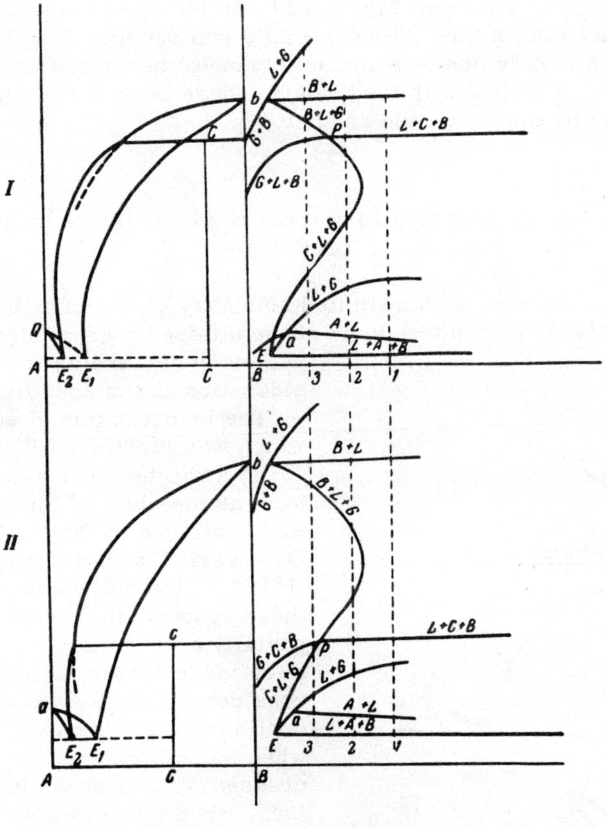

FIGURE 189. Projections of a diagram of a system in which the volatile component forms a compound with the nonvolatile compound

The first of the mentioned curves, which passes through the peritectic point (Figure 188, I), continues into the area of high pressures, and in similarity to the eutectic curve it is only slightly dependent on pressure, i. e., it is almost parallel to the P axis. The other curve passes into the area of low pressures and corresponds to point P_g on the section (Figure 188, IV). Both curves issue from invariant points which correspond to the four-phase equilibrium $S_B + S_C + L + G$. From the figures it is evident that pressure P of a corresponding point falls within the limits $P_2 > P > P_3$. On Figure 189 projections of the diagrams are presented which correspond

302 to two possible variants of sections at the pressure P_2: Figure 189, I, corresponds to Figure 188, II, and Figure 189, II — to Figure 188, III.

On considering systems in which critical phenomena do appear in the presence of a solid phase, three possible cases are noted. In the first two cases the invariant points will be located on three-phase curves and are real points; they may lie either above (first case) or below (second case) the critical point. The diagrams of such systems do not principally differ from those shown on Figure 189. In the third case the invariant points are located in the critical area PQ and become imaginary (Figure 190). There is only one equilibrium curve (of decomposition) of the solid phases C and B, either with the gaseous phase or with the liquid, both of which gradually approach one another.

THE CASE WHEN THE VOLATILE COMPONENT FORMS IMMISCIBLE LIQUIDS WITH THE NONVOLATILE COMPONENT**

Recently, systems with a limited solubility of the volatile component in the nonvolatile one in the liquid phase attained a great significance in petrology and are given wide consideration in the specific literature.

The investigation of such systems, which was started by Büchner (1906, 1918) and other investigators as far back as the start of the 20th century, was continued by Niggli (1946), I. A. Ostrovskii (1950) and Smith and Jones (1958). This obliges us to consider these systems in greater detail, especially as certain essential errors were made during their treatment. In considering these systems we shall confine ourselves only to the case when the solubility of the liquid increases with increase in temperature, i. e., when there is only one upper critical point of miscibility (see page 55).

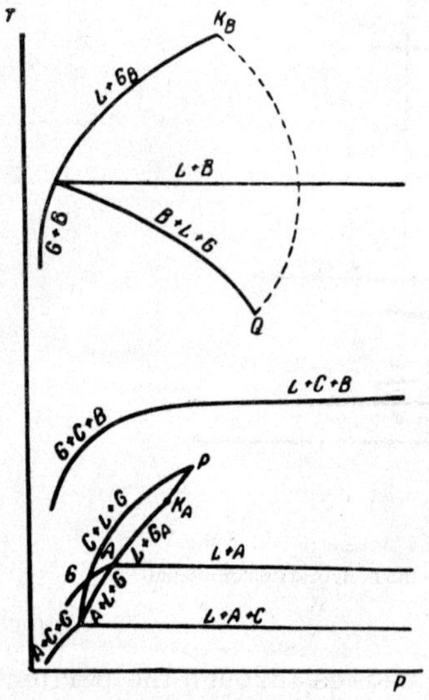

FIGURE 190. A binary system with a compound and critical phenomena. Point B+C+L+G is imaginary

As previously, we shall start by considering systems in which the critical gas-liquid curve does not intersect the three-phase curve, in our case the three-phase curve L_1+L_2+G. At high pressures the corresponding isobaric section (Figure 191, I) is almost identical to that shown on Figure 35, V (see page 58), the only difference being that in this case boiling curves appear above the area of decomposition of liquids. As the position of the decomposition curves of liquids is only slightly dependent on pressure, they will intersect the boiling curve on

decreasing the pressure. This proves that in the mentioned section only one case of three-phase equilibrium L_1+L_2+G (Figure 191, II) is possible, i.e., the first boiling point coincides with the critical point of decomposition of the liquid. On further decreasing the pressure, the line $G+L_1+L_2$ descends and finally intersects the line L_1+L_2+B at a point where four phases L_1, L_2, B, G exist in invariant equilibrium (Figure 192). At lower pressures the decomposition of the liquids becomes impossible, but the
303 crystallization of B from the boiling liquid phase becomes possible, i.e., equilibrium of the type $B+L+G$ (see Figure 191, III).

The projections of the corresponding three-dimensional diagrams are shown on Figure 192.

The location of the three-phase equilibrium curves and invariant points does not call for any special explanation. It should merely be mentioned that the pressure of invariant equilibrium may be rather high; it is not comparable to that of the univariant equilibrium of the type $A+B+L+G$, i.e., the given point may be of real interest to us.

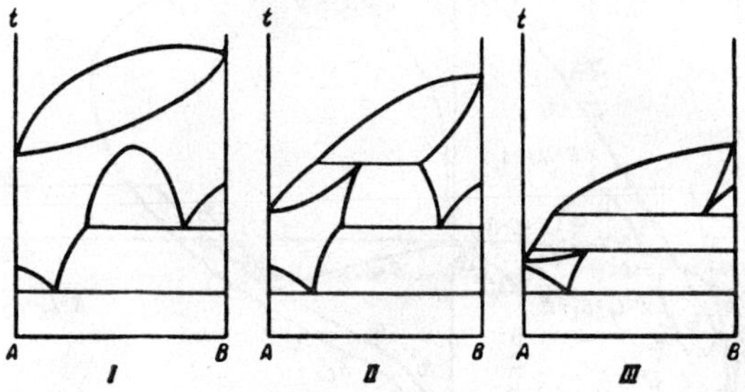

FIGURE 191. Main types of isobaric sections of a binary system with immiscible liquids

It should also be mentioned that on the diagram of the described system two critical curves are present. One of them, which was already encountered previously — the critical curve $L+G$, limits the area of coexistence of liquid and gas on the side of high pressures. According to the accepted condition, this curve does not contact the curves of three-phase equilibrium. The other critical curve — the curve L_1+L_2 — connects the corresponding critical points of isobaric sections at which the compositions of the immiscible liquids become identical. This curve limits the area of immiscible liquids in the field of high temperatures. It issues from point K from which the curve L_1+L_2+G originates, i.e., the gas is in equilibrium with a liquid of critical composition.

As all points on the liquidus-gas critical curve also lie simultaneously on the surface of boiling, it is quite evident that the mentioned curve cannot intersect the surface limiting the immiscible liquids, as at pressures higher than at point K the area of boiling lies above the area of decomposition of liquids. Correspondingly, at high pressures, we shall not obtain a discontinuity in the curve L_1+L_2 as was assumed by certain investigators, nor in the curve $B+L_1+L_2$, with the exception of those cases when the curve L_1+L_2 descends rapidly with decrease in pressure and intersects the curve

$B+L_1+L_2$. On considering the dependence of the solubility of the liquids on pressure, we should first of all mention that according to first approximations the solution occurs without a change in volume, and hence it should be only slightly dependent on pressure. A minor change in volume (either increase or decrease) should occur here, but compression seems to be the more probable (i. e., the volume should decrease somewhat during the formation of the solution), and the increase in pressure should hence facilitate the increase of natural solubility. The principle of mutual compensation of temperature and pressure is thus adhered to, and the critical curve
304 L_1+L_2 will ascend very gradually into the area of high pressures, as shown on Figures 192 and others.

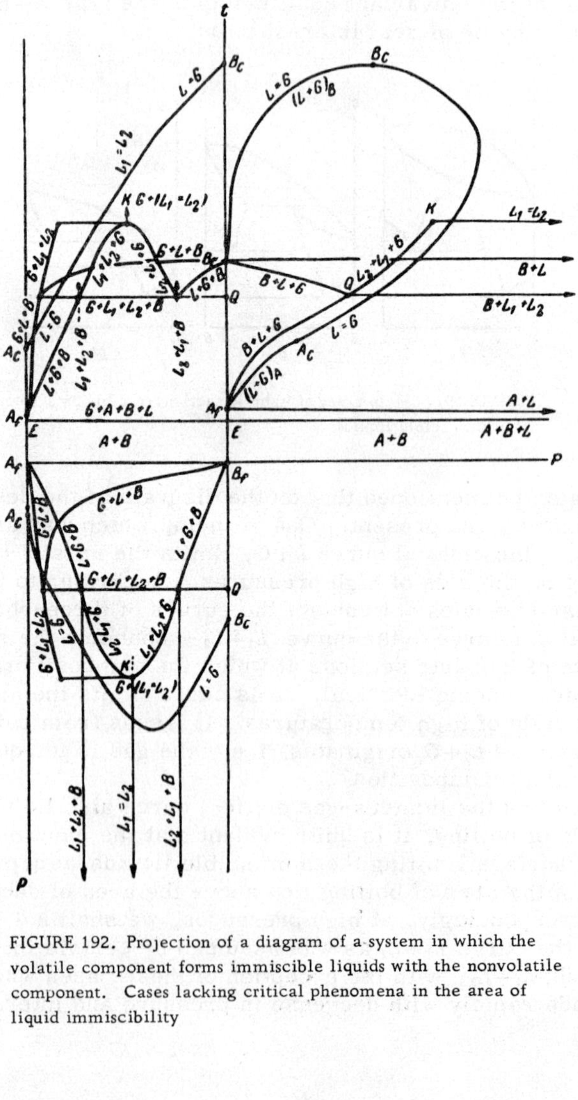

FIGURE 192. Projection of a diagram of a system in which the volatile component forms immiscible liquids with the nonvolatile component. Cases lacking critical phenomena in the area of liquid immiscibility

Let us now consider systems in which critical liquidus-gas phenomena exist in the presence of a third phase. Smith and Jones (1958) distinguished two main types for such diagrams. On Figure 193 a type of diagram is shown in which an intersection of the critical curve $L+G$ with the three-phase equilibrium curve L_1+L_2+G occurs in such a way that the invariant equilibrium $B+L_1+L_2+G$ remains valid and the upper critical point always lies in areas of greater pressure than point K_1 (see Figure 192) and becomes imaginary. Point K on Figure 193 is the lower critical point at which one of the liquids becomes identical to the gaseous state. The critical curve $L+G$ passes directly into the critical curve L_1+L_2. The most objective boundary between G and L_2 on the critical curve is the inflection point of this curve, provided that it exists.

305 At pressures higher than that of the lower critical point P, sections will appear in which the difference between the decomposition into two liquid phases and the boiling of one liquid disappears (Figure 194, I, II, III).

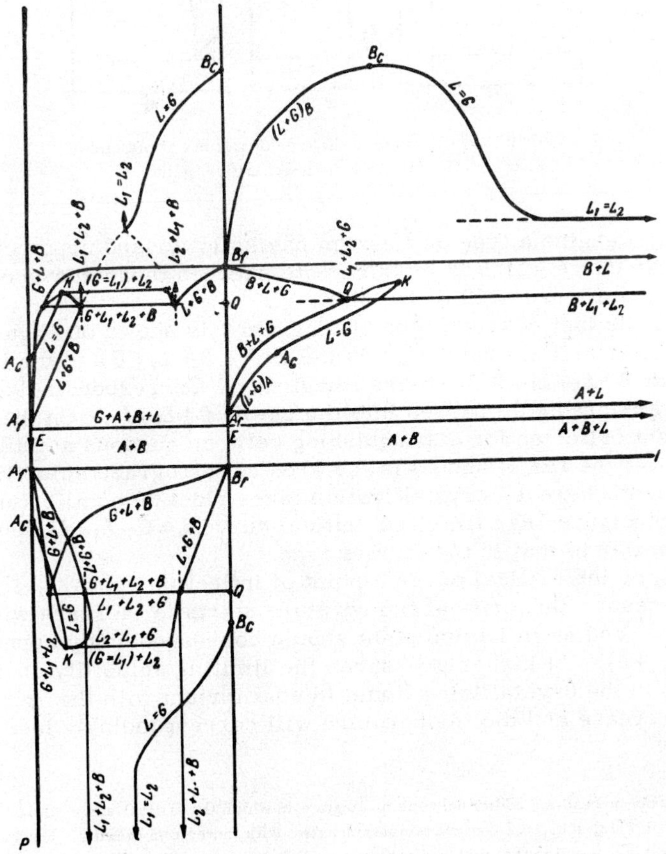

FIGURE 193. Projection of a diagram of a system in which the volatile component forms immiscible liquids with the non-volatile component. The case of critical phenomena in the area of liquid immiscibility and a real point of four-phase equilibrium $B+L_1+L_2+G$

At comparatively low pressures we note a gradual transition of true gas at high temperatures to a liquid at lower temperatures with a characteristic bend in the curve (retrograde solubility) for intermediate fluid solutions.

In the case of very high pressures the second phase will actually be a liquid at all temperatures.

It is interesting to mention that the important system H_2O-CO_2 is related to this type. Here water proves to be a comparatively nonvolatile component (component $\boldsymbol{B}$). The system is considered further on (see page 363).

306

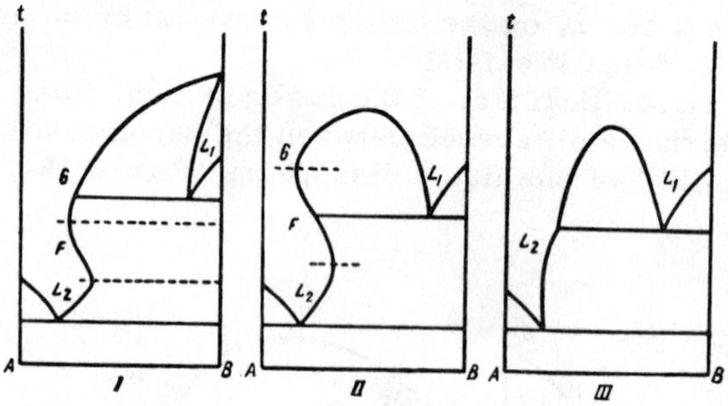

FIGURE 194. Changes in the type of sections at pressures higher than the value for the lower critical point

307 One may imagine a type of diagram having lower and upper critical points on the curve which is intermediate between those shown on the former diagrams (see Figures 192 and 193).

Finally, the last possible type of diagram † is shown on Figure 195. Here the lower critical point lies on the curve $\boldsymbol{B+L_2+G}$ and the invariant equilibrium $\boldsymbol{B+L_1+L_2+G}$ becomes imaginary. Correspondingly, the curve $\boldsymbol{B+L_1+G}$ very gradually passes into the curve $\boldsymbol{B+L_1+L_2}$. On the other hand an important criterion for distinguishing between gaseous and liquid solutions (see Figure 194, I and II) is the area of retrograde solubility which will be absent wherever crystallization takes place in equilibrium with two liquids (see Figure 194, II). The critical curve $\boldsymbol{L=G—L_1=L_2}$ behaves in a manner similar to that in the former type.

If $\boldsymbol{L_1=L_2}$ on the critical curve a point of inflection appears, i.e., above a certain pressure the critical temperature starts to increase with increase of pressure, and an inflection point should correspondingly appear on the curve $\boldsymbol{B+L_1+L_2}$. At higher pressures the limiting solubility of the volatile component in the crystallizing liquid in equilibrium with the "gaseous phase" ($\boldsymbol{L_2}$) will decrease and the temperature will correspondingly increase.

† Smith and Jones present one additional type of diagram in which the critical curve $G=L—L_1=L_2$ is bent sharply upward, i.e., the critical temperature rises with increase in pressure. However, the authors themselves mention that such a diagram is improbable. In a silicate system the checking of this area is practically impossible.

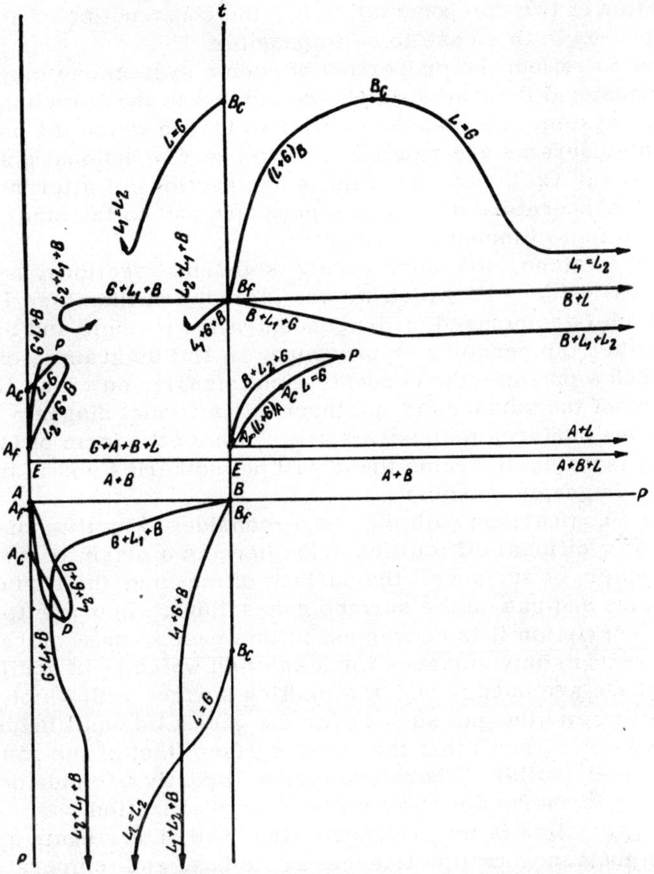

FIGURE 195. Projection of a diagram of a system in which the volatile component forms immiscible liquids with the nonvolatile component. The case of critical phenomena in the area of liquid immiscibility and an imaginary point of four-phase equilibrium $B+L_1+L_2+G$

As is shown further on, some very important systems are related to the mentioned type: r o c k - f o r m i n g   s i l i c a t e s — w a t e r (V. A. Nikolaev's diagram).

TERNARY SYSTEMS WITH VOLATILE COMPONENTS

General comments*

In constructing a complete diagram of a binary system we have made use of all three dimensions for the representation of temperature, pressure and concentration. For an analogous diagram of a ternary system a fourth dimension would be necessary (temperature, pressure and concentration of two components), i. e., the construction of a complete diagram proves in this case to be impossible.

In order to reflect the properties of such a system one may use several three-dimensional diagrams which are related to the complete diagram of the ternary system in a manner similar to that in which the above-considered planar diagrams are related to the three-dimensional diagram of a binary system. On the one hand these are sections of a ternary diagram at constant temperature or constant pressure and on the other — the projections of a four-dimensional diagram.

Isobaric sections, and more rarely isothermal sections, have the form of a trihedral prism with which we are already familiar (see Part I), the base of which is comprised of the concentration triangle and the height of which is either temperature or pressure. A flat diagram is often used in place of such a prism — the concentration triangle, on which lines of intersection of the surfaces of the three-dimensional diagram are projected and the surfaces themselves are shown in the form of isolines. In the case of isobaric diagrams these will be isotherms and in the case of isothermal diagrams — isobars.

From the comparison with the above-considered melting diagrams, we note that additional difficulties arise here as a prism possesses more than one system of surfaces: the surface of melting, the surfaces of boil-
308 ing — liquidus and gas on the surface gas-solidus. In order to obtain a complete description it is convenient in this case to use colored diagrams. Usually, however, only surfaces for a solution which is in equilibrium with the solid phase are used, i. e., the melting surface with which we are already familiar and the gas surface for the gas-solid equilibrium. It is here necessary to recall that the latter surface (that of the gas) is higher than the former (solid). Therefore they will partly coincide on the projection of the isotherm and the course of crystallization.

The ***TX*** projection is the one most often used, i. e., again a trihedral prism with the concentration triangle as its base and temperature as its height. The relation between the mentioned diagram and the isobaric section is the same as for binary diagrams, as is shown further on.

Confining ourselves to these general remarks, we shall pass to the consideration of the most simple ternary systems with volatile components. Here we shall consider two types of systems:

1. A system with one nonvolatile and two volatile components.
2. A system with two nonvolatile and one volatile component.

Firstly we shall consider systems without critical phenomena in crystallizing solutions and shall then pass onto more complicated systems with critical phenomena in the presence of solid phases.

Ternary systems with volatile components lacking critical phenomena in crystallizing solutions

One nonvolatile component and two volatile components*

During the consideration of binary systems we have seen that at high temperatures and relatively low pressures the separated vapor has the composition of the almost pure component A. If several volatile components are included in the system, the phenomena become more complex. Although the separated vapor forms one phase in this case, the composition of this phase is nevertheless not constant but variable. A process of distillation occurs. The composition of the vapor depends on the composition of the corresponding liquid and varies continuously during vaporization. The liquid remaining up to the time of separation of gases is the "residue" of distillation and the total of the separated substances constitutes the "distillate".

Both of them change during the process. The process of distillation may on its own be considered, as in the case of binary systems, either at constant pressure (isobaric distillation) or at constant temperature (isothermal distillation). The process of magmatic distillation, which is encountered in nature, differs essentially from that of distillation under laboratory conditions, where a liquid is distilled on constant heating. In magmatic systems, on the contrary, distillation is accompanied by a loss of heat — mainly during crystallization (retrograde distillation). Therer fore, these concepts should not be confused. When a solution crystallizes the liquid is enriched in the volatile component and the vapor pressure of the latter increases; if it exceeds the pressure of the environment the liquid starts to boil. A vapor of composition close to that of the pure volatile component is separated. Thus, a vapor is separated during crystallization in a manner similar to heating (boiling). We have already noted this in our consideration of binary systems.

We shall now consider a ternary system with volatile components A and A_1 and nonvolatile component B. We should bear in mind the isobaric diagrams of the systems $A+B$ and A_1+B. In each of these we have noted two boiling points — m and m_1 (Figure 196).

309 Let us now consider the concentration diagram of the system A_1AB (see Figure 196), assuming that the system is at constant pressure. The projections of points mm' and m_1m_1' are obtained on the sides of the triangle AB and A_1B as the corresponding points $m(t_1)$, $m'(t_1')$ $m_1(t_2)$ and $m_1'(t_2')$. Points $m(t_1)$ and $m_1(t_2)$ are the terminal points of the curve expressing the concentration of the ternary solution, saturated in component B, which boils at constant pressure. A definite temperature t corresponds to each point on this curve. In the most simple case $t_1>t>t_2$.

[The curves of the vapor composition in equilibrium with the liquid are constructed in a similar manner. These curves are very close to the AA_1 side of the triangle and may even coincide with it.] The composition of the vapor, which corresponds to any of the concentrations of the solution, is represented by a point on such a curve. Vapor $1'$ corresponds to the composition point 1 of the solution, and vapor $2'$ — with composition 2, etc.

Similarly, in the case of the concentration curve $m'(t_1')m_1'(t_2')$, which corresponds to the second, lower boiling point of the thermal solution, we note a correspondence between its points and definite points of vapor composition, e.g., a corresponds to a', b — to b', etc. Let us consider the course of distillation (see Figure 196).

Let the initial composition of the liquid be x. Upon cooling, component B will start to separate and the composition of the liquid changes in the direction $x1$ up to point 1. This point corresponds to a vapor of composition $1'$.

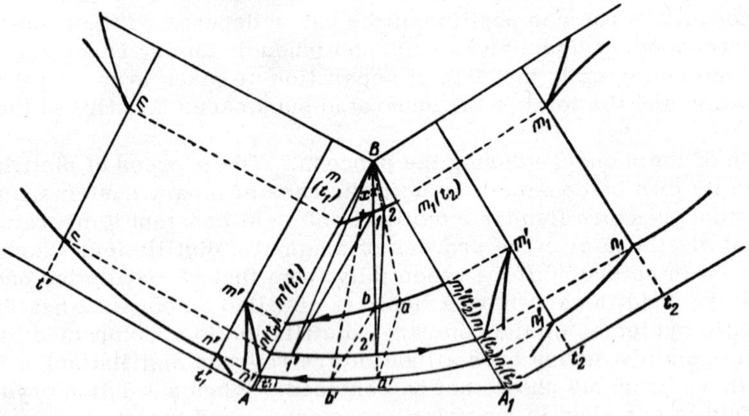

FIGURE 196. Isobaric section of a diagram of a ternary system with one nonvolatile and two volatile components (lacking critical phenomena)

We obtain a triangle $B11'$. On further cooling the composition of the liquid changes along curve $m(t_1)m_1(t_2)$ in the direction indicated by the arrow, and at the same time the composition of the vapor correspondingly changes along the curve $n(t_1)n_1(t_2)$ which usually almost coincides with the AA_1 side of the triangle. [On our figure the scale is markedly distorted.] The apices 1 and $1'$ of the phase triangle are displaced along these curves and the B apex remains constant. Point x naturally always falls within the phase triangle. When x finally reaches the side $B2'$ the liquid reaches composition 2, it will be entirely exhausted and the system will consist of solid $B+$ gas ($2'$). Further cooling and separation of B gives rise to the appearance of a liquid with a composition represented by a point upon the curve $m'(t_1)m_1'(t_2')$ (see Figure 196). This involves the liquefaction of pneumatolytic gases.

310 [If a considerable amount of component B is dissolved in the gaseous phase (shown to scale in our diagram), crystallization from a gaseous solution will take place on the segment $2'a'$ between the end of distillation and the beginning of liquefaction.

Liquefaction starts after the curve is reached. As seen from the sections of binary systems $t_2'>t_1'$ at this stage, i.e., the arrows on the liquefaction curves point in the opposite direction and the entire process moves in the opposite direction, i.e., the last portions of the liquifying gas are enriched in the more volatile component. Liquefaction is terminated after the Bb side of the phase triangle passes through the given point of composition of the system. After this, crystallization of a solid phase from a liquid solution will occur once again up to the eutectic curve, which is very close to the AA_1 side of the triangle and is not shown on the diagram. The course of crystallization of the projection coincides with the segment of crystallization of the gaseous phase (*2′a′*), which is located on a different level on the three-dimensional diagram.]

As is noted, during the separation of a gas, changing its composition from *1′* to *2′*, the ratio $A : A_1$ decreases in it.

Up to this point we have assumed that the gas which separated during the crystallization of component ***B*** is not removed and enters the composition of the system. The diagram changes somewhat in the case when the gas is removed (Figure 197).

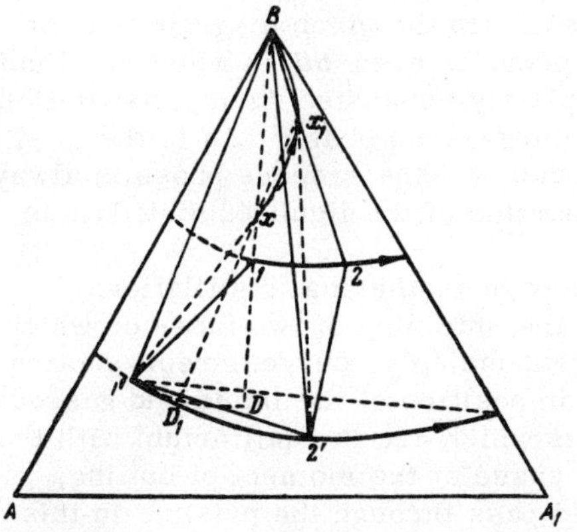

FIGURE 197. Course of distillation in an open ternary system (one nonvolatile, two volatiles)

Let the initial composition of the mixture be represented by ***x***. At the beginning of distillation the system consists of ***B*** and a liquid of composition ***1***; a gas of composition ***1′*** starts to separate. The curves ***1—2*** and ***1′—2′*** correspond to the curves $m(t_1)m_1(t_2)$ and $n(t_1)n_1(t_2)$ respectively on Figure 196.

Thus, at the start of the distillation process the phase triangle will be ***B11′***. If gas ***1′*** is removed, the composition of the system changes in the direction opposite that of ***x1′***, namely as shown by the arrow. The composition of the liquid phase changes at the same time. When, for instance, the liquid phase has the composition ***2*** a gaseous phase of composition ***2′*** may coexist with it. On the removal of gas ***2′*** the composition of the residue ***x***, which is obtained from ***x*** due to the removal of volatile components, changes in the direction opposite to x_12'. As the process of the removal of the gas proceeds continuously, the composition of the residue must change along the curve xx_1B. Should all the distillates of this fractional distillation be collected, their overall composition would be represented by a point D_1, which naturally lies within the segment ***1′—2′***. The overall residue at this moment would be located on the extention of D_1x at the point x_1 where, as we have seen, $2'x_1$ contacts the curve of the residue xx_1B. The overall final distillate ***D*** should lie upon the straight line ***Bx*** if solid ***B*** is the only product remaining at the end of the process.

Thus, during the considered process the compositions of both the liquid and the gas and the general compositions of the residue and distillate change continuously. We may distinguish the following curves on the diagram: liquid compositions ***1—2***, vapor composition ***1′—2′***, distillate composition $1'D_1D$, and general residue composition xx_1B.

311 In the considered system with two volatiles, under an external pressure, the possible separation of one of the volatiles may take place, whereas the other will be held in solution by this pressure, i.e., for the pair ***AB*** there will be a diagram as shown on Figure 179, II, and for the pair A_1B — a diagram of the type shown on Figure 179, I. On the ***AB*** side of

the concentration triangle AA_1B two points are noted — m and m', which correspond to the two boiling points. Concentration curves issue from these two points: of magmatic and hydrothermal solutions respectively (Figure 198). For the pair of components A_1 and B we note no phenomena of boiling and therefore there are no points corresponding to m and m' on Figure 198 on the A_1B side of the triangle. The curves ml and $m'l$ meet at point l, not reaching side A_1B; point l is an inflection point. Mixtures with a relative concentration of volatiles $A_1 : A$ greater than that at point l at the given pressure do not boil at all. [In the given diagram it is of interest to mention the existence of a peculiar area Bll' in which the liquid phase (in a closed system) is not completely exhausted during distillation, and is followed by the liquefaction of the gaseous phase. As in the former case, a reverse liquefaction occurs — the gaseous phase is always enriched in the volatiles and the last portion of the liquified distillate is the most enriched in this component.

Figure 199 provides us with a concept of isothermal distillation.

On this diagram one may also note the unfolding of two faces on which isothermal sections are shown. Correspondingly, curves, representing the composition of the liquid and gaseous phases which are in equilibrium with the solid phase at the moment of boiling, should pass through the prism. In this case, as opposed to binary systems, the pressure will drop (as shown by the arrow), being accompanied by the accumulation of the volatile component.

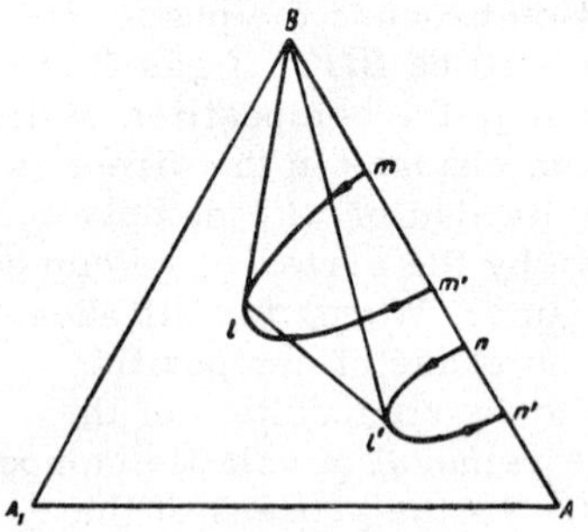

312 FIGURE 198. A second variant of an isobaric section of a diagram of a ternary system with one nonvolatile and two volatile components (lacking critical phenomena)

The course of distillation is shown for the indicated point x. As we have noted from binary sections at high pressures, the system consists of solid and liquid phases. When pressure is lowered and reaches a certain value P_x, which corresponds to the given phase triangle (where usually $P_1 > P_x > P_2$), the formation of the gaseous phase usually starts. As in the case of isobaric distillation, the gaseous phase is continuously enriched in the more volatile component A, and the phase triangle is displaced towards A_1B. In a closed system distillation terminates the moment when the side B of the triangle — gas — passes through the given composition point. On further lowering of pressure, the separation of the solid phase B occurs from the gas (equilibrium solidus — gas).

In the above-considered cases the pressure of the vapor separated from the melt changes continuously to one side with the change in the ratio of concentrations $A : A_1$. Cases may arise when the ternary system A—A_1—B has either a minimum or maximum vapor pressure for a certain intermediate concentration ratio $A : A_1$. Obviously, distillation trends towards the lowering of the vapor pressure. The location of the minimum pressure always determines the direction of this process.

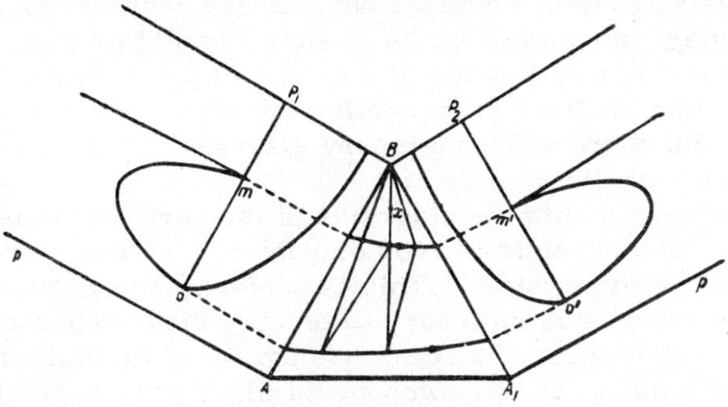

FIGURE 199. Isothermal section of a diagram of a ternary system with one nonvolatile and two volatile components (lacking critical phenomena)

As in the case of isobaric distillation in an open system, i. e., with the removal of gas, the process will not terminate as is shown on Figure 199, but will proceed up to the separation of the gas, which consists of component ***A*** only.]

Two nonvolatile components and one volatile component*

We shall confine ourselves to the consideration of the case where the components do not form compounds. In this case, the nonvolatile components may either form a series of solid solutions, or are incapable of mixing in the crystalline state and yield a eutectic scheme of crystallization.

We shall dwell in greater detail on the system with a eutectic and shall only briefly touch on certain particular systems with solid solutions.

The construction of diagrams of systems with compounds does not entail any special difficulties. They will be directly considered by us further on during the consideration of separate concrete systems.

It is obvious that of the two possible types of diagrams — isobaric, which represents the course of a process at constant pressure and decreasing temperature, and isothermal, which describes the influence of decreasing pressure at constant temperature—the former is the more important. It is necessary to construct a diagram of isobaric cooling of a ternary system in which one component is volatile. After the above statements, this construction does not present any particular difficulties.

If among the three components ***A*** is volatile and ***B*** and ***C*** are nonvolatile, we may construct an isobaric ***TX*** diagram in the usual manner by plotting temperature on a plane perpendicular to the concentration triangle ***ABC***. We are already familiar with the shape of the binary diagrams ***AB*** and ***AC***. This shape was strictly derived from the relations of thermodynamic potentials. Of special interest is the case when a gap occurs in the system during the processes accompanying cooling, i. e., when the process is naturally subdivided into stages which differ from one another by the conditions of crystallization, and does not proceed continuously. In this case the binary diagram has a form as is shown on Figure 179, II. Diagrams constructed on both sides of a trihedral prism (*AB* and *AC*) will be of such a type in a diagram representing the states of a ternary system. On the third face *BC* there will be a binary diagram of two nonvolatile components with a eutectic.

313 Being thus acquainted with the diagrams of the parts of the ternary system, each of which consists of two components, we may now pass on to a diagram of a ternary system. This is performed in a manner similar to that in which we passed from binary to ternary diagrams of crystallization when deriving the ternary eutectic or diagram of crystallization of a system with solid solution of two components which are immiscible in the solid state with the third component, etc.

After constructing binary diagrams on the faces of the prism, we note that for each specific point on these binary diagrams there is a corresponding line, which issues from this point in the ternary diagram. For each line of the partial binary diagram there is a corresponding surface which intersects the face of the prism along the line of the corresponding partial binary diagram. The surfaces intersect along lines which meet at points. If the number of components is increased by one, the number of degrees of freedom is also increased by unity. The invariant system becomes univariant, and the inversion which occurred in the binary system at constant pressure is accompanied by a decrease in temperature in the ternary system.

On Figure 200 this construction is performed: T_B and T_C — are the melting temperatures of components *B* and *C* respectively; T_E — the temperature of their eutectic; T_BT_E and T_CT_E — the crystallization curves of *B* and *C* in the binary dry system, and T_BM_B and T_CM_C — the crystallization curves in the binary systems *B—A* and *C—A*.

Surfaces of incipient crystallization corresponding to *B* and *C* in the ternary system pass through lines T_BT_E and T_BM_B and T_CT_E and T_CM_C respectively; T_EM_E is a cotectic line of simultaneous crystallization of *B* and *C*. Surfaces intersecting along line M_EN_E also pass through M_BN_B and M_CN_C. M_EN_E has the same relation to the eutectic of *B* and *C* as has the line M_BN_B to *B* and M_CN_C to *C*. M_E determines the composition of the liquid phase and N_E that of the corresponding gaseous phase in the system, where the simultaneous eutectic crystallization of *B* and *C* occurs.

The surface $T_CT_EM_CM_E$ is the surface of saturation of the melt with respect to component *C*; $T_BT_EM_BM_E$ — the surface of saturation of the melt with respect to component *B*, which crystallizes from such melts on decreasing the temperature. The surfaces $M_BN_BM_EN_E$ and $M_CN_CM_EN_E$ are surfaces of separation of the gaseous phase from the melt during the crystallization of *B* and *C* respectively.

The entire shaded surface on the figure has the form of an inverted four-sided "roof" (with curved edges). Its edges are projected on the plane of the concentration triangle, as shown on Figure 200. Figure 201 shows this projection separately, and one may follow the course of crystallization and separation of the gaseous phase in various cases. Let us consider two of these cases.

Coposition x_1. Cooling after the attainment of the surface of saturation $T_E T_B M_B M_E$ leads to the separation of component B. The composition of the liquid changes in the direction $x_1 l_1$ up to the eutectic line EM_E; subsequently B and C crystallize simultaneously. The liquid which had started to become enriched in the volatile A during the crystallization of B now becomes even further enriched in this component and reaches a composition M_E. Here a gaseous phase of composition N_E starts to separate. An invariant inversion takes ·lace and the liquid M disappears, being replaced by the solid phases B and C and a gas of composition N. When the liquid is ex-
314 hausted B, C, and the gaseous phase remain in equilibrium. The solid phases B and C crystallize from this gaseous phase on further cooling.

Composition x_2. Cooling leads to saturation with respect to component B and its subsequent separation, the composition of the liquid changing from x_2 to l_2; here a gaseous phase of composition g starts to separate. On further cooling the separation of B is accompanied by a simultaneous change in composition of both the liquid and the gas (vapor), being shifted in the direction indicated by the arrows. When they reach points m_E and n_E crystals of C appear and a simultaneous crystallization of B and C and the separation of a gaseous phase N occur; when the liquid is exhausted, the gaseous phase N and solids B and C remain in equilibrium. The system reaches the same final state as in the first case (composition x_1), but follows a somewhat different path.

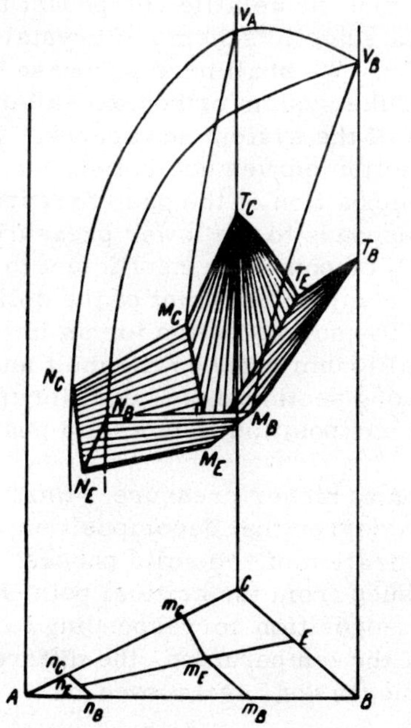

FIGURE 200. A three-dimensional diagram of an isobaric section of a ternary system. One volatile and two nonvolatile components with eutectic crystallization

As shown above, the process proceeds equally in systems both lacking and possessing critical phenomena. The only prerequisite is that the pressure should not be so high as to impede the formation of a separate gaseous phase from the liquid, i. e., if it is lower than the critical pressure at which critical phenomena take place, or lower than the maximum pressure of the gaseous phase reached in systems lacking in critical phenomena.

[On the above-considered diagram (see Figure 201) only phenomena of the formation of the gaseous phase are shown. On further decrease in temperature liquefaction will follow the crystallization of the gaseous
315 phase, as may also be shown on a ternary diagram (Figure 202). Here again a four-phase equilibrium will be attained: solid C, solid B, liquid and gas. It should be mentioned that the temperature at this point will be above the temperatures of liquefaction on the binary diagrams AC and BC, as indicated by the arrows. After liequefaction, crystallization will again take place from the liquid phase along the cotectic line up to the ternary eutectic point, which naturally is of no practical application to magmatic systems.

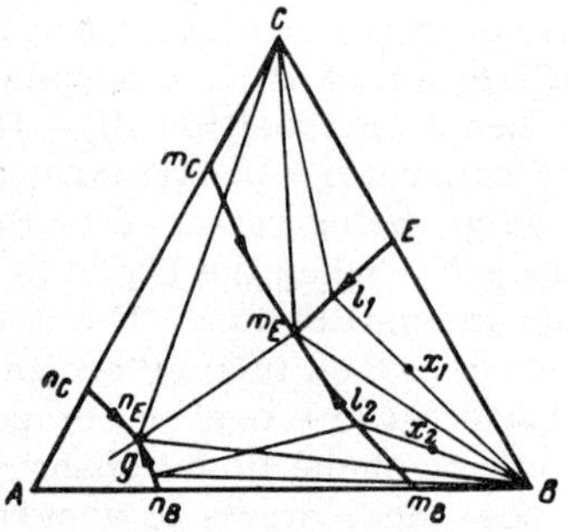

FIGURE 201. Projection of the diagram shown in Figure 200 on the concentration triangle and the crystallization scheme (lacking phenomena of liquefaction)

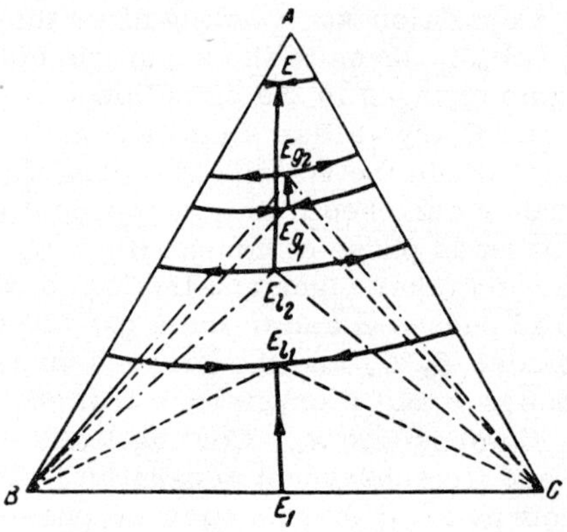

FIGURE 202. Complete projection of the diagram shown on Figure 193 on the concentration triangle

We shall consider a particular case as an example of a system in which nonvolatile components form solid solutions. Components *B* and *C* form solid solutions with a minimum. On addition of the volatile component the decomposition of solid solutions starts and a eutectic scheme of crystallization is established (Roozeboom's Type V). We shall have recourse to this example later on in our consideration of the system orthoclase-albite-water. Figure 203 shows isobaric sections of the system at various pressures on the projection plane. Liquefaction curves are not shown for the simplification of the drawing, and the composition of the gas corresponds to pure component *A*. Figure 203, I, corresponds to the lower pressure. Here the maximum solubility of the volatile component is insufficient to lower the temperature of melting below the composition point of the solid solutions. The course of crystallization of the solid solution for an initial composition d_1 is shown on the figure. Equilibrium between the melt and the solid phase occurs along section d_1d_2, and along section d_2d_3 — equilibrium between the melt, solid phase and gas; the composition of the solid phase changes correspondingly (b_1, b_2, b_3).

Figure 203, II, shows an isobaric section at higher pressure. Here the temperature of melting is lowered to such a degree that decomposition occurs and a curve of simultaneous crystallization of two solid phases appears within the triangle. This curve issues from the critical point *K*, where both solid phases have an identical composition corresponding to the minimum point. On further lowering of the temperature, the difference in composition of the separating phases increases, as is seen on drawing conodes.]

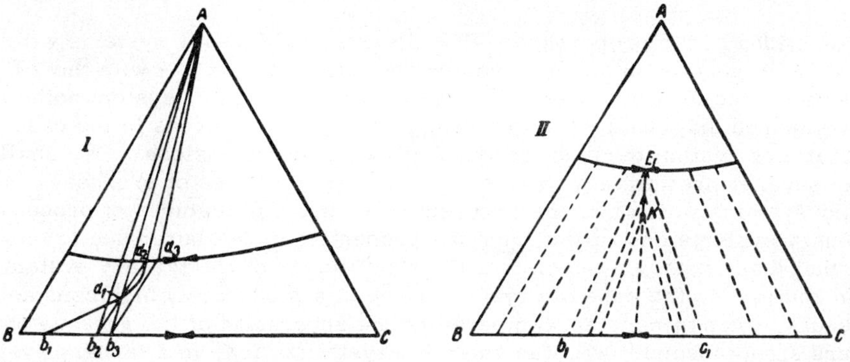

FIGURE 203. A ternary system in which the nonvolatile components form a continuous series of solid solutions with a minimum and limited solid solution with a considerable content of the volatile component: I — isobaric section at lower pressure; II — isobaric section at higher pressure

316 ## Critical phenomena in ternary systems*

In the presence of critical phenomena inversions, analogous to those shown on Figure 184 (see page 298), occur. For a binary system this diagram is three-dimensional. As already mentioned, such a visual diagram for a ternary system (taking into account both temperature and pressure) cannot be shown in three dimensions. A fourth dimension would be required in order to construct such a diagram. One may only present a projection in three dimensions, as is usually the procedure, e. g. , the ***TX*** projection of the four-dimensional diagram, analogous to the projection of a three-dimensional diagram on a plane (see Figure 184), showing only some of the features of the three-dimensional diagram. Such a projection cannot be used directly for the consideration of processes occurring during the cooling of a system, as we use three-dimensional diagrams for binary systems.

[One is obliged each time to construct special isobaric sections already considered for other ternary systems with critical phenomena.]

In this case especially one should avoid errors similar to those resulting from the use of the diagram on Figure 180, ignoring pressure in binary systems (see note on page 289). In ternary systems it is easier to make such errors and much more difficult to discover them.

[In describing critical phenomena, we will start with a system in which two nonvolatile components participate (with a eutectic scheme of crystallization). After considering it in its more general form we shall return to the system with two volatile components.]

Two nonvolatile components and one volatile component*

Some idea of the properties of ***PTX***-diagrams of ternary systems with one volatile may be obtained by comparing isobaric sections with the ***TX*** projection (see Figures 184—186). In the case of a binary system both of the compared diagrams were expressed on a plane, whereas in the case of a ternary system these diagrams will be three-dimensional. We shall make several preliminary comments on the peculiarities of processes in binary systems which become apparent on comparing features of processes in binary systems and follow from the properties of the latter due to the fact that these binary systems are the constituents of the ternary system.

In contrast to the case of crystallization at a moderately high constant pressure, when at constant temperature the separation of the gaseous phase occurs simultaneously with the eutectic crystallization, in a ternary system (the temperature of which decreases during the crystallization of both solid phases and the pressure of which increases) the compositions of both the liquid and gaseous phases change. If critical phenomena are present these compositions approach one another until they coincide and the difference between the liquid and gaseous phases disappears. They are fused into one "fluid" phase at the critical point ***Q***, which is analogous to the same point of a binary system. The indications of the ever-approaching compositions of the liquid and gaseous phases and the continuous transition into a single supercritical "fluid" phase without a marked discontinuity of the process, may serve as proofs of the presence of critical phenomena.

The construction of the ***TX*** projection of a ternary system which presents a three-dimensional projection of a certain diagram — a figure with four dimensions — may be carried out in the same way as used in deriving the isobaric sections of this diagram for the ternary system, by issuing from isobaric sections of the diagram for the binary system.

317 Let us consider the concentration triangle ***ABC*** on which we shall describe the composition of the system and on the perpendiculars of which we shall plot the values of temperatures corresponding to individual points on the ***PTX*** diagram, which we project in the direction of the pressure axis in the ***TX*** plane. We have used this in the construction of planar diagrams — ***TX*** projections in binary systems. In the case of ternary systems we obtain a diagram in the form of a certain three-dimensional figure which is limited laterally by the faces of a prism on which we obtain the corresponding diagrams for the three binary systems ***A—B***, ***A—C***, and ***B—C***, (the constituents of the considered ternary system). From each point of such a planar binary diagram a corresponding line will issue into the ternary diagram: the line of the partial binary diagram will be the trace of the intersection of a certain surface of the ternary diagram with a corresponding surface of a diagram of a partial binary system. These surfaces will intersect along lines which meet at certain points. All these constructions may be carried out as in the case of isobaric sections†.

†[Composition lines of the liquid and gaseous phases are projected onto the concentration triangle, being in equilibrium, at various pressures, with two solids ***B*** and ***C***. At the intersection of these lines with the critical curves Q_1Q_2 and P_1P_2 we obtain the corresponding critical points Q_3 and P_3 in the ternary system, where critical phenomena occur in the presence of two solid phases ***B*** and ***C***. Naturally, the compositions of the liquid and gaseous phases become identical at these points.

In the absence of critcal phenomena the eutectic curves on the ***TX*** projection would continue without a gap up to the invariant point of five-phase equilibrium. On the considered isobaric section (see Figure 205) only separate parts of these curves are shown: for the liquid and gas related to the given pressure].

Figure 204 is a three-dimensional combination of a ***TX*** projection and one of the isobaric sections. The lines of the ***TX*** projection are shown as dark lines and those of the isobaric section — as thin lines. The section is now shown completely for the sake of simplicity of the diagram; it is presented separately on Figure 205. As on Figure 200, the broken line surface is one of crystallization and retrograde boiling. Individual points of isobaric sections are located on lines of the projection, and at the transition of one pressure to another, these points are shifted along lines of the ***TX*** projection; the form of the isobaric section changes correspondingly, preserving, within certain limits, the type of the section and the locations of the areas of states of the system and changing them beyond these limits.

After constructing several isobaric sections of the four-dimensional diagram in a three-dimensional framework, we could have located them, in the mentioned way, in relation to the three-dimensional ***TX*** diagram. This would have provided us with an idea of the four-dimensional diagram itself.

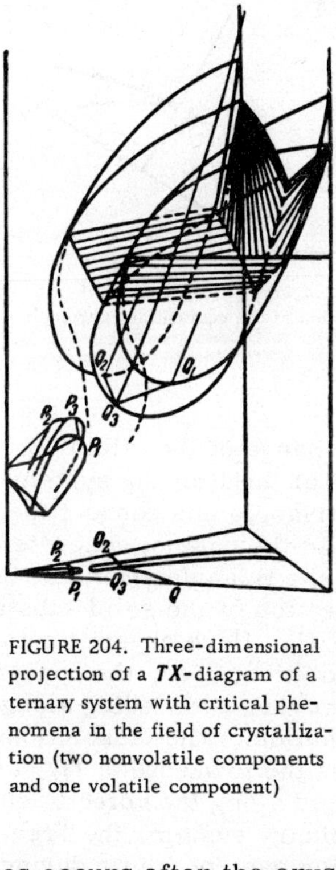

FIGURE 204. Three-dimensional projection of a ***TX***-diagram of a ternary system with critical phenomena in the field of crystallization (two nonvolatile components and one volatile component)

318 [Confining ourselves to the mentioned general considerations, we shall dwell in greater detail on the peculiarities of the above-mentioned isobaric sections. Let us first of all mention that pressure is chosen in such a way in this case that it falls in the interval between the pressures of the two critical points Q_3 and P_3. In such a section boiling may be observed at any ratio of the components ***B*** and ***C*** and no liquefaction will occur, i. e., the transition from the gaseous to the liquid state will be gradual on lowering the temperature. A projection of this section on triangle ***ABC*** is shown on Figure 205. It is self evident that the crystallization up to the boiling point and boiling itself in no way differ from the case shown on Figure 202, i. e., without crystal phenomena, and the change in the course of the process occurs further on. Whereas in the diagram on Figure 202 liquefaction accompanied by the solution of a part of the separated solid phases occurs after the crystallization of two solid phases ***B*** and ***C*** from a gaseous phase, in this case (see Figure 205) the transition to the liquid is gradual.

As we have already noted in the case of binary systems, this gradual liquefaction is reflected on the diagram by a corresponding inflection of the solubility curve. The separation of the solid from the gaseous phase stops at a certain point f and is replaced by its dissolution up to a new point of inflection of curvature f', where the separation of the solid phase starts again. We have already mentioned that these points may be considered as the conditional boundary of the fluid state, i. e., that transitional stage when the solubility of the solid phase in equilibrium with the liquid is not lowered on decreasing the temperature but, on the contrary, is increased. The dissolution of the solid phase along this segment replaces the dissolution at the point of liquefaction in the absence of critical phenomena.

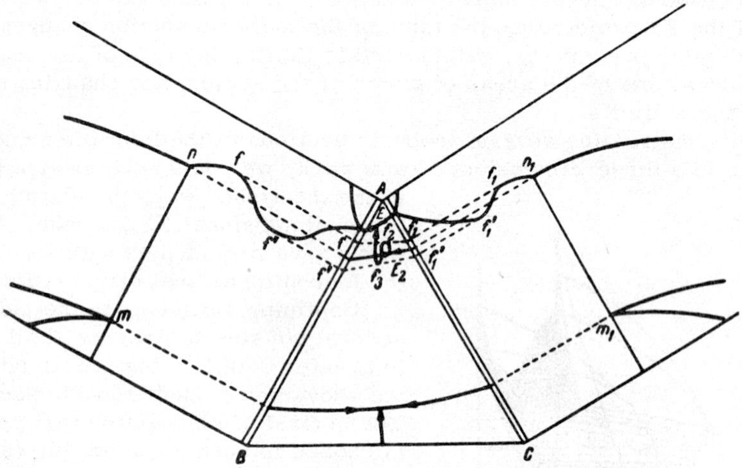

FIGURE 205. Unfolding and projection of an isobaric section of the concentration triangle shown on Figure 197

In order to orientate oneself in the path of change of the "fluid" phase of the ternary system, one may project the points of the limiting systems ***BA*** and ***CA*** onto the sides of the concentration triangle and connect them by means of broken lines within the triangle. In the case of concrete systems the exact location of these curves within the triangle should be deter-
319 mined experimentally. After boiling, the separation of the solid substance from a gaseous phase will occur up to the line ff_1. Here crystallization changes and is replaced by phenomena of dissolution, i. e., the concentration of the fluid will approach the side ***BC***, however, proceeding along a path somewhat different from that of crystallization. The dissolution of the solid phase changes when our point reaches the broken line $f'f_1'$, and here again crystallization proceeds from a true liquid along the cotectic curve f_3E up to the ternary eutectic point. As in a binary system, the "reversion" on the segment $E_2f_2f_3$ corresponds to the jumping over of points during liquefaction (see Figure 202).]

Systems with one nonvolatile component and two volatile components**

Without dwelling on the construction of the ***TX*** projection we shall only consider the most important cases of isobaric sections which correspond to the following types of sections of limiting binary systems:

a) both sections (**AB** and **A_1B**) are taken between the critical points (Figure 206);

b) section **AB** is taken between the critical points, and section **A_1B** has both a boiling point and a liquefaction point (Figure 207);

c) section **AB** is taken between the critical points and continuous crystallization without boiling occurs along section **A_1B** (Figure 208).

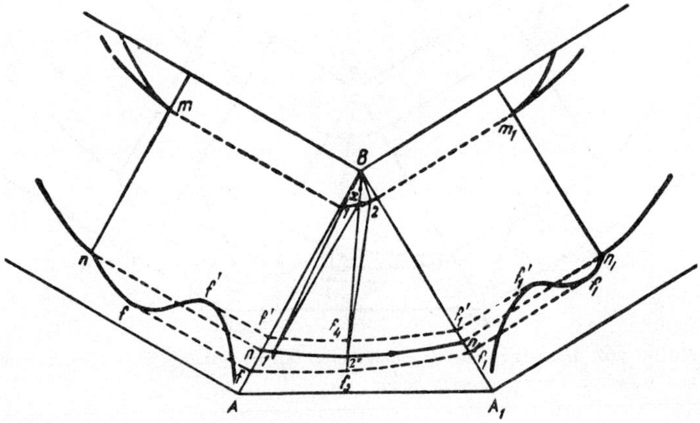

FIGURE 206. Isobaric section of a diagram of a ternary system with one nonvolatile and two volatile components (case a)

In the first of the mentioned cases (Figure 206) we note a boiling line within the triangle, as in the case shown on Figure 196, but the corresponding liquefaction line will be absent and the transition from gas to liquid will be gradual. In order to follow the course of the process in any system similar to the above-considered case, we should insert broken lines corresponding to the boundaries of a specific area of increased solubility of the fluid phase. These curves are drawn by the same method through the points $ff_1f'f_1'$ but no inflection point corresponding to cotectic crystallization will occur on them within the triangle.

Let us consider the course of crystallization in a system the composition of which is shown by x. Crystallization of the melt and boiling with a change in the composition of the gaseous phase does not differ in any way in this case from the course of the process without critical phenomena, as considered above (see Figure 196). Further on, when the entire liquid
320 passes into gas (point 2), we will first obtain crystallization from a gaseous phase, the composition of which changes along the straight line up to the broken line ff_1.

In this case crystallization ceases and is replaced by liquefaction. As more than two phases are in equilibrium in the given case and the composition of the solid phase remains constant, the "reversion" will occur along the same straight line up to the broken line $f'f_1'$ where dissolution is again replaced by crystallization from a liquid. The composition of the liquid phase will be displaced along the same straight line up to the incipient crystallization of one of the volatile components (which is, to all intents and purposes, of no practical significance) and is not shown on the diagram.

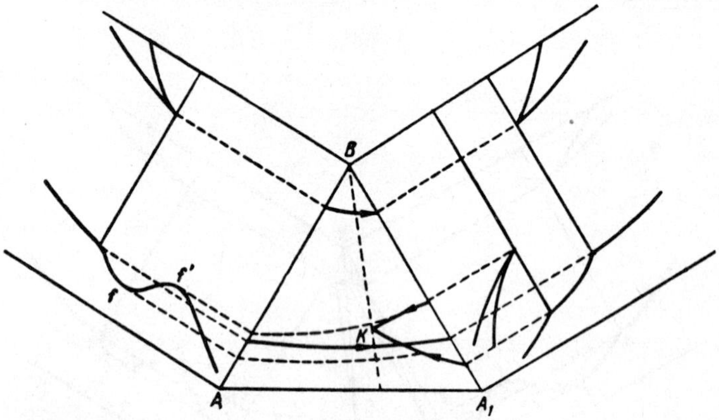

FIGURE 207. Isobaric section of a diagram of ternary system with one nonvolatile and two volatile components (case b)

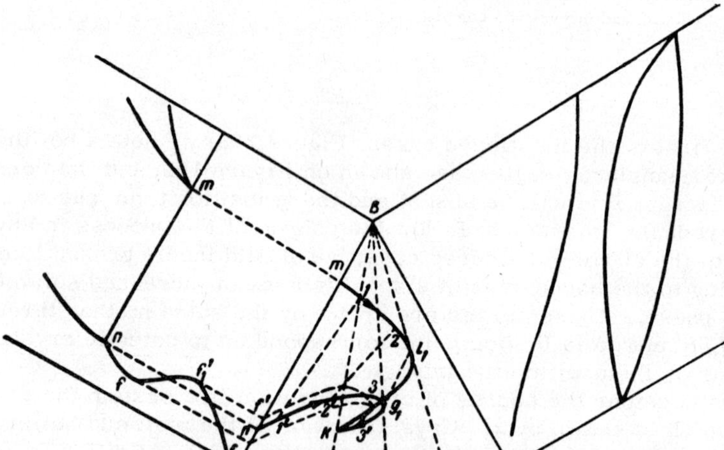

FIGURE 208. Isobaric section of a diagram of a ternary system with one nonvolatile and two volatile components (case c)

The reversion of the composition of the fluid phase along the same path (as in the case of Figure 206, in relation to the segment where the composition of the

321 crystallizing gaseous and liquid phases coincide) is obviously only apparent due to the nature of the projection. The temperature decreases continuously, and if one takes a section of a three-dimensional diagram located within the three-phase prism, a figure is obtained analogous to those shown on the faces of the prism for binary systems. As distinct from these, the given section is however not binary, because on boiling the compositions of the liquid and gaseous phases do not fall in the plane of the section.

In the second case, shown on Figure 207, the phenomenon of boiling throughout the field of the diagram once again does not differ from the previously considered diagrams and the case just mentioned. However, we encounter essential peculiarities further on. Whereas there is a point of liquefaction on side A_1B and two curves correspondingly diverge within the triangular diagram which correspond to the composition of the liquid and gaseous phases in equilibrium with the solid phase during liquefaction, no liquefaction point is located on side AB, i. e., these curves cannot reach side A_1B. The only possible variant in this case is the closure of the area of coexistence of the liquid and gaseous phases (in equilibrium with the solid phase) within the triangle, which occurs at the critical point K, where the compositions of the gas and the liquid become identical.

As a gradual transition from the liquid to the gaseous phase occurs in system AB, we may again note the points f and f' — corresponding to the conventional boundaries of the fluid phase — on this side of the triangle, and may insert broken lines in the triangle. The broken lines will fuse with the lines of liquid and gas which will correspond with the disappearance of the retrograde solubility in the corresponding section.

On drawing the straight line BK we divide the triangle into two fields which differ from one another in the behavior of the system during crystallization. Whereas all compositions lying between A_1B and BK will on cooling pass both the stages of retrograde boiling and liquefaction, the compositions whose points lie within the area ABK will only undergo boiling and subsequently a gradual transition from a gaseous phase to a liquid. In the first case the phenomena are analogous to those considered on Figure 196 (see Russian page 309), and in the second case — to Figure 206. We will not go into details here.

In the third case (see Figure 208) phenomena of retrograde boiling are only encountered on side AB and correspondingly the curve of the boiling crystallizing liquid and the curve of the gas in equilibrium with it pass through the rear of the triangle, and cannot reach the side of the triangle as here the system does not undergo boiling. On Figure 198 (see Russian page 311) we note that in this case, in the absence of critical phenomena, a reversion of curves takes place toward side AB, i.e., the curves of retrograde boiling pass into curves of liquefaction within the triangle. Such a change in course of the curves will also occur in our case, but they cannot reach the side AB, as there is no point of liquefaction in the system, and a gradual transition from the gaseous phase to the liquid takes place on lowering the temperature. Thus, these curves should again terminate within the triangle at point K which is the critical point, as in this case the compositions of the liquid and solid phases are identical.

The intersection of the liquid and gaseous curves is in this case also imaginary as actually the intersection point on the gaseous curve lies above the corresponding point on the liquid curve. Our diagram shows conodes

for the liquid and gaseous phases which are in mutual equilibrium, *1—1′*, *2—2′*, l_1—g_1, *3—3′*. The gaseous phase is always enriched in the most
322 volatile component (naturally, only if the boiling curve of system AA_1 has neither a minimum nor a maximum), and the liquid phase up to the critical point should always be enriched in the dissolved component†.

On drawing straight lines through point *B* and characteristic points of curves, one may divide the entire diagram into several areas. The following are the characteristic points:

1) point l_1, through which passes a tangent to the boiling liquid curve from point *B*. This is essentially an inflection point as beyond it the liquid starts to become enriched in the most volatile component;

2) point g_1, through which passes a tangent to the gaseous curve, which is in equilibrium with the liquid and solid phases, i. e., to the point of inflection of the gaseous curve. The corresponding composition of the gas is in this case in equilibrium with the liquid composition indicated by point l_1;

3) the critical point *K* where the liquid and gaseous curves meet.

The mentioned straight lines divide our concentration triangle into four areas. Compositions located within the area l_1BA_1 will undergo neither boiling nor liquefaction, i. e., the separation of the solid phase will continuously occur from the liquid, and boiling is only possible in unsaturated solutions. Retrograde boiling will take place in the area g_1Bl_1 during crystallization. As this boiling is not complete it will be replaced by the phenomenon of liquefaction in a manner similar to that on Figure 198. In the area KBg_1 both boiling and liquefaction will occur as noted in several other cases. Finally, in the area *ABK* retrograde boiling will occur but instead of liquefaction a gradual transition of the gaseous phase to the liquid occurs. In order to consider the course of crystallization, one should draw auxiliary broken lines which issue from points *ff′* on side *AB* and meet (in similarity to Figure 207) at point *K*.

The cited sections cover all possible combinations of ternary systems with two volatile components. We shall mention that V. A. Nikolaev attaches a special significance to such systems in which the binary system *B—A* corresponds to the type considered by him, lacking an upper critical point, and where the system B—A_1 lacks critical phenomena in the area of crystallization. The first of these corresponds to the equilibrium of a rock-forming silicate with water, whereas the second should characterize equilibrium between the rock-forming silicate and a water-soluble relatively (as compared to rock-forming silicates) volatile salt.

At pressures higher than that at point *M* in the system B—A_1 we will always encounter a section of the third type (see Figure 208), for which the character of the process is markedly dependent on the relation of the volatile components. At a lower content of component A_1 — retrograde boiling will take place followed by a gradual change from the gaseous to the liquid phase; at an intermediate content of — boiling and liquefaction — and at a high content of this component — gradual crystallization without phenomena of boiling. The latter case is scarcely possible in natural systems.

At a pressure lower than that at point *M* in the system B—A_1, but higher than that at the point *P* in the system *B—A*, we pass to a section of the

† It should be mentioned that the diagram of V. A. Nikolaev dealing with this area (1953, Figure 8, page 114 as well as 1947 page 71) is erroneously constructed, as the content of the nonvolatile component (Nikolaev's component *B*) is more enriched in the gaseous phase than in the liquid in the vicinity of the critical point.

second type (see Figure 207), where liquefaction is possible in the area of compositions enriched in component A_1. Finally, at pressures lower than at point P, we have a section lacking critical phenomena, which is of the type shown on Figure 196.

323 CHAPTER XIV

SOME SYSTEMS ILLUSTRATING THE EQUILIBRIUM BETWEEN SILICATES AND VOLATILE COMPONENTS**

THE PROPERTIES OF WATER AT HIGH TEMPERATURES AND PRESSURES

Before passing to the consideration of binary and ternary systems with the participation of water, it is necessary to consider the properties of water itself in areas of high temperatures and pressures.

In Part I we have already presented the general form of the diagram of a unary system with a boiling curve (see Figure 2). Figure 209 shows the boiling curve for water. As we are in this case not interested in the phenomena occurring at low temperatues and pressures, we shall start our curve from 0°, omitting the curves of melting and sublimation and only mentioning that the triple point of equilibrium ice + water + vapor corresponds to a temperature of 0.01° and a pressure of 0.006 atm. It should also be mentioned that at high pressures and low temperatures new polymorphous modifications of ice appear in the system, which are, however, only possible under artificial conditions and are in most cases so unstable that they immediately pass into the usual modifications on changing pressure.

324 As we already know, the boiling curves are terminated at a critical point, which for water corresponds to the temperature $t = 374.1°$ and a pressure $P = 218.4$ atm. At this point the properties of the gas and the liquid become identical, their specific volume being $V_k = 3.1$.

In chemical textbooks we often find the definition according to which the critical temperature is that temperature above which a gas may not be changed to a liquid irrespective of pressure. It should, however, be mentioned that this definition is incomplete and characterizes only one of the properties of the critical temperature. It is correct insofar as that above the critical temperature we will not encounter the curve of two-phase equilibrium on changing the pressure, and hence, we will not observe the liquefaction of gas, i. e., the formation of such a system in which liquid and gas may coexist. However, it in no way precludes the possibility of gradual transition of the gaseous state into a liquid and the existence of a true liquid above the critical temperature. This assumption may be easily proved by the rule of contraries. Actually, in similarity to the way in which we define critical temperature, we may define critical pressure as that pressure above which no amount of heating will change the liquid to a gas. If we consider this definition from one aspect only, we could arrive at the conclusion that a gas cannot exist above the critical pressure. This conclusion would be in direct contradiction to the analogous conclusion arrived at from the first definition.

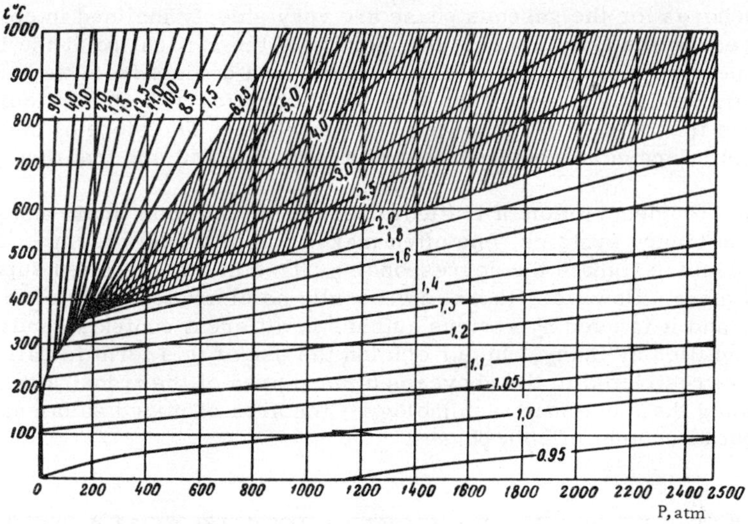

FIGURE 209. *PT* -diagram for water showing isochores (after Kalyuzhnyi). The area of fluids is shaded.

The existing confusion in literature is associated with the fact that no precise definition of the differences between liquid and gaseous states exists for unary systems. We can only readily distinguish liquid from gas when they exist together in a system of two phases. The absence of such a clear-cut distinction is not incidental, as the inversion of the typical liquid at low temperatures and pressures to a typical gas at high temperatures and low pressures may not only be obtained by boiling (on a two-phase curve) but also by bypassing the critical point with a gradual change in properties over the considerable interval of temperatures and pressures.

The incorrectness of the concept of the impossibility of a liquid state above the critical temperature has already been mentioned in the literature and Smith has considered it in great detail (1953). Considering the differences between the liquid and gaseous states with respect to the degree of dissociation of molecules, he justifiably compares the liquid and gas below the critical point with limited solutions (i. e., solution of gas in liquid and liquid in gas), which may coexist on the boundary curve. In the vicinity of the critical point their mutual solubility increases rapidly, and at this point itself the mutual solubility is unlimited. Such continuous solutions of gas in liquid are stable at higher temperatures or pressures, corresponding to a gradual change in the properties themselves.

Until we find an accurate method for determining the "concentration" of the gas and liquid, the best criterion for assessing the state of the system is its specific volume. Figures 209 and 210 show approximation diagrams for water at higher temperatures and pressures (after Smith) upon which isochores are plotted, i. e., lines of equal specific volume. The ideal liquid is considered as being incompressible, i. e., the isochores should be parallel to the pressure axis. In reality, the isochores of liquid water are inclined at a small, but nevertheless visible angle to the pressure axis, no inflections being observed at pressures higher than the critical.

The isochores for the gaseous phase are very steeply inclined in correspondence with the Gas Law.
325 In the vicinity of the critical point the isochores become visibly closer to one another and deviate from the curves, thus characterizing a rapid change in the properties of both the liquid and the gas in the vicinity of the critical temperature. The slope of these isochores is intermediate between that of a typical liquid and that of a typical gas.

During the description of critical phenomena (between liquid and gas) in unary and binary systems, one often makes use of the new term — "fluid", which should designate the corresponding phase in the so-called supercritical state. However, there is generally no clear definition of the term "fluid", and it is used by various authors in different contexts, and is completely omitted by many. In our opinion the use of this term is rational when it is restricted to the above-mentioned area of the transitional state, i. e., using the mentioned terminology — the area of mixed solutions with an essential content of both phases.

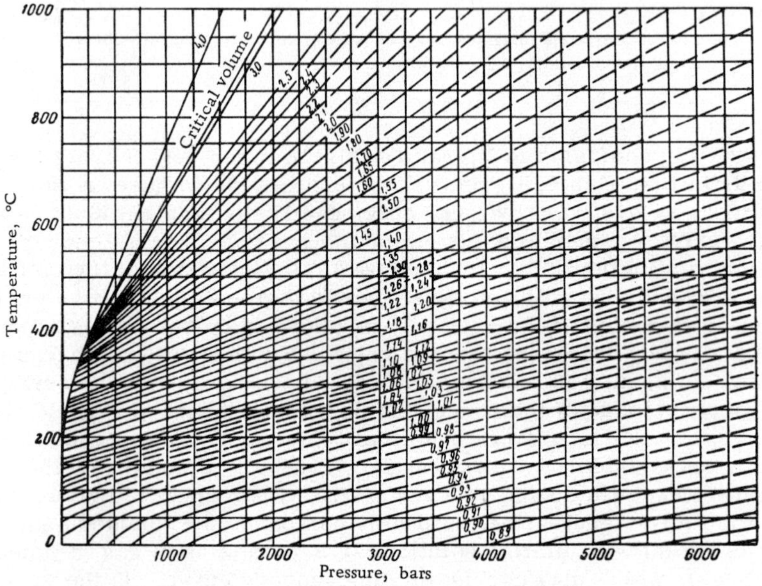

FIGURE 210. Isochores for the system H_2O at high temperatures and pressures (after Smith)

Thus the distinction between the liquid and gaseous states in the unary system may be approached in two ways. The first approximate way is that of the direct conventional division of the fields without introducing an area of a fluid phase. This boundary may be drawn along the isochore corresponding to the critical volume, which in turn corresponds to the character of the behavior of phases at a constant volume (see further on). As seen in Figure 210, on using such a division water may be considered as a liquid at a temperature of 900° if the pressure exceeds 1,750 atm, and at a 1,000 atm — up to 650°. The other way is to distinguish a certain area

of fluids which should begin even below the critical point and well into the area of high temperature and pressure — as is schematically shown in
326 Figure 209. The problem is now to find objective criteria for inserting the mentioned boundaries of the fluid area. Here the solubility in binary systems may be of certain assistance (see page 298).

In connection with the consideration of the properties of water we shall dwell upon one further question which is now of great significance to mineralogy, namely, the phenomena of phase inversions at constant volume. These phenomena may be observed in microscopic inclusions of liquid solutions in minerals, as the volume of these inclusions is practically constant due to the low coefficient of thermal expansion of the minerals as compared to the liquid. The simplest of such inclusions, being pure water at room temperature, consists of two phases: liquid and gas, i. e., the given system will correspond to a two-phase equilibrium of a certain point on the boundary curve. The vapor pressure will in this case be quite low (0.024 atm at 20°). On heating a mineral section in a special furnace placed under the microscope, we may observe the manner in which the gas bubble decreases as the liquid expands, i. e., a certain amount of the gas will pass into the liquid. Obviously the system will change along the two-phase curve up to the point where the bubble disappears.

Assuming that during the formation or filling of cracks of a mineral, a homogeneous solution is usually introduced into the inclusions, the temperature of the inclusion is considered to be not lower than that of homogenization. On further heating, the pressure within the inclusion will decrease rapidly along the corresponding isochore, and usually leads to a breaking up of the mineral section. In order to know the exact temperature at which the inclusion of the solution within the mineral occurred, it is necessary to know the pressure at this point. However, due to the small slope of the isochore the true temperature is usually not very different from the temperature of homogenization, provided that the pressure is not too high.

It is interesting to mention that on cooling such inclusions one may freeze the water in the presence of the gaseous phase, i. e., the point of invariant equilibrium may be reached. This may be of practical importance for determining the concentration of salts in solution according to the lowering of the freezing point.

If the gaseous bubble is very large and the total filling of the inclusion is below the critical value (i. e., the volume of the inclusion divided by the mass of water in the inclusion gives a value greater than 3.1), homogenization proceeds somewhat differently. As in the first case, on heating the gaseous bubble it initially decreases in size and subsequently starts to decrease rapidly in the vicinity of the homogenization temperature. Finally it disappears completely from the liquid. On further heating a decrease in pressure will also take place along the isochore. In this case, however, the isochores have very steep slopes and thus the temperature of capture of the inclusion may be much higher than the temperature of its homogenization.

Thus, by using the method of homogenization of the inclusion, one may determine not only the lower temperature limit at which this inclusion was filled, but also the phase condition of the solution. From the above-mentioned it is clear that the limit in this case will correspond to the isochore of the critical volume, i. e., the solution will become homogenized in the form of a liquid phase, provided that the specific volume of the solution is less than 3.1 — and in the form of a gaseous phase if the specific volume of the solution is greater than 3.1. The area of the fluids is not distinguishable here and a solution which was trapped at a temperature of 700° and a pressure of 1,500 atms, will yield a typical liquid phase on heating.

One should mention that the phenomena of homogenization of inclusions are usually complicated by the presence of other dissolved components
327 (mainly CO_2) and soluble salts. These phenomena are considered on page 364 as well as in other special handbooks and papers (e.g., E. I. Vul'chin, 1951; N. P. Ermakov, 1950; V. A. Kalyuzhnyi, 1955).

THE SIGNIFICANCE OF THE PARTIAL PRESSURE OF THE COMPONENT PARTICIPATING IN THE EQUILIBRIUM WHEN THE GASEOUS PHASE HAS A COMPLEX COMPOSITION

All the systems considered hereafter, containing water or other volatile components, are considered essentially as closed systems, in which the gaseous phase is represented by the same components as the system, i. e., mainly by the corresponding volatile component. In nature we encounter more complex phenomena, as the gaseous phase, which determines the total pressure of the system, is more complex and the considered volatile component has a comparatively low partial pressure as compared to its concentration in the general gaseous phase. Such phenomena are particularly encountered at the Earth surface during the evaporation of aqueous solutions.

We should mention that in this case a distinction should be made between phenomena of boiling and phenomena of evaporation. As any gas will tend to fill all available space irrespective of the presence of other gases, water will evaporate when the partial pressure of the atmospheric water vapor is lower than the corresponding pressure on the curve of two-phase equilibrium at a given temperature, or, as is usually stated, below the saturated vapor pressure of the space. The ratio of this partial pressure P_1 to the pressure of the two-phase equilibrium at the given temperature P_t, as expressed in percent, is termed the relative humidity. The lower the relative humidity the quicker is the rate of evaporation of water. When the relative humidity approaches 100 %, evaporation ceases and the dew point is reached, i. e., condensation of humidity from vapors in the atmosphere as is often observed during the rapid fall of temperature.

Evaporation of aqueous solutions containing salts will occur in a similar manner, but the vapor pressure of such systems is considerably lower than that of pure water and may thus explain the fact that, whereas pure water evaporates at a relative humidity less than 100 %, condensation accompanied by desalination is noted in saline basins, causing the great stability of salt lakes in arid regions. The condensation of water on the surfaces of crystals may commence in this very way. The crystals dissolve in the water, e. g., carnallite.

The considered evaporation phenomena occur only at the surface dividing the liquid and gaseous phases, i. e., in our case, the atmosphere. Therefore, evaporation is comparatively slow and becomes more rapid with the increase in size of the evaporation surface. If temperature is raised to

100°, at which the vapor pressure above the two-phase curve reaches atmospheric pressure, i. e., the total pressure of the gaseous mixture below which all points are located within the liquid phase, gas formation starts below the surface of the liquid. Such a phenomenon is termed boiling. In contrast to evaporation a relatively rapid transition from the liquid phase to the gaseous phase is observed in the case of boiling at constant pressure, if there is a sufficient flow of heat available for phase inversion.

Whereas the process of evaporation is reversible, i. e., both evaporation and condensation may be observed, the process of boiling in such a system is irreversible and may only become reversible in a closed system, where the entire gaseous phase is represented by water. In this case the boundary between evaporation and boiling disappears.

In all the systems considered hereafter, if the processes continue for a sufficient length of time, the corresponding equilibrium should theoretical-
328 ly be obtained, i. e., the reaction should be reversible and should only be dependent (in the presence of a complex gaseous phase) on the partial pressure of the volatile component participating in the given reaction. Moreover, thermodynamic calculations (Yoder, Ann. Rep., 1954; Nikolaev, 1957) show that in these cases the temperature of reaction should even decrease as compared to a simple closed system, i. e., a type of "squeezing" of the volatile component from the solid or liquid phase should occur due to the influence of total pressure.

However, the experimental investigation of these relations has proved very difficult until now and the experiments performed have yielded contradictory results. Thus, they showed that the stability of analcite (Yoder, Ann. Rep., 1954) in an atmosphere of an inert gas (argon) corresponds to the curve obtained for the same pressure of water vapor. An assumption was even made that the molecules of the inert gas may replace water in the lattice of analcite, but this assumption is rather improbable. Data are also available (holding for comparatively low pressures) which show a usual increase of temperature and dehydration of brucite and kaolin (Lebedev, 1958) in an atmosphere of an inert gas. On the other hand, the investigations of Harker (see page 370) on the compositions of carbonates appear to indicate that the obtained equilibrium curves lie between two theoretical variants (approaching the first): the curve as calculated from the partial pressure of carbonic acid, and the curve as calculated from the total pressure of the gaseous phase.

Despite such contradictory experimental results, one may be certain that under natural conditions equilibrial inversion is approached as a rule, i. e., in determining a point on the curve, we should only take into consideration the partial pressures of the volatile components participating in the reaction. This certainly holds true for reactions involving the absorption of volatile components, whereas in the reaction involving the removal of a volatile component (dehydration, decarbonization, etc), the process becomes rather sluggish on increasing total pressure.

In this connection the question may arise as to the possible fluctuations of partial pressure of water vapor during igneous and metamorphic mineral formation. Two points of view exist with regard to this problem: 1) that the mentioned reactions of mineral formation always occur with an excess of water, i. e., the partial water vapor pressure is always sufficiently high (Korzhinskii, 1940); 2) that reactions of mineral formation may occur with a deficiency of water (Yoder, 1952, 1957; Nikolaev, 1947), i.e., the partial water vapor pressure may decrease to almost zero.

According to the first concept, reactions involving the absorption and liberation of water depend mainly on temperature (as illustrated further on the dehydration curves ascend very gradually with increase of pressure), and the corresponding minerals and parageneses are the best geological thermometers.

According to the second concept anhydrous minerals crystallize at considerably lower temperatures in the deficiency of water and the corresponding reactions lose their significance as thermal indicators. According to the adherents to this theory, various metamorphic facies are formed not only when temperature is changed, but also when the amount of water is varied. The existence of series of minerals which are stable in the deficiency of water is also assumed, thus explaining the failure of many experiments (see further the mention of pyrope, page 353).

Without considering this important problem in detail, we shall only mention that, according to the author's opinion, mineralogical data appear to affirm the first point of view. However, during experimental investigations both in laboratory and of the coatings of metallurgical furnaces
329 mineral formation was observed to occur in solid systems without melting and without the presence of water. However, such reactions are only sufficiently intense at temperatures considerably higher than those of metamorphism — above 1,000° (usually above 1,300°). As an example we shall mention the well-known dehydration reaction of kaolinite, which occurs at a temperature of the order of 500° with the formation of amorphous products. Crystallization only starts at a temperature of 930°, accompanied by a characteristic exothermal effect; the size of the crystals is so small that they may scarcely be determined even with the use of X—rays. Crystallization of olivine starts in the same field of temperatures, resulting from the dehydration products of serpentine.

All examples of natural reactions which are claimed to change their direction depending on the excess or deficiency of water, are usually unconvincing, as in the case of the famous Goldschmidt Reaction for the formation of biotite which he presents in the following simplest form:

$$KAlSi_3O_8 + 3(Mg, Fe)\,SiO_3 + H_2O = K(Mg, Fe)_3[OH]_2\boxed{AlSi_3O_{10}} + 3SiO_2$$

Goldschmidt assumed that the existence of biotite-hypersthene granite and gneiss, together with hypersthene and biotite granite and gneiss which were formed under conditions of equal temperature, is a direct proof of the dependence of mineral formation on the water content. In the case of an excess of water the paragenesis of quartz, K-feldspar and biotite is stable, and in the absence of water — that of quartz, K-feldspar and hypersthene; with a deficiency of water all four minerals are stable together.

D. S. Korzhinskii has found an important error in the considerations of Goldschmidt, namely that in the paragenesis of biotite and hypersthene the ratio Mg : Fe is not the same in both minerals (biotite is richer in Fe); this may determine the character of the paragenesis. As we have noted previously, the mentioned relation, which follows directly from the Phase Rule, is an important general principle governing any paragenesis of several femic minerals.

It was also shown (see page 270) that the question of the equilibrium of biotite with K-feldspar and hypersthene is further complicated by the different ratio K:Al in biotite and in the K-feldspar, and finally by the role of hydrogen in the reactions.

The application of experimental data has also proved to be unconvincing. All minerals, which are presumed to be stable in the absence of water, are usually formed simply at higher pressures than those reached in the mentioned experiments (see jadeite, page 348; pyrope, page 357).

However, one may reach complete agreement with Yoder that the partial pressure of water vapor in reactions of mineral formations does at the same time not reach very high values. If the overall pressure often exceeds 12,000 atm (kyanite schist) and even 15,000 atm (eclogite), the partial pressure of water, as may be concluded from the formation of a series of water-bearing minerals (see page 355), usually does not exceed 1,000 to 2,000 atmospheres.

The above-mentioned is, to a considerable extent, related also to the formation of primary igneous minerals, especially hornblende and biotite. There is undoubtedly always more than sufficient water in the magma for the formation of these minerals, and their crystallization depends mainly on the ratio of other components — especially alkalis. The water content of the magma may however have an indirect influence, as its increase
330 lowers the temperature of crystallization of the magma, and this in its turn favors the formation of water-bearing minerals.

SEVERAL SYSTEMS WITH WATER (GENERAL CONSIDERATIONS)

Experimental investigations of systems of this type, which could possibly provide us with a more complete idea of both the qualitative and quantitative aspects of equilibrium phenomena between magma, rock-forming silicates and the gaseous phase, are rather difficult to carry out for technical reasons. Not only must the containing vessels have a sufficiently high refractivity in order to withstand the high pressures involved, but the specific refractivity of the material under investigation must also be sufficiently high in relation to the corrosive action of water and alkaline solutions at high temperatures. The technique of adjustment is in itself most complicated and great difficulties arise when sampling single phases, as the simple quenching method does not give satisfactory results in this case. Therefore, even at present only partial data are available for the majority of important systems and many questions remain unsolved. However, since the year 1926 (when the first edition of this book appeared) numerous interesting data have been accumulated. These data will be presented in brief in this section.

BINARY SYSTEMS OF THE TYPE WATER-SALT (INCLUDING SOLUBLE ALKALINE SILICATES)

At present, relatively complete data are available for a series of systems of the type water-soluble salts. Figure 211 shows three projections of the system H_2O-NaCl, which is characteristic for the case when critical phenomena are absent in saturated solutions. The solubility of the salt in water is so high in this case that the critical curve passes above the curve $B+L+G$, which is continuous and shows P_{max} somewhat below 400 atm. It is somewhat interesting to mention that in the case of other halides — bromides and iodides of sodium — the maximum pressure of the curve falls below 200 and 100 atms respectively, and the maximum pressures for potassium halides are lower than those for sodium.

Far less accurate data are available for the composition of the gas than for the liquidus curve. The gaseous curve is shown on the diagram as a dotted line. Despite the insufficient accuracy of the available figures, there is no doubt that the volatility of the salt is sufficiently high at high temperatures and that the gaseous phase contains a considerable amount of NaCl.

Critical phenomena of crystallizing solutions occur, for instance, in the systems Na_2CO_3-H_2O and Na_2SO_4-H_2O. The lower critical points of both systems differ only slightly from the critical point of water itself. No sufficiently accurate data are yet available for the upper critical points, but one may consider the following characteristic property as being established: whereas the temperature of this point is comparatively only slightly different from that of the lower critical point (falling within the limit of several tens of degrees, the difference in pressure is quite considerable and in all events exceeds 100 atms.

For the system water-silicate the first investigations were carried out by Morey and Fenner as far back as 1917 (the ternary system water-silica-potassium metasilicate). This system will be considered hereafter, showing the characteristic binary sections K_2SiO_3-H_2O and $K_2Si_2O_5$-H_2O (Figures 212 and 213). In both cases the silicates of potassium form incongruently melting compounds with water; in the case of the metasilicate there are two such compounds and in the case of the dimetasilicate — only
331 one compound. The most important properties of the presented system are as follows: the absence of critical phenomena in crystallizing solutions; a comparatively low pressure for the maximum point, which slightly exceeds 10 atms in the first diagram and slightly exceeds 75 atms in the second, i.e., it is considerably lower than for solutions of NaCl. Accurate data on the composition of the gaseous phase are not available.

The presented data, which for some time were the only available experimental data for silicate systems, have exercized a great influence on the development of theoretical ideas on the equilibrium of water with crystallizing magma. By analogy with the given system, it was considered possible that P_{max} in such a system is not very high and in all events attains several hundred atmospheres, i.e., at higher pressures a gradual transition should occur from magmatic melts to hydrothermal solutions. The data on the mentioned systems were also used when considering critical phenomena and, in particular, investigations denied the possibility of critical phenomena in systems of crystallizing magma-water.

As shown by further direct experiments, the available theoretical concepts proved to be quite far from the truth. However, before passing on to the presentation of the results of these investigations, which have not yet led to the construction of complete diagrams, we shall consider binary sections of water-sodium dimetasilicate and water-sodium metasilicate, which differ considerably from the corresponding diagrams for potassium silicates.

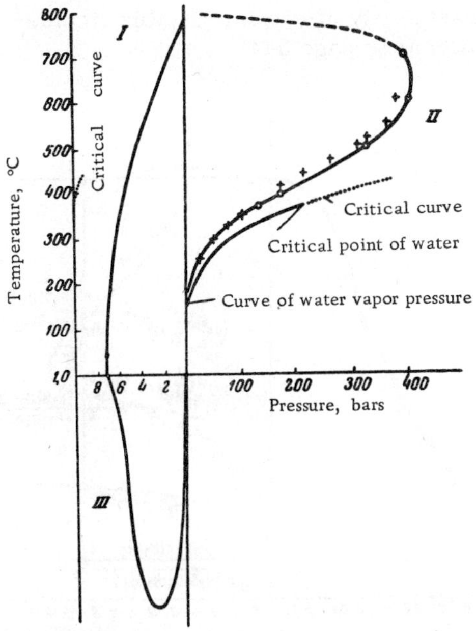

FIGURE 211. Equilibrium diagram of the system $NaCl-H_2O$ (after Morey)

The most complete set of data on equilibrium in the system $H_2O-Na_2O-SiO_2$ are presented in the work of Morey and Hesselgesser published in 1952, i. e., 35 years later than the investigation of the analogous system with potassium silicates. However, even these investigations are far from complete. According to our accepted plan, we shall consider the ternary system in the appropriate section, and shall now consider the peculiarities of the binary sections $Na_2SiO_3-H_2O$ and $Na_2Si_2O_5-H_2O$.

Firstly we shall mention that, unlike the case of potassium silicate, no compounds are formed with water. It should be mentioned that both sections are pseudobinary, as the gaseous phase dissolves a considerable amount of sodium silicate. However, the ratio of silica to sodium oxide in the gaseous and liquid phases differs considerably, thus leading to their change during crystallization. Thus, the compositions of the liquid and gas do not lie in the plane of the corresponding section.

332 Figure 214 shows several data on the relations between the liquid and the gaseous phases of the mentioned section, and Figure 215 shows a ***TP*** projection for the section $Na_2Si_2O_5-H_2O$. Firstly it should be stressed that we are here dealing with a completely different order of pressure values in comparison with solutions of potassium silicates. Investigations were carried out up to pressures of 2,200 atm, but neither the critical point nor the maximum point were reached. The second peculiarity is the rather large
333 solubility of the silicate in the gaseous phase, which here has values of the order of 20 and even 30 %. It should, however, be stressed that with regard to the interpretation of the obtained data, no unanimous opinion is held, and in the previous investigations of Tuttle and Friedman an opinion was

expressed as to the possibility of two immiscible liquids: one of which is rather enriched in water (see page 344).

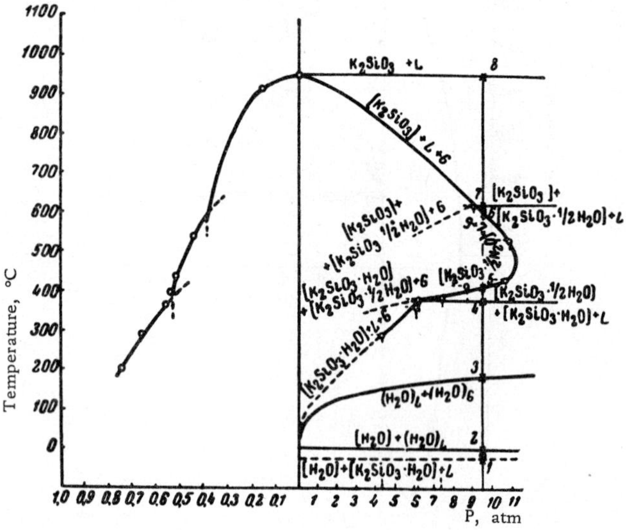

FIGURE 212. Equilibrium diagram of the system K_2SiO_3-H_2O (after Morey)

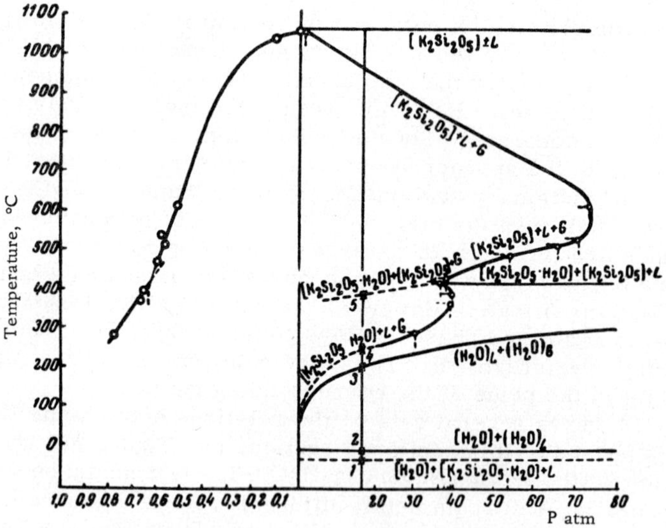

FIGURE 213. Equilibrium diagram of the system $K_2Si_2O_5$-H_2O (after Morey)

However, independent of the latter data, the second phase on Figure 214 is already in reality neither a gas nor a fluid, but a true liquid, i. e., we are here dealing with a definitely immiscible liquid. The approaching of the curves, especially in the case of the metasilicate, allows one to assume the presence of upper critical points at comparatively low temperatures (below 400°), but at very high pressures (greater than 2,500 atm). However, accepting the above mentioned, one may claim that in this case these critical points may be obtained only in such cases when the critical curve $L_1 L_2$ intersects the three-phase equilibrium curve $S+L_1+L_2$.

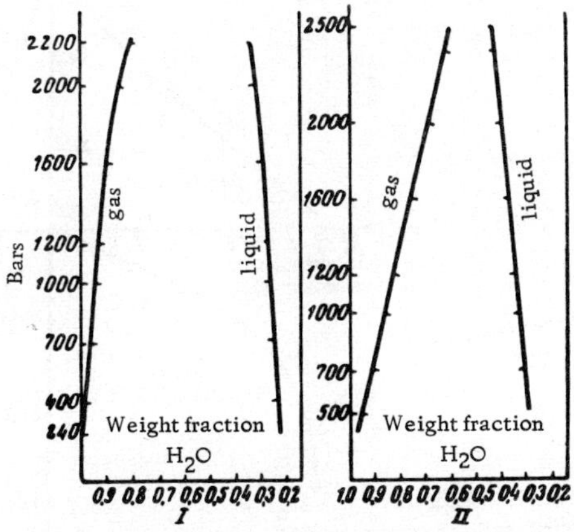

FIGURE 214. ***TX*** projection for binary sections $Na_2Si_2O_5$-H_2O (I) and Na_2SiO_3-H_2O (II) (after Morey and Hesselgesser)

BINARY SYSTEMS WATER— ROCK-FORMING SILICATE

In 1931 Goranson made the first attempt at a direct determination of the solubility of water in silicate melts of granitic composition at high pressures and temperatures, and he immediately discovered a rather important and interesting feature. It was noted that at pressures in the thousands (and not hundreds) of atmospheres range, not only has the maximum pressure (P_{max} or the point Q) not been attained, but the solubility of water proves to be comparatively low. Later, complete investigations were carried out up to pressures of 5,000 atm for the systems K-feldspar-water and albite-water, and other investigators studied analogous systems.

The results obtained are summarized on Figure 216. Despite the fact that a certain disagreement is noted in certain cases and the solubility of water changes somewhat in systems of different silicates, the order of
334 magnitudes nevertheless remains the same. Nowadays, it is quite obvious that even at 5,000 atm the solubility of water in silicate melts only slightly exceeds 10 %, and that a further increase of pressure will have a progressively lessening effect on the solubility of water.

On considering the solubility of similar silicates in water at lower temperatures, we note that in this case also our data are far from complete, but it is nevertheless obvious that this solubility is quite small, being measured in tenths of a percent. It is obvious that the critical temperature and pressure of saturated solutions in this case do not differ in actuality from the same values for pure water, thus determining the location of the lower critical point.

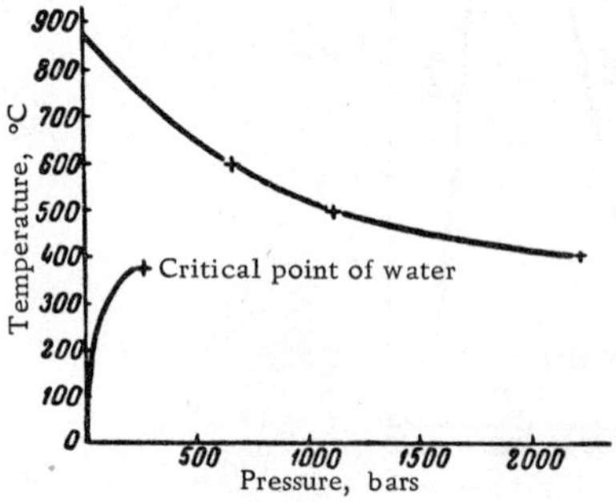

FIGURE 215. ***PT***-projection of the binary system $Na_2Si_2O_5$-H_2O (after Morey)

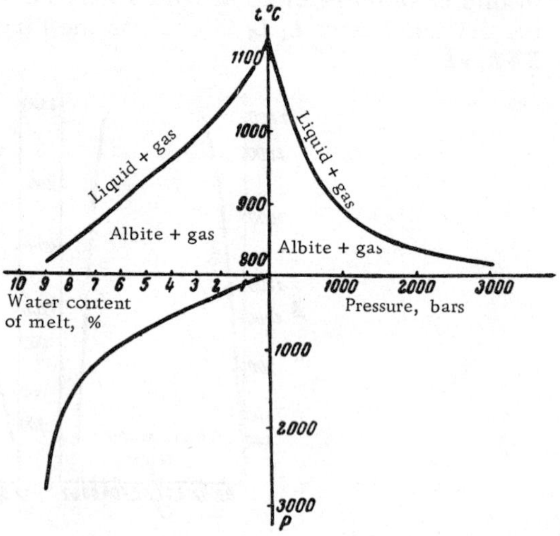

FIGURE 216. Equilibrium diagram for the system albite-water (according to Goranson's data)

The location of the upper critical point ***Q*** has given rise to much controversy. Whereas on the diagram constructed by Bowen and Tuttle (1950) the presence of such a critical temperature is already assumed at relatively low (in comparison with experimental conditions) pressures — of the order of 9,000 atm, V. A. Nikolaev (1945, 1946) expressed in his works a completely novel point of view, assuming that in this case we are dealing with an essentially new form of diagram lacking an upper critical point — the so-called third type. Moreover, Nikolaev considered it possible that at very high pressures the solubility of water in silicate melts reaches a certain maximum and subsequently starts to decrease.

V. A. Nikolaev's Type III diagram is essentially a simple diagram with critical phenomena, showing an upper critical point at infinity. Obviously, the hypothetical inflection of the curve ***G*** is not drawn. However, even in such a form the possible existence of such diagrams is questioned by I. A. Ostrovskii (1950) as well as by other authors, who propose a different interpretation of experimental data — favoring a decomposition into two immiscible liquids.

335 Firstly we shall mention that the configuration of the diagram, as proposed by Bowen, is not very probable as it presents a very abrupt change in compositions at the critical point itself, whereas an increase in the solubility of the gaseous phase is usually already noted over a considerable interval on approaching the critical point. Thus, one should either infer that the curve for the gas has in this case been drawn erroneously or that the upper critical point is in the area at much high pressures (closely conforming to the concepts of Nikolaev). Unfortunately, precise data on

the composition of the second phase and very high pressures are not available, but it is universally accepted that this is an almost pure "water vapor".

Further, on Figure 218, data are presented which illustrate the solubility of silica in water vapor at relatively low temperatures and pressures. V. A. Nikolaev, in mentioning the somewhat high figures presented by other authors (Khitarov et al.), writes "one should expect that at temperatures above 600° C and pressures of the order of 2,000—4,000 atm the solubility of SiO_2 in the supercritical water phase increases several times in comparison with the larger figures shown on Figure 4" (our Figure 218). Even if one considers that we are dealing with a ten- or twenty-fold increase, the content of dissolved silicate in water vapor only attains several per cent, i. e. , it will be quite insignificant within the composition of the melt and may scarcely be taken into account on approaching critical phenomena.

The discovery of a second liquid phase in this or other similar systems can scarcely be anticipated (the possible existence of such a phase was assumed by I. A. Ostrovskii). However, Smith and Jones (1958, see page 306) recently presented a new interpretation of diagrams of the mentioned type, showing that here we are actually dealing with a case of immiscible liquids where the invariant point $B+L_1+L_2+G$ falls in the area of critical phenomena and becomes hypothetical. Hence the curve $B+L_1+G$ passes gradually into $B+L_1+L_2$ and at high pressures we already have liquation and not boiling. The difference between the gas and L_2 is arbitrary, and more complete data on the solubility of silicates in water at high temperatures and pressures are needed for the determination of the boundary in order to reveal the area of retrograde solubility. For the system SiO_2-H_2O considered hereafter one may claim that at least up to a pressure of 700 atm a typical gas is liberated from the melt, which passes into a liquid as temperature is lowered.

Thus, irrespective of what we term the aqueous solutions separated from the silicate melt at different pressures, the type of diagram which determines the limited solubility of water in silicate melts at highest pressures may now be accepted as having been experimentally verified. This is of wide significance for the interpretation of magmatic and post-magmatic processes, as a gradual transition from magmatic melt to post-magmatic aqueous solutions is practically impossible because a well-defined gap is always present here.

The more complex system granite-water (Bowen and Tuttle) is essentially related to the considered type of binary systems and is similar to the binary section of the system quartz-orthoclase-water, where the first two minerals are included in eutectic relations.

The diagram of such a system, constructed in three projections on Figure 217, shows that at a pressure of 4,000 atm melting already starts at a temperature of 640° and a water content exceeding 9 % in the melt. This corresponds to a depression of the melting point of more than 300°, i. e. , more than 30° per 1 % water.

336 The cited experimental data allow one to approach the consideration of conditions of formation of granitic magmas. It is necessary to note that several other considerations, mentioned hereafter (temperature of decomposition of K-Na feldspars, formation temperature of muscovite, etc) show that the actual final temperature of granites is somewhat below 700°,

and the water vapor pressure apparently does not exceed 1,500 to 2,000 atm.

As is known, during the last decade the so-called theory of granitization has gained marked recognition. This theory assumes the metasomatic origin of a part or the majority of granites as the result of an interaction between ascending solutions and various rocks without melting phenomena. Without going into the physicochemical regularities of metasomatism, we cannot criticize this theory in detail and should only mention that the mentioned ideas contradict the anchieutectic relations in typical granites. On the other hand, a vast amount of material has accumulated which affirms the geological possibility of the replacement of various rocks by granites. This led D. S. Korzhinskii (1955) to the concept of magmatic granitization.

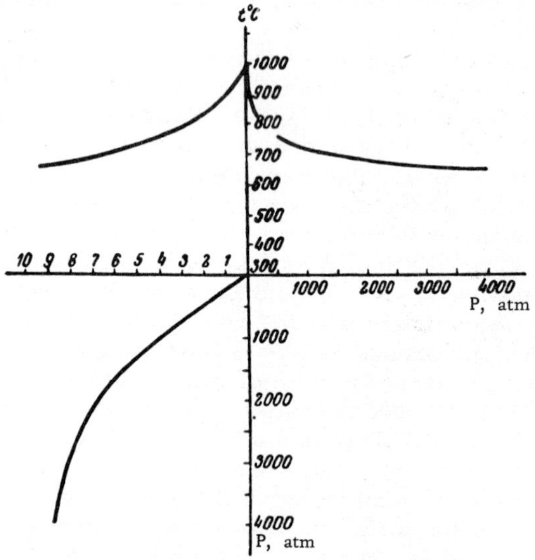

FIGURE 217. Melting diagram of the pseudo-binary system granite-water (according to data of Bowen and Tuttle)

Assuming that the granitic melt in zones of magmatization is formed only due to the increase in temperature, the melting may occur only in the case of the presence in the rock of all components of the granitic eutectic, i. e., rocks will melt, already approaching the composition of granite. Moreover, when the temperature corresponding to the point on the curve of Figure 217 is reached, only a small part of the rock will melt, depending on the amount of water present. Further melting will occur when temperature is increased, either along the binary curve of the isobaric section, if the water vapor pressure remains constant, or (in greater likelihood) along the curve $S+L+G$, if one assumes a decreasing partial pressure of H_2O. Thus, Bowen and Tuttle mentioned that at a 2 % water content the melting of granite, which started at a pressure of 4,000 atm and 640°, yields only 20 % liquid at this temperature and finally terminates at 830°.

337 D. S. Korzhinskii assumes, in opposition to the mentioned schemes (as in the theory of metasomatic granitization), that the melting occurs under the action of hot ascending solutions. This broadens the range of composition of melting rocks, as the solutions carry with them alkalis, and in some cases alkali aluminates. Melting occurs at the same time with an excess of water. At constant temperature all constituents of the granitic eutectic are completely melted and the resulting melt will be saturated in water. The subsequent portions of the melt will diffuse through the formed magma without dissolving in it. These solutions are termed by D. S. Korzhinskii as "transmagmatic solutions".

We shall mention that prior to the appearance of the mentioned term many investigators essentially affirmed the possibility of movement of volatile components through the magma, e. g., during the formation of alkaline rocks (Zavaritskii, 1937).

D. S. Korzhinskii assumes that "transmagmatic solutions" are capable of dissolving the excess unmelted parts of the rock and removing them, thus purifying the granitic magma. Being saturated with alkalis and in equilibrium with the magma, these solutions maintain the potential of alkalis in the magma at a certain constant level, thus causing a special behavior — the so-called "mobility of alkalis".

New data on the solubility of water in rock-forming silicates led A. N. Zavaritskii (1947) to a new concept of the genesis of granite pegmatites. This author mentioned that one can no longer speak of a specific "hydro-igneous" pegmatitic magma, the concept of which was based on the assumption of unlimited miscibility of water in the melt. A. N. Zavaritskii writes: "Thus we arrive at a complete negation of a specific pegmatitic magma or a specific pegmatitic melt.

The formation of the main minerals of a pegmatite is ascribed by us to the recrystallization of minerals under the action of a separated "volatile phase", which is in equilibrium with these minerals. These phenomena of recrystallization occur in an almost closed system.

Negating the existence of a specific pegmatitic magma or a specific pegmatitic melt, we may, however, speak of a pegmatite-forming solution which causes recrystallization. This solution is the same solution in which elements of ore deposits are transported from the magma.

During the initial moments after separation it is saturated with minerals of the parent rock, but is nevertheless undersaturated with respect to the rare specific pegmatite minerals and ore minerals. Therefore, during these initial moments they have not yet separated from it.

The further behavior of such solutions is known.

The separated volatile phase — gaseous solution — difuses through the country rocks and, due to the complex composition of the solution, it undergoes fractional distillation. The composition of the solution changes, which is a factor of "pegmatitization", and as an outcome of this change in the composition of the solvent, the dissolved substances start to separate from it on the one hand, and the minerals which were in equilibrium with it become unstable in the solution and undergo dissolution and replacement by new minerals on the other hand".

The system changes from a closed to an open one due to the removal of certain constituents from the solutions. This is reflected by mineral formation: the metasomatic replacement of minerals occurs.

338 Finally, A. N. Zavaritskii (1950) explains the genesis of aplites by removal of part of the volatiles from the magma during the formation of fractures in the roof of the magma chamber, resulting in a considerable oversaturation which leads to a fine-granular texture.

The system SiO_2-H_2O

In the above-presented systems data are available on the solubility of water in melts of rock-forming silicates, and it is usually only mentioned that the composition of the gaseous phase practically corresponds to that of pure water.

More or less complete data on solubility of silicate in water (both in the liquid and in water vapor) are available only for the system SiO_2-H_2O, which merits special consideration.

Figure 218 presents various data on the solubility of SiO_2 in H_2O at different temperatures and pressures after Kennedy (1950) with amendments introduced by Morey and Hesselgesser (1957). Three-phase equilibrium curves were plotted on the diagram with the lower critical point ***P*** almost coinciding with the critical point of water. As seen from the given diagram, the solubility of SiO_2 increases with temperature and pressure, but is usually measured in tenths or hundredths of a per cent. The so-called retrograde solubility is readily discernible on the diagram and corresponds to the gradual change of properties of the solution from gas to liquid. Attention should be drawn to the fact that such retrograde solubility, i. e., increase of solubility on decrease in temperature, also occurs at pressures somewhat lower than the critical value, i. e., after phenomena of liquefaction.

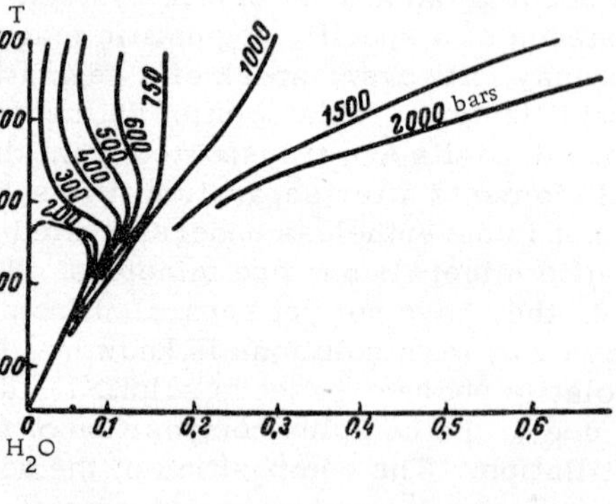

FIGURE 218. Equilibrium diagram of the system SiO_2-H_2O (after Kennedy, with amendments by Morey)

Thus here we clearly discern that a gradual accumulation of properties of a gas occurs in the liquid not only in the area where the coexistence of two phases (gas + liquid) becomes impossible, but also in the region where it is truncated by a phase transition on the boiling curve.

The *TP* projection of the system SiO_2-H_2O(Figure 219) was recently presented in full by I. A. Ostrovskii et al. (1959).

The three-phase equilibrium curve solid + liquid + gas first passes into the field of cristobalite then dridymite, and already at pressures of 1,400 atm it enters the field of **α**-quartz. Here it may be traced up to a pressure of 4,000 atm where the melting temperature approximately equals 1,100°. Stewart arrived at almost identical results (Ann. Rep., 1957) on tracing this curve up to pressures of 5,000 atm. It is interesting to mention that within the pressure range of 2,000 to 5,000 atm the temperature drops from 1,130 to 1,065°, i.e., by only 65°.

TERNARY SYSTEMS WITH THE PARTICIPATION OF WATER

The system K_2SiO_3-SiO_2-H_2O

We have already considered the limiting systems and the binary sections of the given system as investigated by Morey and Fenner (1917).
339 Figure 220 shows the usual diagram of the given system with the isotherms for the melting surface. In addition to the above-considered compounds $K_2SiO_3 \cdot H_2O$ and $K_2SiO_3 \cdot 0.5 \cdot H_2O$ for the system K_2SiO_3-H_2O and $K_2Si_2O_5 \cdot H_2O$ for the system $K_2Si_2O_5$-H_2O (all of which melt incongruently), there is one additional compound — an acid disilicate of potassium, $KHSi_2O_5$, which melts congruently. The area in the vicinity of the H_2O apex and a small strip which directly adjoins the side SiO_2-H_2O are as yet uninvestigated.

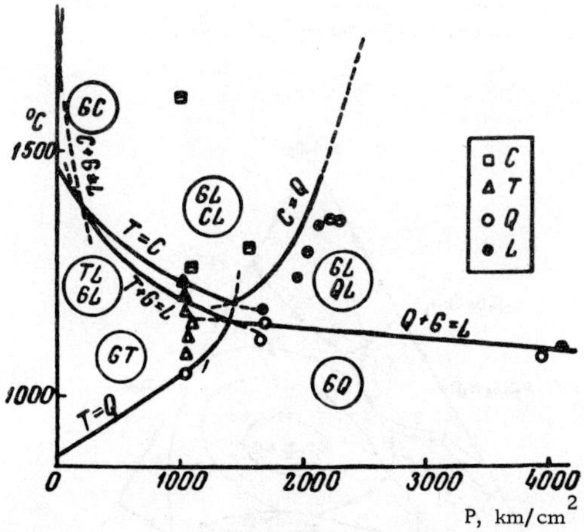

FIGURE 219. **PT**-projection of the system SiO_2-H_2O (after I. A. Ostrovskii, et al.)

C — cristobalite; T — tridymite; Q — quartz; *L* — liquid

The process of separation of the gaseous phase from crystallizing solutions at various pressures is illustrated on Figure 221, where together

with lines bordering areas of separation of single phases, isobars are also shown which represent the vapor pressure in saturated solutions. Using
340 this diagram one may construct an isobaric section similar to the one shown on Figure 222 for any given pressure, drawing the corresponding isobar which will indicate the composition of the boiling saturated solutions. In the area of higher vapor pressures the gaseous phase will be stable, whereas in the area of lower vapor pressures the liquid phase will be stable. The composition of the gaseous phase is not represented here as it is considered to be close to that of pure water.

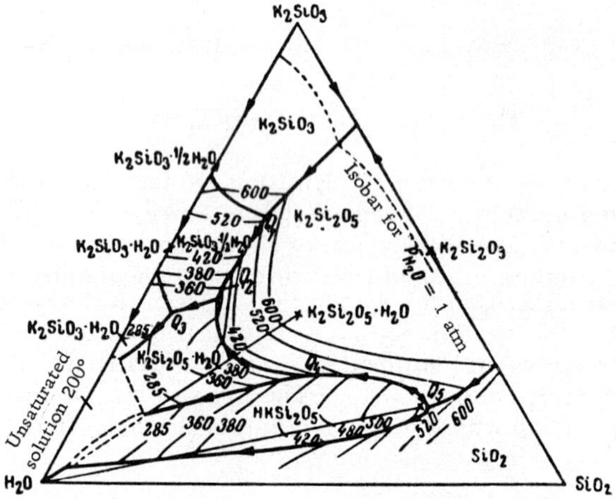

FIGURE 220. The system SiO_2-K_2SiO_3-H_2O. Isotherms of crystallization (after Morey and Fenner)

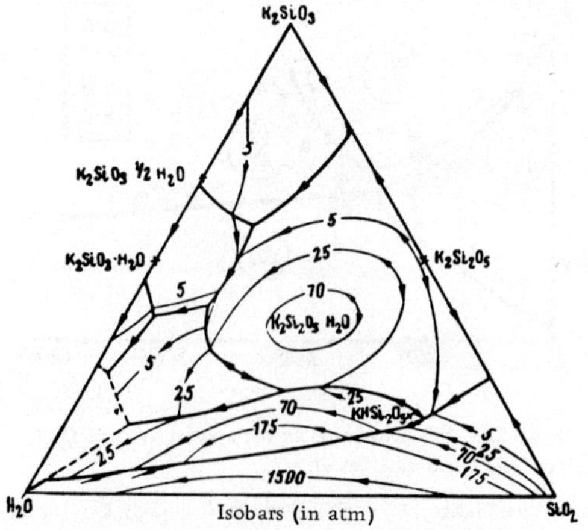

FIGURE 221. The system SiO_2-K_2SiO_3-H_2O. Isobars of crystallization (after Morey and Fenner)

For the sake of elucidation on Figure 222, I, a corresponding isobaric diagram is given for a pressure of 5 atm and for Figure 222, II — for 70 atm. Obviously, in order to obtain an area of gaseous existence the lines of crystallization must be truncated (they are shown by broken lines on the diagram). Using such diagrams, it is easy to trace the behavior of the system at any composition at the given pressure. Thus, for point *x* (Figure 222, I) crystallization commences with the separation of a compound $K_2Si_2O_5$. When the composition reaches the boundary line, phase $K_2SiO_3 \cdot 0.5 \cdot H_2O$ joins in the crystallization. At point ***a*** boiling takes place and a gas, similar in composition to pure water, will be in equilibrium with the solid phases. Further on, on lowering temperature, liquefaction will recommence and the forming liquid will have a composition corresponding to a point ***b***. Liquefaction is accompanied by a considerable dissolution of both solid phases. After the disappearance of the gaseous phase cotectic crystallization will occur up to the invariant point, at which the following reaction takes places:

$$K_2Si_2O_5 + L \rightarrow K_2SiO_3 \cdot 0.5H_2O + K_2Si_2O_5 \cdot H_2O.$$

As our system is located within the triangle $K_2Si_2O_5$-$K_2SiO_3 \cdot 0.5H_2O$-$K_2Si_2O_5 \cdot H_2O$, crystallization will terminate at this peritectic point.

From the configuration of the diagram it is obvious that for compositions the crystallization of which is terminated at point *P*, i. e., which lie within the triangle K_2SiO_3-$K_2Si_2O_5$-$K_2SiO_3 \cdot 0.5H_2O$ and to the right of the isobar, no boiling of crystallizing solutions will occur.

For an isobar with a pressure of 70 atm we have a considerably larger area of compositions showing continuous crystallization without boiling.
341 This is the particular property of all compositions which lie above the upper tangent drawn from point $K_2Si_2O_5$ to the closed isobar in the center of the triangle. In the lower part of the triangle boiling will occur at certain compositions the crystallization of which starts with SiO_2 or $KHSi_2O_5$. Thus, for instance, for a composition shown by point *x* crystallization starts with the separation of phase $KHSi_2O_5$. Further on, quartz will join in the crystallization and the composition of the liquid will change along the boundary curve, up to the isobar where boiling will occur (***a***). When liquefaction commences the liquid corresponds to point ***b***, after which the joint crystallization of the two mentioned phases will continue up to a temperature of 200° (the system has not been investigated beyond this point).

With regard to the general properties of the system, attention should be drawn to the existence of a maximum pressure lying upon the line H_2O-$K_2Si_2O_5$ (on the line passing through the H_2O apex and the compound which is formed by nonvolatile components), which feature is characteristic of such systems. As always (see Figure 202, page 314), the lowest pressure for boiling solutions is observed on cotectic lines as is reflected by the corresponding inflection of the isobars.

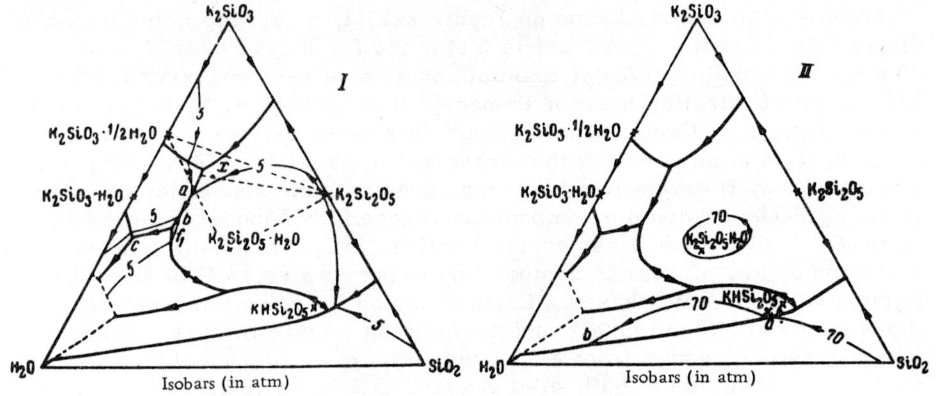

FIGURE 222. Isobaric sections of the system SiO_2-K_2SiO_3-H_2O

I — pressure of 5 atm; II — pressure of 70 atm

342 ## The system Na_2SiO_3-SiO_2-H_2O

The investigation of the given system is as yet far from complete and not all the data obtained are sufficiently convincing. However, the material is so interesting that it merits consideration.

Firstly we shall mention that in contrast to the analogous system with potassium silicate, only anhydrous phases are possible in this case, and no corresponding hydrosilicates are formed. Thus the system is considerably less complex. On the (SiO_2) (Na_2SiO_3) side we have two eutectic points, between quartz and sodium dimetasilicate and between sodium dimetasilicate and sodium metasilicate respectively. From these points two cotectic curves issue towards the middle of the triangle as is shown on Figure 223. Investigations were carried out within the 200–400° C temperature range.

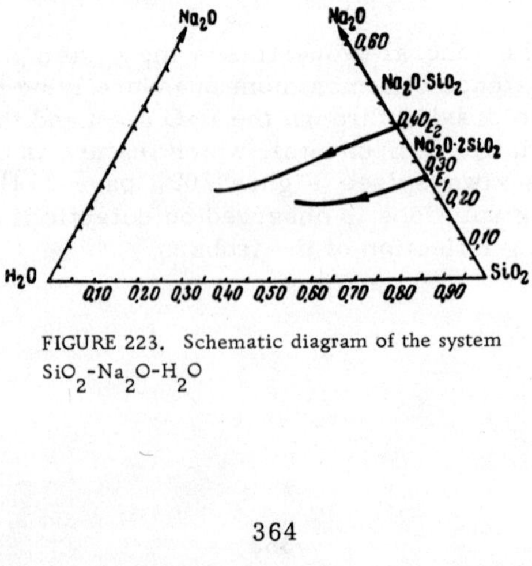

FIGURE 223. Schematic diagram of the system SiO_2-Na_2O-H_2O

The most complete investigations are those of Morey and Hesselgesser (1952) and we shall therefore start our considerations by presenting their results. All these investigations were carried out at a temperature of 400° and at pressures of up to 2,500 atm. The diagram which was constructured (Figure 224) was isothermal and polybaric, i. e., it applies to a given temperature and shows the dependence of the composition of the gaseous phase on the pressure which was plotted along the top of the prism.

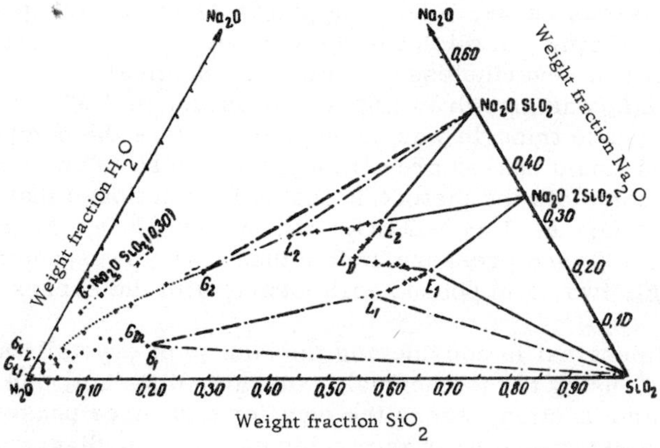

FIGURE 224. Isothermal section of the diagram SiO_2-Na_2O-H_2O at 400°C (after Morey and Hesselgesser)

Points E_1 and E_2 on the given diagram are invariant points and correspond (at constant temperature) to equilibria of four phases: quartz, sodium dimetasilicate, liquid (E_1) and gas, and sodium dimetasilicate, sodium metasilicate, liquid (E_2) and gas. The mentioned points correspond to the composition of the liquid, and points G_{E_1} and G_{E_2} — to the corresponding composition of the gas. The coordinates of the invariant point E_1 are as follows: pressure — 240 atm; gas composition — 0.998 H_2O, 0.001 Na_2O
343 and 0.001 SiO_2; liquid composition — 0.229 H_2O, 0.200 Na_2O and 0.571 SiO_2. It is self-evident that the point E_1 lies on the cotectic curve of the polythermal melting diagram (see Figure 223) and at the same time on the isobaric boiling curve (on a diagram of the type shown on Figure 222). The invariant point E_2 was determined with less accuracy. Its pressure was accepted as being 380 ± 20 atm and its composition was determined by extrapolation.

To the left of points E_1 and E_2 curves of three-phase equilibrium issue: liquid, gas, quartz (E_1L_1); liquid, gas, dimetasilicate (E_1L_D); liquid, gas, dimetasilicate (E_2L_D) and liquid, gas, metasilicate (E_2L_2). As in the case of these curves we are also dealing with an equilibrium between a liquid and a solid phase, and the melting surface is almost unchanged on varying pressures, the curves themselves coincide with the position of the isotherm of the melting surface at 400° (see Figure 223). On the given diagram these curves are polybaric as they show the change of pressure of the gaseous phase in all boiling solutions.

Two branches of the curve, which correspond to the liberation of dimetasilicate from boiling solutions, meet in the vicinity of the line $Na_2Si_2O_5$-H_2O which characterizes a binary section. In relation to the liquid this point is only insignificantly shifted from the mentioned line (the ratio $Na_2O:SiO_2$ is 0.65 instead of 0.66), but in relation to the point G_{Di} of the gaseous composition the ratio is already 0.61, i. e., the section is not strictly binary for boiling solutions, although in general the relations in the gaseous phase already approach those of the liquid phase.

The liquid composition E_1L_1 on the curve corresponds to the composition G_{E_1} G_1 on the gaseous curve, where the points L_1 and G_1 are given for a pressure of 2,500 atm. In this case the compositions are fairly close to one another but are nevertheless by no means identical.

The curve E_2L_2 was also drawn up to a pressure of 2,500 atm and a corresponding phase triangle was constructed. Here the compositions of the gaseous and liquid phases are already quite close to one another.

In order to complete the picture it should be mentioned that in the triangles ($Na_2O \cdot SiO_2$) E_2 ($Na_2O \cdot 2SiO_2$) and ($Na_2O \cdot 2SiO_2$) E_1 (SiO_2) at a temperature of 400° and pressure higher than that corresponding to the points E_1 and E_2, two solid phases will coexist with the corresponding liquid.

It is self-evident that in considering the course of crystallization at any given pressure, using the isothermal, polybaric diagram, one should construct an isobaric section, where the equilibria of three phases become invariant and are expressed by phase triangles. On our diagram such phase triangles are constructed for a pressure of 2,500 atm.

Before passing on to the consideration of possible critical phenomena and the immiscibility of liquids at lower temperatures, the character of the solutions, which are termed gaseous by the authors, should be considered. As follows from the mentioned considerations (see page 325) at a temperature of 400° and a pressure of 2,000 atm we shall obtain neither a gaseous nor a fluid phase, but a typical liquid phase. Thus, boiling phenomena do not occur here but typical immiscibility of liquids which is also characterized by the high solubility of the nonvolatile components.

According to Morey and Hesselgesser, in the case of the binary systems dimetasilicate-water and metasilicate-water, critical points may be reached in these sections at comparatively high pressures but low temperatures (below 400° for dimetasilicates and somewhat higher for metasilicates). The critical curve should pass through these points but cannot, however, reach the side (H_2O) (Na_2O), where critical phenomena are absent. The critical curve also does not reach the cotectic curve dimetasilicate-quartz, which is characterized by the continuous nature of crystallization. Thus,
344 the critical area is a closed one within the concentration triangle. The second critical area, which is also a closed one, is adjacent to the side (H_2O)(SiO_2), upon which the upper critical point is absent.

Taking into consideration the fact that under the mentioned conditions it will be the equilibrium L_1+L_2, and not $L+G$, that will be achieved, the upper critical point may appear provided that the critical curve L_1+L_2 is lowered on increasing the pressure and will intersect the curve $S+L_1+L_2$. The possibility of such phenomena should be affirmed by further investigation.

Investigations at lower temperatures were carried out by Tuttle and Friedman (1948) but are less complete, particularly as pressures were not measured. The most important results of these investigations were the conclusions arrived at on the existence of immiscible liquids with the area of liquid immiscibility having a lower critical point. Immiscibility was absent at 200°, but was noted at 250°, and at higher temperatures (up to 350°) the difference in composition of the liquids increased.

An idea of the character of the solutions may be obtained from the pseudo-binary section (Figure 225) which was constructed from the data of the authors' investigation.

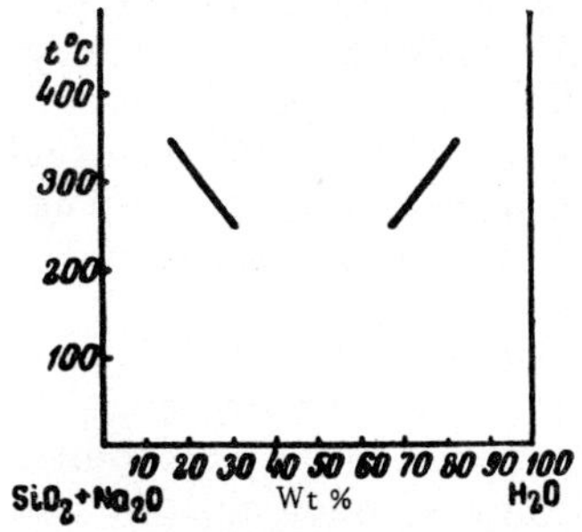

FIGURE 225. Schematic diagram for the composition of immiscible solutions in the system SiO_2-Na_2O-H_2O (according to data of Tuttle and Friedman)

As already mentioned, in the experiments of Tuttle and Friedman (1948) pressure was not measured, and one is thus not able to plot the points in space. However, if the presented considerations are correct, binary sections of the mentioned system should be related to the type shown on Figure 193 (see page 305), i. e., an invariant point $S+L+L_2+G$ should be present at comparatively low temperatures of the order of 250°. This seems to contradict the data of Morey and Hesselgesser — and in all events those sections shown on Figure 214. The investigations of V. A. Kalyuzhnyi also indicate the possible existence of such a point. During the investigation of certain natural inclusions in quartz, he succeeded in observing the appearance of a second liquid on heating in the presence of a gaseous bubble, thus leaving no doubts as to the character of the determined phase. The composition of the solution was not investigated in this case and one may only assume that it is close to the considered system.

Further investigations should elucidate the as yet unsolved question and make the character of the diagrams and the sections more accurate. However, even at present it is clear that the relation between water and sodium silicates has an intermediate position between the system H_2O-potassium silicate (which is analogous to systems with soluble salts) and the ratio between water and rock-forming silicates.

The system $NaAlSi_3O_8$-$KAlSi_3O_8$-H_2O

If the earlier considered system for the explanation of magmatic crystallization was of only theoretical significance, as similar compositions could only be realized in post-magmatic solutions, the investigation of this system is rather important for the understanding of the crystallization of the main rock-forming minerals.

345 Investigations were carried out by Bowen and Tuttle (1950) within the pressure range up to 2,000 atm. The solubility of water in the silicate

solution proved to be quite small, i. e., these experiments affirmed the results of Goranson's investigations. The authors considered the following figures as being optimum: 7.5 % at P = 1,000 atm and t = 854°; and 9 % at P = 2,000 atm and t = 785°. All determinations were made on melts which correspond to the minimum point of feldspars.

As already mentioned previously, in a dry system there is a continuous miscibility between feldspars with a minimum point, which approximately corresponds to 65 % sodium component and a temperature of 1,063°, but the diagram is complicated by the presence of a wide area of leucite crystallization, within which the system is no longer binary. On Figure 115 (see page 180), together with the section corresponding to a dry system, sections are given for a system with water at pressures of 1,000 and 2,000 atm. Firstly it is evident that the leucite field is markedly diminished, and on further increase of pressure it should wedge out, as was proved by Goranson at a pressure of 2,000 atm. It is evident from the sections that initially the increase in water vapor pressure leads to a sharp drop in the temperature of melting, which later is arrested, corresponding to the character of solubility of water. The authors considered the minimum point to be somewhat shifted toward the composition of approximately 70 % of sodium component. However, such shifting cannot be considered as proved, as the point itself is determined with a low degree of accuracy both in the dry system and in the system with water. The position of the maximum on the decomposition curve also negates such a shift.

The provided sections only reveal the relations of the feldspars without considering water. In order to consider the courses of crystallization triangular diagrams should be constructed — separate diagrams for temperature and pressure. However, as we have already noted, the projections of these diagrams are often presented on the same concentration triangle, the description of which is quoted from Bowen and Tuttle.

Actually, the "composition triangle" here presented is really only a part of that triangle, and it is constructed in an unusual manner, which requires some description. Although it is customary to represent ternary compositions on an equilateral triangle, it is well known that one can use a triangle of any shape whatever with coordinates drawn parallel to its sides. For the present purpose it is convenient to use a triangle in which the water ordinate is greatly exaggerated, and, accordingly, we have constructed an isosceles triangle with an altitude five times that of an equilateral triangle. Water is plotted at its apex, and one of the feldspars at each end of its base. Of this triangle we show only the lower portion in Figure 226 because it is the only portion required to represent the experimental results. The proportion of water is read in the ordinary manner with the aid of lines parallel to the base, but, as it is desirable to know the feldspar proportions apart from water content, instead of drawing coordinates parallel to the other two sides of the triangle, lines radiating from the H_2O apex are drawn. With their aid the feldspar proportions represented by any point can be read directly, whatever the water content may be.

The isobars and isotherms are then plotted with the aid of the data on which the diagrams in Figure 115 (see page 180) are based and also the

determined water contents of the liquids (glasses), including Goranson's determinations. The significance of these curves is simply stated. The isobar of 1,000 kg/cm^2 for example, represents the composition of all liquids that are saturated with feldspar crystals and with water vapor at
346 a pressure of 1,000 kg/cm^2, and an isotherm, for example, that for 950°, represents the composition of all liquids that are just saturated with feldspar crystals and water vapor at 950°. Tie-lines may now be added, joining the composition of any liquid represented by a point on an isobar with the particular feldspar in equilibrium with it. These tie-lines are given only for the 1,000 kg/cm^2 isobar. An attempt to show them for more than one isobar would result in a confusion of lines. Figure 115 supplies this information also, for horizontal lines on Figure 115 are such tie-lines.

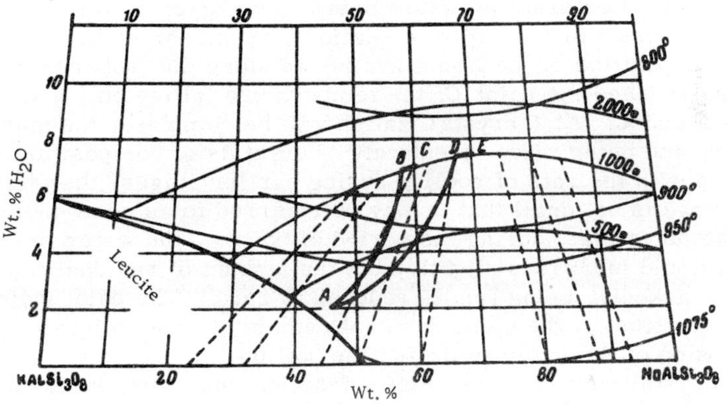

FIGURE 226. Portion of equilibrium diagram for the system $KALSi_3O_8$-$NaAlSi_3O_8$-H_2O (according to Bowen and Tuttle)

The diagram (Figure 226) is now ready for use to describe the course of crystallization of a liquid (magma) of any composition in the system. We shall bring out the principles involved by discussing crystallization of only one composition.

Composition ***A*** contains $NaAlSi_3O_8$ and $KAlSi_3O_8$ in the relative proportions 45 : 55 with 2 per cent of H_2O. We shall assume that a liquid of that composition exists under an external pressure of 1,000 kg/cm^2 and at an elevated temperature, say 1,060°. Crystallization of this liquid will begin, if it is cooled, at approximately the temperature of the isotherm passing through ***A***. It will not be exactly at the temperature of that isotherm, because, as will be recalled, the isotherms locate a saturation surface, the liquids being saturated with both water vapor and crystals. At 1,000 kg/cm^2 this liquid ***A*** is not saturated with water vapor, the saturation pressure being about 400 kg/cm^2. Accordingly, the temperature of initial crystallization at the higher pressure of 1,000 kg/cm^2 will be raised slightly above that indicated by the saturation surface. This effect of pressure in raising the crystallization temperature in the condensed system is small. Therefore, the liquid will begin to crystallize at a temperature slightly higher than that indicated by interpolation between isotherms on the saturation surface; but the actual difference is very small, and, for all practical purposes, isotherms on the saturation surfaces can be used.

Crystallization will therefore begin at approximately 1,000°. The first crystals formed will have the composition ***Ab_{14}***; and, if the perfect equilibrium type of crystallization prevails, the first part of the course of crystallization is given by the curve ***AB***. At ***B***, which is a point on the 1,000 kg/cm^2 isobar, the liquid is now saturated with water vapor. The temperature, now indicated precisely by the isotherm through ***B***, is, by
347 interpolation, 850°. The position of the point ***B*** is determined by the tie-line (three-phase boundary) which passes through ***A***. The mass now consists of liquid ***B***, which contains nearly 7 percent H_2O, and crystals of **Ab_{40}**. The proportion of liquid to crystals is approximately 30 : 70.

The assumed external pressure being 1,000 kg/cm^2, further abstraction of heat will result in boiling (vesiculation), accompanied by crystallization, and the composition of the liquid now moves along the isobar until the liquid has reached the point ***C***, the temperature falling only a very little between ***B*** and ***C***. At ***C*** crystallization will be complete, the mass having boiled dry and being made up entirely of crystals of composition **Ab_{45}**. Compared with the rate of cooling during earlier stages, that during the boiling stage is so small that it may be referred to as semi-isothermal. It is to be noted that, during the period of boiling, the water content of the residual liquid increases slightly, but the amount of residual liquid decreases markedly, being finally reduced to zero. The point indicating the total composition of the mass does not remain at ***A*** but migrates to the feldspar side of the diagram during the boiling.

With crystallization of the perfect fractionation type, initial formation of crystals will be as above, but thereafter the liquid will follow the course ***ADE***, boiling beginning at ***D***, a significantly lower temperature than ***B***, and both boiling and crystallization being completed at ***E***, which is the point of minimum temperature on the isobar (840°). The crystals in this case are continuously zoned, with compositions ranging from **Ab_{14}** to **Ab_{70}**".

The curves of decomposition of solid solutions are of particular interest in these systems. The existence of these curves was predicted some time ago on the basis of interpretation of mineralogical data and by analogy with other systems of solid solutions of potassium and sodium (see page 177). During the devitrification of glasses of the composition of alkali feldspars under a considerable water vapor pressure, it was found, by means of X-ray studies, that two phases were present. Accurate calculation allows for the determination of the formed feldspars according to these data, i. e., for the given temperature the location of the points upon the decomposition curves shown on Figure 115 may be found (see page 180).

The critical point at which the decomposition curves meet, forming one curve with a maximum point, corresponds to a temperature of 660° at an approximate composition of 55 % albite component. The location of this point determines the upper boundary of decomposition of feldspars. With regard to the composition of the maximum point, one should note that the accuracy of its determination is doubtful. In reality one may claim that on further lowering of the melting curve (e. g., as the result of addition of another neutral, fusible substance) the intersection of this curve with the curves of decomposition should eventually occur, and during the initial moments the minimum point on the melting curve should probably coincide with the critical point of decomposition. Thus, a difference of 15 % for the mentioned points with such small differences in temperature is barely possible.

According to Yoder, Stewart and Smith (Ann. Rep., 1957), at 5,000 atm the decomposition curve of solid solutions intersects the melting curve at t = 698° and a composition range of $Or_{34}Ab_{66}$-$Or_{58}Ab_{42}$. Thus, the decomposition curve of solid solutions (solvus) in this case is located considerably higher than as stated by previous investigators (40° higher). The authors assume that the intersection occurs to one side of the minimum point, for which they give the following values: T = 695°, composition = $Or_{29}Ab_{71}$. As the temperature difference is very small, the accuracy of determination of the points may be doubted.

348 The system $NaAlSi_3O_8$-$NaAlSiO_4$-H_2O

This system was investigated by Mackenzie (Ann. Rep., 1954) up to a pressure of 1,000 atm, the obtained ratios differing slightly from those of the dry system. The eutectic point in this case corresponds to the composition $Ab_{72}Ne_{28}$† instead of $Ab_{76}Ne_{24}$ — in the absence of water, the temperature being lowered from 1,068° to 870°.

The limiting solid solution of nepheline, enriched in silica, in comparison with the standard formula at a temperature of 750°, approximately corresponds to the composition $Ne_{75}Ab_{25}$, which is rather dubious (see page 185).

The compositions corresponding to $Ab_{95}Ne_5$ have undoubtedly been obtained during crystallization of nepheline, i. e., solid solutions of nepheline and albite are practically absent. In the same system at lowered temperatures, Yoder (Ann. Rep., 1954) investigated the equilibrium curve of analcite, shown on Figure 227. The temperature of decomposition was given for fairly high pressures below 600°, thus giving rise to serious doubts as to the possibility of the magmatic crystallization of analcite in teschenites and other alkali basaltoids.

During all the above-mentioned experiments the synthesis of jadeite was not achieved. In this mineral, as distinct from other phases of this system, aluminum occurs in six-fold coordination, thus causing a sharp change in properties and primarily a considerable increase in specific gravity. As the reaction of jadeite formation should be accompanied by a decrease in volume, an assumption was made some time ago that this mineral is formed at high pressures. After the completion of Yoder's first experiments, an opinion was expressed that much higher pressures are necessary for the formation of jadeite and that even the calculated figures of Adams are low (see page 187). On the other hand one of the investigators at high pressures — Yoder — proposed the hypothesis that, as in a series of other cases (formation of pyrope, see further on page 354), the crux of the question lies in the water regime, in the deficiency

† Ab — albite; Ne — nepheline.

of which anhydrous phases are formed which are unstable at the corresponding water vapor pressure under experimental conditions.

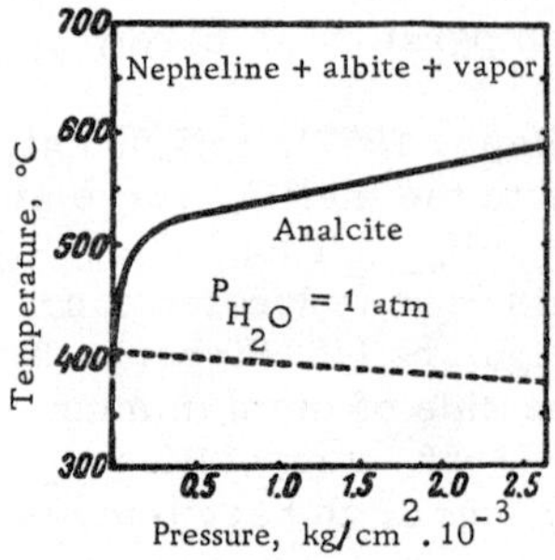

FIGURE 227. Decomposition curve of analcite in an atmosphere of inert gas (after Yoder). The broken line represents the theoretical curve of the partial pressure of H_2O equal to 1 atm

Later experiments by Robertson, Birch and MacDonald (1957) affirmed the correctness of the first point of view showing that the reaction of formation of jadeite occurs at much higher pressures. Whereas several experiments were carried out by these authors in an atmosphere of nitrogen, experiments were also formed in the presence of water vapor, thus providing data on the area of stability of jadeite.

Figure 228 shows the general diagram of the considered system. The equilibrium curve analcite-jadeite is shown schematically as only the location of the nodal point is known and not the slope of the curve.

Recently the location of the decomposition curve of albite into jadeite and quartz was determined and found to lie above the coesite curve. The decomposition curve of albite into analcite and quartz has also been investigated (not shown on Figure 228). It diverts from the corresponding point of equilibrium of four minerals towards low pressures.

249 Finally, in order to complete the picture, it should once more be mentioned that at pressures of the order of 10,000 atm and above, it has been established that on lowering the temperature a polymorphous inversion of nepheline occurs with the formation of the heavy modification, which has an olivine-like structure, i. e., a six-fold coordination of aluminum in similarity to jadeite. However, the stability field of this modification has not been established with any accuracy and is not shown on Figure 228.

The system $CaMgSi_2O_6$-$CaAl_2Si_2O_8$-H_2O

Figure 229 shows the crystallization diagram for the system diopside-anorthite at a water vapor pressure of 5,000 atm according to data of Osborn (Ann. Rep., 1954). As seen from comparison with the crystallization diagram of the "dry" system, presented on the same figure, not only a lowering of temperature occurs in this case, but also a considerable shifting of the eutectic point. The melting point of anorthite is lowered to a greater extent than that of diopside, with the result that under these conditions anorthite, the melting temperature of which is 150° higher than that of diopside, will melt at a lower temperature than diopside. This causes the shifting of the eutectic point to a composition of approximately 73 % anorthite component instead of 42 % when water is absent, the temperature of the eutectic point being lowered from 1,274° to 1,095°.

The mentioned peculiarities allow for the assumption that for the same composition of the initial magma, depending on pressure, the first crystallizing phase will in some cases be diopside (at high pressures) and in others — anorthite (at low pressures). Correspondingly, during crystallization differentiation, ultrabasic rocks will form in some cases and in others — anorthosites.

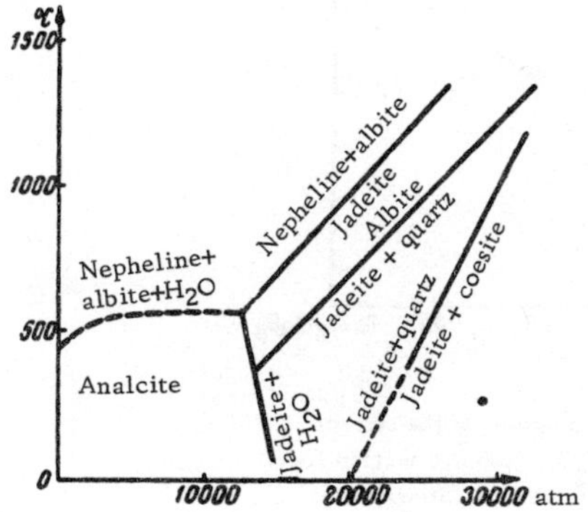

FIGURE 228. Schematic diagram of the system $NaAlSiO_4$-SiO_2-H_2O (according to data of Robertson, Birch and MacDonald)

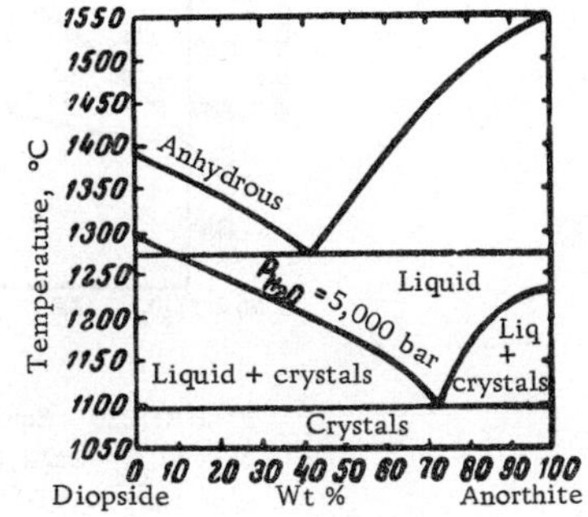

FIGURE 229. The system diopside-anorthite-H_2O. Isobaric section at a pressure of 5,000 atm (after Osborn)

The system $NaAlSi_3O_8$-$CaAl_2Si_2O_8$-H_2O

Figure 230 presents, according to preliminary data (Yoder, Stewart and Smith, Ann. Rep., 1956), the equilibrium diagram of plagioclases at a water vapor pressure of 5,000 bar. The liquidus curve, as compared with the data of the "dry" system, is shifted by more than 300°.

The location of the solidus has not yet been determined accurately, i. e., the influence of water upon the distance between the solidus and liquidus curves is not clear. In the vicinity of the anorthite composition β-aluminite
350 crystallizes, the stability field of which is also vague. In other experiments it appears that the hexagonal modification $CaAl_2Si_2O_8$ was obtained as an unstable phase.

At higher pressures and lower temperatures anorthite reacts with water to form lawsonite $CaAl_2Si_2O_8 \cdot 2H_2O$ — a mineral characteristic of the jadeite facies.

The system $KAlSi_3O_8$-$CaAl_2Si_2O_8$-H_2O

Figure 231 presents, according to preliminary data, the diagram of the system $KAlSi_3O_8$-$CaAl_2Si_2O_8$-H_2O at a pressure of 5,000 bar (Ann. Rep., 1956).

As mentioned previously, under these conditions orthoclase melts congruently, its melting point being below 900°. In comparison with the "dry" system $KAlSi_3O_8$-$CaAl_2Si_2O_8$-SiO_2, the content of the anorthite component in the eutectic is considerably higher, although it generally remains low (of the order of 12 %). The limiting compositions of solid solutions are An_2Or_{98} and $An_{96}Or_4$.

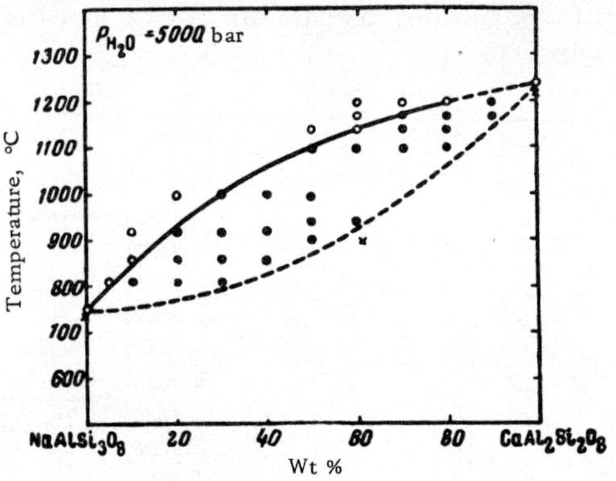

FIGURE 230. Equilibrium diagram of the system $NaAlSi_3O_8$-$CaAl_2Si_2O_8$-H_2O. Isobaric section for a pressure of 5,000 atm (after Yoder, Stewart and Smith)

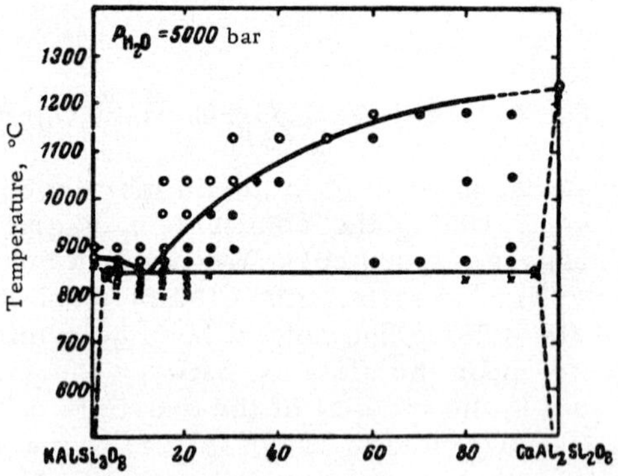

FIGURE 231. The system $KAlSi_3O_8$-$CaAl_2Si_2O_8$-H_2O. Isobaric section for a pressure of 5,000 atm (after Yoder, Stewart and Smith)

351 ## The system MgO-SiO_2-H_2O

The ternary system MgO-SiO_2-H_2O was investigated by Bowen and Tuttle in 1949. Figure 232, I, presents a ***TP***-diagram on which univariant curves were plotted corresponding to the dehydration reaction of individual minerals from their parageneses. These curves are analogous to the corresponding decomposition curves of compounds of binary systems, as projected on the ***TP*** plane, but in our case of a ternary system there are not two but three solid phases with water vapor. Between the curves are located corresponding, stable three-phase parageneses, one of the phases of which is water vapor. Figure 232, II, shows the corresponding paragenetic diagrams.

Certain complimentary investigations were carried out later on, which concern the decomposition curve of brucite, the location of which curve was

only approximately established by the authors and has given rise to many controversies. This curve does not correspond to thermodynamic data,
352 and according to its location it differs sharply from other curves by virtue of its steeper slope — actually intersecting these curves. On Figure 232 this curve is plotted according to recent data, but even here it rises more steeply than should be anticipated from analogy with other curves.

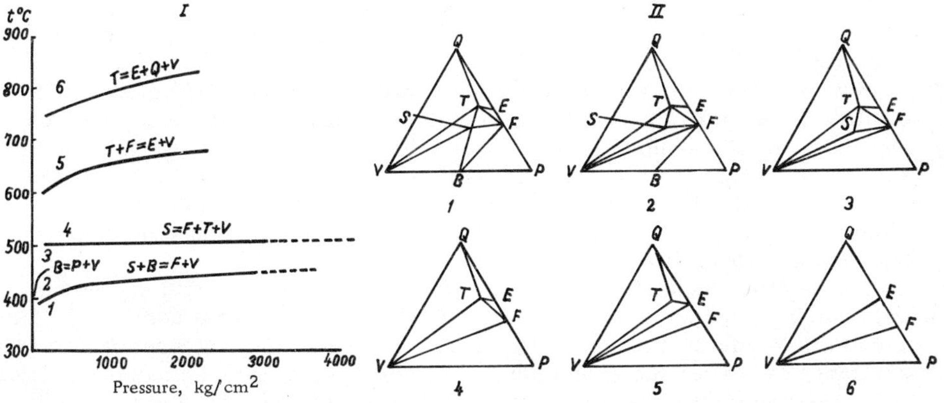

FIGURE 232. I — the system $MgO-SiO_2-H_2O$. **PT**-projection for temperatures up to 900° (after Bowen and Tuttle). Brucite decomposition curve (after Lebedev); II — paragenetic diagrams for the system $MgO-SiO_2-H_2O$

Investigations of the considered system showed firstly that the dehydration curve of minerals rises very slightly on increasing pressure, the largest rise being observed at a pressure of 100 atm, after which the curve becomes almost parallel to the abscissa. This character of dehydration curves of silicates during metamorphic reactions was assumed by Korzhinskii some time ago and was considered by him as a proof of the liquid state of water in these cases. Experimental data have affirmed the correctness of Korzhinskii's conclusions that the temperature of dehydration is only slightly dependent on pressure, and showed at the same time that this also occurs during the liberation of water in the form of a gaseous phase which gradually attains the properties of a liquid (see page 333).

The considered system is mainly of interest to the petrology of igneous rocks, as it provides accurate data on the upper temperature limit of the serpentinization reaction. Metamorphic serpentinization of olivine with the formation of brucite at low pressures (of the order of 500 atm) occurs at a temperature of 400° and rises up to 450° at pressures greater than 2,000 atm. Experiments on the serpentinization of natural olivines, containing 10 % fayalite component, showed that in this case the temperature of serpentinization is lowered by almost 80°.

The upper temperature limit of stability of serpentine, i. e., the temperature of serpentinization with removal of magnesium, is approximately equal to 500° and does not rise up to 550° even at pressures greater than 3,000 atm. The equilibrium curve in this case has an especially gentle slope.

At present data on the solubility of water in melts of ultrabasic composition are not yet available. However, one may assume with some assurety that this solubility is scarcely larger than in a melt corresponding to the composition of an acid magma, i. e., lower than 10 %. Making the corresponding assumption that 1 % of water lowers the melting temperature by approximately 40°, we may claim that even at very high water vapor pressures the temperature of ultrabasic magma remains higher than 1,000°. On the basis of these data, Bowen and Tuttle arrived at a conclusion, conforming Bowen's earlier concepts on the general impossibility of existence of ultrabasic magma, as the contacts of ultrabasic intrusions do not furnish any data supporting these high temperatures.

Without going into the complex and controversial question of the origin of ultrabasic rocks, it should nevertheless be mentioned that we are confronted with an undoubted case of an ultrabasic magma with a rather insignificant content of alkalis and alumina during the formation of meimechites and kimberlites—rocks with a glassy matrix. As inclusions of phlogopite are sometimes found in these rocks, the magma temperature was lower than 1,100° (see page 355).

On the other hand one may reach complete agreement with another of these authors' conclusions, namely that these experiments completely disprove the formerly expressed hypothesis of certain petrographers as to the existence of serpentine in magma; i. e., on the primary igneous formation of serpentine. One may claim that at the highest pressures not only the dissociation curve of serpentine, but also that of talc, will not intersect the melting curve, and hence the crystallization of these minerals from the melt is impossible, as is completely affirmed by the absence of these minerals in the corresponding effusives.

353 Special mention should be given to the fact that under the mentioned experimental conditions, i. e., at a pressure of 3,000 atm and a temperature of 1,000°, antophyllite was not obtained by these authors. The stability field of this mineral remains unknown, and it may only be assumed that it forms at higher pressures.

The system Al_2O_3-SiO_2-H_2O

The first, comparatively complete data on the investigation of this system were published by Roy and Osborn in 1952—1954 for pressures up to 2,000 atm. They mainly investigated the limits of stability of hydrosilicates, while showing that the decomposition curves are even more gentle than for magnesium silicates and that they correspondingly lie in the area of lower temperatures. We shall only mention that the decomposition of kaolinite, up to a pressure of 3,000 atm, only slightly exceeds 400° and for pyrophyllite — 570°. The temperature of decomposition, of diaspore in the system Al_2O_3-H_2O is also only slightly higher than 400°. It is interesting to note that, starting from a temperature of 420°, the stable anhydrous phase proved to be mullite according to preliminary data.

As recently as 1955 D. Roy succeeded in synthesizing andalusite at temperatures of 450° to 650° and pressures of approximately 600—2,000 atm. The synthesis of andalusite was facilitated by increasing the pressure as well as by the admixture of MgO. In the absence of MgO andalusite was only obtained with the aid of seeding. Data on the sillimanite-kyanite inversion curve, as well as on the synthesis of andalusite at the expense of mullite, are presented on page 192.

The system $MgO-Al_2O_3-SiO_2-H_2O$, chlorites and the pyrope problem

This system was investigated by Yoder (1952), and D. Roy and R. Roy (1955) and in addition complementary data on the stability of chlorite were provided by Nelson and R. Roy (1958). The investigations were carried out within the pressure range up to 2,100 atm and at temperatures up to 990°.

The question of chlorite stability is of the greatest interest. This problem has not yet been finally solved, but the available material is of great interest from the mineralogical point of view.

The upper stability limit of chlorite has been determined fairly accurately and proves to be approximately at a temperature of 700° for clinochlore and somewhat lower for chlorites of other compositions, as shown on Figure 233.

Chlorites with compositions varying over the broadest range become stable at temperatures lower than 650°.

The lower stability limit of chlorites is known with a much lesser degree of accuracy. As shown on the same figure, there are reasons for believing that, already below 400°, chlorites with a normal structure are unstable and a new type of chlorites — the so-called septechlorites — appear, the layer thickness of which is of the order of 7 Å, whereas in normal chlorites this distance is approximately 14 Å. Septechlorites are structurally analogous to serpentines.

Judging from the presented data, the temperature of formation of chlorite-bearing metamorphic rocks is much higher than is at present assumed by many authors. Thus, Barth (1956) accepts an upper temperature limit of 400° for the epidote-amphibolite facies, and a lower temperature limit of as low as 100° for the greenschist facies. Taking into account the fact that the iron content should somewhat lower the temperature of decomposi-
354 tion, we may accept 600° as the upper temperature limit of the epischists of Grubenmann (which include both mentioned facies), and 400° as the lower temperature limit of regional metamorphism (without essential metasomatism). The mentioned assumptions should be verified by carrying out detailed investigations of the distribution of septechlorites in various rocks, as well as by adding to the accuracy of the equilibrium diagram, as it is still somewhat inexact.

With regard to the area of low-temperature equilibria, we shall mention that montmorillonite is the characteristic solid phase up to a temperature of 400°, although already at temperatures above 250° no purely magnesian end member of the series is obtained and a certain amount of alumina is necessary. Montmorillonites with compositions varying over a fairly narrow range were obtained up to a temperature of 480°.

Among the high temperature minerals it is very interesting to note the stability area of magnesian cordierite, which appears at a comparatively low temperature, namely above 500°. In his investigations Yoder drew attention to the fact that he was not successful in obtaining pyrope. He assumed that this failure was not due to the fact that pyrope is stable at very high pressures but only to the presence of an excess of water, whereas pyrope should, according to his opinion, be formed within the same temperature range as chlorites in the deficiency of water. Using this as a basis, Yoder arrives at broad generalizations on the possible formation of various metamorphic facies at the same temperature, depending on the excess or deficiency of water. He also denies the specific role of high pressures in metamorphism, stating that "pressure should be divested of its mystic power and relegated to a minor role".

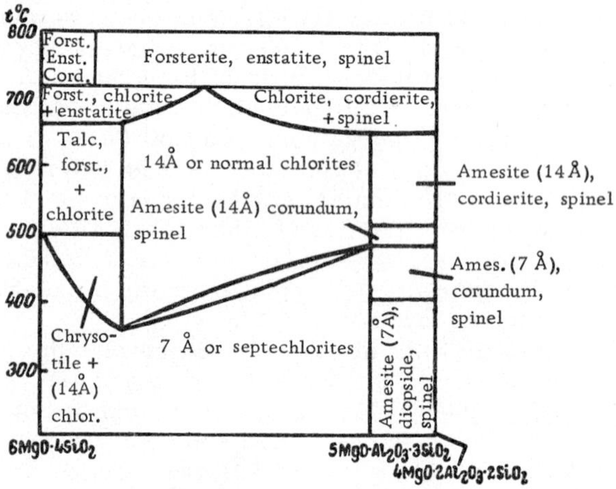

FIGURE 233. Pseudobinary section of the system $MgO-Al_2O_3-SiO_2-H_2O$, illustrating the stability of chlorites (after Nelson and Roy)

As seen from the above mentioned, by including water as a usual, i.e., inert (according to D. S. Korzhinskii) component, he reverts to the old concept of Goldschmidt and develops it.

The validity of Yoder's concept was proved by himself and other authors almost immediately after the publication of the cited paper.

Pyrope was successfully synthesized at pressures above 20,000 atm, and recently even its decomposition curve was outlined, being close to that
355 of kyanite (page 192). Nevertheless, (as is often the case) the generalization, which was based on false foundations, is preserved in literature and has a negative influence on the physicochemical interpretation of parageneses, thus undermining from the correctness of the conclusions arrived at.

THE HYDROTHERMAL SYNTHESIS OF CERTAIN MINERALS OF MULTICOMPONENT SYSTEMS

Muscovite

The synthesis and dehydration curve of muscovite were investigated by

De Vries and Roy (1953)†, as well as by Yoder and Eugster (1953). The equilibrium curve on Figure 234 shows that, with the exception of an area of very low pressures (up to 300 atm), the decomposition temperature of muscovite varies from 630° at 300 atm to 720° at 2,000 atm water pressure. This permits one to accept a temperature of 700° as the boundary between high and moderate temperature zones in crystalline schists.

The curve of the granitic eutectic is plotted on the same figure according to data of Bowen and Tuttle. The corresponding curves should intersect at a pressure of approximately 1,500 atm, i. e., above this pressure muscovite should be thermally stable with the melt, corresponding to the granitic eutectic. However, the mentioned fact can scarcely be regarded as proof of the primary origin of muscovite in nature, and the observation of textures still remains the essential criterion.

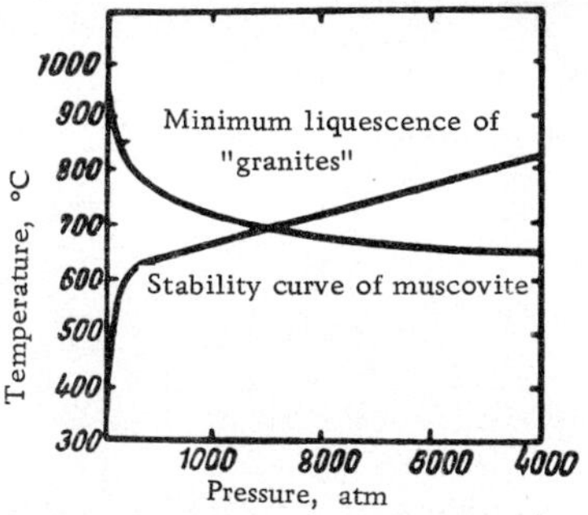

FIGURE 234. Stability curve of muscovite

The most accurate evidence for the possibility of the primary genesis of any mineral is the finding of this mineral in the form of porphyrite inclusions in an unaltered glassy rock. As muscovite has never been found in such a form despite the thorough investigation of rhyolites with a very high content of phenocrysts, the probability of the primary genesis of muscovite remains small.

Other micas

The conditions of formation of ferric biotite are described further on (see p 360). Figure 235 shows the dehydration curves of phlogopite and paragonite, also according to the data of Yoder and Eugster (Ann. Rep., 1954). In similarity to other magnesium silicates, as compared to aluminosilicates, phlogopite decomposes at much higher temperatures than muscovite, thus making biotite a stable mineral in magmatic rocks and during high temperature metamorphism. It is desirable that the relations between biotites of varying composition and SiO_2 be investigated in the near future, as this is of great significance to paragenetic analysis. From mineralogical observations it follows that at high temperatures (above the decomposition temperature of muscovite) magnesium biotite is an undersaturated silicate, and may exist together with quartz provided that it contains not less than 40 % to 50 % iron component (Korzhinskii, 1946; Sobolev, 1949), whereas at lower temperatures it is phlogopite which is in stable coexistence with quartz.

356 The dehydration curve of paragonite lies below the stability curve of muscovite. Therefore, this mineral is only stable under conditions of low-temperature metamorphism. The study of the equilibrium conditions of paragonite with muscovite and albite has not yet been carried out at lower temperatures.

† Unpublished; quoted from Roy and Tuttle (1956).

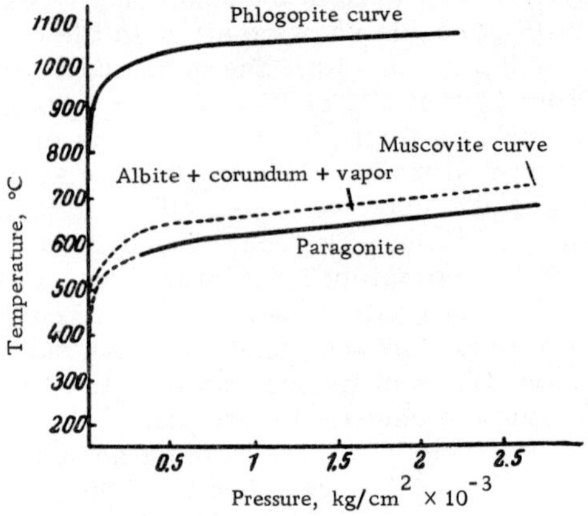

FIGURE 235. Stability curves of paragonite and phlogopite (after Yoder and Eugster)

Amphiboles

The synthesis and study of the dehydration curves of magnesium amphiboles were carried out by Boyd (1954—1956). Figure 236 shows a part of the decomposition curve of tremolite (Ann. Rep., 1954). This curve obviously represents the lower limit of stability of orthorhombic and monoclinic pyroxenes with quartz, which proves to be 850° for a pressure of 1,000 atm. With regard to the temperature of formation of tremolite, it may scarcely be regarded as being this high, as in the case of more complex systems amphiboles richer in alumina are formed.

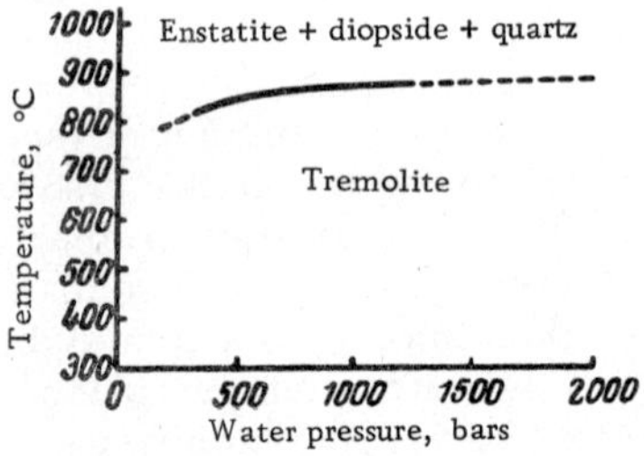

FIGURE 236. Decomposition curve of tremolite (after Boyd)

Figure 237 shows the curve limiting the stability field of aluminum-rich amphiboles of the pargasite type (after Boyd. — Ann. Rep., 1955—1956). As anticipated, the temperature of decomposition of such hornblendes is much higher than that of tremolite. In order to solve the problem of the relations between primary and secondary amphiboles in igneous rocks, as well as in crystalline schists, an investigation of amphiboles of variable composition of the tremolite-pargasite series should be carried out in the future in the presence of plagioclase or simply in an excess of alumina.

At present various minerals of the garnet group have been synthesized. However, in the majority of cases the equilibrium boundaries have not yet been elucidated, and one may only claim that they are rather different from one another.

357 Spessartite (not containing calcium) and uvarovite may crystallize directly from a melt at normal pressures. Calcium garnets of the grossularite type were synthesized, and their water content increases on lowering of temperature. Figure 238 shows the equilibrium diagram after Yoder (1950). Anhydrous grossularite was synthesized at a temperature of 800° and a pressure of 2,000 atm (Yoder, H., Ann. Rep., 1954).

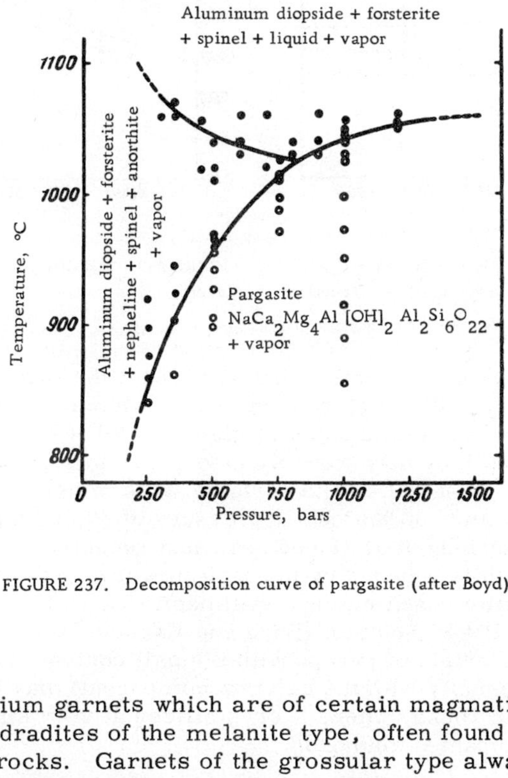

FIGURE 237. Decomposition curve of pargasite (after Boyd)

The only calcium garnets which are of certain magmatic origin are the titanium-rich andradites of the melanite type, often found as phenocrysts of leucite-bearing rocks. Garnets of the grossular type always appear to be metasomatic or metamorphic, but they are stable at low pressures (in contact with hypabyssal rocks).

The successful synthesis of almandine and pyrope was also carried out comparatively recently. Almandine was obtained at comparatively low pressures, starting from approximately 1,000 atm, but the reaction is sufficiently vigorous only at pressures above 7,000 atm. Figure 239 shows the corresponding equilibrium curve after Yoder (1959); the crosses indicate synthetic almandine and the circles with crosses — the stability field of natural almandine.

358 If one completely accepts (including the broken line) the presented equilibrium curve, one may assume that almandine is stable up to normal

pressures at temperatures lower than 800°. However, it is well known that this garnet is not formed even in contact hornfelses in andalusite zones of deeper intrusions and in schists. As already mentioned, almost no syntheses of almandine have been carried out at low pressures, and it may be assumed that the area of its stability has a different limit (see Figure 240).

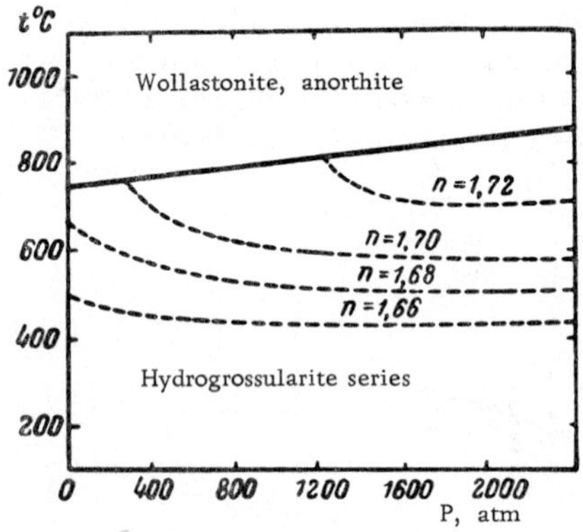

FIGURE 238. Schematic stability diagram of grossularite and hydrogarnets (after Yoder)

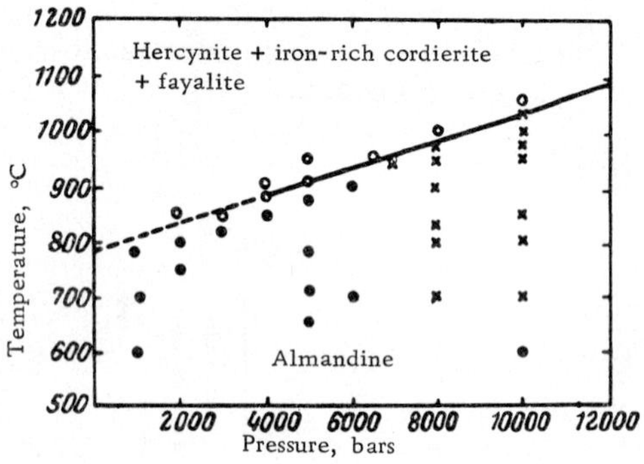

FIGURE 239. Stability curve of almandine (after Yoder)

Incidentally, we shall mention that at present the igneous origin of almandine (usually with a high content of the calcium component) may be considered as proved beyond doubt in glassy rhyolites and dacites (phenocrysts), and hence they may possibly also occur in granites. The synthesis of pyropes is the most difficult to perform. This mineral was obtained by Coes at a temperature of 900° and a pressure of 30,000 atm. Later on the synthesis was carried out at 20,000 atm, and recently an area of stability was outlined which is bounded by the melting curve and the decomposition curve (into enstatite, saphirite and sillimanite ?), with an intersection point at t = 1,510° and P = 21,600 atm (Boyd and England).

The magmatic origin of pyrope with a small content of the almandine component (and usually with the calcium component) may be assumed in certain ultrabasic rocks, where it crystallizes at very high pressures and sometimes accompanies diamonds.

Chloritoid and staurolite

The synthesis of chloritoid was carried out by Halferdahl (Ann. Rep., 1956) at a temperature of 675 ± 25° and a pressure of 10,000 atm. The stability field of chloritoid has not yet been determined, but there are indications that the attempts at attaining it at a pressure of 2,000 atm and a temperature of below 600° yielded negative results. It may also be noted that, in similarity to other silicates containing iron, the stability of chloritoid is also dependent on the redox potential of the gaseous mixture.

Starting from a pressure of 7,000 atm (at a temperature of 650°) staurolite is formed during the decomposition of chloritoid. The synthesis of staurolite was accomplished recently. At pressures below 7,000 atm staurolite was not obtained. The decomposition of natural chloritoid at 2,000 atm (t = 600°) resulted in the formation of cordierite and hercynite, and not almandine, which again raises doubts as to the possible formation of the latter at such comparatively low pressures.

Figure 240 presents preliminary data on the decomposition curve of chloritoid at various pressures (Ann. Rep., 1957).

359 Tourmaline

The synthesis of iron-bearing tourmaline was accomplished by M. Michel-Lévy (1949) at a pressure of 500 atm and a temperature of 500°. Accurate data on the pressure and temperature of the experiments carried out by other authors who synthesized tourmaline, are absent.

A great number of experiments on the synthesis of tourmaline from various aluminosilicates in the presence of boric acid, magnetite, NaCl and water were carried out by Frondel and Collete (1957). The temperature of experiments ranged from 350° to 550° and the corresponding pressures from 2,000 to 700 atm.

Tourmaline was obtained from many minerals; certain silicates which are rich in calcium (grossularite) and alkalis (orthoclase, albite, biotite, etc) constitute an exception.

Epidote and zoisite

The first positive results on the synthesis of epidote and zoisite were obtained at relatively high pressure (2,000 atm) and a temperature of 800°. However, one may claim that these minerals are stable at lower pressures and temperatures. The increase in iron content undoubtedly lowers the lower pressure limit.

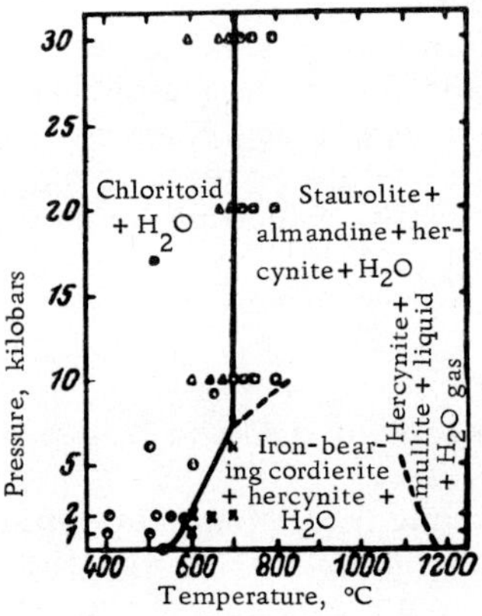

FIGURE 240. Stability curve of chloritoid (after Halferdahl)

The schematic diagram of epidotization of Ramberg's plagioclases (see Barth, 1956) does not show true temperatures. According to Ramberg, epidotization of anorthite starts at 450°, and that of oligoclase ($Ab_{80}An_{20}$) is lowered to 300°. There is, however, every reason for assuming that the beginning of this reaction is related to the temperature range of approximately 650—550° (see page 354).

Oxidation-reduction reactions occurring in the presence of a gaseous mixture are of the greatest importance in nature. One may now regard as established the fact that, with the exception of reactions occurring at the surface and in its close vicinity, free oxygen is absent in such gaseous mixtures, and the oxidizing agents are carbon-dioxide and water. As all components of the gas are in equilibrium, it is sufficient to consider only one of their reactions. D. S. Korzhinskii (1940) showed the significance of Boudouard's reaction, which is considered in brief further on. At present interesting experiments are being carried out on the synthesis of iron-bearing silicates in equilibrium with water and hydrogen, i. e., essentially an oxidation reaction which is connected with the dissociation of water (see also page 372); the results of these investigations are presented in brief in this chapter.

Aegirine-water equilibrium

Equilibrium between aegirine and water vapor at pressures up to 5,000 atm was investigated by I. A. Ostrovskii in 1957.

It showed that under these conditions aegirine melts incongruently. As
360 distinct from the "dry" system, the new solid phase is in this case not hematite but magnetite. The equilibrium diagram is shown on Figure 241.

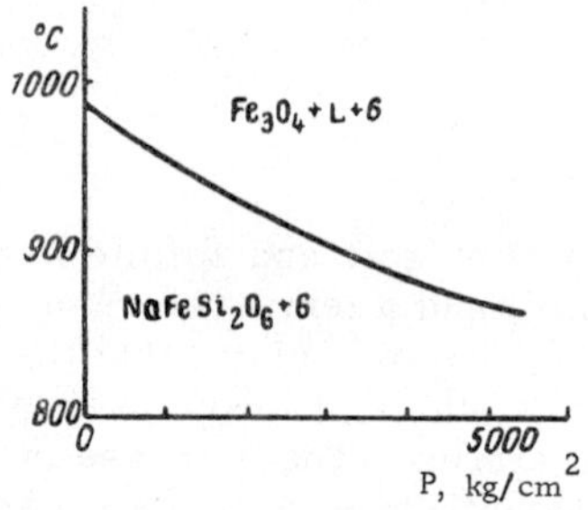

FIGURE 241. The system $NaFeSi_2O_6$-H_2O (after Ostrovskii)

It is of interest to mention that at the roof of the autoclave, where the temperatures are somewhat lower than in the main body, crusts and druses of aegirine crystals accumulated without inclusions of magnetite, having formed from the gaseous phase.

These experiments were carried out in the absence of hydrogen or in the presence of only very small amounts of it. On the addition of a considerable amount of hydrogen, aegirine disappears and minerals containing ferrous oxide appeared: fayalite, alkaline amphibole, etc.

The equilibrium of biotite in the presence of H_2O and H_2

I. A. Ostrovskii (1956, 1957) synthesized ferric biotite and determined the ratio $H_2 : H_2O$ during the decomposition reaction of biotite with the formation of magnetite. This ratio proved to be approximately equal to 1 : 16 at a temperature of 715° and a pressure of 4,000 atm.

American investigators have adopted a somewhat different approach to the determination of this equilibrium. Eugster in 1956 (Ann. Rep., 1955–1956) presented a preliminary equilibrium diagram for ferric biotite

(annite) as dependent on pressure and partial pressure of oxygen at an overall pressure of 2,000 atm. Obviously, as the result of the dissociation reaction of water — $2H_2O = 2H_2 + O_2$, a certain equilibrium is established between hydrogen and oxygen, and the values of their partial pressures are interrelated: an increase of the hydrogen pressure in the gaseous mixture should retard the dissociation of water, thus lowering the partial pressure
361 of oxygen and lowering the oxidizing ability of the gaseous mixture. Thus, there is a strict correspondence between the presented values of "oxygen pressure" and the partial pressure of the hydrogen in the mixture.

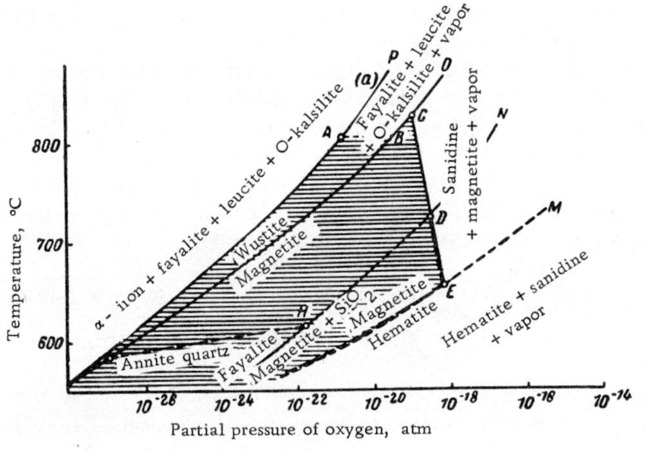

FIGURE 242. The stability field of ferric biotite (annite) as dependent on temperature and partial pressure of oxygen (after Eugster)

The diagram represented on Figure 242 shows the upper stability limit of ferric biotite at approximately 800°. According to I. A. Ostrovskii, "at a total pressure of 4,000 kg/cm^2 and a partial hydrogen pressure approximately equal to 1/12 of the total pressure, the temperature of dissociation of biotite is 840° ± 20""".

Equilibrium of fayalite with H_2O and H_2

The equilibrium of fayalite with water and hydrogen was investigated by I. A. Ostrovskii (1957). Unexpectedly, it was shown that at a water vapor pressure of 5,000 atm fayalite does not melt even at a temperature of 900°, and only some data on the conditions of equilibrium of the following reaction were obtained:

$$3Fe_2SiO_4 + 2H_2O \rightleftarrows 2Fe_3O_4 + 3SiO_2 + 2H_2.$$

At a temperature of 800° and a pressure of 4,000 atm the equilibrium pressure of hydrogen is approximately 1/8 of the total pressure of the water-hydrogen gaseous mixture.

In his conclusions, I. A. Ostrovskii (1957) states "the obtained data allow for the expression of certain preliminary assumptions on the pressure limits of hydrogen in a gaseous phase which is in equilibrium (at approximately 4,000 atm) with natural granitic melts within the temperature range of 700—800°. The fact that in granites biotite does not decompose into a mixture of orthoclase and magnetite allows for the assumption that the partial pressure of hydrogen in them is higher than 1/16 of the total pressure. On the other hand, the normal association in granites consists of magnetite and quartz, and fayalite is not characteristic. Therefore, the partial pressure of hydrogen should not exceed 1/8 of the total pressure of the water-hydrogen gaseous mixture".

However, it should be added that the equilibrium between the ferric biotite, fayalite, quartz and potassium feldspar is undoubtedly also dependent on the concentration of K_2O in the system. In reality, in all natural granitic biotites the molecular ratio $Al_2O_3 : K_2O$ is much higher than unity, and hence the reaction with the formation fayalite instead of biotite is accompanied by the absorption of K_2O, e. g., according to the following scheme (iron oxide is not taken into account):

$$10KFe_{2.4}Al_{0.4}[OH]_2\boxed{AlSi_3O_{10}} + 2K_2O + 24SiO_2 \rightleftarrows$$
$$\rightleftarrows 12Fe_2SiO_4 + 14KAlSi_3O_8 + 10H_2O.$$

Thus, the decrease in the concentration of K_2O in the melt, all other conditions being equal, will facilitate the formation of fayalite (or orthorhombic pyroxene instead of ferromagnesium biotites). In general the maximum possible limit of hydrogen pressure of the gaseous mixture (which corresponds to the lowest oxidation potential of water and Boudouard's reaction with a mixture of water and carbon dioxide), according to Ostrovskii, is much higher than the presented figure — approximately 2/3 of the total pressure, but may only be attained in the presence of metallic iron. The finding of native iron in basic rocks is very rare, i. e., the mentioned limit is only attained in very rare instances.

The synthesis of sodium-iron amphiboles

The synthesis of sodium-iron amphiboles (riebeckite type?) was carried out by I. A. Ostrovskii in equilibrium with H_2O and H_2 in 1956. Quantitative
362 data are lacking, but a transition from the sodium-iron amphibole to fayalite was observed, the formation of which, according to the above-presented data, corresponds to a partial hydrogen pressure not less than 1/8 of the total pressure". The fayalite-alkali amphibole inversion temperature in a melt with the participation of a gaseous phase decreases slightly from 725° to 695° as pressure is increased from 125 to 1,000 kg/cm^2". According to Ostrovskii, this provides one with an idea of upper temperature limit of stability of alkali amphiboles. An opinion has also been expressed that with a lower content of hydrogen cossyrite becomes stable, and that aegirine is only stable in the absence of hydrogen. However, the relations between the mentioned minerals are probably more complex, as in nature the

paragenesis of riebeckite with aegirine is known, whereas cossyrite is only found very rarely.

Eugster indicates a somewhat higher temperature of decomposition for magnesium riebeckites, increasing with increasing oxygen pressure, as shown on Figure 243.

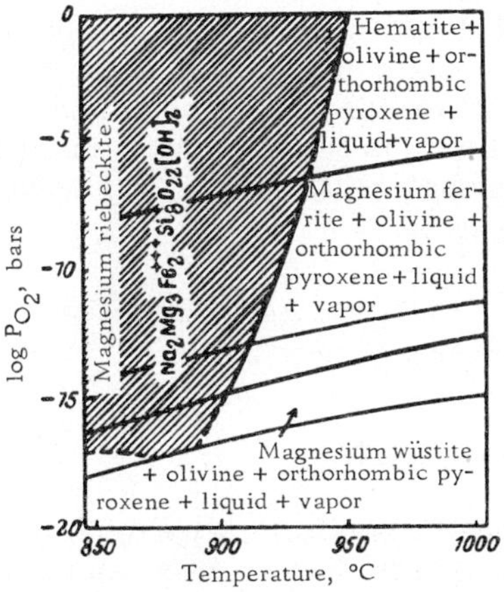

FIGURE 243. Equilibrium conditions of magnesium riebeckite (after Eugster)

THE SYNTHESIS AND INVESTIGATION OF FLUORINE-BEARING SILICATES

The synthesis of fluorine-bearing silicates and their investigation constituted a most important step in the study of silicates containing volatile components in general. The particular technique used in such experiments is due to the fact that fluorine has a comparatively low volatility, and melting is achieved at normal atmospheric pressure. Under these conditions it is difficult to present a complete equilibrium diagram of the system as the fluorine is vaporized. The mentioned peculiarities allowed for the wide development of experimental works long before the inception of true hydrothermal synthesis.

Of greatest interest in this field are the investigations of D. P. Grigor'ev, which were carried out over a period of approximately ten years (1935—1944). The principal idea behind these investigations is the fact that fluorine may completely replace the hydroxyl radical in a series of water-bearing minerals; the corresponding fluorine varieties may be synthesized on melting. In practice, Grigor'ev succeeded in synthesizing fluorine micas and fluorine amphiboles. Both phlogopite and ferromagnesium biotite were obtained as well as tremolite and hornblende. In addition, varieties which differ in composition (in addition to fluorine content) from natural minerals were obtained, and in particular amphiboles with a higher calcium content as well as biotite containing calcium.

The reactional relations obtained by Grigor'ev are of specific interest to petrology, as well as the data on the immiscibility of silicate melts. He succeeded in obtaining reaction rims of amphibole around pyroxene and of biotite around amphibole. These reactional relations are undoubtedly primary, i. e., they were formed as a result of a reaction between an earlier separated solid phase and a melt. As is known from similar reaction rims in igneous rocks which were used by Bowen as the basis for
363 constructing his reaction scheme, doubts were expressed as to their primary igneous origin.

Grigor'ev also succeeded in obtaining the decomposition of a fluorine-bearing silicate melt into immiscible liquids in the area of melts in which a continuous miscibility in the liquid state was noted in the absence of fluorine (for further details see page 207). Lank's criticism of Grigor'ev concerning these experiments should scarcely be considered valid.

A synthesis of fluorine-bearing amphiboles was carried out by Bowen and Schairer (1935a) at almost the same time as the experiments of Grigor'ev. Using Grigor'ev's data, several authors later synthesized various varieties of micas, in particular fluor-muscovite and lithium mica. These experiments were applied in the artificial mica industry.

Among other experiments on the synthesis of fluorine-bearing silicates, one should mention the achieving of the synthesis of minerals of the chondrodite group (Rankama, 1947). The experimental conditions are in this case even closer to those pertaining in nature, as the corresponding minerals have a high fluorine content, which may not be completely replaced by hydroxyl ions.

SYSTEMS WITH THE PARTICIPATION OF CO_2

The system H_2O-CO_2 †

The given system is of great significance to mineralogy owing to the extraordinary role of water and carbon dioxide in postmagmatic solutions. Nevertheless, a sufficiently complete investigation of equilibrium in the area of high temperatures and pressures was only carried out fairly recently by N. I. Khitarov and S. D. Malinin (1958). We shall use their results for constructing the diagram (Figure 244).

The given system is characterized by a limited solubility in the liquid phase and is related to the first type with critical phenomena (see Figure 193, page 305). Owing to the very low solubility of water in carbon dioxide at low temperatures, the lower critical point almost coincides with the critical point of pure carbon dioxide. The point of four-phase equilibrium (compare with Figure 193, page 305) is somewhat lower than zero and is of no real interest to us. In addition, its representation cannot be of assistance as, according to scale, the composition of the second liquid and of the gaseous phase are identical.

The gradual transition of the critical curve L_1+G into the curve L_1+L_2. is of special interest. Correspondingly, the upper critical point is not only not observed up to a pressure of 700 atm, but it is probably completely absent. The authors of the investigation mentioned that already for the "isotherms of 200 and 250° the merging of the branches with increase of pressure may not occur, the curves passing through corresponding maxima and minima". We are in complete agreement with the mentioned assumption (see page 305) and therefore show an inflection of the critical curve in the area of higher pressure on Figure 244.

Isothermal sections are correspondingly shown on the ***PX*** and ***TX*** projections on this figure (directly according to Khitarov and Malinin) as well as isobaric sections compiled from their data. The consideration of the diagram on Figure 244 allows one to arrive at the conclusion that, whereas an unlimited solubility of carbon dioxide in water occurs at temperatures above the critical temperature of water, at lower temperatures, in particular below 300°, this solubility becomes quite limited, even at very high

† Written in conjunction with V. A. Kalyuzhnyi.

pressures, and a phase separation should occur, where not only water
but also carbon dioxide has a liquid character at considerable pressures
364 exceeding 700 atm. In the case of salt solutions the decomposition tempera-
tures of liquids should undoubtedly be raised.

The direct observations of phase relations of the considered system may be carried out on micro-inclusions of solutions in minerals. Here, at temperatures lower than 32°, we often note all three phases — water, liquid CO_2 and a gaseous bubble: carbon dioxide will markedly predominate in the gas. On heating, homogenization of the two phases of CO_2 is observed even before the temperature of 32° is reached, and subsequently at high temperatures (usually of the order of 300° or more) a complete homogenization of the inclusion is observed. The position of points on the projections for two such inclusions shows that they fall almost exclusively on the isobaric section for 580 atm (see Figure 244), thus permitting the determination of the pressure at the moment of homogenization.

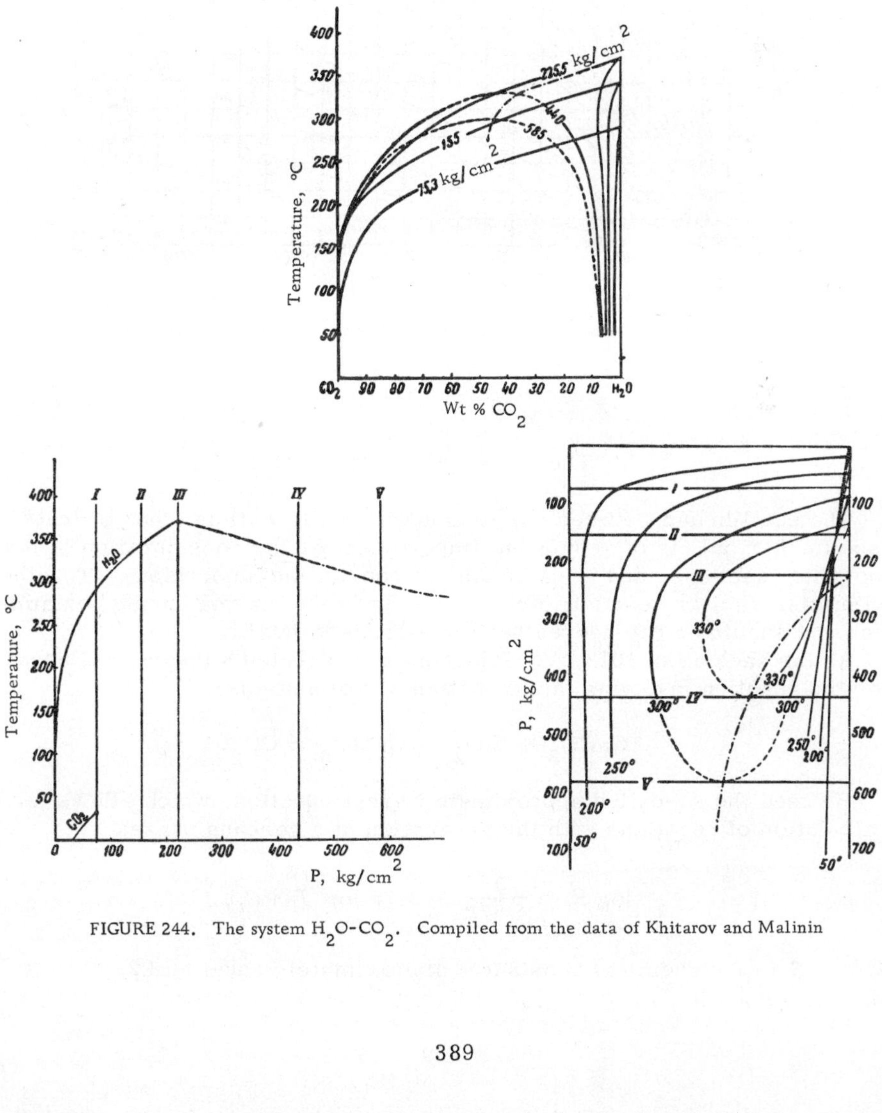

FIGURE 244. The system H_2O-CO_2. Compiled from the data of Khitarov and Malinin

During the formation of minerals at a temperature of 300° and less,
365 two types of inclusions may be noted: water with dissolved carbon dioxide, and carbon dioxide. An approximate determination of temperature and pressure may be arrived at from the point of intersection of isochores for the unary system (Kalyuzhnyi, 1955).

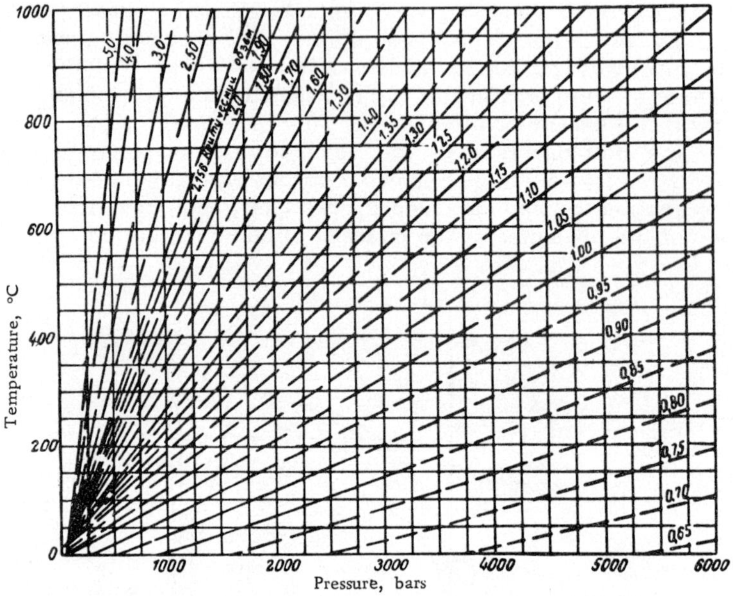

FIGURE 245. Isochores of the unary system CO_2 (after Smith)

The system $CaO\text{-}SiO_2\text{-}CO_2$

The equilibrium between calcite and silica, as well as other silicates (see further on), is of very great importance for the consideration of metamorphic reactions, and in particular of contact metamorphism. Together with this, similar reactions occur directly in the magma during contamination and should be considered here in sufficient detail.

As far back as in 1912, V. Goldschmidt presented a theoretical equation for the reaction involving the formation of wollastonite:

$$CaCO_3 + SiO_2 \rightleftarrows CaSiO_3 + CO_2.$$

He used the so-called approximate Nernst equation, which allows for the calculation of reactions with the separation of a gaseous phase:

$$\log P = \frac{Q}{4.57T} + 1.75 \log T + C_1,$$

where C_1 is the chemical constant — approximately equal to 3.2.

The heat of dissociation was accepted according to the value of D. Chernobaev and S. P. Vologdin — Q = 25,300 cal/mole. The obtained curve (Figure 246) was widely used by mineralogists up to recent times.

366 Later on several attempts were made at introducing amendments to this curve, among which we shall mention the experimental data of Kröger and Illner (1936), new calculations by Danielsson (1950) and, lastly, the latest experimental investigations of Harker and Tuttle (1956).

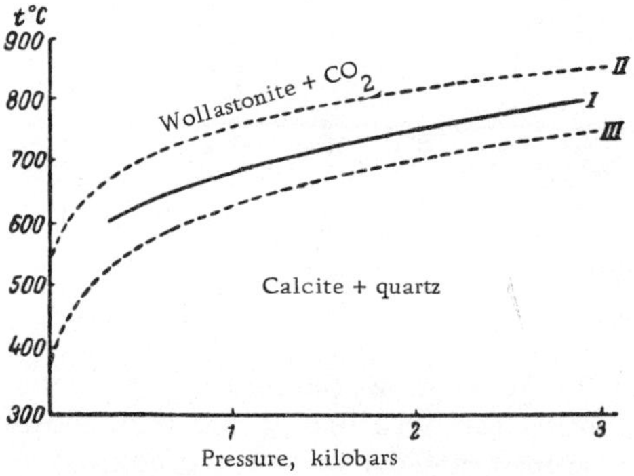

FIGURE 246. Stability curve of calcite according to Harker and Tuttle (I); according to data of Goldschmidt (II); and Danielsson (III)

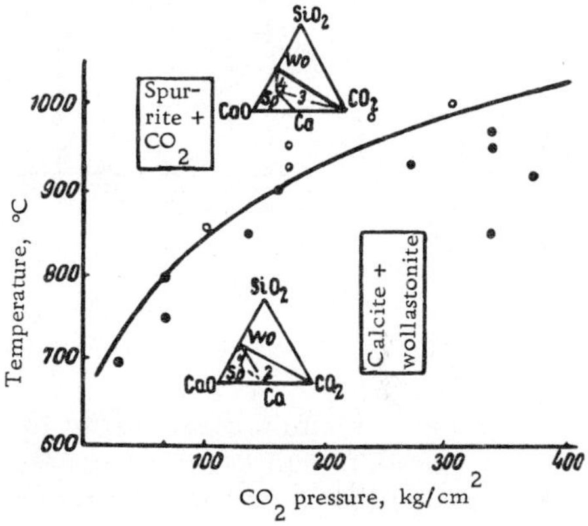

FIGURE 247. Curve of formation of spurrite (after Tuttle and Harker)

Results of investigations of the latter authors are shown on Figure 246. Using the same approximation of the Nernst equation, they determined the heat of dissociation as 23,400 cal/mole, which allowed them to calculate the extrapolation of the curve into the field of lower pressures. The curve thus obtained lies somewhat lower than Goldschmidt's curve.

A complete equilibrium diagram of the considered system is not yet available, but several of the already clarified data are of interest.

According to Eitel (1923), on the melting diagram of the system Ca_2SiO_4-$CaCO_3$ at high pressures a carbonate-silicate, $2Ca_2SiO_4 \cdot CaCO_3$, appears, which melts incongruently; according to its composition it is an analogue of spurrite, but it differs in its properties. The inversion temperature of natural spurrite in this carbonate-silicate is approximately 1,200°.

367 Tuttle and Harker (1957) indicated the lack of convincing evidence in Eitel's data concerning the fact that the obtained compound is not a calcium orthosilicate. They succeeded in synthesizing spurrite directly from a mixture of calcite and quartz glass in the presence of CO_2. On lowering the temperature the decomposition of spurrite into wollastonite and calcite occurred. The corresponding diagram is shown on Figure 247.

As is known, carbonate-silicates of other compositions are also found among natural minerals: tilleyite $Ca_5[Si_2O_7][CO_3]_2$ and scawtite $Ca_6Si_4O_{11}[CO_3]_3$.

D. S. Korzhinskii (1935) calculated the equilibrium curve $Ca_2SiO_4 + CO_2 \rightleftarrows CaCO_3 + CaSiO_3$, for which the heat of reaction was accepted as being equal to 31,400 cal. The curve is shown on Figure 248.

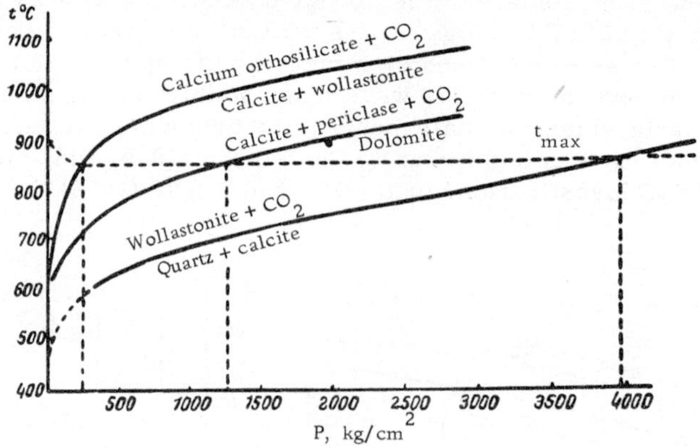

FIGURE 248. Diagram illustrating the depth facies of marbles (after Korzhinskii)

After the calculation of the formation curve of wollastonite, Goldschmidt used the obtained data for considering the temperature of metamorphism, assuming that, starting from certain comparatively low pressures, the temperature of formation of wollastonite is higher than 700°. He was not able to consider this curve as a geological manometer, as he accepted the upper temperature limit of metamorphism as being very high (1,500°, where intersection of the wollastonite curve with the line of maximum temperature could not be expected).

Later on the developments in petrography called for a revision of this limit, resulting in the considerable lowering of the maximum temperature of metamorphism (up to 850—900°). While studying the marbles of the Aldan platform, which are undoubtedly related to the high-temperature crystalline schist facies, bearing both quartz and calcite in equilibrium with one another (as was established from analyses of marbles of other Archaen complexes), D. S. Korzhinskii arrived at the conclusion that these rocks were formed at such high CO_2 pressures that the temperature formation of wollastonite was higher than the temperature of metamorphism. According to Goldschmidt's curve, a temperature of 850° corresponds to a pressure of 4,000 atm, which is the limiting pressure for the formation of wollastonite in marbles. According to recent data, this limit should be somewhat higher (greater than 5,000 atm).

368 The mentioned data were used by Korzhinskii as a basis for his theory of depth facies (see Figure 248). He related the Aldan marbles with their stable calcite-quartz paragenesis in the high-temperature facies to the deepest facies. Later this facies was subdivided according to criteria of stability or instability of grossularite. This, however, met with strong opposition. To the most shallow facies are related marbles with larnite Ca_2SiO_4 (calcium orthosilicate), which thus far have only been determined in one locality on the globe (Ireland), as well as marbles with more common (quite rare) mineral — spurrite, merwinite and others, which have since been discovered in more than ten localities. In the USSR such rocks have been described in the basin of the lower Tunguska (Sobolev, 1935) in contact

with the intrusion of gabbro-diobases (trapps). In addition to a small CO_2 pressure, an increased temperature of metamorphism is necessary, as is the case at contacts of hypabyssal basic intrusions. Between the above-mentioned two facies, Korzhinskii distinguishes marbles with monticellite and galenite and those with periclase. The upper stability limit of periclase corresponds to the intersection of the dolomite dissociation curve with the line of maximum temperature (approximately 1,200 atm).

As already mentioned previously, the dissociation reactions, and in all events the reactions of carbonatization, are not dependent on the total pressure, but on the partial pressure of CO_2. By terming the corresponding facies as depth facies, Korzhinskii accepts the hypothesis† that the partial pressure of CO_2 increases in proportion to depth in metamorphic processes.

The objections found in the literature are either directed against the mentioned hypothesis or are connected with a misunderstanding, when the quartz-calcite paragenesis is considered in marbles metamorphosed at comparatively low temperatures (in these cases it is obviously also stable at low CO_2 pressures).

An analogous interpretation of the reaction with the liberation of CO_2 for assessing the pressure, without resource to the data of Korzhinskii, was carried out by Bowen in 1940. He drew attention to the difference between progressive metamorphism (liberation of carbon dioxide) and regressive metamorphism (absorption of carbon dioxide) during the course of the reaction. Whereas the reaction of decarbonitization depends on the total pressure (in his opinion), the reaction of carbonitization is dependent on the partial pressure of carbon dioxide. As mentioned on page 328, such a difference scarcely exists for conditions of equilibrium and in particular for prolonged processes of mineral formation. However, the available experimental data are as yet insufficient, and the question is still open to dispute.

The system $NaAlSiO_4$-$CaCO_3$-CO_2

Figure 249 shows the equilibrium of the system $NaAlSiO_4$-$CaCO_3$ at a CO_2 pressure of approximately 110 atm according to Eitel (1922). The cancrinite, which, in contrast to natural cancrinite, does not contain water, melts incongruently and crystallizes together with calcite at the eutectic point.

Eitel assumes that the mentioned diagram is in good agreement with the concepts on the formation of primary calcite in nepheline-bearing syenites. However, at the mentioned pressure calcite cannot crystallize together with nepheline, with the exception of the case when nepheline is preserved in the form of unaltered relicts surrounded by a margin of cancrinite in the mesostasis of cancrinite and calcite.

369 In real nepheline-bearing syenites, which contain calcite (see for example the Botogolsk rocks; Sobolev, 1947), nepheline predominates over calcite

† This hypothesis is expressed in the works of D. S. Korzhinskii (1932, 1935 and 1940).

and reaction rims of cancrinite are undoubtedly of post-magmatic origin. At the same time the formation of garnet reaction rims around calcite (temperatures above 600°—700°) oblige one to assume that calcite, which undoubtedly existed at a sufficiently high temperature, is actually found in the magma and is a xenogenic mineral.

The forementioned considerations do not apply to the controversial question of the possible primary origin of calcite in carbonatites.

The system $CaCO_3$-$MgCO_3$-CO_2

Figure 250 shows the dissociation curves of the most important carbonates according to the most recent data of Harker and Tuttle (1955). The figure shows that, whereas the dissociation of calcite calls for quite high temperatures (usually not attained in nature) even at comparatively low pressures the decomposition of dolomite and magnesite occurs at considerably lower temperatures which are within the normal temperature range of magmatic and metamorphic processes.

Already at atmospheric pressure the dissociation temperature of calcite equals 898° and increases rapidly as pressure is increased; the melting temperature is 1,340° at a pressure of 110 atm. Hence, it is obvious that calcite (in the absence of free SiO_2) is usually quite stable in contact with igneous rocks, and the question of its primary crystallization in alkaline rocks may be raised — and in all events of the preservation of calcite xenoliths trapped in magma.

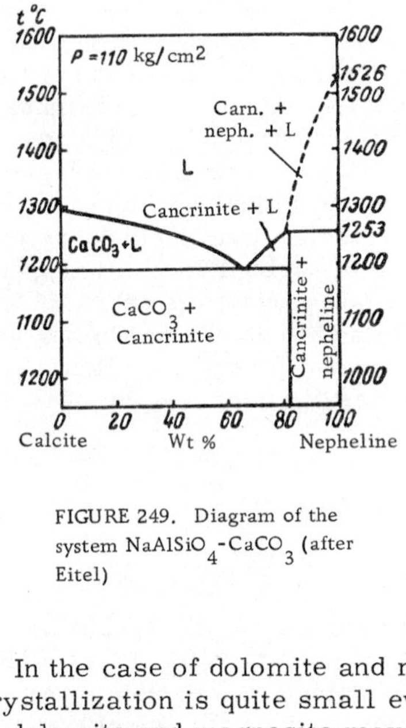

FIGURE 249. Diagram of the system $NaAlSiO_4$-$CaCO_3$ (after Eitel)

370 In the case of dolomite and magnesite, the probability of their primary crystallization is quite small even at high pressures. The stability curves of dolomite and magnesite may provide us with an idea of the temperature of formation of periclase in metamorphic rocks. Accepting the upper temperature limit of metamorphism (with the exception of specific cases of hypabyssal contacts of basic rocks) as approximately 850°, we note that in dolomite periclase may form up to a pressure of 1,200 atm. In magnesite the formation of periclase is possible up to 2,000 atm, but in comparison with dolomites such rocks are rather rare.

We shall here mention that the dissociation curve of siderite is situated considerably lower than the magnesite curve, and that an admixture of iron in magnesium carbonates lowers the dissociation temperature.

In 1958 Harker carried out an interesting experiment on the stability of magnesite at different partial pressures of CO_2 in the presence of an inert gas — argon. The author arrived at the conclusion that the temperature of dissociation depends mainly on the partial pressure of CO_2 and not on the total pressure. However, it is worth mentioning that this conclusion was arrived at on the basis of a comparatively small number of experiments, whereas from another series of experiments it follows that the temperature of dissociation also increases with increase of total pressure. Thus, the question (of mineralogical importance) of the role of partial pressure of CO_2 in the total pressure of the gaseous mixture during the dissociation of a carbonate, as in the case of the dissociation of minerals containing water (see page 348, data on analcite), calls for further investigation.

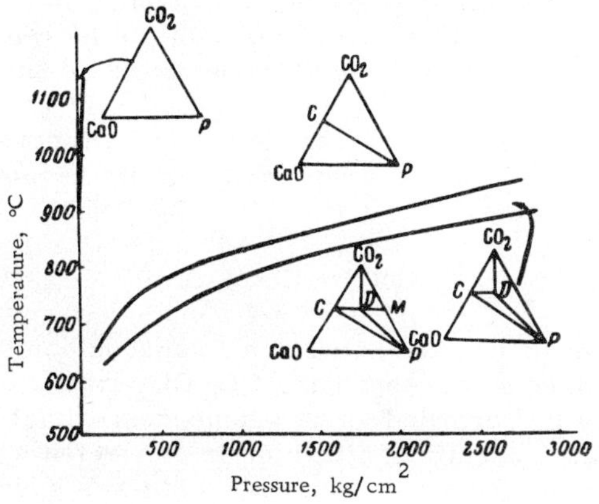

FIGURE 250. The system MgO-CaO-CO_2. Decomposition curves of magnesite (M) and dolomite (D)

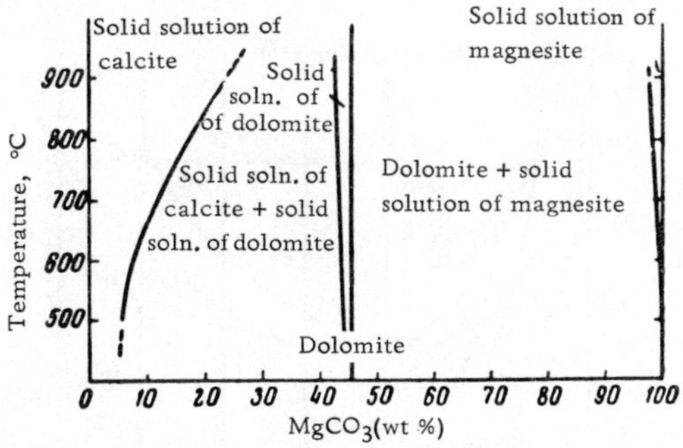

FIGURE 251. Solid solutions in the system $CaCO_3$-$MgCO_3$ (after Harker and Tuttle)

In the above-mentioned experiments of Harker and Tuttle the stability of solid solutions in the considered system was also determined. The results are shown on Figure 251. Whereas the solubility of the calcium component in magnesite and dolomite is fairly small even at high temperatures, the solubility of the magnesium component in calcite increases rapidly with increase in temperature, and hence the composition of calcite in the presence of dolomite may serve as an indication of the temperature of its formation.

Carbon monoxide and Boudouard's reaction

It has already been mentioned previously that, due to the absence of free oxygen in the gaseous mixture, phenomena of oxidation are connected with the interaction of other gaseous components, in particular water. The
371 oxidizing ability of the latter decreases with the increase in content of free hydrogen, thus determining the character of the reaction (see page 359).

As carbon dioxide is also present in the endogeneous gaseous mixture, the oxidation reaction may be associated with its decomposition, according to the following scheme:

$$CO_2 \rightarrow CO + O.$$

As equilibrium is attained between all gaseous components, each $H_2 : H_2O$ ratio will have a corresponding $CO : CO_2$ ratio, i. e., either one of these values may be regarded as an independent variable. However, the $CO : CO_2$ ratio has a certain limit which is connected with the so-called Boudouard reaction, the significance of which was specially stressed by D. S. Korzhinskii (1940) and A. G. Betekhtin (1949) in connection with the reactions of mineral formation.

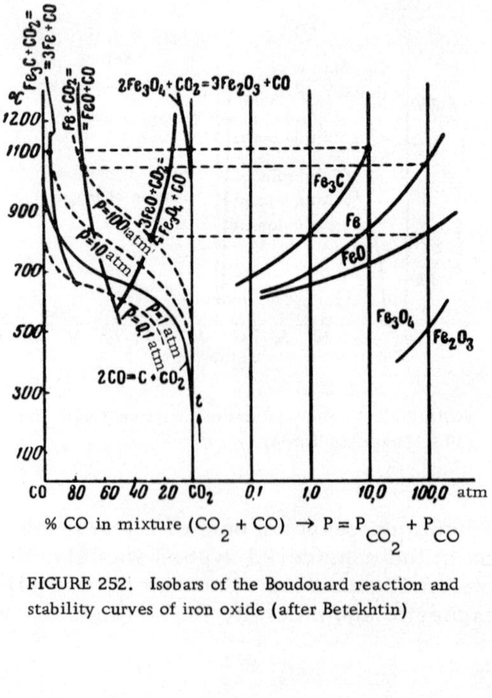

FIGURE 252. Isobars of the Boudouard reaction and stability curves of iron oxide (after Betekhtin)

When the CO content in the gaseous mixture reaches its lower limit, a type of auto-oxidation occurs according to the scheme:

$$2CO \rightarrow CO_2 + C,$$

which is termed the Boudouard reaction. Obviously, the reaction corresponds to the lowest oxidizing ability of the gaseous mixture, and to the highest hydrogen content.

As the volume of solid carbon may, to all intents and purposes, be accepted as zero, the Boudouard reaction proceeds with a considerable reduction in volume, and an increase of pressure should favor the formation of carbon dioxide. Thus, on increasing pressure the limiting content of carbon monoxide decreases, and the lower limit of oxidation potential of the mixture correspondingly increases. Therefore at high pressures of the gaseous mixture of carbon dioxide the appearance of native iron becomes impos-
372 sible at great depths. Obviously, this only holds true up to the moment when all the CO_2 is reduced to C.

Boudouard's reaction is accompanied by considerable liberation of heat. Correspondingly, the increase in temperature will hinder the formation of CO_2 and lead to an increase in the limiting content of CO_2 and a lowering of the lower boundary of oxidation potentials.

Boudouard's reaction and its significance in the oxidation and reduction of iron have been thoroughly studied in connection with metallurgical processes. Figure 252 shows isobars of the Boudouard reaction (after A. G. Betekhtin) as well as curves of reduction of iron oxides by carbon monoxide. The points of intersection of the mentioned curves with the isobars show the corresponding equilibrium conditions. Thus, while the formation of native iron at a total pressure of 10 atm is possible above temperatures of 820°, at 100 atm this reaction already calls for temperatures greater than 1,000°.

It is self-evident that the lowering of the oxidizing ability of the gaseous mixture up to the Boudouard reaction, should be connected with the activity of reducing agents. Hydrogen may be such a reducing agent if its content is higher than the limit of the corresponding Boudouard reaction. Very often in nature various hydrocarbons of organic or endogenous origin probably serve as such reducing agents. The hydrocarbons themselves also yield carbon on oxidation, which is formed according to the scheme:

$$CH_4 + CO_2 \rightarrow 2C + 2H_2O.$$

In this case the Boudouard reaction should constitute an intermediate stage.

Hydrogen sulfide (as well as other sulfides) is also an important component of the gaseous mixture and plays an active role in oxidation and reduction reactions. The separation of sulfides is directly connected with these reactions, as well as the acidity of solutions, which increases markedly during the formation of sulfuric acid from hydrogen sulfide. This process is most evident in those cases when endogenous solutions meet ascending waters containing free oxygen. This results in a marked oxidation reaction and an increase in acidity, which also results in the change of character of metasomatism. The potassium feldspathization is replaced by alunitization. All these reactions are most significant from the point of view of mineralogy (Betekhtin, 1949, 1953, et al.). However, as they occur in the post-magmatic stage, their consideration does not fall within the scope of this book.

* * *

The last fifteen years were marked by great advances in the experimental investigation of silicate systems with volatile components. For the first time it became possible to simulate natural conditions in the laboratory, by carrying out experiments at high temperatures and pressures; initially of the order of 2,000—5,000 atm and today up to pressures of at least 40,000 atm which already correspond to the pressures in the deeper parts of the Earth's crust. The results obtained, which were presented in brief and discussed in this chapter, allow one to consider from a new angle the elucidation of several questions of theoretical petrology: the problem of magma formation, the relations between magma and post-magmatic solutions, the determination of temperature and pressure of the most important reactions of mineral formation. If at present many problems are as yet incompletely solved, it is certain that in the near future wider, more systematic experimental material will become available, which will serve as a reliable basis for the elucidation of the physicochemical conditions of igneous and metamorphic processes. This, in its turn, will provide a solution to the most important problems of recent geology.

BIBLIOGRAPHY

Publications in Russian

Betekhtin, A. G. O vliyanii rezhima sery i kisloroda na parageneticheskie sootnosheniya mineralov v rudakh (The Influence of the Sulfide and Oxide Regimes on the Paragenetic Relations of Minerals in Ores). — Izvestiya AN SSSR, seriya geologicheskaya, No. 3. 1939.

Betekhtin, A. G. Gidrotermal'nye rastvory, ikh priroda i protsessy rudoobrazovaniya (Hydrothermal Solutions, Their Nature and Processes of Ore Formation). — Sbornik "Osnovye problemy v uchenii o magmatogennykh rudnykh mestorozhdeniyakh". Izdatel'stvo AN SSSR. 1953.

Bilibin, Yu. A. Petrografiya Illymakhskogo intruziva (The Petrography of the Illmakh Intrusion). — Gosgeolizdat. 1947.

Botvinkin, O. K. Fizicheskaya khimiya silikatov (Physical Chemistry of Silicates). — Promstroiizdat. 1935.

Chirvinskii, P. N. Srednii khimicheskii sostav glavnykh mineralov izverzhennykh, metamorficheskikh i osadochnykh porod (Average Chemical Composition of the Predominant Minerals of Igneous, Metamorphic and Sedimentary Rocks). — L'vovskoe geologicheskoe obshchestvo. 1953. (First edition 1928).

Ermakov, N. P. Issledovaniya mineraloobrazuyushchikh rastvorov (Investigations of Mineral-Forming Solutions). 1950.

Fedorov, E. S. Vazhnyi shag nauchnoi petrografii (Important Undertaking in Petrography). — Zapiski Gornogo instituta, Vol. 6. 1916.

Fersman, A. E. Pegmatity (Pegmatites). Vol. I, 3rd ed. —Izdatel'stvo AN SSSR. 1940.

Fischer, R. Nesmesimost' rasplavov soderzhashchikh okisly tyazhelykh metallov, silikaty i fosfaty i znachenie nesmesimosti dlya geokhimii i ucheniya o rudnykh mestorozhdeniyakh (Immiscibility of Melts Containing Oxides of Heavy Metals, Silicates and Phosphates and the Significance of Geochemistry for the Study of Ore Deposits). — Sbornik "Eksperimental'nye issledovaniya v oblasti petrografii i rudoobrazovaniya". Izdatel'stvo IL. 1954 (original reference not found).

Galakhov, F. Ya. Izuchenie glinozemistoi oblasti alyumosilikatnykh sistem (Study of the Alumina Field in Aluminosilicate Systems). — Avtoreferat dissertatsii, Leningrad. 1958.

Gerasimov, A. P. Geologicheskoe stroenie Mineralovodskogo raiona (The Geological Structure of the Mineral'nye Vody District). — Trudy TsNIGRI, Vol. 93. 1937.

Ginzberg, A. S. Eksperimental'naya petrografiya (Experimental Petrography). — Izdatel'stvo Leningradskogo Gosudarstvennogo Universiteta. 1951.

Grigor'ev, D. P. O roli ftoristykh khloristykh i vol'framovokislykh soedinenii pri iskusstvennom poluchenii magnezial'nykh slyud (The Role of Fluoride, Chloride and Molybdate Compounds in the Preparation of Magnesian Micas). — Zapiski Vserossiiskogo mineralogicheskogo obshchestva, Part 34, No. 2. 1935.

Grigor'ev, D. P. O tak nazyvaemykh "temperaturakh plavleniya" nekotorykh mineralov (The So-called "Temperatures of Melting" of Certain Minerals). — Zapiski Vserossiiskogo mineralogicheskogo obshchestva, Part 66, No. 4. 1937.

Grigor'ev, D. P. Sintez i issledovanie biotita (Synthesis and Investigation of Biotite). — Doklady AN SSSR, Vol. 20, No. 5. 1938.

Grigor'ev, D. P. Iskusstvennoe poluchenie amfibolov (Artificial Preparation of Amphiboles). — Zapiski Vserossiiskogo mineralogicheskogo obshchestva, Part 68, No. 2. 1939.

Grigor'ev, D. P. O reaktsionnom printsipe v petrogenezise (The Reaction Principle in Petrogenesis). — Sovetskaya geologiya, No. 12. 1939.

Grigor'ev, D. P. Sintez i issledovanie flogopita (Synthesis and Investigation of Phlogopite). — Doklady AN SSSR, Vol. 93, No. 2. 1944.

Grigor'ev, D. P. and E. V. Iskyul'. Differentsiatsiya nekotorykh rasplavov kak rezul'tat obrazovaniya dvukh nesmeshivayushchikhsya zhidkostei (Differentiation of Certain Melts as the Result of the Formation of Two Immiscible Liquids). — Izvestiya AN SSSR, otdel matematicheskikh i estestvennykh nauk. 1937.

Kalyuzhnii, V. A. Zhidkie vklyucheniya v mineralakh kak geologicheskii barometr (Liquid Inclusions in Minerals as a Geological Barometer). — Mineralogicheski sbornik L'vovskogo geologicheskogo obshchestva, No. 9. 1955.

Karapet'yants, M. Kh. Khimicheskaya termodinamika (Chemical Thermodynamics). — Goskhimizdat. 1953.

Khitarov, N. I., A. B. Slutskii and R. V. Arsen'eva. Sintez i kharakteristika koezita — ustoichivoi modifikatsii kremnezema pri vysokikh davleniyakh (Synthesis and Characteristics of Coesite — the Stable Modification of Silica at High Pressures). — "Geokhimiya", No. 8. 1957.

Korzhinskii, D. S. Termodinamika i geologiya nekotorykh metamorficheskikh reaktsii s vydeleniem gazovoi fazy (Thermodynamics and Geology of Certain Metamorphic Reactions Involving the Separation of a Gaseous Phase). — Zapiski Vserossiiskogo mineralogicheskogo obshchestva, Part 64, No. 1. 1935.

Korzhinskii, D. S. Faktory mineral'nykh ravnovesii i mineralogicheskie fatsii glubinnosti (Factors of Mineral Equilibria and Mineralogical Plutonic Facies). — Trudy IGN AN SSSR, No. 12, petrologicheskaya seriya, No. 5. 1940.

Korzhinskii, D. S. Printsip podvizhnosti shchelochei pri magmaticheskikh yavleniyakh (The Principle of Mobility of Alkalis During Igneous Phenomena). — Sbornik, posvyashchennyi akademiku D.S. Belyankinu. Izdatel'stvo AN SSSR. 1946.

Korzhinskii, D. S. Otkrytye sistemy s vpolne podvizhnymi komponentami i pravilo faz (Open Systems with Completely Mobile Components and the Phase Rule). — Izvestiya AN SSSR, seriya geologicheskaya, No. 2. 1949.

Korzhinskii, D. S. Problemy petrografii magmaticheskikh porod, svyazannye so skvoz'magmaticheskimi rastvorami i granitizatsiei (Problem of the Petrography of Igneous Rocks Connected with Trans-magmatic Solutions and Granitization). — In: "Magmatizm i svyaz' nim poleznykh iskopaemykh". Izdatel'stvo AN SSSR. 1955.

Korzhinskii, D. S. Vyvod termodinamicheskikh potentsialov sistem s podvizhnymi komponentami (The Derivation of Thermodynamic Potentials of Systems with Mobile Components). — Doklady AN SSSR, Vol. 106, No. 2. 1956.

Korzhinskii, D. S. Fiziko-khimicheskie osnovy analiza paragenezisov mineralov (Physicochemical Basis of the Analysis of the Paragenesis of Minerals). — Izdatel'stvo AN SSSR. 1957.

Kurnakov, N. S. Vvedenie v fiziko-khimicheskii analiz (Introduction to Physicochemical Analysis). Fourth edition, edited by V. Ya. Anosova i M. A. Klochko.—Izdatel'stvo AN SSSR. 1940.

Lebedev, V. I. Nekotorye rezul'taty izucheniya kaolina i brusita pri pomoshchi krivykh nagrevaniya pod davleniem (Certain Results of Investigations of Kaolinite and Brucite Under Pressure by Means of Heating Curves). — Trudy Pyatogo soveshchaniya po eksperimental'noi i tekhnicheskoi mineralogii i petrografii. Izdatel'stvo AN SSSR. 1958.

Leontovich, M. A. Vvedenie v termodinamiku (Introduction to Thermodynamics). — Gostekhteoretizdat. 1950.

Levinson-Lessing, F. Yu. Uspekhi petrografii v Rossii (Advances in Petrography in Russia). — Izdanie geologicheskogo komiteta, 1923 (see also "Izbrannye trudy", Vol. 2. Izdatel'stvo AN SSSR. 1949).

Levinson-Lessing, F. Yu. O svoeobraznom tipe differentsiatsii v variolite Yalguty (A Peculiar Type of Differentiation in the Variolites of Yalguta). — Trudy Petrograficheskogo instituta AN SSSR, No. 5. 1935 (see also "Izbrannye trudy", Vol. 1. Izdatel'stvo AN SSSR. 1949).

Levinson-Lessing, F. Yu. Problema genezisa magmaticheskikh porod i puti k ee razresheniyu (Problems of Genesis of Igneous Rocks and the Methods of Solving Them). — Izdatel'stvo AN SSSR. 1934 (see also "Izbrannye trudy", Vol. I. Izdatel'stvo AN SSSR. 1949).

Lodochnikov, V. N. Metody izobrazheniya mnogokomponentnykh sistem (Methods of Representing Multicomponent Systems). — Izvestiya Instituta fiziko-khimicheskogo analiza, Vol. 2, No. 2. 1924.

Lodochnikov, V. N. Serpentiny i serpentinity il'chirskie i drugie i petrologicheskie voprosy, s nimi svyazannye (Serpentines and Serpentinites of Il'chirsk and Elsewhere and the Petrological Questions Connected with Them). — Trudy TsNIGRI, No. 38. 1935.

Nikitin, V. D. Diagramma ravnovesiya sistemy $MgO-SiO_2$ (Equilibrium Diagram of the System $MgO-SiO_2$). — Izvestiya Sektora fiziko-khimicheskogo analiza, AN SSSR, Vol. 16, No. 3. 1947.

Nikolaev, V. A. O kriticheskikh yavleniyakh v binarnykh sistemakh silikat-voda (Critical Phenomena in Silicate-Water Binary Systems). — Izvestiya AN SSSR, seriya geologicheskaya, No. 5. 1945.

Nikolaev, V. A. Novaya diagramma ravnovesiya binarnoi sistemy tipa silikat-voda (A New Equilibrium Diagram of A Binary System of the Type Silicate-Water). — Doklady AN SSSR, Vol. 51, No. 9. 1946.

Nikolaev, V. A. O nekotorykh osobennostyakh reaktsii metamorfizma s uchastiem vody i uglekisloty (Certain Peculiarities of the Reaction of Metamorphism With the Participation of Water and Carbon Dioxide). — Izvestiya AN SSSR, seriya geologicheskaya, No. 4. 1947.

Nikolaev, V. A. O troinykh sistemakh s letuchimi komponentami i printsipe ogranichennoi rastvorimosti letuchikh v silikatnykh rasplavakh (Ternary Systems with Volatile Components and the Principle of Limited Miscibility of Volatiles in Silicate Melts). — Zapiski Vserossiiskogo mineralogicheskogo obshchestva, Part 75, No. 4. 1947.

Nikolaev, V. A. O primenenii termodinamiki k nekotorym petrologicheskim protsessam (The Application of Thermodynamics to Certain Petrological Processes). — Zapiski Vsesoyuznogo mineralogicheskogo obshchestva, Part 86, No. 2. 1957.

Ostrovskii, I. A. Diagramma ravnovesnykh sostoyanii v sisteme silikat-voda dlya sluchaya ogranichennoi smesimosti i vozmozhnoe istolkovanie dannykh Goransona (Diagram of Equilibrium Conditions in the System Silicate-Water in the Case of Limited Miscibility and a Possible Interpretation of Goranson's Data). — Doklady AN SSSR, Vol. 72, No. 3. 1950.

Ostrovskii, I. A. Issledovanie po mineraloobrazovaniyu v nekotorykh silikatnykh rasplavakh pod davleniem vodyanogo para i vodoroda (Investigations of Mineral Formation in Certain Silicate Melts Under Water Vapor and Hydrogen Pressure). — Trudy Instituta geologii rudnykh mestorozhdenii, petrologii, mineralogii i geokhimii AN SSSR, No. 1. 1956.

Ostrovskii, I. A. Usloviya mineraloobrazovaniya v nekotorykh vodnosilikatnykh sistemakh pri vysokikh davleniyakh i temperaturakh (Conditions of Mineral Formation in Certain Hydrosilicate Systems at High Temperatures and Pressures). — Trudy Instituta geologii rudnykh mestorozhdenii, petrologii, mineralogii i geokhimii AN SSSR, No. 9. 1957.

Ostrovskii, I. A., G. P. Mishina and V. M. Povilaitis. PT-proektsiya sistemy kremnezem-voda (PT-Projection of the System Silica-Water). — Doklady AN SSSR, Vol. 126, No. 3. 1959.

Piip, B. I. Klyuchevskaya sopka i ee izverzheniya v 1944—1945 gg. i v proshlom (The Klyuchevskaya Volcano and Its Eruption in 1944—1945 and in the Past).—Trudy Laboratorii vulkanologii AN SSSR, No. 11. 1956.

Petrov, D. A. Troinye sistemy (Ternary Systems). — Izdatel'stvo AN SSSR. 1953.

Polkanov, A. A. O kontaktovom vzaimodeistvii diabaza i gneisogranita na Kol'skom poluostrove (The Contact Interaction of Diabase and Granite-Gneiss on the Kola Peninsula). — Trudy Sankt-Peterburgskogo obshchestva estestvoispytatelei, Vol. 44, Nos. 4—6. 1913.

Rosenbusch, H. Opisatel'naya petrografiya (Descriptive Petrography). ONTI. 1934 (original reference not found).

Schreinemakers, Nonvariantnye, monovariantnye i divariantnye ravnovesiya (Invariant, Monovariant and Divariant Equilibria). — Izdatel'stvo IL. 1948 (original reference not found).

Sobolev, V.S. Petrologiya trappov Sibirskoi platformy (Petrology of Trapps of the Siberian Platform). — Trudy Arkticheskogo instituta, Vol. 43. 1936.

Sobolev, V.S. Osobennosti magmaticheskikh proyavlenii i metallogenii platform na primere formatsii sibirskikh trappov (Peculiarities of Igneous Manifestations and Metalogeny of Platforms as Illustrated by the Siberian Trapps). — Tezisy doklada MGK, Moskva. 1937.

Sobolev, V.S. Petrografiya Botogol'skogo shchelochnogo massiva (Petrography of the Bogotol'sk Alkaline Massif). — Sbornik "Botogol'skoe mestorozhdenie grafita i perspektivy ego ispol'zovaniya". OGIZ, Irkutsk. 1947.

Sobolev, V.S. Vvedenie v mineralogiyu silikatov (Introduction to the Mineralogy of Silicates). — Izdatel'stvo L'vovskogo Universiteta, 1949.

Sobolev, V.S. O roli davleniya pri mineraloobrazovanii (The Role of Pressure in Mineral Formation).—In: Mineralogicheskii sbornik L'vovskogo geologicheskogo obshchestva, No. 9. 1955.

Sobolev, V.S. Dopolneniya k diagrammam plavkosti sistem nefelin-kaliofilit-kremnezem i ortoklaz-albit-anortit (Amendments to the Melting Diagrams of the Systems Nepheline-Kaliophyllite-Silica and Orthoclase-Albite-Anorthite). — Mineralogicheskii sbornik L'vovskogo geologicheskogo obshchestva, No. 10. 1956.

Sobolev, V.S. and A.I. Tsvetkov. O termicheskikh effektakh na krivykh nagrevaniya distena i andaluzita (Thermal Effects on Heating Curves of Kyanite and Andalusite). — Geokhimiya, No. 4. AN SSSR. 1956.

Sobolev, V.S. O fazovykh sootnosheniyakh v sisteme nefelin-kal'silit-anortit-kremnezem (Phase Relations in the System Nepheline-Kalsilite-Anorthite-Silica). Mineralogicheskii sbornik L'vovskogo geologicheskogo obshchestva, No. 13. 1959.

Soboleva, O.S. Diagrammy ravnovesiya smeshannykh kristallov s vodnymi rastvorami (Equilibrium Diagrams of Mixed Crystals with Aqueous Solutions). — Naukovi zapysky L'vivskohu Derzhavnohu Universitetu, Vol. 21, No. 3. seriya khimichna. 1952.

Tsvetkov, A.I. Izomorfnye zameshcheniya v gruppe besshchelochnykh piroksenov (Isomorphous Substitution in the Group of Alkali-free Pyroxenes). — Trudy IGN, No. 138, petrograficheskaya seriya, No. 41. 1951.

Vul'chin, E.I. Fiziko-khimicheskaya interpretatsiya termometricheskikh izmerenii metodom gomogenizatsii vklyuchenii (Physicochemical Interpretations of Thermometric Measurements by the Method of Homogenization of Inclusions). —Mineralogicheskii sbornik L'vovskogo geologicheskogo obshchestva, No. 5. 1951.

Zavaritskii, A.N. Fiziko-khimicheskie osnovy petrografii izverzhennykh gornykh porod (Physicochemical Principles of Igneous Petrography). —Nauchnoe khimiko-tekhnologicheskoe izdatel'stvo, Leningrad. 1926.

Zavaritskii, A. N. Lamprofiry i rodstvennye im porody na gore Magnitnoi (Lamprophyres and Related Rocks on Mt. Magnitnaya). — Trudy Petrograficheskogo instituta AN SSSR, No. 5. 1935.

Zavaritskii, A. N. Shchelochnye gornye porody Ishima (Alkaline Rocks of Ishim). — Trudy Petrograficheskogo instituta, Nos. 7—8. 1936 (see also "Izbrannye trudy", Vol. 2, Izdatel'stvo AN SSSR. 1958).

Zavaritskii, A. N. Petrografiya Berdyaushskogo plutona (Petrography of the Berdyaush Pluton). — Trudy TsNIGRI, No. 96. 1937. (see also "Izbrannye trudy", Vol. 2, Izdatel'stvo AN SSSR. 1958).

Zavaritskii, A. N. Pereschet khimicheskikh analizov izverzhennykh gornykh porod (Recalculation of Chemical Analyses of Igneous Rocks). — Izdatel'stvo AN SSSR. 1941.

Zavaritskii, A. N. Osnovnoi vopros fizicheskoi khimii protsessa obrazovaniya pegmatitov (The Principal Question of the Physical Chemistry of Pegmatite Formation). — Izvestiya AN SSSR, seriya geologicheskaya, No. 5. 1944.

Zavaritskii, A. N. O pegmatitakh, kak obrazovaniyakh promezhutochnykh mezhdu izverzhennymi gornymi porodami i rudnymi zhilami (Pegmatites as Intermediate Formations between Igneous Rocks and Ore Veins). — Zapiski Vserossiiskogo mineralogicheskogo obshchestva, Part 76, No. 1. 1947 (see also sbornik "Osnovnye problemy v uchenii o magmatogennykh rudnykh mestorozhdeniyakh". Izdatel'stvo AN SSSR. 1953).

Zavaritskii, A. N. Vvedenie v petrokhimiyu izverzhennykh gornykh porod (Introduction to Petrochemistry of Igneous Rocks). 2nd Edition. — Izdatel'stvo AN SSSR. 1950.

Zavaritskii, A. N. O vysokotemperaturnykh plagioklazakh (High-temperature Plagioclases). — Zapiski Vsesoyuznogo mineralogicheskogo obshchestva, Part 80, No. 3. 1951.

Zavaritskii, A. N. et al. Novye diagrammy dlya opredeleniya sostava vysokotemperaturnykh plagioklazov (New Diagrams for the Determination of the Composition of High-Temperature Plagioclases). — Zapiski Vsesoyuznogo mineralogicheskogo obshchestva, Part 87, No. 5. 1958.

Publications in Other Languages

[Foreign Part]

Abelson and oth. Researches in Geochemistry, New York. 1959.

Adams, L. H. A Note on the Stability of Jadeite. —Am. Jour. Sci. Vol. 251 and 299—308. 1953. [Russ. Transl. 1954].

Andersen, O. The System SiO_2-Forsterite-Anorthite. —Amer. Journ. Sci. 4 ser., Vol. 39. 1915.

Andersen, O. and N. L. Bowen. Das binäre system MgO-SiO_2. — Zeitschr. anorg. Chem., B. 87. 1914.

Annual Report of the Directory of the Geophysical Laboratory.—Year-book of the Carnegie Institution of Washington, No. 53. 1953—1954, No. 55. 1955—1956, No. 56. 1956—1957.

Backlund, H. Die Umgrenzung der Svekofenniden.—Bull. Geol. Inst. Upsala, Vol. 27. 1937.

Backlund, H. G. The Granitization Problem.–Geol. Mag. Vol. 83, pp. 105–117. 1946. [Russian translation. 1950].

Barth, T. F. W. Theoretical Petrology. Willey. New York. 1952. [Russ. transl. 1956].

Bowen, N. L. The Binary System $Na_2Al_2Si_2O_8$-$CaAl_2Si_2O_8$. –Amer. Journ. of Sci., 4 ser., Vol. 34. 1912.

Bowen, N. L. The Melting Phenomena of the Plagioclase Feldspars. – Amer. Journ. of Sci. (4), Vol. 35. 1913.

Bowen, N. L. The Ternary System Diopside-Forsterite-Silica. —Amer. Journ. of Sci., Vol. 34. 1914.

Bowen, N. L. The Crystallization of Haplobasaltic, Haplodioritic and Related Magmas. –Amer. Journ. of Sci, 4 ser., Vol. 236. 1915.

Bowen, N. L. System Diopside-Forsterite-SiO_2. —Zs. anorg. Chem., B. 90. 1915.

Bowen, N. L. The System Diopside-Anorthite-Albite. —Amer. Journ. Sci., 4 ser., Vol. 40. 1915.

Bowen, N. L. Diffusion in Silicates Melts. —Journ. Geol., Vol. 29. 1921.

Bowen, N. L. The Behaviour of Inclusions in Igneous Magma.— Journ. Geol., Vol. 30 (suppl.). 1922.

Bowen, N. L. The Reaction Principle in Petrogenesis. –Journ. Geol., Vol. 30. 1922.

Bowen, N. L. The Evolution of Igneous Rocks. —Princeton Univ. Press, Princeton N. Y. 1928. [Russian translation. 1934].

Bowen, N. L. "The Broader Story of Magmatic Differentiation Briefly Told". —Ore Deposits of the Western States, 106—128, Amer. Ins. Min. Met. Eng. New York. 1933. [Russ. transl. 1937].

Bowen, N. L. The System SiO_2-Diopside-Leucite. —Amer. Journ. of Sci., Vol. 233. 1937.

Bowen, N. L. Recent High-Temperature Research on Silicates and its Significance in Igneous Geology. –Amer. Journ. of Sci. Vol. 233. 1937.

Bowen, N. L. Progressive Metamorphism of Siliceous Limestone and Dolomite.—Journ. Geol., Vol. 48, No. 3. 1940.

Bowen, N. L., and Greig, J. W. The System Al_2O_3-SiO_2. —Amer. Ceramic. Soc. Journ., Vol. 7. 1924.

Bowen, N. L. and G. Morey. The Melting of Potash Feldspar. —Amer. Journ. of Sci. (5), Vol. 4. 1922.

Bowen, N. L. and J. F. Schairer. "The System FeO-SiO_2". —Amer. Journ. Sci. 214, 177—213. 1932. [Russ. transl. 1937].

Bowen, N. L. and J. F. Schairer. Grünerite from Rockport, Massachusetts, and a Series of Synthetic Fluoramphiboles. –Amer. Mineral., Vol. 20. 1935.

Bowen, N. L. and J. F. Schairer. The System MgO-FeO-SiO_2 — Amer. Journ. of Sci., Vol. 229. 1935.

Bowen, N. L. and J. F. Schairer. The System SiO_2-Anorthite-Leucite. Amer. Journ. of Sci., Vol. 233. 1937.

Bowen, N. L. and J. F. Schairer. The System SiO_2-Fayalite-Nepheline. –Amer. Journ. of Sci., Vol. 233. 1937.

Bowen, N. L., J. F. Schairer and E. Posnyak. The System SiO_2-CaO-FeO. —Amer. Journ. of Sci., Vol. 226. 1933.

Bowen, N. L., J. F. Schairer and H. U. Willems. The System SiO_2-Na_2SiO_3-Fe_2O_3. –Amer. Journ. of Sci., Vol. 220. 1930.

Bowen, N. L., and O. F. Tuttle. "The System MgO-SiO_2-H_2O". Bull. Geol. Soc. Amer., 60, 439—460. 1949. [Russ. transl. 1950].
Bowen, N. L., and O. F. Tuttle. "The System $NaAlSi_3O_8$-$KAlSi_3O_8$-H_2O".—Journ. Geol. 58, 489—511. 1950. [Russ. transl. 1950].
Büchner, E. Flüssige Kohlensäure als Lösungsmittel.—Zeitschr. phys. Chemie, B. 54. Hf. 6. 1906.
Büchner, E. Die heterogenen Gleichgewichte von H. B. Roozeboom (Systeme mit zwei flüssigen Phasen). Teil 2, Hf. 2. Braunschweig. 1918.
Coes, L. J.—Science, 118, 131. 1953.
Clark, S. P., E. C. Robertson and F. Birch. Experimental Determination of Kyanite-Sillimanite Equilibrium Relations at High Temperatures and Pressures.—Amer. Journ. of Sci., Vol. 255. 1957.
Daly, R. A. Igneous Rocks and the Depth of the Earth.—McGraw-Hill, New York. 1933. [Russ. transl. 1936].
Danielsson, Allan. Das Calcit Wollastonitgleichgewicht.—Geochim. et Cosmochim. Acta, Vol. I. 1950.
Day, A. L., J. P. Iddings and E. T. Allen. The Isomorphism and Thermal Properties of the Feldspars.—Publ. Carnegie Inst. Wash. No. 31. 1905.
Eckermann, V. The Genesis of the Rapakivi Granites. —Geol. Fören. Förhandl. 59. 1937.
Eckermann, V. The Rapakivi Facts, an Answer to a Reply.—Geol. Fören, Förhandl., 60. 1938.
Eitel, W. Über das binare System $CaCO_3$. Ca_2SiO_4 und den Spurrit.—Neues Jahrb. Min. Beilage, B. 48. 1923.
Eitel, W. The Physical Chemistry of the Silicates.—The University of Chicago Press. Chicago, Illinois USA. 1954.
Epstein, P. S. Textbook of Thermodynamics.—John Willey and Sons, New Yok. 1937. [Russ. transl. 1948].
Eskola, P. The Mineral Facies of Rocks. Norsk. geol. Tidskrift. B. VI. 1920.
Fenner, C. N. The Stability Relations of the Silica Minerals.—Amer. Journ. Sci. 36, 331—384. 1913. [Russ. transl. 1937].
Fenner, G. The Residual Liquids of Crystallizing Magmas.—Miner. Mag. Vol. 134. 1931.
Ferguson, J. B. and H. E. Merwin. The Melting Points of Cristobalite and Tridymite.—Amer. Journ. of Sci., ser. 4, Vol. 46. 1918.
Ferguson, J. B. and H. E. Merwin. Wollastonite and Related Solid Solutions in the Ternary System CaO-MgO-SiO_2.—Amer. Journ. of Sci., ser. 4, Vol. 48. 1919.
Ferguson, J. B. and H. E. Merwin. The Ternary System CaO-MgO-SiO_2.—Amer. Journ. of Sci., 4, Vol. 48. 1919.
Foster, W. R. Synthetic Sapphirine and its Stability Relations in the System MgO-Al_2O_3-SiO_2.—Amer. Ceramic. Soc. Journ. Vol. 33. 1950.
Foster, W. R. High-Temperature X-ray Diffraction Study of the Polymorphism of $MgSiO_3$.—Amer. Ceramic Soc. Journ. Vol. 34. 1951.

Franco, R. R. and J. F. Schairer. Liquidus Temperatures in Mixtures of the Feldspars of Soda, Potash and Lime.—Jour. Geol. Vol. 59, pp. 259-267. 1951. [Russ. transl. 1954].

Frondel, C. and R. L. Collete. Synthesis of Tourmaline by Reaction of Mineral Grains with $NaCl-H_3BO_3$ Solution, and its Implications in Rock Metamorphism.—Amer. miner. Vol. 42, Nos. 11 and 12. 1957.

Gibbs, J. W. On the Equilibrium of Heterogeneous Substances.—Trans. Conn. Acad. Vol. 111, 1876—1878.

Gillingham, T. E. The Solubility and Transfer of Silica and Other Non-volatiles in Steam.—Econ. Geol., Vol. 43, pp. 241—272. 1948. [Russ. transl. 1950].

Goldschmidt, V. Die Kontaktmetamorphose im Kristiangebiet.—Vid. Skr. Mat. Natur. kl., Nr. 1. 1911.

Goranson, R. W. Silicate-Water Systems.—Amer. Journ. of Sci. Vol. 25A. 1938.

Goranson, R. W. Physical Effects of Extreme Pressures.—Repr. Scient. Monthly, Vol. 51. pp. 524—535. 1940. [Russ. transl. 1954].

Greig, J. W. Formation of Mullite from Kyanite, Andalusite and Sillimanite.—Amer. Journ. of Sci. No. 61, Vol. 212. 1926.

Greig, J. W. "Immiscibility in Silicate Melts".—Amer. Journ. of Sci. (5), 13. 1—44, 133—154. 1927. [Russ. transl. 1937].

Greig, J. W. On the Evidence which has been Presented for Liquid Silicate Immiscibility in the Laboratory and the Rocks of Agate Point.—Amer. Journ. of Sci. ser. 5, Vol. 15. 1928.

Greig, J. W., and T. F. W. Barth. The System $Na_2O \cdot Al_2O_3 \cdot 2SiO_2$ (Nephelite, Carnegieite) — $Na_2O \cdot Al_2O_3 \cdot 6SiO_2$ (Albite).—Amer. Journ. of Sci. ser 5, Vol. 35. 1938.

Greig, J. W., H. E. Merwin and E. S. Shepherd.—Amer. Journ. of Sci. New., Vol. 225. 1933.

Griggs, D. T. and G. C. Kennedy. A Simple Apparatus for High Pressures and Temperatures.—Amer. Journ. of Sci. Vol. 254, No. 12. 1956.

Grout, F. Petrography and Petrology. New York—London. 1932.

Hannay, J. B. and J. Hoggarth.—Proceedings of the Royal Society. A. Mathemat. and Physical Sci., Vol. 29. 1879.

Harker, A. Metamorphism. Methuen, London. 1939. Russ. trans.

Harker, R. J. The System $MgO-CO_2$-A and the Effect of Inert Pressure on Certain Types of Hydrothermal Reaction. Amer. Journ. of Sci. Vol. 256, No. 2. 1958.

Harker, R. J. and O. F. Tuttle. Studies in the System $CaO-MgO-CO_2$. Part I - The Thermal Dissociation of Calcite, Dolomite and Magnesite. Amer. Journ. of Sci. Vol. 253, No. 4. 1955.

Harker, R. J. and O. F. Tuttle. Experimental Data on the P_{CO_2} T-curve for the Reaction: Calcite + Quartz $\leftrightarrows$ Wollastonite + Carbon Dioxide.—Amer. Journ. of Sci., Vol. 254, No. 4. 1956.

Hess, H. H. Chemical Composition and Optical Properties of Common Clinopyroxenes. Amer. Miner. Vol. 234. 1949.

Keith, M. L., and O. F. Tuttle. "Significance of Variation in the High-Low Inversion of Quartz".—Amer. Journ. of Sci. Bowen vol. 203—280. 1952. [Russ. transl. 1952].

Kennedy, G. C. "Equilibrium Between Volatiles and Iron Oxides in Igneous Rocks". —Amer. Journ. of Sci. 246, 529—549. 1948. [Russ. transl. 1950].

Kennedy, G. C. Pyrophyllite-Sillimanite-Mullite Equilibrium Relations to 20,000 Bars and 800°C.—Geol. Soc. Amer. Bull. Vol. 66. 1955.

Kennedy, G. C. "Some Aspects of the Role of Water in Rock Melts" Crust of the Earth, Geol. Soc. Amer. Spec. Paper 62, 489—503. 1955. [Russ. transl. 1957].

Kennedy, G. C. The Brucite-Periclase Equilibrium.—Amer. Journ. of Sci. Vol. 254, No. 9. 1956.

Kuno, H. Ion Substitution in the Diopside-Ferropigeonite Series of Clinopyroxenes. —Amer. Miner., Vol. 40, Nos. 1 and 2. 1955.

Kröger, C. and K. Illner. Über die Systeme Alkalioxyd — CaO-Al_2O_3-SiO_2-CO_2. Die Gleichgewichte in System CaO-SiO_2-CO_2.—Zeitschr. anorg. und allg. Chemie, B. 229. 1936.

Laar van, J. J. Die Thermodynamik einheitlicher Stoffe und binarer Gemische. Groningen und Batavia. 1936.

Laves, F. "Phase Relations of the Alkali Feldspars".— Journ. Geol. 60, 436—450, 549—574. 1952. [Russ. transl. 1956].

MacDonald, G. J. F. Quartz-Coesite Stability Relations at High Temperatures and Pressures. Amer. Journ. of Sci., Vol. 254, No. 12. 1956.

MacKenzie, Wm. S. The Crystalline Modifications of $NaAlSi_3O_8$. —Amer. Journ. of Sci. Vol. 255. 1957.

Michel-Levy, M. Synthesis of Tourmaline and Jeremeievite,–C. R. Acad. Sci. (Paris), Vol. 228. 1949.

Morey, G. W. The Solubility of Solids in Gases. —Econ. Geol., Vol. 52, No. 3. 1957.

Morey, G. W. The System Water-Nepheline-Albite: A Theoretical Discussion. —Amer. Journ. of Sci. Vol. 255. 1957.

Morey, G. W. and N. L. Bowen. The Melting of Potash Feldspar.—Amer. Journ. of Sci. (5), Vol. 4. 1922.

Morey, G. W. and C. N. Fenner. The Ternary System H_2O-K_2SiO_3-SiO_2. —Amer. Chem. Soc. Journ., Vol. 39. 1917.

Morey, G. W. and J. M. Hesselgesser. The Solubility of Some Minerals in Superheated Steam at High Pressures.—Econ. Geol. 46, 821—835. 1951. [Russ. transl. 1954].

Morey, G. W. and J. M. Hesselgesser. "The System H_2O-Na_2O-SiO_2 at 400° C".—Amer. Journ. of Sci. Bowen Vol. 343—372. 1952. [Russ. transl. 1954].

Nelson, B. W. and R. Roy. Synthesis of Chlorites and their Structural and Chemical Constitution. Amer. Miner. Vol. 43, No. 7 and 8. 1958.

Neumann, F. Über die Stabilitatsverhältnisse de Modifikationen im Polymorphen System Al_2SiO_5.—Zeitschr. anorg. und allgem. Chem. B. 145, Hf. 3—4. 1925.

Niggli, P. Die Leichtflüchtigen Bestandteile im Magma. Leipzig. 1920.

Niggli, P. Das Magma und Seine Produkte. Leipzig. 1937. [Russ. transl. 1946].

Oftedahl, C. Studies on the Igneous Rock Complex of the Oslo Region. IX. The Feldspars.—Norsk. Vidensk. -Akad. Oslo Skr.. I. Math.-Naturv. Kl., No. 3, p. 63. [Russ. transl. 1952].

Osborn, E. F. and J. F. Schairer. The Ternary System Pseudowollastonite-Akermanite-Gehlenite.—Amer. Journ. of Sci. (5), Vol. 39. 1941.

Osborn, E. F., and D. B. Tait. The System Diopside-Forsterite-Anorthite.—Amer. Journ. of Sci. Bowen, vol. 413—434. 1952. [Russ. transl. 1954].

Perret, F. The Vesuvius Eruption of 1906.—Carnegie Inst. Wash. Publ. No. 339. 1924.

Perrin, R. and M. Rouboult. On the Granite Problem. Journ. Geology, Vol. 57, 357—379. 1949. [Russ. transl. 1950].

Rankin, G. A. Das Ternäre System Kalziumoxyd-Aluminiumoxyd-Siliziumoxyd.—Zeitschr. Anorg. Chem. B. 92—93. 1915.

Rankin, G. A. and H. E. Merwin. The Ternary System SiO_2-Al_2O_3-MgO.—Amer. Journ. of Sci., Vol. 45. 1918.

Read, H. H. Meditations on Granite.—Geologists' Assoc. Proc. Vol. 54, pp. 64—85. 1943. [Russ. transl. 1949].

Robertson, E. C., F. Birch, and G. F. J. MacDonald. Experimental Determination of Jadeite Stability Relations to 25,000 bars.—Amer. Journ. of Sci., Vol. 255. 1957.

Roedder, E. W. Low Temperature Liquid Immiscibility in the System K_2O-FeO-Al_2O_3-SiO_2.—Amer. Min. 36, 282—286. 1951. [Russ. transl. 1954].

Roedder, E. W. A Reconnaissance of Liquidus Relations in the System K_2O-2SiO_2-FeO-SiO_2.—Amer. Journ. of Sci., Bowen vol., 435—456. 1952. [Russ. transl. 1954].

Roozeboom. Über die Löslichkeit von Mischkrystallen, Speziell Zweier Isomorpher Körper.—Zeitschr. f. Phys. Chem. 504, 8. 1891.

Roozeboom. Erstarrungspunkte der Mischkrystalle Zweier Stoffe.—Zeitschr. f. Phys. Chem., 385, 30. 1899.

Roy, R. and E. F. Osborn. Studies in the System Alumina-Silica-Water, Problems of Clay and Laterite Genesis, Symposium. Am. Inst. Min. Metal. Eng. 76—80. 1952. [Russ. transl. 1954].

Roy, R. and E. F. Osborn. The System Al_2O_3-SiO_2-H_2O.—Amer. Miner., Vol. 39, Nos. 11—12. 1954.

Roy, D. M. Hydrothermal Synthesis of Andalusite.—Amer. Miner. Vol. 39, Nos. 1—2. 1954.

Roy, D. M. and R. Roy. Synthesis and Stability of Minerals in the System MgO-Al_2O_3-SiO_2-H_2O.—Amer. Miner., Vol. 40, No. 3 and 4. 1955.

Roy, D. M. and R. Roy. A Re-Determination of Equilibria in the System MgO-H2O and Commets on Earlier Work.—Amer. Journ. of Sci., Vol. 255, No. 8. 1957.

Roy, R. and O. F. Tuttle. Investigations under Hydrothermal Conditions. Physics and Chemistry of the Earth. Vol. I.—Pergammon Press, London and New York. 1956.

Schairer, J. F. The Alkali Feldspar Join in the System $NaAlSiO_4$-$KAlSiO_4$-SiO_2.—Journ. Geol. 58, 512-517. 1960. [Russ. transl. 1952].

Schairer, J. F. and N. L. Bowen. Preliminary Report Equilibrium Relations Between Feldspathoids, Alkali Feldspars and Silica.— Trans. Amer. Geophys. Union. 16 th. Amer. Meeting. 1935.

Schairer, J.F. and N.L. Bowen. The System Leucite-Diopside-Silica.— Amer. Jour. of Sc., Vol. 245, 193-204. 1938. [Russ. transl. 1950].

Schairer, J. F. and N. L. Bowen. The Binary System $CaSiO_3$-Diopside and the Relations between $CaSiO_3$ and Akermanite.—Amer. Journ. of Sci. ser. 5, Vol. 40. 1942.

Schairer, J. F. and N. L. Bowen. Melting Relations in the System Na_2O-Al_2O_3-SiO_2 and K_2O-Al_2O_3-SiO_3. –Amer. Journ. of Sci., Vol. 245. 1947.

Schairer, J. F. and N. L. Bowen. The System Anorthite Leucite-Silica,–Bull. Soc. Geol. Finlande, 20, 67—87. 1947. [Russ. transl. 1950].

Schairer, J. F. and N. L. Bowen. The System K_2O-Al_2O_3-SiO_2. — Amer. Journ. of Sci., Vol. 253, XII. 1955.

Schairer, J. F. and N. L. Bowen. The System Na_2O-Al_2O_3-SiO_2.— Amer. Journ. of Sci., Vol. 254, No. 3. 1956.

Schairer, J. F. and K. Yagi. The System FeO-Al_2O_3-SiO_2.—Amer. Journ. of Sci., Bowen Volume, 471-512. 1952. [Russ. transl. 1954].

Schaller, W. T. "Pegmatites" Ore Deposits of the Western States 144—151,-Amer. Inst. Min. Met. Eng. New York. 1933. [Russ. transl. 1937].

Shepherd, E. S. and G. A. Rankin. Die Binären Systeme von Tonerde mit Kieselsäure, Kalk und Magnesia. –Zeitschr. f. Anorg. Chem. B. 68. 1910.

Smits, A. The Course of the Solubility Curve in the Region of Critical Temperatures of Binary Mixtures.—Proc. Royal Soc. Amsterdam, Vol. 6. 1903.

Smith, J. V. and O. F. Tuttle. The Nepheline-Kalsilite System: I. X-Ray Data for the Crystalline Phases. Vol. 255, No. 4. 1957.

Smith, F. G. Review of Physicochemical Data of Supercritical Fluids.— Econ. Geol. 49, 331—332. 1953. [Russ. transl. 1954].

Smith, F. G. and R. E. Jones. The Solubility of Solid Gases. — Econ. Geol., vol. 53, No. 3. 1958.

Spencer, E. The Potash-Soda Feldspars. Parts I and II.—Min. Mag. 24, 453-494, 25, 81-118. 1937-38. [Russ. transl.].

Tammann, G. Lehrbuch der heterogenen Gleichgemichte.—Brunswick, Fn. Vieweg and Sohn. 1924. [Russ. transl. 1952].

Thompson, U. B. Jr. The Thermodynamic Basis for the Mineral Facies Concept. –Amer. Journ. of Sci., Vol. 253, No. 2. 1955.

Tilley, C. E. Nepheline-Alkali Feldspar Parageneses.—Amer. Journ. of Sci., Vol. 252, No. 2. 1954.

Tiley, C. E. The Leucite Nepheline Dolerite of Meiches, Vogelsberg, Hessen. – The Amer. Miner. Vol. 43, Nos. 7—8. 1958.

Towkeieff, J. Differentiation in Basalt Lava. Island Magee, Co. Antriw. Geol. Mag. Vol. 71, No. 845. 1934.

Tsuboi, S. On the Course of Crystallization of Pyroxenes from Rock-Magmas. –Jap. Journ. Geol. Geogr., Vol. 10. 1932.

Tuttle, O. F. and N. L. Bowen. High Temperature Albite and Contigous Feldspars. –Journ. Geol. 58, 572—584. 1950. [Russ. transl. 1952].

Tuttle, O. F. and N. L. Bowen. Origin of Granite in the Light of

Experimental Studies in the System $NaAlSi_3O_8$-$KAlSi_3O_8$-SiO_2-H_2O.—Mem. Geol. Soc. Amer. No. 74. 1958.

Tuttle, O. F. and I. I. Friedman. Liquid Immiscibility in the System H_2O-Na_2O-SiO_2. Jour. Am. Chem. Soc. 70, 919—926. 1948. [Russ. transl. 1950].

Tuttle, O. F. and R. I. Harker. Synthesis of Spurrite and the Reaction Wollastonite + Calcite ⇆ Spurrite + Carbon Dioxide.—Amer. Journ. of Sci., Vol. 255, No. 3. 1957.

Tuttle, O. F. and J. V. Smith. The Nepheline-Kalsilite System. II. Phase Relations.—Amer. Journ. of Sci., Vol. 256, No. 8. 1958.

Vogt. Physikalisch-chemische Gesetze der Kristallisationsfolge in Eruptivgesteinen.—Tscherm. Min. Petr. Mitt. B. 24 und 25, 1905—1906.

Vogt. The Physical Chemistry of the Crystallization and Magmatic Differentiation of Igneous Rocks.—Journ. of Geol., Vol. 29. 1921; Vol. 30. 1922; Vol. 31. 1923.

Vogt. Die physikalisch-chemische Gesetze der magmatischen Differentiation- Videns. Skrifter I. Mat. -Naturv. Klasse, No. 17. 1923.

Vogt. The Physical Chemistry of the Magmatic Differentiation of Igneous Rocks— Videns. Skrifter I. Mat. -Naturv. Klasse, No. 15. 1924.

Waals, J. D. van der. and Kohnstamm. Lehrbuch der Thermostatik. Leipzig. 1927. [Russ. transl. 1936].

Wager, L. R. and W. A. Deer. Geological Investigations in East Greenland. Part III. The Petrology of the Skaergard Intrusion, Kangerdlungssuaq Greenland.—Meddel. om Gronland, B. 105, No. 4. 1939.

Wahl, W. Beiträge zur Geologie der Präkämbrischen Bildungen im Gouvernement Olonez 3. Die Gesteine der Westküste des Onega-Sees.—Fennia 24, No. 3. 1908.

Wahl, W. Eutectics and the Crystallization of Igneous Rocks. —"Bull. Comm. Geol. Finlande". No. 87. 1929.

Winchell, A. H. Studies in the Feldspar Group.—Journ. Geol. Vol. 33, No. 7. 1925.

Winchell, A. N. and H. Winchell. Elements of Optical Mineralogy.—New York. John Willey and Sons. 1951. [Russ. transl. 1953].

de Wys, E. C. and W. R. Foster. The System Diopside-Anorthite-Akermanite.—Miner. Mag., Vol. 31, No. 240, III. 1958.

Yoder, H. S. Stability Relations of Grossularite.—Journ. Geol. Vol. 58, No. 3. 1950.

Yoder, H. S. High-Low Quartz Inversion up to 10,000 bars.—Trans. Amer. Geophys. Union, 31, 827—835. 1950. [Russ. transl. 1959].

Yoder, H. S. Change of Melting Point of Diopside with Pressure.—Journ. Geol. 60, 364—374. 1952. [Russ. transl. 1954].

Yoder, H. S. The MgO-Al_2O_3-SiO_2-H_2O System and the Related Metamorphic Facies.—Amer. Journ. of Sci., Bowen Volume, 569—627. 1952. [Russ. transl. 1954].

Yoder, H. S. Role of Water in Metamorphism. Crust of the Earth. —Geol. Soc. Amer. Spec. Paper, 62, 505—524. 1955. [Russ. transl. 1957].

Yoder, H. S. and H. P. Eugster. Phlogopite Synthesis and Stability Range.—Geochim. et Cosmochim. Acta, Vol. 6. 1954.

Translator's Appendix

Zavaritskii's Numerical Characteristics for Petrochemical Calculations

Zavaritskii's numerical characteristics are parameters which can be calculated from the chemical analysis of a rock. They reflect the quantitative molar relations between the rock constituents.

Zavaritskii's mathematical scheme for the calculation of his parameters is based on a standard procedure. The resulting values can be recalculated so as to give the source data (weight percent). The mathematics involved is relatively simple, and the parameters obtained clearly reflect the most important chemical characteristics of the analyzed magmatic rock.

The numerical characteristics allow for the clarification of the following properties of rocks:

1) the ratio of salic to femic constituents as reflected by the chemical differences between the aluminosilicates and "simple" silicates (olivines, pyroxenes, etc);

2) the relative saturation of the rock with respect to silica;

3) the general composition of the feldspars, i. e., the ratio of alkali-aluminosilicates to calcium-aluminosilicates;

4) the compositional relations within the alkali-aluminosilicates and "simple" silicates.

In order to express quantitatively these diagnostic properties, Zavaritskii makes use of two groups of values, namely:

i) Principal numerical characteristics, used for expressing the first three mentioned properties;

ii) Complementary numerical characteristics, used for expressing the fourth-mentioned property.

As already stated, these parameters are based on molar ratios, i. e., the chemical weight percentages are divided by the respective molecular weights, prior to any further calculation. In order to simplify subsequent calculations the number of constituents taken into consideration is reduced by adding MnO to FeO and TiO_2 to SiO_2. Iron is calculated as the bivalent ion. Hydrogen (as water) is not taken into account (the primary water content of the rock generally changes radically during secondary processes). Oxygen is eliminated from the calculation since its amount is dependent on the number of metal atoms and their valence. The remaining constituents are therefore Si, Al, Fe, Mg, Ca, Na and K.

The principal numerical characteristics consist of the following values — a, c, b and s; by definition $a + b + c + s = 100$.

These parameters have the following significance:

a is the relative number of the alkali-metal atoms bound to aluminum in the aluminosilicate mineral lattices.

c is the relative number of calcium atoms bound with aluminum in the aluminosilicates. Rocks deficient in alumina, relative to alkali-metals, will contain sodium ferrisilicates instead of the "normal" calcium aluminosilicates.

b is the relative number of all remaining metal atoms (excluding silicon and titanium) which constitute the "simple" silicates and are not included in the aluminosilicates.

s is the relative number of silicon atoms (including titanium).

This group of parameters is obtained from the chemical analysis by the following procedure:

1) weight percentages, as stated in the chemical analysis, are converted to molar proportions using suitable tables;

2) the values of silica and titania are added to give the value S;

3) the alkalis are summed: if $(K_2O + Na_2O) < Al_2O_3$, this sum is multiplied by two to give the value A;

4) if $(K_2O + Na_2O) > Al_2O_3$, alumina (multiplied by two) will be the limiting value of A. The alkali excess over alumina $[(K_2O + Na_2O) - Al_2O_3]$ multiplied by two (Na_2O') is then calculated to obtain $\overline{C}$;

5) if CaO exceeds the alumina excess with respect to alkalis $[CaO > Al_2O_3 - (K_2O + Na_2O)]$, the difference is defined as the value C;

6) if there is an excess of alumina over calcium plus alkalis $[Al_2O_3 > CaO + Na_2O + K_2O]$, C will equal CaO. The alumina excess is then multiplied by two and is then added to the value B;

7) the molar proportion of Fe_2O_3, multiplied by two, is added to that of FeO (+ MnO) in order to obtain FeO'. When dealing with rocks oversaturated in alkalis, the value $\overline{C}$ should be subtracted from FeO' (for each sodium atom in excess of alumina one iron atom should be reserved in order to form sodium ferrisilicates);

8) in the rare cases when $\overline{C} > FeO'$ (there is not sufficient iron to combine with the excess sodium for the formation of ferrisilicates), the difference $(\overline{C} - FeO')$ is defined as Na';

9) the difference (CaO — C) is calculated to give the excess of calcium, i. e., CaO';

10) the parameter B is obtained by the summation of FeO + MgO + CaO'. Al_2O_3' (multiplied by two) and Na' are added to B if they are present;

11) the values A, C, B and S (or A, $\overline{C}$, B and S) thus obtained are recalculated to percentages of their sum to give the principal numerical characteristics a, c, b and s (or a, $\overline{c}$, b and s) respectively.

The complementary numerical characteristics consist of the following eight parameters: f', m' and c' which reflect the role played by iron, magnesium and calcium, respectively, in "simple" silicate minerals; n is a measure of the relative amount of sodium molecules in the alkali aluminosilicates. The amount of titanium in the rock is represented by the parameter t, while the relative importance of ferric iron included in the "simple" silicates is expressed by the parameter φ. Rocks rich in alumina yield the value a' instead of c', which reflects the surplus of alumina left after combination with all the calcium atoms in the aluminosilicates. This alumina excess is included, as already explained, in group B. When $\overline{C} > FeO'$, the complementary numerical characteristic n' is obtained, which reflects the relative amount of sodium left over in B after combining the maximum possible amount of sodium in aluminosilicates and ferrisilicates.

Using Zavaritskii's numerical characteristics a magmatic rock may be related to one of the following four types:

i) rocks with a normal composition — the principal numerical characteristics are a, c, b and s and the complementary parameters are f', m', and c' and n, t and φ;

ii) rocks oversaturated in alumina — the principal parameters are d, $\bar{c}$, b and s and the complementary parameters are a', f' and m' and n, t, φ;

iii) rocks slightly oversaturated with respect to alkalis — the principal parameters are a, $\bar{c}$, b and s and the complementary parameters are f', m' and c' and n, t and φ;

iv) rocks markedly oversaturated with respect to alkalis — the principal parameters are a, $\bar{c}$, b and s and the complementary parameters are n', m' and c' and n, t and φ.

The degree of saturation with respect to silica is represented by the parameter Q, i. e., the relative number of silicon atoms. If the value of Q is positive, it indicates the possible occurrence of free silica (as quartz) in the rock. Rocks composed of silica-saturated minerals, but lacking free silica, will give Q values approximating zero, while undersaturated rocks will yield negative values of Q. The Q values are readily obtained from principal numerical characteristics (according to the ratio of silicon to metal atoms in the rock-forming silicates) as follows:

1) $Q = s - (3a + 2c + b)$ — for normal rocks;

2) $Q = s - (3a + 2\bar{c} + b)$ — for rocks containing sodium ferrisilicate.

The relation between Zavaritskii's parameters and the commonly used Niggli values is here provided.

In the case of normal rocks, $a : c : b : s = 2\,alk : [al - alk] : [fm + c - (al - alk)] : si$.

In the case of rocks oversaturated with respect to alumina, $a : c : b : s = 2alk : c : [fm + 2\,(al - alk) - 2c] : si$.

In the case of rocks oversaturated with respect to alkalis, $a : \bar{c} : b : s = 2al : 2\,[alk - al] : [fm + c - 2\,(alk - al)] : si$.

The complementary numerical characteristics may be expressed in Niggli values as follows:

normal rocks — $f' : m' : c' = fm\,(1 - mg) : fm.\,mg : [c - (al - alk)]$;

rocks oversaturated with respect to alumina — $a' : f' : m' = [(al - alk) - c] : fm\,(1 - mg) : fm.\,mg$;

rocks oversaturated with respect to alkalis — $f' : m' : c' = [fm - 2(alk - al)]\,(1 - mg) : [fm - 2\,(alk - al)]\,mg : c$.